Solar Cell Materials

Developing Technologies

Wiley Series in Materials for Electronic and Optoelectronic Applications

www.wiley.com/go/meoa

Series Editors

Professor Arthur Willoughby, *University of Southampton, Southampton, UK*

Dr Peter Capper, *SELEX Galileo Infrared Ltd, Southampton, UK*

Professor Safa Kasap, *University of Saskatchewan, Saskatoon, Canada*

Published Titles

Bulk Crystal Growth of Electronic, Optical and Optoelectronic Materials, Edited by P. Capper

Properties of Group-IV, III–V and II–VI Semiconductors, S. Adachi

Charge Transport in Disordered Solids with Applications in Electronics, Edited by S. Baranovski

Optical Properties of Condensed Matter and Applications, Edited by J. Singh

Thin Film Solar Cells: Fabrication, Characterization and Applications, Edited by J. Poortmans and V. Arkhipov

Dielectric Films for Advanced Microelectronics, Edited by M. R. Baklanov, M. Green and K. Maex

Liquid Phase Epitaxy of Electronic, Optical and Optoelectronic Materials, Edited by P. Capper and M. Mauk

Molecular Electronics: From Principles to Practice, M. Petty

CVD Diamond for Electronic Devices and Sensors, Edited by R. S. Sussmann

Properties of Semiconductor Alloys: Group-IV, III–V and II–VI Semiconductors, S. Adachi

Mercury Cadmium Telluride, Edited by P. Capper and J. Garland

Zinc Oxide Materials for Electronic and Optoelectronic Device Applications, Edited by C. Litton, D. C. Reynolds and T. C. Collins

Lead-Free Solders: Materials Reliability for Electronics, Edited by K. N. Subramanian

Silicon Photonics: Fundamentals and Devices, M. Jamal Deen and P. K. Basu

Nanostructured and Subwavelength Waveguides: Fundamentals and Applications, M. Skorobogatiy

Solar Cell Materials

Developing Technologies

Edited by

Gavin Conibeer
School of Photovoltaic and Renewable Energy Engineering,
University of New South Wales, Australia

Arthur Willoughby
Emeritus Professor, Engineering Materials Research Group,
Faculty of Engineering and the Environment,
Southampton University, UK

WILEY

This edition first published 2014

Registered office
John Wiley & Sons Ltd, The Atrium, Southern Gate, Chichester, West Sussex, PO19 8SQ, United Kingdom

For details of our global editorial offices, for customer services and for information about how to apply for permission to reuse the copyright material in this book please see our website at www.wiley.com.

Library of Congress Cataloging-in-Publication Data

Solar Cell Materials: Developing Technologies / [edited by]
Gavin Conibeer, Arthur Willoughby.
pages cm – (Wiley series in materials for electronic & optoelectronic applications)
Includes bibliographical references and index.
ISBN 978-0-470-06551-8 (hardback)
1. Photovoltaic cells–Materials. I. Conibeer, Gavin, editor of compilation.
II. Willoughby, Arthur, editor of compilation.
TK8322.P4584 2014
621.3815′42–dc23

2013037202

A catalogue record for this book is available from the British Library.

ISBN: 9780470065518 (13 digits)

Set in 10/12pt Times by Aptara Inc., New Delhi, India

1 2014

This book is dedicated to our wives, Julie and Jenni, without whose support it would not have been possible.

Contents

Series Preface

WILEY SERIES IN MATERIALS FOR ELECTRONIC AND OPTOELECTRONIC APPLICATIONS

This book series is devoted to the rapidly developing class of materials used for electronic and optoelectronic applications. It is designed to provide much-needed information on the fundamental scientific principles of these materials, together with how these are employed in technological applications. The books are aimed at (postgraduate) students, researchers and technologists, engaged in research, development and the study of materials in electronics and photonics, and industrial scientists developing new materials, devices and circuits for the electronic, optoelectronic and communications industries.

The development of new electronic and optoelectronic materials depends not only on materials engineering at a practical level, but also on a clear understanding of the properties of materials, and the fundamental science behind these properties. It is the properties of a material that eventually determine its usefulness in an application. The series therefore also includes such titles as electrical conduction in solids, optical properties, thermal properties, and so on, all with applications and examples of materials in electronics and optoelectronics. The characterization of materials is also covered within the series in as much as it is impossible to develop new materials without the proper characterization of their structure and properties. Structure-property relationships have always been fundamentally and intrinsically important to materials science and engineering.

Materials science is well known for being one of the most interdisciplinary sciences. It is the interdisciplinary aspect of materials science that has led to many exciting discoveries, new materials and new applications. It is not unusual to find scientists with a chemical engineering background working on materials projects with applications in electronics. In selecting titles for the series, we have tried to maintain the interdisciplinary aspect of the field, and hence its excitement to researchers in this field.

ARTHUR WILLOUGHBY
PETER CAPPER
SAFA KASAP

List of Contributors

Tayebeh Ameri, Konarka Technologies Austria GmbH, Austria

Hamed Azimi, Konarka Technologies Austria GmbH, Austria

Daniel Bellet, Laboratorie des Matériaux et du Génie Physique, CNRS, France

Edith Bellet-Amalric, CEA-CNRS-UJF group 'Nanophysique et Semiconducteurs', CEA/INAC/SP2M, France

Christoph J. Brabec, Konarka Technologies Austria GmbH, Austria and Konarka Technologies Germany GmbH, Germany

Gavin Conibeer, School of Photovoltaic and Renewable Energy Engineering, University of New South Wales, Australia

S. Delbos, Institut de Recherche et Développement sur l'Energie Photovoltaïque (IRDEP), France

N. J. Ekins-Daukes, Department of Physics, Imperial College, London, UK

Martin A. Green, School of Photovoltaic and Renewable Energy Engineering, University of New South Wales, Australia

J. F. Guillemoles, Institut de Recherche et Développement sur l'Energie Photovoltaïque (IRDEP), France

Claudia Hoth, Konarka Technologies Germany GmbH, Germany

N. Naghavi, Institut de Recherche et Développement sur l'Energie Photovoltaïque (IRDEP), France

M. Paire, Institut de Recherche et Développement sur l'Energie Photovoltaïque (IRDEP), France

R. E. I. Schropp, Energy research Center of the Netherlands (ECN), Solar Energy, The Netherlands and Department of Applied Physics, Eindhoven University of Technology, The Netherlands

Andrea Seemann, Konarka Technologies Germany GmbH, Germany

Roland Steim, Konarka Technologies Germany GmbH, Germany

J. Vidal, Institut de Recherche et Développement sur l'Energie Photovoltaïque (IRDEP), France

Arthur Willoughby, Faculty of Engineering and the Environment, University of Southampton, UK

1 Introduction

Gavin Conibeer[1] and Arthur Willoughby[2]
[1] *School of Photovoltaic and Renewable Energy Engineering, University of New South Wales, Australia*
[2] *Faculty of Engineering and the Environment, University of Southampton, UK*

1.1 INTRODUCTION

The environmental challenges to the world are now well known and publicised, and all but a small minority of scientists accept that a reduction on dependence on fossil fuels is essential for addressing the problems of the greenhouse effect and global warming. Everyone is aware of the limited nature of fossil-fuel resources, and the increasing cost and difficulty, as well as the environmental damage, of extracting the last remnants of oil, gas and other carbonaceous products from the earth's crust.

Photovoltaics, the conversion of sunlight into useful electrical energy, is accepted as an important part of any strategy to reduce this dependence on fossil fuels. All of us are now familiar with the appearance of solar cell modules on the roofs of houses, on public buildings, and more extensive solar generators. Recently, the world's PV capacity passed 100 GW, according to new market figures from the European Photovoltaic Industry Association (14 February 2013), which makes a substantial contribution to reducing the world's carbon emissions.

It is the aim of this book to discuss the latest developments in photovoltaic materials which are driving this technology forward, to extract the maximum amount of electrical power from the sun, at minimal cost both financially and environmentally.

1.2 THE SUN

The starting point of all this discussion is the sun itself. In his book 'Solar Electricity' (Wiley 1994), Tomas Markvart shows the various energy losses to the solar radiation that occur when it passes through the earth's atmosphere (Figure 1.1).

The atmosphere also affects the solar spectrum, as shown in Figure 1.2.

A concept that characterises the effect of a clear atmosphere on sunlight, is the 'air mass', equal to the relative length of the direct beam path through the atmosphere. The extraterrestrial spectrum, denoted by AM0 (air mass 0) is important for satellite applications of solar cells. At its zenith, the radiation from the sun corresponds to AM1, while AM1.5 is a typical solar spectrum on the earth's surface on a clear day that, with total irradiance of 1 kW/m^2, is used for the calibration of solar cells and modules. Also shown in Figure 1.2 are the principal absorption bands of the molecules in the air. AM1.5 is referred to frequently in a number of the chapters in this book, and readers should be aware of its meaning.

Solar Cell Materials: Developing Technologies, First Edition. Edited by Gavin Conibeer and Arthur Willoughby.

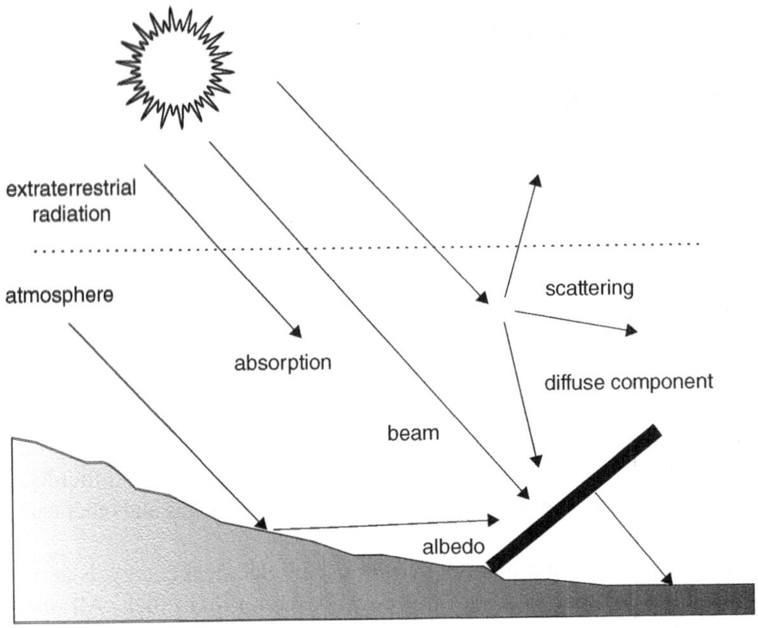

Figure 1.1 Solar radiation in the atmosphere. (Reproduced with permission from Markvart, 2000. Copyright © 2000, John Wiley & Sons, Ltd.)

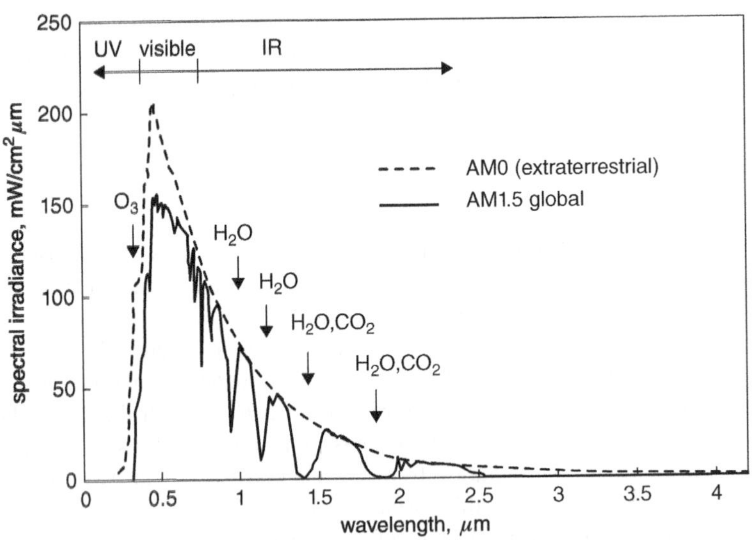

Figure 1.2 The solar spectrum. (Reproduced with permission from Markvart, 2000. Copyright © 2000, John Wiley & Sons, Ltd.)

1.3 BOOK OUTLINE

The book starts with a clear exposition of the fundamental physical limits to photovoltaic conversion, by Jean-Francois Guillemoles. This covers the thermodynamic limits, the limitations of classical devices, and develops this theme for more advanced devices. The identification of device parameters used in other chapters can also be found in this chapter.

Material parameters, of course, also require a thorough understanding of characterisation tools, and the second chapter, by Daniel Bellet and Edith Bellet-Amalric, outlines the main material characterisation techniques of special interest in solar cell science. X-ray analysis, electron microscopy, ion-beam techniques and spectroscopy characterisation methods are discussed, including Raman, X-ray photoelectron and UV/Visible spectroscopy, which are rarely detailed in such a materials book.

The next chapter, by Martin A Green, concentrates on developments in crystalline silicon solar cells. Despite the fact that silicon is an indirect-bandgap semiconductor, and therefore is a much less efficient absorber of above-bandgap light than direct-gap semiconductors (such as GaAs), silicon is still the overwhelming choice for solar cell manufacture. As the second most abundant element in the earth's crust, with a well-established technology, the chapter explores recent developments that have produced low-cost devices with efficiencies approaching the maximum physically possible.

Amorphous and microcrystalline silicon solar cells, are next reviewed by Ruud E I Schropp. These thin-film technologies are finding many exploitable applications with their lower usage of absorber materials and use of foreign substrates.

Turning next to direct-bandgap semiconductors, Nicholas J Ekins-Daukes outlines recent developments in III-V solar cells. III-Vs give the highest efficiencies of any solar cell materials. But despite their large absorption coefficients for above-bandgap light, the materials are relatively expensive, and often difficult and rare to extract from the earth's crust. Their place in the technology is assessed, together with recent advances.

Chalcogenide thin-film solar cells are next reviewed by Miriam Paire, Sebastian Delbois, Julien Vidal, Nagar Naghavi and Jean-Francois Guillemoles. Cu(In Ga)Se_2 or CIGS cells have made impressive progress in recent years with the highest efficiencies for thin-film cells, while Cu_2ZnSn $(S,Se)_4$ or CZTS or kesterite uses less-rare elements than CIGS, and so has significant potential for large-scale production.

The field of organic photovoltaics (OPV) has become of great interest since the efficiency achieved rapidly increased from around 1% in 1999, to more than 10% in 2012 (Green 2013). The chapter by Claudia Hoth, Andrea Seemann, Roland Steim, Tayebeh Amin, Hamed Azimi and Christoph Brabec reviews this novel technology, concentrating on the state-of-the-art in realising a photovoltaic product.

Lastly, one of us (Gavin Conibeer) looks to the future, by outlining third-generation strategies that aim to provide high conversion efficiency of photon energy at low manufacturing cost. Approaches covered include multiple energy level cells (such as tandem cells and multiple exciton generation), modification of the solar spectrum (such as by down- and upconversion), and thermal approaches (such as thermophotovoltaics and hot-carrier cells). The emphasis in all these approaches is on efficiency, spectral robustness, and low-cost processes using abundant nontoxic materials. The book ends with some concluding remarks by the editors, looking to the future in this rapidly developing field.

Finally, no book in this very extensive field can claim to be complete. To explore the field further, readers are recommended to consult 'Thin Film Solar Cells' by Jef Poortmans and Vladimir Arkipov (Wiley 2006), a companion volume in this Wiley Series on Materials for Electronic and Optoelectronic Applications, which includes such areas as dye-sensitised solar cells (DSSCs), in the chapter by Michael Gratzel. We hope that this book, with its emphasis on technological materials, will be of use to all who are interested in this field.

REFERENCES

Markvart, T., 'Solar Electricity' Wiley, Chichester 2000.

Poortmans, J., and Arkipov, V., '*Thin Film Solar Cells*' Wiley, Chichester 2006.

Green, M.A., Emery, K, Hishikawa, Y., Warta, W., and Dunlop, E.D., Solar cell efficiency tables (version 41), Progress in Photovoltaics: *Research and Applications*, 21 (2013) p. 1–11.

2 Fundamental Physical Limits to Photovoltaic Conversion

J.F. Guillemoles
Institut de Recherche et Développement sur l'Energie Photovoltaïque (IRDEP), France

2.1 INTRODUCTION

Where to stop the quest for better devices? What does better mean? The conversion efficiency arises prominently in this respect.

More efficient devices, everything kept equal, would first translate into cheaper solar electricity. Are there limits to reducing the cost of PV electricity? In 2012, modules were sold 0.5–0.7 €/W and the cost of solar electricity is around 20 cts/kWh. In the longer term, development of photovoltaics (PV) has to be based on a major technological breakthrough regarding the use of processes and materials at very low cost, or/and on the engineering of devices offering far higher performance, harvesting most of the available solar energy. Two approaches are targeted at this issue today: the first aims at low-cost materials and low-cost processes to reduce the surface cost of PV devices, possibly sacrificing some of the device efficiency, and the second, aiming at the maximal possible efficiency, at the same cost as today's modules (see Figure 2.1). There is a major difference between these two approaches: the conversion concepts, the materials and the processes.

If we think in terms of the manufacturing costs of PV modules, the target aimed at requires that the system needs to produce 1 MWh (comprising about 0.2 m^2 of high-end c-Si modules lasting 25 yr) cost less than €30 for parity with the base load or €120 for grid parity. For a very low-cost device, for instance based on polymers or organic–inorganic hybrids, with an expectation for conversion efficiencies on a par with those achieved by the amorphous Si line (on the grounds of similar structural disorder and a low carrier mobility) and shorter life durations, the budget is €7.5/m^2 (5-year life duration with 5% efficiency, including power electronics and installation), closer to the cost of structural materials than of functional electronic materials. Finally, for profitable electricity production, we need to pay attention to the system costs. Thus, one sees that it might be extremely difficult to attempt to reduce production costs far beyond what is currently being obtained with inorganic thin-film systems.

This chapter will deal with the scientific issues behind the photovoltaic conversion process, keeping in mind what would make a difference to having this technology more widely used.

The first of these questions is of course the efficiency of the processes. Since the appearance of the first PV devices, the question of the conversion efficiency limits arose, and for a good reason: not only does it have high scientific and technological visibility, it is also

Solar Cell Materials: Developing Technologies, First Edition. Edited by Gavin Conibeer and Arthur Willoughby.

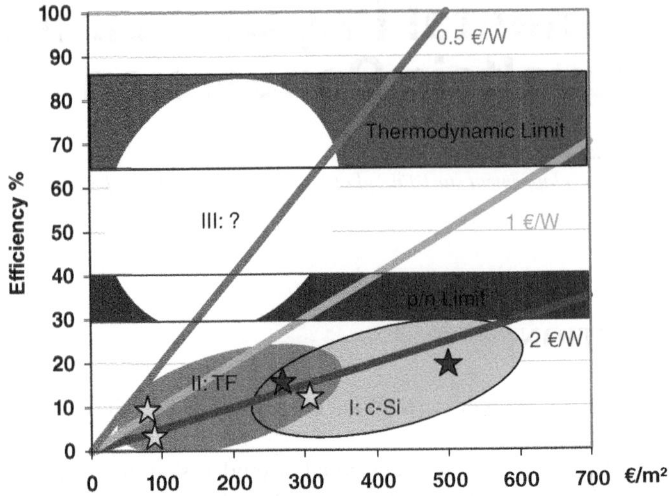

Figure 2.1 Relation of the cost per watt of solar energy to the surface cost of manufacturing solar devices (modules) and the device efficiency. The light gray, dark gray, and white oval regions represent the ranges found for crystalline silicon (first-), thin-film (second-), and third-generation solar technologies, respectively. The white zone marks the anticipated range for very high efficiency devices. For comparison, limiting efficiencies derived from thermodynamic constraints are also indicated as horizontal bands (low range: no concentration, high range: maximal concentration). Stars indicate industrial production costs as they could be estimated in 2010 from available data: filled stars for c-Si modules and hollow stars for thin-film technologies. (Adapted with permission from Green, 2003. Copyright © 2003, Springer.)

one of the major factors in lowering the cost of generating solar electricity. Interestingly, this question of efficiency limit took quite a bit of time before being settled [Landsberg and Badescu, 1998].

The paper of Schockley and Queisser, devising an approach based on a detailed balance approach of photovoltaic conversion is still one of the most quoted papers on PV, yielding the limit of single-junction, standard PV devices.

This question has also been approached on a more general basis, using thermodynamics (Landsberg and Tonge, 1980, Parrott 1992, De Vos 1992) to give device-independent or even process-independent limits (Section 2.1). These limits are essentially related to the source (the sun) characteristics and to the conditions of use (e.g. ambient temperature). Perhaps more useful, and practical, limits have been proposed for defined processes.

In very general terms, photovoltaic conversion in its simpler form supposes several steps:

1. solar photon transferred to the active part of the system;
2. absorption of the photons and energy transfer to the electronic system;
3. selective extraction of electrons to contacts (2 at least);
4. channelling of e-free energy to useful load whose impedance is adjusted.

These steps are illustrated in Figure 2.2 and describe PV process as it is working in all working devices, with nonessential modifications for organic PV (in which electron and holes are coupled as excitons) and multijunction cells (where the incident spectrum on a cell is a part of the total solar spectrum).

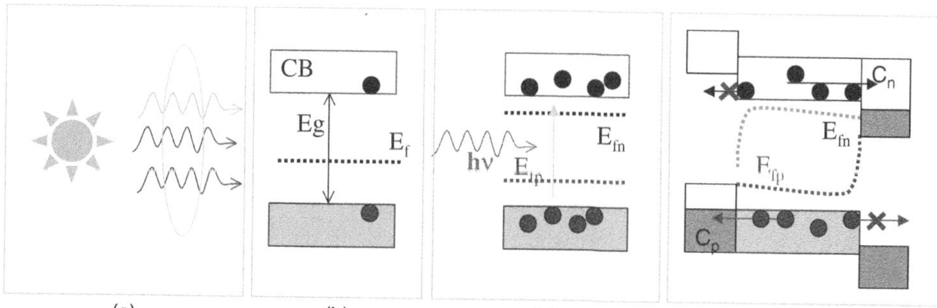

Figure 2.2 Photovoltaic action proceeds in 3 steps. (a) photons have to be collected and coupled to the converting system; (b) the converting systems contains occupied (VB) and empty states (CB) separated by a gap between which light induces transitions are allowed; (c) upon photon absorption, two populations of charge carriers are created; (d).

In this process, the work per absorbed photon is equal to the electromotive force between the electrodes (i.e. the voltage) times the elementary charge, that is the difference in the quasi-Fermi levels in the two contacts. In the absence of current extraction and when the mobility is high enough this quasi Fermi level (QFL) separation is that of electrons and holes in the absorber (see Section 2.4 for a more complete discussion).

Because the QFL is generated by the incoming flux, it increases with the light flux. This can also be understood as a larger generation rate per unit volume will create a larger density of electron–holes pairs, and therefore a higher conduction-band electron QFL and a lower valence-band electron QFL (that is a larger chemical potential of holes in the VB). This large QFL separation can be obtained in different ways, everything else being equal:

- by decreasing the recombination pathways for photogenerated carriers, for instance increasing the carrier lifetime;
- by increasing the generation of electron–hole pairs, for instance by concentration of the incident solar influx;
- by decreasing the generation volume, for instance by thinning down the cells, which requires light trapping to keep the total generation constant.

A large chemical potential can be seen as a large partial pressure: this helps the extraction of generated carriers and therefore a larger free energy per carrier can be collected, whereas, whatever the concentration of electron–hole pairs, their potential energy is always the same, near E_g. The collection of carriers depends on the chemical potential of the carriers in the contacts, that is, *in fine*, of the external conditions, and for instance the load into which the solar cell will deliver power.

The maximum power is delivered when the load impedance matches the differential impedance of the generator (as is true by the way for any generator).

Indeed, if the device has a current–voltage characteristic $I(V)$, the power $P = I \cdot V$ is maximum for

$$\frac{dP}{dV} = 0 \text{ and, as } \frac{dP}{dV} = I + \frac{dI}{dV} \cdot V \tag{2.1}$$

this yields:

$$R_{\text{Load}} = \frac{I}{V} = -\frac{dI}{dV}, \tag{2.2}$$

which is the relation announced.

The current at $V = 0$ (short-circuit) is noted I_{sc}, while the voltage at $I = 0$ (open circuit) is noted V_{oc}.

One can write a relation such that:

$$\eta = \frac{I_{sc} \cdot V_{oc} \cdot \text{FF}}{P_{inc}} \tag{2.3}$$

where η is the efficiency, P_{inc}, the solar incident power and FF, the fill factor, is a number close to 0.85 and slowly dependent on the working point for an ideal cell.

Importantly, each carrier has to be collected at a specific contact, that is, ideally, the contact should be selective for one of the carriers, and prevent collection from the other. Selective contacts usually take the form of barriers for one of the contacts, as for instance in a p/n junction.

There are general relationships based on thermodynamics that fundamentally limit the efficiency of conversion of light into work. They have been discussed extensively in the literature and are presented in Section 2.2.

Photovoltaic conversion, as sketched in Figure 2.2, starts with a good coupling between the conversion device and the sun. We therefore start to explore limits to photon management. This limit is essentially given by the 2nd thermodynamic principle: a PV device does not become hotter than the sun (Section 2.2). It is then important to know how good the absorption of the material can be as this determines the dimensions and geometry of the device as its thickness has to be the best compromise between being

- large enough so that light is efficiently absorbed;
- small enough so that excited carriers are transported to the contacts before recombination.

Progress in material science has been extremely fast in the past decades, and that knowledge may help us to get closer to the above limits. So, it becomes more important to understand precisely what these are. We therefore look in more detail at the working of a standard (even if somewhat idealised) PV device (Section 2.3).

The next question is then: can we come up with practical approaches to get closer to the thermodynamic limits? A variety of such approaches have been proposed, although they sometimes come at the expense of increased device complexity. They fall into several categories that will be described below (Section 2.4).

2.2 THERMODYNAMIC LIMITS

Thermodynamics sets the most fundamental limits to any energy-conversion process via its two fundamental laws: energy conservation and maximum entropy for closed systems.

The thermodynamic framework captures perfectly the fact that the useable energy (called work, W) that can be extracted from a body is only a fraction of its total internal energy (E).

2.2.1 The Sun is the Limit

Common sense tells us that the energy source is certainly essential in delving into the limits of the extractible power, but thermodynamics teaches us that since we deal not with energy creation but energy conversion, we have to look at both the source and the sink. Our sink is the earth, whose ambient temperature is around 300 K, which we will take as the sink temperature.

How should we characterise the source? The sun emits a considerable power as electromagnetic radiation (4×10^{23} kW) into space burning some 10^{15} g/s of hydrogen in nuclear reactions and converting some 5000 t/s of matter into pure energy. This radiation is, to a good approximation, thermal and described as blackbody radiation of a temperature of 5800 K (often approximated as 6000 K in the literature). The resulting solar spectrum (Figure 2.3) is rather broad, with most energy being radiated in the near-infrared and in the visible ranges.

As was discussed in Section 2.1, the conversion efficiency of a solar cell depends on the incoming flux: more flux (of the same spectral density) means more power out. In principle, light from the sun can be concentrated using, for instance, lenses and mirrors. By such systems, the solid angle under which the cell sees the sun changes increases by the concentration factor, due to the conservation of étendue.

The maximum concentration of solar flux is limited to a factor of $C_m = 42\,600$. Indeed, the maximum concentration factor is 4π divided by the solid angle of the sun, which is when the exposed surface of the cell can see the sun in all directions. This can also be understood from another point of view: at higher concentration factors, the focal point would have a higher radiation temperature than the source, which is thermodynamically impossible: the image of an object cannot be hotter than the object itself. Otherwise, one could reversibly transport heat from a cold body toward a hot body, something forbidden by the second law of thermodynamics.

$$\eta_{\mathrm{Carnot}} = \frac{W}{E_s} = 1 - \frac{T}{T_s}$$

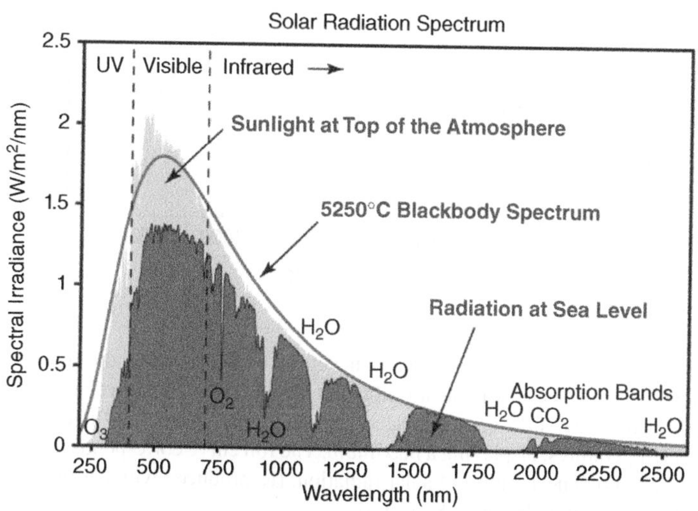

Figure 2.3 Solar spectrum at the top of the atmosphere and at sea level.

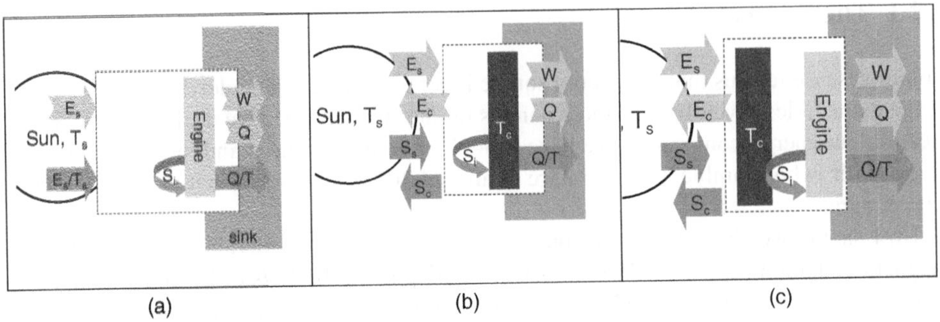

Figure 2.4 Systems for calculating various thermodynamic limits (a) for the Carnot efficiency, (b) for the Landsberg efficiency and (c) for the endoreversible efficiency. T_s is the source temperature and T is the sink temperature. Energy exchanges are represented with light gray arrows and entropy exchanges (or creation: S_i) with dark gray arrows. E is radiant power, Q is heat flux, and W is work.

2.2.2 Classical Thermodynamics Analysis of Solar Energy Conversion

The first step in computing thermodynamic efficiencies is to define the inputs and outputs, that is to define the conversion system.

A first such system is composed of the sun's surface and the conversion engine (Figure 2.4a). The surface of the sun receives energy and entropy fluxes from the sun's interior

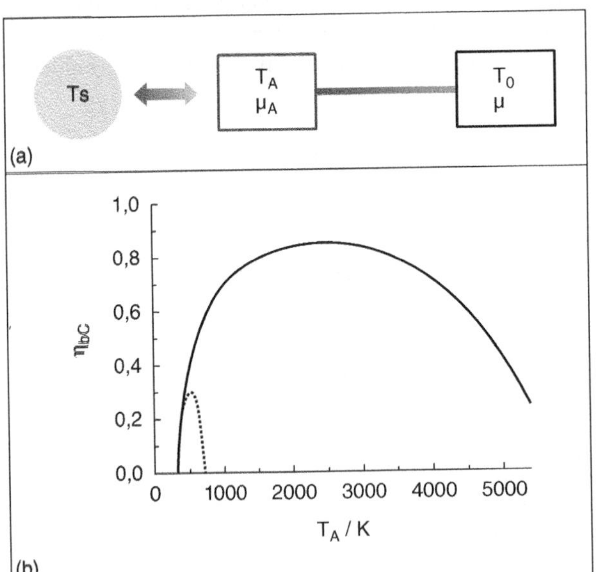

Figure 2.5 Efficiency of an endoreversible solar cell. (a) schematics of the device, showing that solar heat is radiatively exchanged with an absorber at temperature T_A while a Carnot engine operating between T_A and T_0 produces work. T_0 is connected to a cold reservoir. (b) the efficiency is computed for various absorber temperatures, displaying a maximum. The full curve corresponds to full concentration and the dotted one to no concentration of solar radiation. (Reproduced with permission from Wurfel, 2005. Copyright © 2005, John Wiley & Sons, Ltd.)

where nuclear reactions take place. Energy and entropy are then transferred radiatively to the device that rejects heat and work at ambient temperature T. Classically (see Green 2003 for a recent discussion), the limiting efficiency of such a process is obtained when the internal entropy generation is negligible and given by:

$$\eta_{\text{Carnot}} = 1 - \frac{T}{T_s} \tag{2.4}$$

This is the Carnot efficiency of the process, with a value of 95%. Of course, such an efficiency does not count as a loss of the fraction of power radiated by the cell to the sun, but this is not the point (the latter is a very small quantity). The main issue is that to have negligible internal entropy generation, the converter and the sun's surface should have almost the same temperature, otherwise the radiative transfer between the two would result in finite entropy generation. But then, the amount of work produced is infinitesimally small: most of the sun's power is recycled, which misses the purpose.

A more exact calculation taking into account the entropy of radiation to the sun, first done by Landsberg, gives:

$$\eta_{\text{Landsberg}} = 1 + \frac{1}{3}\frac{T^4}{T_s^4} - \frac{4}{3}\frac{T}{T_s} \tag{2.5}$$

that is, an efficiency of 93.3%. This corresponds (Figure 2.4b), as above, to a vanishingly small power extracted, as above, as the solar cell should be near open-circuit voltage for near-zero-entropy generation, except in the case where the system are nonreciprocal, as discussed by Ries and later by Green. Normally, optical systems are reciprocal, that is symmetric by time inversion, and this implies that if a solar cell gets light from the sun from a solid angle Ω_a, the light it emits (and we will see that solar cells must emit light in operation) should also reach the sun with a probability proportional to Ω_a. Yet systems exist that are nonreciprocal [Ries, 1983, Green, 2003], and in principle the light they emit could be sent to another device to be converted. In this way, a process can be conceptually conceived where the conversion is reversible, and the conversion device only emits thermal radiation (at ambient temperature) back to the sun. Although this seems to be an academic exercise, the Landsberg limit is effectively the highest possible limit for power extraction achievable, at least in principle. The Landsberg limit is achieved under maximal concentration, and calculation for various illumination have also been made [Brown, 2003] and give for instance a value of 73.2% for AM 1.5.

Figure 2.4c proposes yet another way for solar energy conversion using a Carnot engine using a reciprocal system. The sun heats up a blackbody absorber to a temperature T_a, and a Carnot engine extracts work while heat is dumped in the heat sink at temperature T. The thermodynamic analysis of this endoreversible system (De Vos, 1992) led to the formula:

$$\eta_{\text{Endo}} = \left(1 - \frac{T}{T_a}\right)\left(1 - \frac{C}{C_m}\frac{T_a^4}{T_s^4}\right) \tag{2.6}$$

where C is the solar concentration and C_m is the maximal possible solar concentration ($C_m = 46\,200$).

Maximal power is extracted at maximal concentration for an optimal temperature of $T_a = 2480$ K. The efficiency is then 85.4%. The endoreversible efficiency is quite dependent on the concentration ratio as illustrated in Figure 2.4.

The efficiency can be marginally increased to 86.8% if the light is split into quasi-monochromatic beams (e.g. using dichroic mirrors), each of them being converted by an endoreversible system such as the one above. The gain comes from the fact that the blackbody absorbers emit broad band light that can be partially recycled (Green, 2003).

Do we know of Carnot-like engines for the conversion of light into electricity? Actually yes: as we will see later, solar cells (somewhat idealised of course) are good approximate realisations of such engines when they are illuminated by a monochromatic light. We will also discuss another type of Carnot-like engine: the hot-carrier solar cell.

All the above description has been done using only one thermodynamic variable (temperature), but can be generalised to cases where other variables change and need to be taken into account. This would be the case for photochemical conversion, including a special case: that of photovoltaic conversion (De Vos, 1992).

2.3 LIMITATIONS OF CLASSICAL DEVICES

Beyond the general limits set by thermodynamics and optics that were reviewed above, there should be more specific bounds on practical devices. The most important of them is the first that has been developed: the semiconductor diode. It was first made as a Schottky diode (Se diodes [Palz, 2010) and later as a planar silicon p/n homojunction [Schockley and Queisser, 1961], followed by planar heterojunctions (GaAs), convoluted heterojunctions (Cu_2S), p-i-n structures (a-Si:H) and more recently interpenetrated (dye solar cells) or bulk heterojunctions (polymeric cells). They all share a common feature that limits their performance: absorption of light, generally done in one single material, results in only one type of excitation of the absorber that can be converted. This excitation, whose energy is close to the low-energy edge of the optical absorption threshold (the bandgap, E_g), is converted after significant relaxation in the absorber has occurred. This material parameter, the bandgap, is central to the understanding of the fundamental losses in PV conversion by solar cells. Essentially, these losses originate from the fact that one is trying to convert a broadband spectrum (blackbody radiation) with a system having a single characteristic energy (E_g): one is bound to find the best compromise.

To avoid being too specific on the system of conversion, some approximations are generally used when evaluating various approaches. A real device will be nonideal in many ways:

- Charge transport always entails ohmic losses, but these can be made negligible if the carrier mobility is high enough (so that the total ohmic drop is less than kT/q).
- Photon transport entails optical losses such as reflection or parasitic absorption. These can be made very small in practice with careful design and use of antireflection coatings.
- Heat-transport issues: because of power dissipation in the above losses and others (e.g. carriers thermalisation), photovoltaic devices become warmer under operation, and this is detrimental to the conversion efficiency. Yet, in principle, the device could be efficiently coupled to a thermostat and its temperature uniform and fixed at a given value (for instance the standard 25 °C value), so variations in temperature are ignored.

2.3.1 Detailed Balance and Main Assumptions

Detailed balance is a principle dating from the early part of the 20th century [Onsager, 1931] ensuring thermodynamic consistency of the formulation of the problem. It specifically requires that each microscopic process (carrier scattering, light-matter interaction, etc.) in a system is in equilibrium with its reverse when in equilibrium. This gives a relationship between the rates of a process and of its reverse and leads to the very useful assumption that this relation is still valid out of equilibrium. It can also be understood as the time-inversion invariance of quantum mechanics that imposes that the probability of a scattering event $I_i> \rightarrow I_f>$ and of its reverse be the same (this is not ensured though when rotations or magnetic fields are present). Kirchoff's law relating absorption and emission probabilities of a system pertains to this type of law.

Detailed balance was first used to assess the limits of conversion efficiency of solar cells by Shockley and Queisser [Schockley and Queisser, 1961], in conjunction with other useful assumptions. Applied to solar cells, the detailed balance requires the photon absorption rate for each pair of states in equilibrium to balance the photon emission. It is then assumed that away from equilibrium the microscopic probabilities of scattering from an initial to a final state are the same, only the occupation of the initial states changes and modifies the rates.

To describe a solar cell in a very general way, many simplifying assumptions have been made, and they help reveal the nature of the photovoltaic action and the limitations of the processes. In a first set of assumptions, we will assume that transport is not limiting (mobilities and thermal conductivities are high, and optical losses minimal, i.e. all photons above threshold are used in the conversion process).

Moreover, the conversion process can be nonideal for other reasons:

- If a fraction of the photons of a given energy that can be absorbed by the converter is not used to create an electron–hole pair, this is clearly suboptimal.[1]
- If carriers can recombine with nonradiative channels (an electron and a hole recombine without emitting a photon). Such recombinations contribute to heat dissipation, and are therefore also nonoptimal.
- If in a given material, several absorption processes are possible for a given photon (e.g. intraband and interband). Normally, one of them will enable more output power from the device, therefore the other processes act as parasitic absorption losses.

In a second set of assumptions, we will assume the optical absorption of the converting material has the following properties:

1. The absorptivity of the active material is either 0 (under the absorption threshold) or 1 (above). That this is indeed optimal has been discussed in details [Marti and Araujo, 1994].
2. Nonradiative recombination is neglected. Indeed, it has been found to be very small in some systems. Note that radiative recombination is mandatory for consistency (detailed balance rule) and has to be accounted for.

---

[1] note that this can't be compensated by the reduced radiative recombination that would be entailed, according to detailed balance, by a reduced absorptivity as shown by [Marti and Araujo, 1994].

3. For each photon, one assumes that only one process is allowed, namely, the most efficient process for energy conversion. This is a convenient simplifying assumption, even in the case where two processes give the same power yield.

If one is interested in the limit, it is only necessary to take into account fundamental loss mechanisms, i.e. those that cannot be made vanishingly small, even in principle. All of the above could, conceptually at least, be eliminated.

2.3.2 p-n Junction

In the radiative limit, the global photon balance of the solar cell imposes that when an electron–hole pair is created (a photon is absorbed), it either recombines (and gives a photon) or this electron–hole pair is extracted to do some work in the external circuit. It can be expressed as:

$$I = q \cdot (\Phi_{\mathrm{abs}} - \Phi_{\mathrm{em}}) \tag{2.7a}$$

$$\Phi_{\mathrm{abs}} = \int A(E) \cdot \dot{N}_{\mathrm{inc}}(E) \cdot \mathrm{d}E \tag{2.7b}$$

$$\Phi_{\mathrm{em}} = \int \dot{N}_{\mathrm{em}}(E) \cdot \mathrm{d}E \tag{2.7c}$$

where $A(E)$ is the absorptivity (equal to the emissivity: $A(E) = \varepsilon(E)$, according to Kirchhoff's law), N_{em} and N_{inc} are, respectively, the photon density of emitted and incident photons. When photons from the cell issued from recombination can be emitted in the full hemisphere above the device, N_{em} is given by [Wurfel, 1982]:

$$\dot{N}_{\mathrm{em}}(E) = \frac{2\pi}{h^3 c^2} \varepsilon(E) \frac{E^2}{e^{(E-\mu)kT} - 1} \tag{2.8a}$$

where μ is the electron–hole quasi-Fermi-level separation in the device, equal to the chemical potential of the radiation, and is generally assumed constant across the device, in which case it can be equated to the electric bias V if measured in eV (see below).

In a more general case, one should introduce the emitted radiated beam étendue, a conserved quantity during propagation that can be interpreted as its entropy [Markvart, 2008]. For a ray with directions within a small solid angle $\delta\omega$ passing through an area δA at an angle with the normal θ, propagating in a medium of refractive index n, the element of étendue δe is defined as

$$\delta e = n^2 \cos\theta \cdot \delta\omega \cdot \delta A \tag{2.8b}$$

The emitted flux within this étendue is then [Markvart, 2008]

$$\dot{N}_{\mathrm{em}}(E) = 2 \frac{\delta e}{h^3 c^2} \varepsilon(E) \frac{E^2}{e^{(E-\mu)kT} - 1} \tag{2.8c}$$

To be still more general, and more realistic, the absorptivity should be replaced by the quantum efficiency (number of electrons collected per photon absorbed). In the context of limit efficiencies and ideal devices, the two can be equated.

The Fermi level is given by – at equilibrium, and in the Boltzmann approximation – as a function of the equilibrium carrier concentration n°:

$$E_{\mathrm{f}} = E_{\mathrm{f}^\circ} + kT \ \ln(n) \tag{2.9a}$$

The quasi-Fermi level concept generalises this relationship in nonequilibrium for each carrier population, with n the actual (nonequilibrium) carrier population, as

$$\begin{aligned} E_{\mathrm{fn}} &= E_{\mathrm{f}} + kT \ \ln(n/n^\circ) \\ E_{\mathrm{fp}} &= E_{\mathrm{f}} + kT \ \ln(p/p^\circ) \end{aligned} \tag{2.9b}$$

By definition,

$$\mu = E_{\mathrm{fn}} - E_{\mathrm{fp}}, \tag{2.9c}$$

and is a local quantity. Whereas:

$$qV = E_{\mathrm{fn}}^{\mathrm{n}} - E_{\mathrm{fp}}{}^{\mathrm{p}} \tag{2.9d}$$

is the difference at the n and p contacts. If the mobility is large, μ does not vary much and can be approximated everywhere by the external voltage. These expressions can be easily generalised in a more general case using Fermi distributions.

The power extracted from the solar cell reads:

$$P = I \cdot V = \mu \cdot (\Phi_{\mathrm{abs}} - \Phi_{\mathrm{em}}(\mu)) \tag{2.10}$$

That is, *in fine*, (finally) to say that the electrical work can be related simply to the work done by the photons as the product of the chemical potential of the photons times the net flux absorbed. As we will see later, this can be generalised to more complex systems, and this greatly facilitates their analysis.

With P_{inc}, the incident power, given by

$$P_{\mathrm{inc}} = \int \dot{N}_{\mathrm{inc}}(E) \cdot \mathrm{d}E \tag{2.11a}$$

one can define the efficiency as:

$$\eta = \frac{P}{P_{\mathrm{inc}}} \tag{2.11b}$$

where P is the electrical power taken at maximum power point (see Section 2.1).

Using assumptions of Section 2.3.1, it is plain that the delivered power of the solar cell depends only on the spectrum of incident light, the device temperature and the bandgap of the absorber (as the value at which the absorptivity goes from 0 to 1, in the above approximation).

Solving equations (2.7) to (2.11) gives the ideal (maximal) efficiency of a solar cell (e.g. Green, 2003, Wurfel, 2007). One can, moreover, quantify from these equations the contribution of various losses: the fraction of the incident power that is not absorbed, the fraction that is re-emitted, the fraction that ends up as heat in the absorber (excess kinetic energy of the initially created electron–hole pairs) and the fraction dissipated in exchange of carrier collection. While other terms are obvious and have been described many times, the latter requires possibly more explanation. One of the descriptions is given in [Green, 2003]: there is still a difference between the energy of an electron–hole pair thermalised in the absorber (that is $E_g + 3kT$) and the useful work extracted (that is μ) from that electron–hole pair. This loss depends on the amount of absorbed photons and has been discussed for a cell illuminated by a blackbody at temperature Ts [De Vos and Pauwels, 1981, Landsberg, 1998, Hirst et al., 2011].

From the point of view of electrons, driving the photogenerated charges to the contacts has some entropy cost: because of the finite mobility, there should be a drop in the quasi-Fermi level separation. This entropy production rate can in part be related to the Joule effect [Parrot, 1992], but has even in ideal devices with very high mobilities, an irreducible component related to the entropy per photogenerated carrier pair and to the net recombination current [De Groot and Mazur, 1984], that stems from the imbalance of a chemical reaction:

$$\frac{\mathrm{d}S_{\mathrm{rec}}}{\mathrm{d}t} = \oiiint \frac{1}{T} \int \left(E_g + 3 \cdot kT - \left(E_{\mathrm{fn}} - E_{\mathrm{fp}}\right)\right) \cdot (G(E) - R(E, \mu)) \cdot \mathrm{d}E \cdot \mathrm{d}V \qquad (2.12)$$

which is zero at equilibrium because of detailed balance. Here, $G(E)$ and $R(E)$ are, respectively, the generation and recombination rates of electron–hole pairs of energy E. But as:

$$I = \int \mathrm{d}E \oiiint (G(E) - R(E, \mu)) \cdot \mathrm{d}V \qquad (2.13)$$

this is just the missing amount. Why is it called collection losses? The free energy collected from charge carriers is vanishingly small under a small departure from equilibrium although each electron–hole pair has an internal energy (potential + kinetic) of $E_g + 3kT$. Actually, carriers photogenerated under low illumination conditions are very dilute (large entropy per carrier), leading to a penalty in useful work that can be extracted from them as compared to when carriers are more concentrated: more entropy per carrier has to be dumped in the contacts.

The point of view of photons, also here, is more fundamental. The photon spectrum reradiated by the cell differs from that incident. More specifically, at maximum power point, the flux re-emitted is globally less intense than the one received (the flux difference is the extracted current). The flux re-emitted is also in general spread over a larger solid angle than the incident one. And finally, the emitted photon spectrum is different from the incident one. All three terms are discussed in [Markvart, 2008] and can be related to a change in photon entropy and to internal entropy creation during the photovoltaic conversion process.

Using relation (1.3), we can make a semiquantitative argument concerning the behaviour of η with P_{inc}, keeping the incident light spectrum unchanged. It could be naively expected that η should be independent of incident power, but this is not the case. Indeed, while the short-circuit current is proportional to the photon flux in a very large range (that is, keeping the spectrum identical, also to P_{inc}), and while FF increases only slightly (as long as the series resistances can be neglected of course), slower than $\ln(V_{oc})$, V_{oc} increases logarithmically as $\ln(P_{inc}/t)$, with t the cell thickness.

This stems from the usual exponential dependence of the recombination rate with the cell voltage. Specifically, in the radiative limit explored here, $R(E,\mu) \sim R^{\circ}(E)\cdot\exp(qV/kT)$, and is independent of position. Using equation (2.13), with $I = 0$ gives the desired relationship that V_{oc} is proportional to the logarithm of the volume integral of generation rate, i.e. to the log of the incident photon flux.

Finally, η increases essentially as $\ln(P_{inc}/t)$, yielding a gain close to 3% efficiency points per decade in solar cells: a good part of the journey to high efficiency starts with solar concentration. Indeed, looking at the progress made in terms of efficiency by the best triple junction (GaInP/GaAs/GaInAs, ~36% without concentration and ~43% at ×500 suns) devices as compared to the best single-junction devices (GaAs, ~28%), it appears that about half of the gain stems from the concentration factor while the other half comes from adding 2 other junctions.

One should also note that the important factor is P_{inc}/t, so that as long as the vast majority of photons are absorbed, reducing the cell thickness is as effective as increasing concentration for efficiency gains (but without the trouble associated with a more precise tracking of the sun). The optimal cell thickness depends on the ability to absorb most solar light within a small volume, and therefore both on the light-trapping strategies and on the absorptivity of the absorbing material. In the ergodic limit (complete randomisation of light rays) of light trapping, the effective absorptivity of the device can be increased up to a factor $4n^2$ as compared to the absorptivity of a slab [Yablonovitch, 1982]. Recently (Yu et al., 2010, Atwater and Polman, 2010), it has been shown that the absolute limit in terms of light trapping is somewhat larger than the ergodic limit, even if the absolute limit is not known yet.

For a device at 300 K illuminated by a standard AM1.5 spectrum, the efficiency and the contribution of the various losses are represented in Figure 2.6.

2.3.3 The Two-Level System Model

Omitting the losses due to electron–hole thermalisation and to nonabsorption of photons, an ideal solar cell illuminated is very close to being a Carnot engine. This can be realised by using photons near the bandgap of the solar cell. A good model for this situation is to represent the solar cell by a two-level system. If we now consider a two-level systems irradiated by photons matching exactly the transition, in the radiative limit we obtain:

$$I = q\cdot(N_{abs} - N_{em}(E)) = q\frac{2\pi E^2}{h^3c^2}A(E)\left(\overline{f}_{inc} - \frac{1}{e^{(E-\mu)/kT} - 1}\right) \tag{2.14a}$$

$$\overline{f}_{inc} = \frac{C_m - C}{C_m}\frac{1}{e^{E/kT} - 1} + \frac{C}{C_m}\frac{1}{e^{E/kT_s} - 1} \tag{2.14b}$$

$\overline{f}_{inc}$ is the incident photon average occupation factor, assuming a blackbody source diluted by a factor C/C_m (dilution due to sun–earth distance, C is the actual concentration used, while

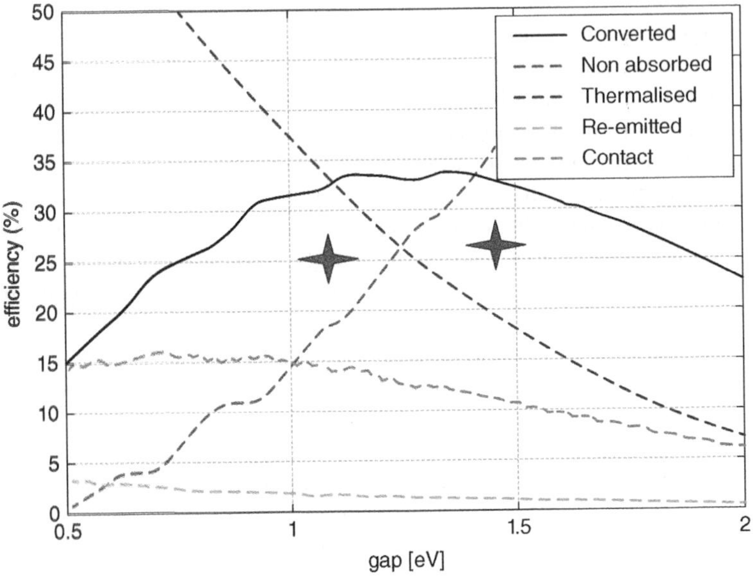

Figure 2.6 Efficiency of solar conversion by a single absorber as a function of its absorption threshold(gap) for a device at 300K illuminated by an AM1.5G spectrum and repartition of the various losses as explained in the text: the fraction of the incident power that is not absorbed,the fraction that is re-emitted as a result of reciprocity,the fraction that represents the excess kinetic energy of the initially created electron–hole pairs and the fraction dissipated in exchange of carrier collection. The stars represent two examples of very efficient cells actually made: c-Si (25%) and GaAs (28.8%) technologies.

$C_m = 46\,200$ is the maximally achievable concentration), and P_{inc} (to which the electrical power is to be compared to obtain the efficiency) writes as:

$$P_{inc,BB}(E, T_s) = \frac{C}{C_m} \frac{2\pi}{h^3 c^2} \frac{E^3}{e^{E/kT_s} - 1} \tag{2.15}$$

Clearly, there is no thermalisation in this case. Under maximal concentration, maximum efficiency is achieved at V_{oc} where

$$\mu_{oc} = q \cdot V_{oc} = E_g \cdot (1 - T/T_s) \tag{2.16}$$

showing its relationship to a Carnot device.

For a solar cell, maximal efficiency is achieved maximising equation (2.10) (as in equation (2.1)), and differentiation gives, after some algebra, the transcendental equation:

$$\overline{f}_{inc} = f_M \cdot \left[1 + (1 + f_M) \cdot \left(E/kT - \mathrm{Ln}\left(\frac{1 + f_M}{f_M}\right)\right)\right] \tag{2.17a}$$

where f_M is the emitted photon occupation factor at maximum power point, given by

$$f_M = \frac{1}{\exp\left(\frac{E - \mu_M}{kT}\right) - 1} \tag{2.17b}$$

and $\mu_M = q \cdot V_m$ is the quasi-Fermi level splitting at MPP. The efficiency at maximum power point (MPP) is then just given by $\eta = I_m \cdot V_m / P_{inc} = I_m \cdot q V_m / (E \cdot I_{ph})$, where I_{ph} is the photogenerated current and:

$$I_m = q \frac{2\pi E^2}{h^3 c^2} \left(\bar{f}_{inc} - f_M \right) = I_{ph} \left(1 - f_M / \bar{f}_{inc} \right) \tag{2.17c}$$

so that η is a function of E and f_{inc} alone (since f_M is an implicit function of E and f_{inc} as well).

Inclusion of a bandwidth yields a correction on account of thermalisation of electron–hole pairs in the band (carrier point of view). It can also be seen as modification of the incident spectrum (zero chemical potential and temperature T_s for a blackbody) to an emitted photon spectrum (Bose–Einstein distribution with finite chemical potential and ambient temperature): this is akin to cooling the photons while producing work [Markvart, 2008]. For a zero bandwidth, the cooling is isentropic [Wurfel, 1997].

To measure how far a device is from ideality, as suggested by equation (2.16) $E_g - V_{oc}$ [Guillemoles et al., 2008], or rather $E_g - q \cdot V_m$ [Nayak et al., 2011] have been used as landmarks for device quality.

2.3.4 Multijunctions

A tandem stack is made up of separate cells each with a different energy gap that absorb photons at different energy levels [Henry, 1980, Green, 2003]. The stack has the highest bandgap material at the top of the stack and allows lower-energy photons to pass through to the next cell, shown in Figure 2.7.

This arrangement of cells gives the stack a higher voltage than what would be obtained by a single diode as each photon is absorbed by a bandgap closer to its energy level and delivers a higher free energy.

A tandem stack can be in series (constrained, Figure 2.7c) or a combination parallel and series connections (Unconstrained, Figure 2.7b). In a tandem cell the number of distinct interband energy transitions is equal to the number of cells.

In the radiative limit and for ideal cells again, the power extracted from each cell can be expressed as:

$$P = \sum_n I_n \cdot V_n = \sum_n \mu_n \cdot (\Phi_{abs,n} - \Phi_{em,n}(\mu_n)) \tag{2.18}$$

where again the electrical work performed by the solar cell can be equated to the free energy change of the photons. For the series-constrained device, there is an additional constraint on the currents (all equal) that essentially translates in that photocurrents of all cells should be almost equal. The difference between optimised series connected and multiterminal device efficiencies are actually rather small (less than a percent) [Brown, 2003]. The main issue is more an increased spectral sensitivity for the series stack.

Because two-level solar cells could be ideal devices, it is tempting to consider solar energy conversion after the spectrum has been split into thin slices and consider an infinite stack of two-level junctions to examine the limit of solar energy conversion.

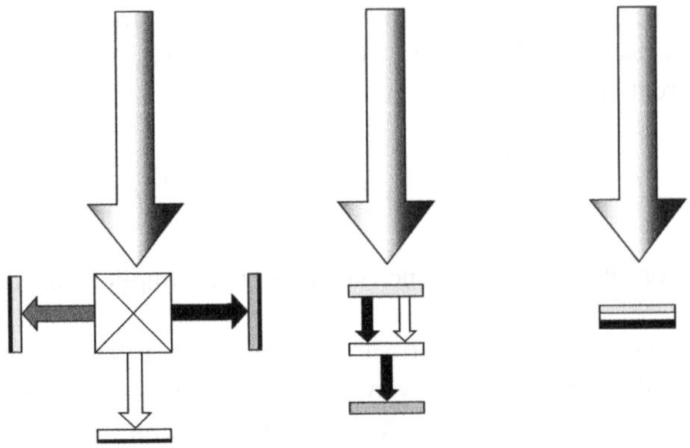

Figure 2.7 From left to right: (a) An optical device separates sunlight by means of filters in three beams "gray" (blue), "white" (green) and "black" (red), which are converted by three cells whose bandgaps were adapted for these three spectral bands. One can simplify the device as indicated in (b) noticing that the "gray" cell does not absorb the wavelengths higher than its threshold of absorption. It can thus be used as a filter for "white" and "black" cells. In the same way, the "white" cell lets the "black" beam pass. It is necessary, of course, that the substrates of the first two cells are transparent. Finally, in (c) a transparent electrical contact between the cells is realised. One can achieve this stack by deposition of the various cells on the same substrate. To avoid creation of a reverse diode at the interface between two cells (a diode p-n if diodes n-p are piled up), one connects them by a tunnel junction. In case (c) the cells are connected in series, whereas in the cases (a) or (b) one remains free to use the electricity produced in the most effective way. (© European Physical Society, EDP Sciences, 2010.)

Computing the optimal combination of gaps can become quite intricate when many single devices are associated [Rudd et al., 2005], so, let us consider asymptotic formulas, valid for a large number of devices.

An approximate method in the case of series tandem cells is to use the cumulative photon flux from the chosen spectrum and divide it into $n + 1$ sections to try to equally distribute the number of photons between transitions. This method is most applicable to constrained tandems that have the same current through all levels, and it predicts correctly the larger half of the energy transitions. These starting values are also quite useful for unconstrained tandems, as in the limit of large level number, the two have increasingly similar parameters. This type of approach, based on the characteristics of the spectrum is quite useful from a practical point of view and provides a better understanding of the devices.

This is confirmed using previously found values for multiple levels cell (1 to 40 levels, blackbody, full concentration case) and curve fitting, an interpolation formula was found for the smaller, average and larger energy threshold values. The actual points and the calculated values are shown in Figure 2.8 and Table 2.1. These equations are rather accurate for

Table 2.1 Tandem starting value equations.

Equations for Tandem BB MC	x = Number of Levels
$J = 5.8E7^*(x + 1)^{-1.04}$	$E_{ave} = 1.11x^{-0.064}$
$E_{min} = 1.02x^{-0.558}$	$E_{max} = 0.709\ln(x) + 1.04$

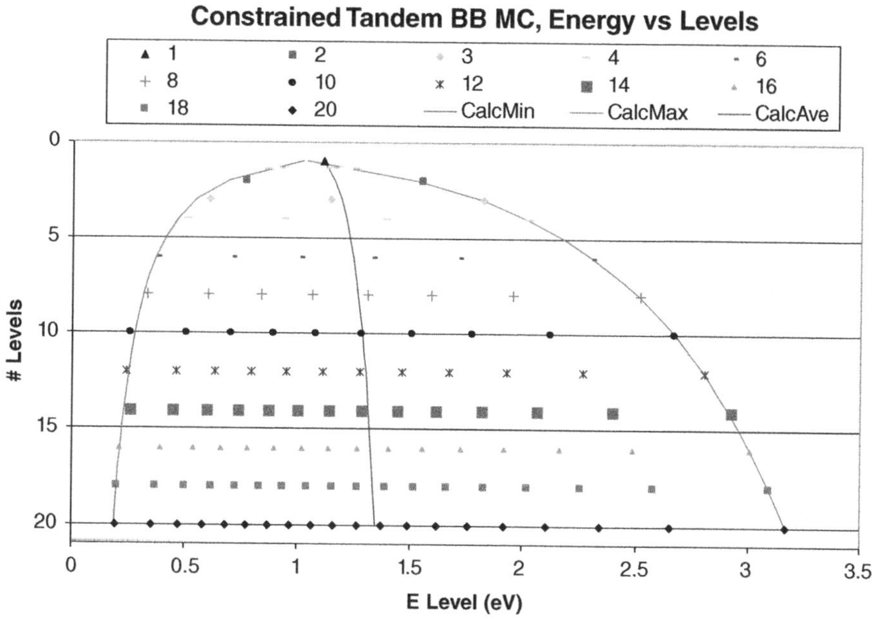

Figure 2.8 Optimised energy levels of tandem cells under blackbody maximum concentration. The lines show the calculated values generated by equations. (Reproduced with permission from Rudd, 2005. Copyright © 2005, Australian Solar Council.)

blackbody spectrum considered ($T - 6000$ K, full concentration). These three values were then used to generate starting values for all energy levels for an arbitrary number of levels under the same spectrum. The lower half of the levels is quite satisfactorily generated using this approach. However, the higher energy levels are at the moment not predicted with enough accuracy, and the approach is very sensitive to the spectrum, as are the fitting coefficients used in the interpolation formula.

The efficiency of a stack of N junctions itself is given to a good approximation by an asymptotic formula given by Brown [Brown, 2003], where η^{∞} is the efficiency of the infinite stack under the considered illumination, and "a" is a constant $\sim$1 whose exact value depends and if multiterminal or Series connected celles are considered:

$$\eta(N) = \frac{\eta^{\infty}}{1 + \frac{a}{N}}, \quad \eta^{\infty} : 68\% \text{ (AM1.5) and } 86.8\% \text{ } (\times 46\,200) \tag{2.19}$$

It has been proposed [Green, 2003] that nonreciprocal systems (without time-inversion symmetry, such as when magnetic fields or rotations are involved) could yield even higher conversion efficiencies, close to Carnot efficiency. A detailed discussion of these systems is beyond the scope of this chapter, but the interested reader can consult the sources [Ries, 1983, Green, 2003, Brown, 2003].

It should be noted that the actual devices have incompressible, finite, losses (optical, electrical) upon addition of a junction in a stack. For instance, there is a penaly for each additional tunnel junction (voltage drop) as it is not ideal. There should also be a penalty in terms of transmittance for each additional cell that impacts all cells below it. Finally, there will also be a cost penalty because of additional processing and also (very importantly) reduced yield. Overall, in terms of cost (€/W), it is unlikely that a large gain can be achieved beyond 3 junctions, even in the case of large concentration ratios ($\times 1000$).

2.4 FUNDAMENTAL LIMITS OF SOME HIGH-EFFICIENCY CONCEPTS

The design of optimal devices rests on the following, previously exposed, assumptions, within the detailed balance model:

1. for each photon, one assumes that only one process is allowed, namely, the most efficient process for energy conversion;
2. the absorptivity of the active material is either 0 or 1 in each energy range;
3. nonradiative recombination is neglected.

Many conversion concepts have been proposed with these restrictions. They are summarized in Table 2.2, together with their limiting efficiency and practical achievements. Following the approach used above for multijunction cells, one could consider an infinity of two-level systems, each of them tuned to a specific solar photon energy. As each photon energy of the solar spectrum can be used independently of any other, and its conversion can be optimised separately, this is setting an upper limit of the conversion efficiency of the solar spectrum. Solving equation (2.17) for each wavelength yields the maximal free energy extractible from the incoming photon flux at this same wavelength. The result is plotted in Figure 2.9 for

Table 2.2 Efficiency limits

Type of cell	Efficiency range AM1.5 at 46 000× (%)	Notes
Tandem, $N > 100$	68–87	Extremely complex, 43% achieved for 6 junctions
Hot carriers	67–86	Device not demonstrated in practice
Impact ionisation (all E_g mult.), TPV	44–86	Extremely constrained electronic band structure. Some demonstrators (TPV)
Thermionic (thermoelectric)	54–85	In practice, 23% (12%) achieved
Tandem, $N = 3$	49–64	In practice, 44% achieved (×970)
Upconversion	48–63	Very small gain demonstrated in practice
Intermediate level	48–63	Device not demonstrated in practice
Tandem, $N = 2$	43–56	In practice, 32% achieved
Downconversion	39–52	No gain demonstrated in practice
Impact ionisation (2 E_g)	38–52	Small gain demonstrated in practice (5%)
Simple p–n junction	31–41	28.8% under AM 1.5

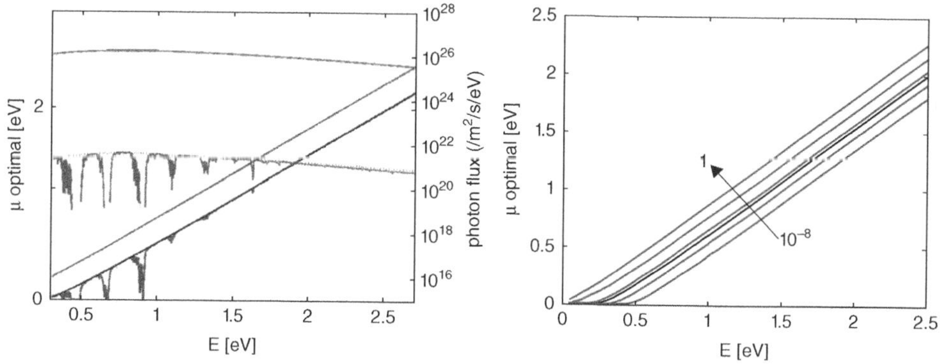

Figure 2.9 Left, optimal extractible free energies (in eV) extractible from the photovoltaic conversion of several solar spectra (respectively 6000 K Blackbody radiation, AM0 and AM1.5, central curve) as a function of the converted photon energy. Right, same curve represented for a 6000 K blackbody spectrum at its full intensity or attenuated in intensity by a factor 10^{-2}, 10^{-4}, 10^{-6} and 10^{-8} (in gray). The black curve corresponds to attenuation by 1/46 200, i.e. the intensity level received on earth.

representative solar spectra (blackbody 6000 K, AM0 and AM1.5G). For a given spectrum, this gives the highest possible free energy each (net) absorbed photon can yield. This value depends significantly of the amount of the flux of photons performing the work! When the spectrum is close to that of blackbody radiation, this free energy varies about linearly with the photon energy.

For a given conversion process, the extractible free energy at room temperature (i.e. work) is bounded by the values obtained above, that is the sum over all energies of the net photon flux multiplied by their optimal free energy. For a specific device, the sum only proceeds over the absorbed photons, while the optimal free energy to be considered depends on the conversion process. For instance, in the case of multijunction solar cells, the output voltage of each cell in the stack (optimally equal to the chemical potential of the emitted photons) is determined in accordance with the bandgap of the cell in which it is absorbed, rather than with the absorbed photon energy, but for a photon flux corresponding to the total flux of photons absorbed by the cell.

Again, the power delivered by the stack can be written from the electrons or photons point of view (equation (2.18)), from which the maximum efficiency can be obtained:

$$P = \sum_n I_n \cdot V_n = \sum_n \mu_n \cdot (\Phi_{\mathrm{abs},n} - \Phi_{\mathrm{em},n}(\mu_n)) = \int \mathrm{d}E \cdot \mu(\mathrm{E}) \cdot (\Phi_{\mathrm{abs}}(E) - \Phi_{\mathrm{em}}(\mu(E))) \tag{2.20}$$

2.4.1 Beyond Unity Quantum Efficiency

An alternative to the above approach (multijunctions) to increase efficiency, is to look for processes where the quantum yield e-/photons is not 1 (one photon gives one electron–hole pair), as assumed above. For instance, an electron–hole pair of sufficient energy might conceivably yield several electron–hole pairs by a process that is the inverse of Auger recombination: impact ionisation. Obviously, the current would be increased as compared to the single-junction case of the same gap (because of a better use of high-energy photons). Still another possibility is to have several photons able to promote an electron–hole pair, so that several subgap photons could create an electron–hole pair at the bandgap energy. In this case also, the current is increased as compared to the single-junction case of the same gap (because of a better use of low-energy photons).

In both cases, we have interaction channels between electron–hole pairs eh that enable them to exchange energy:

$$\mathrm{eh1} + \mathrm{eh2} \Leftrightarrow \mathrm{eh3} \tag{2.21}$$

Because of energy conservation, for these 3 pairs:

$$E_1 + E_2 = E_3 \tag{2.22}$$

If we now consider the entropy production by a process such as equation (2.22), it contains a term proportional to the affinity of the reaction A (De Groot 84) and to the recombination rate, as in equation (2.12). In the conditions of quasireversibility, close to equilibrium, where a minimum of entropy will be produced, we should therefore also have:

$$A = \mu_1 + \mu_2 - \mu_3 \sim 0 \tag{2.23}$$

so that the chemical potential of the photons emitted by the absorber may depend on the photon energy, contrary to the single-junction case, but also so that all electron–hole pairs of a given energy are created by the same process, and share the same chemical potential.

Such interaction channels can be generalised with four and more pairs. A particularly important case with four interacting pairs is given by the process:

$$\mathrm{eh1} + \mathrm{eh2} \Leftrightarrow \mathrm{eh3} + \mathrm{eh4} \tag{2.24}$$

because it describes also intraband collisions leading to equipartition of energy.

When $\mu(E)$ is continuous and smooth, $\mu(E)$ can be shown to be a linear function of E (Wurfel, 1997). States are often, in materials, grouped in bands between which carriers interact very rapidly, and are therefore in equilibrium. If they are moreover in equilibrium with the lattice (room temperature), the chemical potential is constant over this group of states. The chemical potential of the photon connecting two such groups of states (a photon is connecting only one pair of states, according to our assumption, an assumption often made, even if not always explicitly, in calculations of limiting efficiencies) is also constant over the corresponding energy range.

Let us now proceed to a generalisation of the above formalism, in the case where the quantum yield (QY) of a photon is not unity, i.e. whether a photon can create more than one electron–hole pair or several photons are required to create the pair.

Then, it holds for the current that:

$$I = \int \mathrm{d}E \cdot q \cdot \mathrm{QY}(E) \cdot (\Phi_{\mathrm{abs}}(E) - \Phi_{\mathrm{em}}(E, \mu(E))) \tag{2.25}$$

where $\mu(E)$ is a photon chemical potential that now depends on the photon's energy. For 2-terminal devices, as will be considered in the following,

$$P = I \cdot V = \int \mathrm{d}E \cdot \mathrm{QY}(E) \cdot q \cdot V \cdot (\Phi_{\mathrm{abs}}(E) - \Phi_{\mathrm{em}}(E, \mathrm{QY}(E) \cdot q \cdot V))$$
$$P = \int \mathrm{d}E \cdot \mu(E) \cdot (\Phi_{\mathrm{abs}}(E) - \Phi_{\mathrm{em}}(E, \mu(E))) \qquad (2.26\text{a\&b})$$

that is, on average, $q \cdot V \cdot QY(E) = \mu(E)$. We will see below that in some cases, this equality holds at every energy.

2.4.1.1 Impact Ionisation/Multiple Exciton Generation

The absorption of photons having energy more than twice that of the gap makes it possible to consider other mechanisms, where the excess energy can be used to create a second electron–hole pair. This phenomenon is called 'impact ionisation' (Figure 2.10). Devices using impact ionisation can have practical interest, provided that the process is effective already in the vicinity of the physical threshold ($\sim 2E_g$), otherwise the gain is likely to be quite small (Landsberg et al., 1993). The concept can be generalised to a photon of 3 times or more the bandgap energy producing 3 (or more) electron–hole pairs.

From the carriers' point of view, each photon of energy between n and $(n+1)$ times E_g produces n pairs, so that the generation in this range is strongly increased, as compared to the case treated in the previous section. But it is not only a generation pathway that has been added, but also a new recombination channel: n electron–hole pairs can now also recombine together by emitting a photon (in the radiative limit) of the corresponding energy.

Therefore, generalising the process of equation (2.21) for the creation of n pairs with a single photon of sufficient energy, one obtains the generalisation of equation (2.23):

$$\mu(n \cdot E_g) = n \cdot \mu(E_g) = n \cdot q \cdot V \qquad (2.27)$$

where V, the output voltage, is equal to the chemical potential of a pair at the bandgap separation since electrons are extracted from the bottom of the conduction band and holes

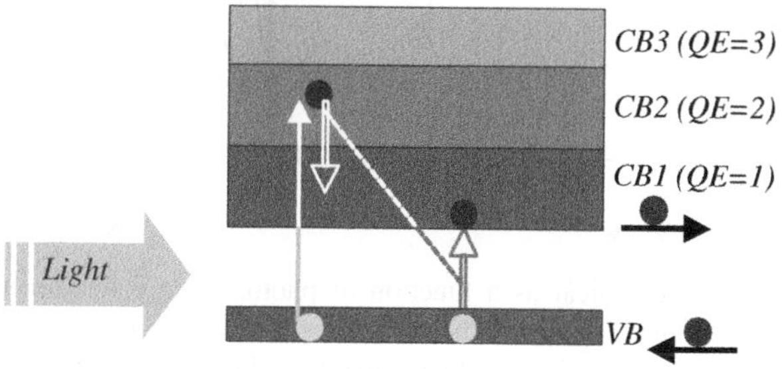

Figure 2.10 Processes in impact ionisation. VB is the valence band and CB 1–3 the conduction bands. For each range of width E_g, absorption in the band enables the creation of a number of electron–hole pairs that is a multiple of E_g as shown by the white arrows. Black arrows are for carrier current flow.

from the top of the valence band. Because n, in the above equation, is the quantum yield of the (possibly high energy) photon, it can be also written:

$$\mu(E) = \mathrm{QY}(E) \cdot q \cdot V \tag{2.28}$$

This equation has been commented several times (DeVos et al., 1998, Landsberg and Badescu, 2002), including in the case when $\mu(E)$ is not varying in steps, but is continuous, and we refer the interested reader to the corresponding papers.

Finally, with I_n denoting current from transitions of energy between n and $(n+1)$ times E_g, the output power of the device can be written:

$$\begin{aligned} P &= \sum_n I_n \cdot V = \sum_n q \cdot V \cdot \mathrm{QY}(n \cdot E_g) \cdot (\Phi_{\mathrm{abs},n} - \Phi_{\mathrm{em},n}(q \cdot V \cdot \mathrm{QY}(n \cdot E_g))) \\ &= \sum_n \mu_n \cdot (\Phi_{\mathrm{abs},n} - \Phi_{\mathrm{em},n}(\mu_n)) \end{aligned} \tag{2.29}$$

Interestingly, the first equality denotes the power from the carrier's point of view, and the second one, from the photon's point of view. Now, if the quantum yield is as high as physics allows ($\mathrm{QY}(E) < E/E_g$), it is possible to find a cell voltage so that the chemical potential gets close to that of the optimum chemical potential of Figure 2.9 and Figure 2.11, especially so in the case of maximal solar concentration, and with a small gap. Then, we can expect the output power of the device to be very close to that of the ideal multijunction solar cell. This is indeed found in the studies, with a maximal efficiency of 85.9% found for a gap of 0.048 meV. Under diffuse illumination, the optimal E_g was found around 0.76 eV, giving a peak efficiency of 43.6% (De Vos et al., 1998, Green, 2003).

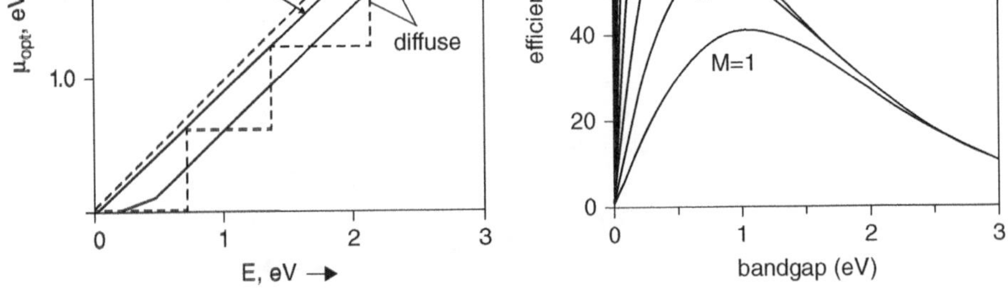

Figure 2.11 Left, optimal chemical as a function of photon energy for direct and diffuse light (continuous lines, see also Figure 2.9), and the best matches possible with a staircase function (according to Green, 2003). Right, device efficiency under maximal solar concentration as a function of the absorber's bandgap and number of bands from which impact ionisation is effective. (Reproduced with permission from De Vos, 98. Copyright © 1998, Elsevier.)

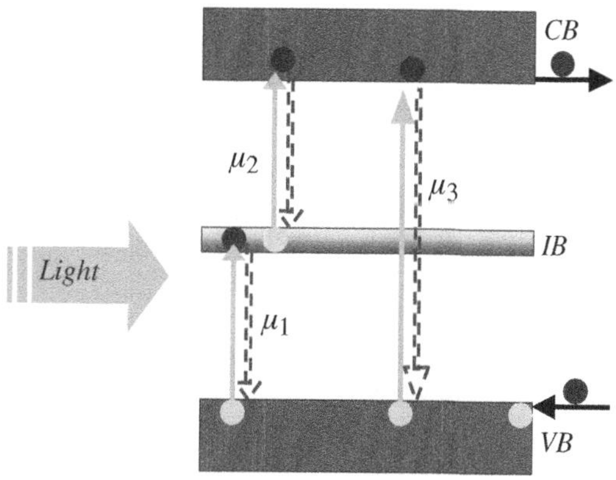

Figure 2.12 processes in intermediate band solar cells. VB is the valence band, IB the intermediate band or levels and CB the conduction band. Upwards arrows show the 3 possible generation processes, dashed arrows mark the three recombination processes and black arrows are for carrier current flow. The μ_i are the photon chemical potentials corresponding to the transitions.

2.4.1.2 Intermediate States/Bands (ILSC)

Proposals have been made to better utilise the low-energy part of the incident solar spectrum by using intermediate level semiconductors (ILSC) either with localised states [Wolf, 1960, Wurfel, 1993, Keevers and Green, 1994] or delocalised ones [Kettemann and Guillemoles, 1995, Luque and Marti, 1997]. It has been shown that such concepts have efficiency expectations up to 49% in an optimal device [Green, 2003], comparable to those of triple-junction devices (51%) under Air Mass 1.5 (AM1.5) standard terrestrial solar radiation [Green, 2003]. Indeed these intermediate level materials also have three possible optical transitions that may be adjusted to optimally use the solar spectrum. Contrary to multijunction devices, only one absorber material is needed, a fact that limits optical losses, tunnel junction losses and would help to design simpler, possibly cheaper, elaboration schemes. However, approaching those values would require rather low nonradiative recombination rates as compared to the radiative ones [Keevers and Green, 1994], a fact that has so far impeded the development of ILSC devices: without a specific mechanism to prevent them, nonradiative processes are far more efficient than radiative ones, especially those that involve the intermediate levels.

To start with, let's consider a simple case, as in Figure 2.12 and Figure 2.13, where the intermediate band (IB) or levels are distributed over a narrow energy interval located in the middle of the bandgap, and where transitions proceed at the same rate between the valence band (VB) and IB and between IB and the conduction band (CB), as required when there is no current extraction through the IB.

In this case, clearly, $\mu_1 = \mu_2$, and $qV = \mu_3 = 1/2 \cdot \mu_1$. Because one needs 2 photons to get an extractable pair, the QY of each transition is $^1/_2$. One has again that $qV = \mathrm{QY}(E) \cdot \mu(E)$.

The current in the cell is the sum of that obtained from transitions between VB and CB and either VB and IB or IB and CB (both have to be equal).

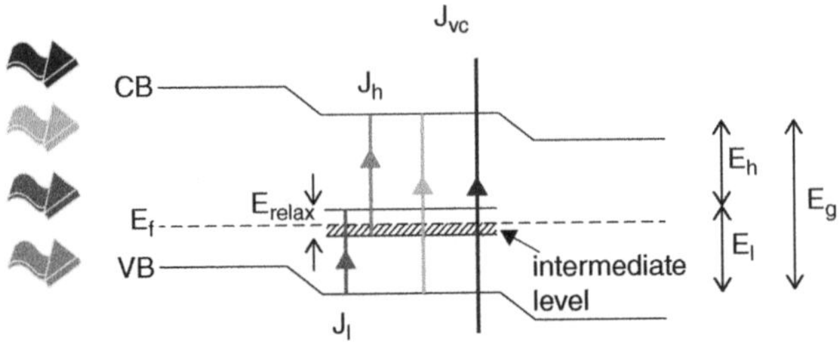

Figure 2.13 Photon processes in an intermediate band solar cell. (Reproduced with permission from Conibeer, 2007. Copyright © 2007, Elsevier.)

Again, it holds that the cell output power can be written either as a product of tension times current (the current is the photon flux times the quantum efficiency) or a sum over all energies of the photon chemical potential times the photon flux. Identifying the 2 means that

$$\mu(E) = q \cdot V \cdot \mathrm{QY}(E)$$

whenever a single process occurs per photon.

This reads again:

$$\begin{aligned} P &= \int \mathrm{d}E \cdot \mu(E) \cdot (\Phi_{\mathrm{abs}}(E) - \Phi_{\mathrm{em}}(E, \mu(E))) \\ &= \int \mathrm{d}E \cdot \mathrm{QY}(E) \cdot q \cdot V \cdot (\Phi_{\mathrm{abs}}(E) - \Phi_{\mathrm{em}}(E, \mathrm{QY}(E) \cdot q \cdot V)) \end{aligned}$$

We can observe that $\mu(E)$ has exactly the same variation as in the impact ionisation case, provided that the bandgap of the intermediate band absorber is twice that of the MEG absorber and that the quantum yield of the MEG absorber is limited to 2. Therefore, we expect the two systems to have the same efficiency.

This can be generalised to the case when the IBSC is not symmetric. In that case, μ_1 and μ_2 need not be equal (they refer to different photon populations), and a quantum yield for each photon is difficult to define. The relationship between μ_1 and μ_2 is still given by the fact that the currents between VB and IB or IB and CB have both to be equal. In the cases (favourable) where the current generated (i.e. the number of absorbed photons) are almost equal in the 2 transitions starting or ending at the intermediate band, it holds that:

$$\mu_1 = \frac{1}{2}\{q \cdot V + kT \ \ln(J_2^\circ / J_1^\circ)\}$$

where J_i° is the transition's saturation current. When the IBSC system is not symmetric, one gets an additional degree of freedom to adjust better to the optimal utilisation of the full solar spectrum and therefore to obtain higher efficiencies. The maximum efficiency under maximal concentration (blackbody 6000 K) was found to be 63.1% [Luque and Marti, 1997].

Such device concepts have been generalised to the case where several intermediate bands could contribute, a case that has been investigated in [Brown, 2003]. It was found that it becomes very difficult to arrange the transitions so that photons are efficiently used over the full solar spectrum. The case where specific selection rules, e.g. spin related, apply to the transitions, was also investigated in [Olsson et al., 2009].

2.4.2 Beyond Isothermal Conversion: Hot-Carrier Solar Cells (HCSC)

In the analysis of Section 2.3, another implicit assumption was that all processes in the device were isothermal, and occurring at room temperature. What if this assumption were relaxed?

A well-known case of nonisothermal conversion is given by thermoelectricity. Indeed, both solar cells and thermoelectric generators are semiconducting p/n junction devices. In a single-junction device maximum current and voltage are both unequivocally determined by the bandgap in the Shockley–Queisser model in the case of isothermal conversion process. But if the (normally) thermalised part of the absorbed solar energy in the device could be converted by a Seebeck effect, it could contribute an additional voltage as [Ketteman and Guillemoles, 2002]:

$$\Delta V = (S_n - S_p)\Delta T$$

where S_n and S_p are the Seebeck coefficients of the n and p regions, respectively.

Unfortunately, the creation of a large temperature gradient in semiconductors is precluded by their relatively high thermal conductivity. An alternative design has been proposed [Ross and Nozik, 1982], dubbed "hot-carrier solar cells", whereby excess photon energy is redistributed among the electron–hole pairs, which thereby get heated to a temperature above ambient, and extracted through energy selective contacts, which limits the heat flow and the entropy production.

This is because the heat (Q) and current (I) flows through the contacts are given by [Le Bris and Guillemoles, 2010]:

$$I = \sum \kappa \cdot \left(f\left(\frac{E-\mu_H}{kT_H}\right) - f\left(\frac{E-\mu}{kT}\right)\right) \quad Q = \sum E\kappa \cdot \left(f\left(\frac{E-\mu_H}{kT_H}\right) - f\left(\frac{E-\mu}{kT}\right)\right)$$

where the sums are carried out over all contributing conduction channels and where f is here the Fermi–Dirac occupation function and κ the transmission coefficient. For a perfectly energy selective contact pair, defined by the energy separation E_{ext}, and working close to reversibility:

$$f\left(\frac{E_{ext}-\mu_H}{kT_H}\right) \approx f\left(\frac{E_{ext}-\mu}{kT}\right)$$

that is, with $\mu = q \cdot V$, the relationship between the thermodynamic parameters of the hot electron–hole gas reads:

$$q \cdot V = E_{ext}\left(1 - \frac{T}{T_H}\right) + \mu_H \cdot \frac{T}{T_H}$$

Needless to say, putting to work the thermal energy dissipated upon absorption of photons requires a careful control of all causes of thermal dissipation, mainly by interaction with

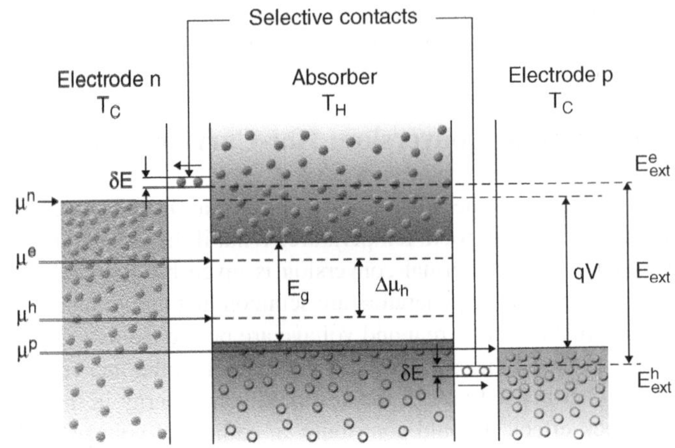

Figure 2.14 Band diagrammatic scheme of a hot-carrier solar cell (HCSC). Electron–hole pairs are created in the absorber producing a gas at temperature T_H, and chemical potential $\Delta\mu_H$, and extracted via two energy selective contacts (energy width δE) to cold n and p contacts at ambient temperature T_c. In this system, the bandgap E_g determines the photons absorbed (hence, the photocurrent) while the extraction energy, E_{ext}, determines the photovoltage. [Courtesy of A. Le Bris.]

the absorber atomic lattice [Le Bris et al., 2011A and B] and by heat transport to the cold contacts [Le Bris and Guillemoles, 2010].

The Ross and Nozik model [Ross and Nozik, 82] follows from the following balance equations, for current and for energy:

$$J = J_{abs} - J_{em}$$
$$P_{el} = P_{abs} - P_{em} - dQ/dt$$

where J_{abs} (J_{em}) is the number of absorbed incident photons (emitted photons, respectively), P_{abs} (P_{em}) is the *f* absorbed incident power (emitted power, respectively), and $P_{th} = dQ/dt$ is a heat-loss term (to the lattice by electron–phonon interaction) that will be neglected here in the context of limiting efficiency, but is otherwise of paramount importance. In the case of an energy selective contact:

$$P_{el} = J \cdot \Delta E = \left[P_{abs} - P_{em} - \frac{dQ}{dt} \right]$$

$$P_{abs} = A \cdot f_s \cdot \int_{E_g}^{\infty} \frac{E^3 \cdot dE}{\left(\exp\left(\frac{E}{kT_s}\right) - 1\right)} + A \cdot (f_{em} - f_s) \cdot \int_{E_g}^{\infty} \frac{E^3 \cdot dE}{\left(\exp\left(\frac{E}{kT_a}\right) - 1\right)}$$

$$P_{em} = A \cdot f_{em} \cdot \int_{E_g}^{\infty} \frac{E^3 \cdot dE}{\left(\exp\left(\frac{E-\mu_h}{kT_h}\right) - 1\right)}$$

The above equations form a complete set from which the operation of a hot-carrier solar cell can be modeled and its efficiency computed (Figure 2.15).

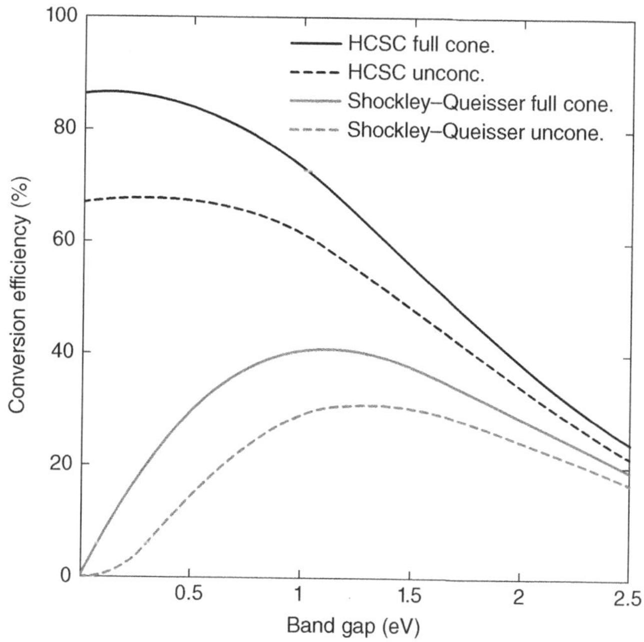

Figure 2.15 Plot of ideal efficiencies for hot carrier solar cells (thickines) compared to standard single junctions (thin lines), either under unconcentrated light (dashes) or fully concentrated light (continuous line).

This system also fits the developments presented above, and especially the photon representation of the power delivered (as in equation (2.26b)) will be quite useful.

Let us define:

$$\hat{\mu}(E) = E\left(1 - \frac{T}{T_H}\right) + \mu_H \cdot \frac{T}{T_H} = (E - E_{ext})\left(1 - \frac{T}{T_H}\right) + qV$$

This photon chemical potential yields the same Bose distribution of photon and spectral power at ambient as the Bose distribution at μ_H and T_H. For photons, only the occupation factor has a physical meaning, but the useable work per photon (when photons are emitted by a hot body and observed at a different temperature, one has to be careful) at room temperature is given by the above equation [Green, 2003]. Again, this linear relationship of $\mu(E)$ can fit quite nicely (for a proper choice of μ_H and T_H) the optimal $\mu(E)$ curve of Figure 2.9b, especially at high concentration, therefore, very high efficiencies can be expected for such a conversion concept. Indeed, with a limiting efficiency of 85.4% under maximal concentration and 67% under diffuse light, the limiting efficiencies are close to that of the infinite stack of junctions.

Note that there is a fundamental limit to the current that can be transported by a selective-energy contact, because of the Heisenberg relationship:

$$\Delta E \cdot \Delta t \approx \hbar$$

When $\Delta E \sim \delta E \sim kT$, $\Delta t \sim 0.1$ ps and the maximum current per conduction channel is still $\sim$1 μA, quite a large quantity, actually. At maximum concentration, the current would be in the kA/cm^2 range, which corresponds to a transmission channel density of $10^9\ cm^{-2}$.

2.4.3 Beyond the Single Process/ Photon: Photon Conversion

There are still other possibilities to approach limit conversion efficiencies. An important one relies on photon interconversion processes, enabled by materials with nonlinear optical properties.

These processes are not described with the same simplifying assumptions as above, since the same photons are used in different conversion processes (in the photon conversion device and in the photovoltaic part itself). This precludes the use of the previous approach, yet it is noteworthy that similar efficiencies were computed with these approaches and with approaches previously described that bear resemblances with them.

- photon addition,
  In the of "upconversion" or "photon addition" approaches, the photons whose energy is too weak to be used directly by a photodiode could be converted by nonlinear processes into a lower number of photons of larger energy.

  Photon addition is similar in concept to intermediate band (it is actually an intermediate band device of very high radiative efficiency operated at open circuit and whose re-emitted photons are used by a standard single junction).

  This type of devices has been studied in detail [Trupke et al., 2002a]. Its limiting efficiencies are 63.1% under maximum concentration. This limit is remarkably similar to the one found under the same illumination for ILSC.
- photon cutting,
  Another principle, 'photon cutting' consists, conversely, in absorbing photons of high energy in a luminescent converter to emit several photons of lower energy, to increase the current output of the photodiode. Photon cutting is similar in a way to impact ionisation (it can be viewed as an MEG device of very high radiative efficiency operated at open circuit and whose re-emitted photons are used by a standard single junction). This type of devices has been studied in details [Trupke et al., 2002b]. Its limiting efficiencies are 52% under maximum concentration. Again, this value is remarkably similar to the one found under the same illumination for MEG, (with only up to 2 excitons produced from a single photon).
- TPV, and selective emitters
  In a last concept, in the thermophotovoltaïc configuration, solar energy heats a radiator at high temperatures (>1000 K typically). The energy reradiated by the radiator is then converted by a photovoltaic cell whose gap is adapted to radiated energy (Harder, 2003). This system can be used to convert heat into electricity. In the thermophotonic configuration [Green, 2000, Harder and Green, 2002], the radiator is an online polarised junction p/n that makes it possible to increase the quantity of radiation emitted. If the radiator is a blackbody, its operation is described by Figure 2.5b.

  TPV is similar in a way to hot carrier (TPV is actually equivalent to a HCSC of very high radiative efficiency operated at open circuit and whose re-emitted photons are used by a standard single junction). The limiting efficiencies are similar to those of HCSC [Wurfel et al., 2005].

2.5 CONCLUSION

Limiting efficiencies for solar energy conversion is bound by the 2nd law of thermodynamics, as expected. Some devices already get close to the allowed limits. These are now well known and have been summarized in Table 2.2

Today, solar cells convert this energy using semiconductor-based diodes in a process where the photon flux creates an overpressure (as compared to equilibrium) of photogenerated carriers in the absorber. Albeit efficient, this process could be superseded in alternative devices, in principle at least, where other physical phenomena would contribute to solar energy conversion (carrier multiplication, photon conversion, hot-carrier collection, etc.).

It is noteworthy that all approaches proposed fall in different classes depending on the quantum yield of intermediate converting processes. These quantities turn out to be determining for the final limiting efficiency.

Finally, adopting the dual point of view of electrons or of photons, an important relationship relating device output voltage, photon chemical potential and quantum yield of intermediate processes is obtained.

NOTE

This chapter was submitted for publication in September 2012.

REFERENCES

H.A. Atwater and A. Polman, *Nature Mater.* **2629** (2010) doi: 10.1038.

A. Brown, PhD, UNSW 2003.

G.J. Conibeer, *Mater. Today* **10** (2007) 42–50.

S.R. De Groot and P. Mazur, "*Non Equilibrium Thermodynamics*", Dover Pub, New York. 1984.

A. De Vos, *Endoreversible Thermodynamics of Solar Energy Conversion*, Oxford University Press, Oxford, 1992.

A. De Vos et al., *Solar Energy Mater. Solar Cells* **51** (1998) 413–424.

A. De Vos and H. Pauwels, *Appl. Phys.* **25** (1981) 119–125.

M.A. Green, *Proceedings 16th EC Photovoltaic Solar Energy Conference*, eds James and James (2000) 51.

M.A. Green, *Third Generation Photovoltaics: Advanced Solar Electricity Generation*, Springer: Berlin, 2003.

J.F. Guillemoles, J.P. Connolly, O. Ramdani, O. Roussel, D. Guimard, V. Bermudez, N. Naghavi, P.P. Grand, L. Parissi, J. Kurdi, J. Kessler, O. Kerrec, and D. Lincot, *J. Nano Res.* **4** (2008) 79.

J.F. Guillemoles, *Europhys. News* **41/2** (2010) 19–22.

N.-P. Harder and M.A. Green, *Semicond. Sci. Technol.* **18** (2003) S270–S278.

N. Harder, *Semicond. Sci. Technol.* **18** (2003) S151–S157.

C.H. Henry, *J. Appl. Phys.* **51** (1980) 4494.

L.C. Hirst, et al., *Prog. Photovolt.: Res. Applic.* doi: 10.1002/pip.1024, **19** (2011) 286–293.

M.J. Keevers and M.A. Green, *J. Appl. Phys.* **75**, (1994) 4022–4031.

S. Kettemann and J.F. Guillemoles, *Physica E* **14** (2002) 101–106.

S. Kettemann and J.F. Guillemoles, *Proc. Of the 13th Photovolt. Sol. Energy Conference*, Nice, 1995 119.

P.T. Landsberg and P.T. Badescu, *Prog. Quantum Electron.* **22** (1998) 211–230.
P.T. Landsberg, H. Nussbaumer and G. Willeke, *J. Appl. Phys.* **74** (1993) 1451.
P.T. Landsberg and V. Badescu, *J. Phys. D: Appl. Phys.* **35** (2002) 1236–1240.
A. Le Bris and J.-F. Guillemoles, *Appl. Phys. Lett.* **97** (2010) 113506.
A. Le Bris, L. Lombez, S. Laribi, G. Boissier, P. Christol and J-F. Guillemoles, *Energy Environ. Sci.* 2012, DOI: 10.1039/C2EE02843C.
A. Le Bris, J. Rodière, C. Colin, S. Collin, J.-L. Pelouard, R. Esteban, M. Laroche, J.-J. Greffet, and J.-F. Guillemoles, J. Photovolt. **2** (2012), 506–511, doi: 10.1109/JPHOTOV.2012.2207376.
P.T. Landsberg and G. Tonge, *J. Appl. Phys.* **5** (1980) R 1–19.
A. Luque and A. Marti, *Phys. Rev. Lett.* **78** (1997) 5014.
T. Markvart, *J. Opt. A-Pure Appl. Opt.* **10** (2008) 015008.
A. Marti and G.L. Araujo, *Sol. Energy Mater. Solar Cells* **43** (1994) 203–222.
P.K. Nayak, J. Bisquert, and D. Cahen, *Adv. Mater.* **23** (2011) 2870–2876.
P. Olsson, C. Domain and J.F. Guillemoles, *Phys. Rev. Lett.* **102** (2009) **102**:227204.
L. Onsager, *Phys. Rev.* **37** (1931) 405.
W. Palz "*Power for the World: The Emergence of Electricity from the Sun*," Pan Stanford Publishing, 2010, DOI 10.4032/9789814303385.
J.E. Parrott, *Solar Energy Mater. Solar Cells* **25** (1992) 73685.
H. Ries, *Appl. Phys. B* **32** (1983) 153.
R.T. Ross and A.J. Nozik, *J. Appl. Phys.* **53** (1982) 3813.
J.W. Rudd, G.J. Conibeer and J.F. Guillemoles "User Friendly Software for Calculation of Limiting Cell Efficiencies for Research on Multijunction Solar Cells" *Proceedings of the Proc. 43rd Australia and NZ Solar Energy Society Conf.*, 2005.
W. Shockley and H.J. Queisser, *J. Appl. Phys.* **32** (1961) 510.
T. Trupke, M.A. Green, and P. Wurfel, *J. Appl. Phys.* **92**, (2002) 1668–1674.
T. Trupke, M.A. Green and P. Würfel, *J. Appl. Phys.* **92**, 7 (2002) 4117–4122.
M. Wolf, *Proc. IRE* **48** (1960) 1246–1263.
P. Wurfel, "*Physics of Solar Cells*" 2nd edn, Wiley, Wiley VCH, Weinheim, 2007.
P. Wurfel, et al., *Prog. Photovolt.* 2005.
P. Wurfel, *J. Phys. C: Solid State Phys.* **15** (1982) 3967.
P. Wurfel, *Solar Energy Mater. Solar Cells* **29** (1993) 403–413.
P. Wurfel, *Solar Energy Mater. Solar Cells* **46** (1997) 43–52.
E. Yablonovitch, *J. Opt. Soc. Am.* **72** (1982) 899.
M. Yamaguchi, "Super-High Efficiency III-V Tandem and Multijunction Cells", in *Clean Electricity from Photovoltaics*, M.D. Archer, R. Hill, eds., Imperial College Press, London, 2001; Vol. 1, Chapter 8, pp. 347–375.
Zongfu Yu, Aaswath Raman, and Shanhui Fan, " Fundamental limit of nanophotonic light trapping in solar cells", PNAS 2011 (early edition), www.pnas.org/cgi/doi/10.1073/pnas.1008296107.

3 Physical Characterisation of Photovoltaic Materials

Daniel Bellet[1] and Edith Bellet-Amalric[2]
[1] *Laboratorie des Matériaux et du Génie Physique, CNRS, France*
[2] *CEA-CNRS-UJF group 'Nanophysique et Semiconducteurs', CEA/INAC/SP2M, France*

3.1 INTRODUCTION

The purpose of the present chapter is to provide an introductory and short summary related to some material characterisation techniques of particular interest in solar cell science. Due to the length of the present chapter, only a few characteristic techniques are described and discussed. Some general remarks dealing with the correspondence between photovoltaic materials characterisation needs and physical techniques are briefly presented in Section 3.2. The following sections deal with chosen characterisation techniques: X-ray (Section 3.3), electron microscopy (Section 3.4) and spectroscopic (Section 3.5) methods.

Generally speaking, a better understanding of a material is based on the complementarity between different characterisation methods. This coupling is very important and will be illustrated with recent examples relating to materials specifically designed for photovoltaic applications. Clearly, a much larger number of examples could have been chosen. In the author's view, this chapter should obviously not be a detailed catalogue of all the techniques currently used but should rather more focus on a pedagogical presentation, based with concrete examples within the solar cell domain. Another goal is to emphasise on how a chosen characterisation technique has been recently developed or improved allowing then a more efficient approach of the needs on "new photovoltaic materials" e.g. thin films, nanostructures and organics. In parallel to the present contribution, general and interesting textbooks on material characterisation can be found in the literature [Zhang et al., 2009], [Brandon and Kaplan, 2008]. Finally, the present chapter is very complementary with reports on advanced characterisation techniques for solar cells, especially for thin films: see for example [Abou-Ras et al., 2011a], [Poortmans and Arkhipov, 2007] [Durose et al., 2004].

3.2 CORRESPONDENCE BETWEEN PHOTOVOLTAIC MATERIALS CHARACTERISATION NEEDS AND PHYSICAL TECHNIQUES

While fabrication of thin films or functional nanostructures is an outstanding challenge, the ability to characterise microstructure or nanostructure morphology, spatial order, chemical

Solar Cell Materials: Developing Technologies, First Edition. Edited by Gavin Conibeer and Arthur Willoughby.

composition, etc. and to correlate these characteristics to physical properties is an equally important activity. One of the main goals of research into characterisation techniques is to achieve, at the pertinent scale, understanding of the chemical and physical attributes of the structure, so that morphology and/or chemical composition effects on physical properties may be understood. To do so, various experimental techniques that offer high spatial and elemental resolution are often used, including for instance scattering (as X-ray diffraction, see Section 3.3), microscopy (electronic (see Section 3.4) as well as optical), spectroscopic methods (related in Section 3.5) or scanning probe analysis (such as atomic force microscopy). These statements hold for any material science problem; below we focus on materials dedicated to solar cells applications.

The examples chosen in this chapter to illustrate some appropriate use of characterisation techniques, deal both with organic and inorganic materials for solar cell applications. The three main phenomena that should occur in series within a solar cell are: photon absorption, carrier separation and carrier collection. Any feature in the bulk and at the material interface that could lower the efficiency of any of these three previous phenomena is of importance and should then be thoroughly considered and characterised. The important characteristic lengths playing a role in a solar cell are, among others: absorption length, diffusion length and specimen thickness. But the atomic arrangement also plays a key role since the main electronic properties depend on its chemical and structural nature, including any heterogeneity or disorder.

To be more specific, let us focus on solution-processed organic solar cells that are attracting substantial world-wide attention due to their potential as a low-cost photovoltaic technology. The active thin layer in such solar cells has to be simultaneously thick enough (~150–200 nm) to absorb most of the light and have internal structures small enough (~10 nm) to allow captured excitons to travel from the site of charge separation to the carriers collection location to ensure electricity generation [Swaraj et al., 2010]. One of the field's fundamental challenges is to create such structures in a controlled manner. Therefore, a good knowledge and understanding of the structural and electronic properties, both in the bulk and at the interfaces, are of crucial importance and should be characterised in detail. In such materials the important interface is between electron-donor and -acceptor materials where the charge separation of the exciton takes place. This problem will be briefly addressed later in this chapter.

3.3 X-RAY TECHNIQUES

X-ray methods are very powerful tools and are nearly always used when investigation of materials is concerned (either in thin films or bulk shape) for any photovoltaic applications. This analysis is nondestructive and provides quantitative information on the crystalline phases. With a suitable choice of X-ray source and optical geometry, X-ray diffraction (XRD) analysis can be applied to a very wide range of scientific material problems, such as crystallinity, phase determination, crystallite size, surface and interface roughness, texturation (orientations), etc. Many interesting books derive X-ray physics with fundamental and applied aspects for a wide community of material scientists, chemists, biologists or physicists (see for instance [Cullity and Stock, 2011]; [Als-Nielsen and McMorrow, 2001]; [Willmott, 2011]; [Guo, 2011]; [Pavlinsky, 2008]; [Warren, 1990]). In the following, essential features are reported concerning the main characteristics of X-ray techniques as well as a few related examples. Generally speaking, it is important to keep in mind that only elastic scattering from electrons is generally considered. The amplitude of the signal is very dependent upon *Z*, the atomic number. The order of magnitude of X-ray energy used for scattering experiments strongly depends on the material under investigation and should in principle be lower than the material absorption energy to avoid fluorescence. In a laboratory scale Cu-Kα radiation is often used (energy of 8 keV). For such an energy, the penetration depth drastically depends upon the incident angle and the material (*Z*): it can vary from zero to several tens of micrometres. Moreover, X-ray diffraction averages data over an area (up to few cm^2) much larger than the nano-objects under study.

Crystalline properties are often to be considered in solar cell materials. For instance, a lot of efforts concerning thin-film polysilicon technologies are directed towards producing a material with higher crystallographic quality than the current ones [Beaucarne and Slaoui, 2006]. Indeed it is believed that enhancing the crystallographic quality will increase the efficiency of these solar cells [Beaucarne and Slaoui, 2006].

3.3.1 X-Ray Diffraction (XRD)

The crystal phase identification of unknown materials is often performed by powder XRD. A powder can be considered here as an aggregation of fine grains (often also called crystallites) with one or several crystal structures having random crystallographic grain orientations. The diffracted intensity is proportional to the sum of X-rays reflected from all the grains giving rise to the diffraction peaks called Bragg peaks. Comparing experimental data in terms of peak positions and peak intensities with data base (such as the International Centre for Diffraction Data, ICDD) enables identification of the nature of the different materials or the phases of the investigated powder. X-ray powder diffraction is very often used in material science and not only to identify the crystal structure but also for strain analysis (shift of peak position, peak broadening), degree of crystallisation (peak intensity) or mosaicity or grain size (peak broadening). Let us focus briefly on this last feature.

XRD data can be linked with the coherent X-ray length that can be related to small crystalline regions. Indeed, when X-ray diffraction occurs on a material having small coherent domains, the XRD peak widths are wider. Due to the inherent consequences of the Fourier transforms, the smaller the nanocrystallites the larger the associated X-ray Bragg peaks. These domains correspond to the crystallites. In a polycrystal they are often associated with the grains (however, in some cases grains can also be constituted of crystallites due to the presence of twin boundaries), with precipitates in the bulk and with coherent crystallites in a single crystal porous material such as porous silicon [Lomov et al., 1995]. Line broadening due to domain size is not a trivial issue. This stems as well from the fact that in polycrystals the grain size is not at all uniform and grains can have different shapes [Mittemeijer and Scardi, 2004].

The Scherrer approximation is valid in crystalline powders where grains are assumed to be free from faulting and strain so that the peak FWHM ($\Delta\theta$) is only due to the crystallite size (L). This gives rise to the following approximate relation:

$$\Delta\theta = (K \cdot \lambda)/(L \cdot \cos\theta) \tag{3.1}$$

where λ is the X-ray wavelength and θ the angular position of the diffracted peak. K is a factor close to unity that depends upon the shape of particles ($K = 0.92$ for spherical particles). The Scherrer formula was derived for cubic systems but it can be generalised as an approximation of the average grain size, only perpendicular to the reflecting planes, also for noncubic lattices. A relation such as equation (3.1) cannot, for instance, be applied to deduce grain size along the plane that is parallel to the substrate, as is often reported in the literature.

In photovoltaics, the Scherrer approximation has been, for instance, used to determine the crystallite size of nanoparticles for the tandem solar cells where nanocrystallites are associated with quantum confinement [Parola et al., 2012] although for these very thin layers grazing XRD (see Section 3.3.2) is more appropriate [(Huang et al., 2007); (Hao et al., 2009); (Surana et al., 2012)]. The crystallite size is indeed a crucial parameter since it strongly influences the energy bandgap of the nanocrystals that act as quantum dots.

Such analysis is often corroborated with transmission electron microscopy (TEM) (Section 3.4.3) observations and gives rise, in general, to reasonable agreements. In addition, an oxide or amorphous shell can be observed by TEM (or by using Raman spectroscopy [Parola et al., 2012]); it cannot be deduced from XRD experiments, showing the complementarity of these techniques. Indeed, TEM imaging works in real space while XRD is intrinsically in reciprocal space; the first method is direct, destructive and local while the second one is nondestructive and averages information over a large sample volume. In addition, XRD often needs inputs for interpreting the data, and TEM or other techniques are therefore very useful for such a task.

A single crystal can be defined by having its crystallographic axis in the same direction everywhere. On the other extreme, a perfect powder exhibits randomly oriented grains with no preferential crystallographic orientations. In between, numerous materials possess a polycrystalline structure consisting of crystal grains with different orientations that might not be randomly oriented. Texture can be defined as the distribution of crystallographic orientations of a polycrystalline sample (for instance a specimen having random crystallographic orientations is said to have no texture). This was defined a long time ago [Harris, 1952] and depends on the percentage of grains having a given orientation. The texture coefficient is defined (and measured) as the ratio between the observed Bragg peak area divided by the area of the same Bragg peak associated with a perfect powder of the same material. Texture is seen in almost all engineered materials, and can have a great influence on material properties [Kocks et al., 2000]. An example of PV applications concerns the texture of transparent conductive oxides (TCOs). TCOs can exhibit large electrical conductivity and high optical transmittance and be used as front transparent electrode in solar cells (see Section 3.5.5 for more details). The texture of TCO films can strongly influence the grain-boundary nature, which partially affects the carrier mobility since a grain boundary can scatter the free electron carriers. Therefore, the texture of one TCO (fluorine-doped SnO_2) has recently been investigated as a function of growth temperature and sample thickness [(Consonni et al., 2012), (Consonni et al., 2013)]. It has been shown, for instance, that a texture transition can occur in these thin polycrystalline films from <101> to <100> and <301> crystallographic orientations when the growth temperature is raised.

Figure 3.1 exhibits results dealing with XRD and materials currently used for solar cells. It illustrates the potential of such technique and the different types of information usually extracted. Figure 3.1a shows a schematic XRD set-up. Figure 3.1b exhibits a XRD pattern of CdS film on ITO/glass substrate [Toyama et al., 2006]. The assignments of the crystal planes were performed on the basis of the lattice spacings of hexagonal CdS and ITO. Strain as well as the degree of preferred orientation was also extracted from this data, as reported in Figure 3.1b [Toyama et al., 2006]. Figure 3.1c exhibits a high resolution XRD reciprocal space mapping around the $(\bar{1}\bar{1}5)$ reflection of CdSe/ZnSe layer grown on GaAs substrate

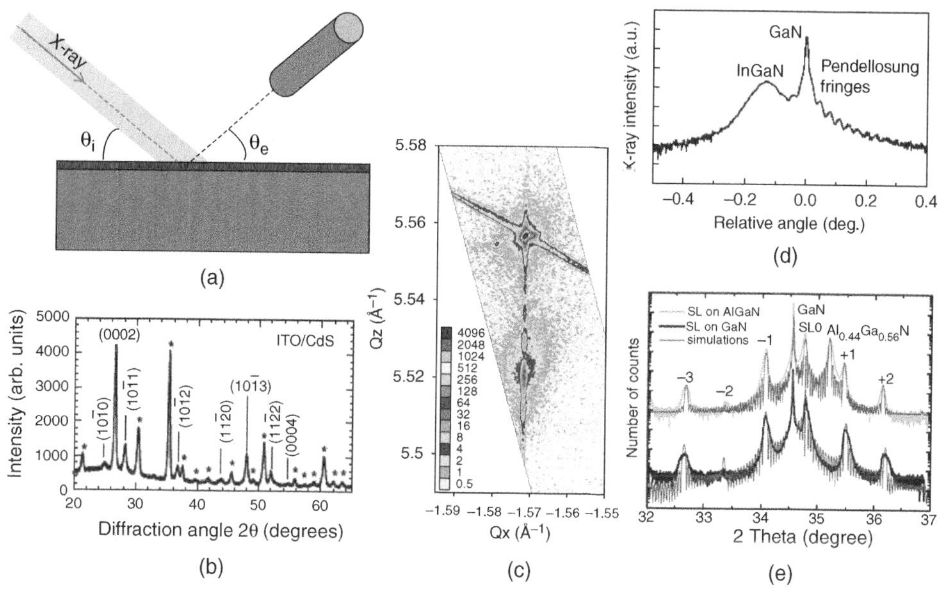

Figure 3.1 (a): Schematic representation of an X-ray diffraction setup. (b): X-ray diffraction pattern of CdS deposited on the ITO/glass substrate, from [Toyama et al., 2006]. The diffraction peaks labelled are assigned to hexagonal CdS (*hkil*) crystal planes and other peaks labelled asterisk are attributed to ITO. (c): High-resolution X-ray diffraction reciprocal space mapping around the ($\bar{1}\bar{1}5$) reflection of self-organised CdSe/ZnSe quantum dots. (Adapted with permission from Robin, 2005. Copyright © IOP Publishing, 2005.) For clarity two scales were used: one for the diffuse background and another one for the tops of the peaks. (d): High-resolution X-ray $2\theta-\omega$ scanning profile of the (0002) reflection from GaN/InGaN/GaN heterojunctions grown on *c*-plane sapphire. (Reproduced with permission from Zhang, 2008. Rights managed by AIP Publishing LLC.) (e): High-resolution X-ray $2\theta-\omega$ scanning profile of the (0002) reflection from a GaN/$Al_{0.44}Ga_{0.56}N$ superlattice grown on $Al_{0.44}Ga_{0.56}N$ (top) and on GaN (bottom) and the corresponding simulations. (Reproduced with permission from Koskar, 2011. Rights managed by AIP Publishing LLC.)

[Robin et al., 2005]. The fact that ZnSe and GaAs reflections appear both at the same inplane momentum q_x value is the proof that the II-VI layer is perfectly pseudomorphic with the GaAs substrate [Robin et al., 2005]. Figure 3.1d shows a high-resolution X-ray diffraction profile (i.e. parallel to the q_z direction) of the (0002) reflection from p-GaN/i-InGaN/n-GaN solar cell heterojunctions grown on c-phase sapphire [Zheng et al., 2008]. The presence of fringes (called Pendellösung) attests to the high structural quality of the InGaN/GaN heterojunction [Zheng et al., 2008]. Figure 3.1d corresponds to a single layer on a substrate. Figure 3.1e deals with a multilayer: XRD scans of the (0002) reflection of the GaN/$Al_xGa_{1-x}N$ (with $x = 0.44$) superlattices [Kotsar et al., 2011]. The simulations are superimposed on the experimental scans. The main superlattice reflections (labelled SL0 in Figure 3.1e) presents asymmetric shape with a tail toward larger (smaller) angles for the structures grown on GaN (AlGaN) that can be associated with a gradual misfit relaxation, in agreement with reflection high energy electron diffraction (RHEED) and TEM observations [Kotsar et al., 2011].

Generally speaking, the three main advantages of XRD stem from the statistical data often related to a large specimen volume, the extreme sensitivity of XRD data upon any modification in terms of growth conditions, annealing, etc. and finally from the ease to perform in situ experiments (such as annealing for instance). These three features explain why X-ray techniques are so often used in material science.

It is also worth noting that X-rays are also attractive for imaging since they can provide high-resolution images thanks to their wavelength being much shorter than the visible range. Moreover, two other features are important: the X-ray penetration into specimens enables the examination of thick samples and their chemical sensitivity provides interesting prospects of elemental mapping. If we add the achievements of synchrotron radiation [Willmott, 2011], which has been much applied in nanoscience [Guo, 2011], one can reasonably foresee promising prospects in the near future for X-rays use in photovoltaics materials especially for high-resolution and in situ experiments [Boscherini et al., 2010].

3.3.2 Grazing-Incidence X-Ray Diffraction (GIXRD)

Grazing-incidence X-ray diffraction (GIXRD) is now a mature technique that has been extensively applied to problems where polycrystalline thin films, powders or nanostructures are concerned [Tanner et al., 2004]. Indeed, GIXRD is a very useful method for clarifying crystalline information when dealing with very thin diffracting specimen. The use of very small incidence angle α_i (a few tenths of degrees in general), as schematically depicted in Figure 3.2a, increases the passing length of X-ray through the specimen and at the same time decreases the indepth X-ray penetration and thus lowers the diffracted contribution from the substrate. X-rays penetrate only a few nanometres into the sample for α_i values below the critical angle (α_c) that is defined by:

$$\alpha_c = \lambda \cdot \sqrt{\frac{\rho \cdot r_0}{\pi}} \tag{3.2}$$

where ρ is the average density of electrons, and r_0 is the classical free-electron radius. Therefore, an extremely high surface sensitivity can be attained by using GIXRD. This is illustrated in Figure 3.2b where the calculated penetration length and the square of the electrical field are plotted versus the incident angle for silicon and considering the CuK_α radiation ($\lambda = 0.154$ nm). Moreover, by adjusting the angle of incidence, the irradiated volume of the sample can be chosen. Typically, the penetration depth is a few tens to a few hundreds of nanometres and the X-ray beam surface area impinging on the specimen is of a few square centimetres. The experimental curve is usually recorded at a fixed incident angle α_i and by varying the 2θ angle in a range where several Bragg peaks are present (typically 20–80 degrees for instance).

Surface-sensitive diffraction methods such as GIXRD can reveal crystalline properties of very thin films of a few tens of nanometres such as the thin contacts used for CdTe cells [Durose et al., 2004] or in thin films containing Si or Ge nanocristallites, surrounded by an oxide, which should behave as a rather large bandgap semiconductor due to quantum confinement. Figure 3.2c exhibits a typical GIXRD of such thin films that are 30, 50 and 75 nm thick [Surana et al., 2012]. Clearly, GIXRD can improve the signal and minimise the contribution of the Si substrate. The good crystallinity of the quantum dots is then confirmed and one can use the Scherrer equation to deduce the dot size (of about 5 nm in the present case) [Surana et al., 2012]. Another important example concerns semiconducting material composed of

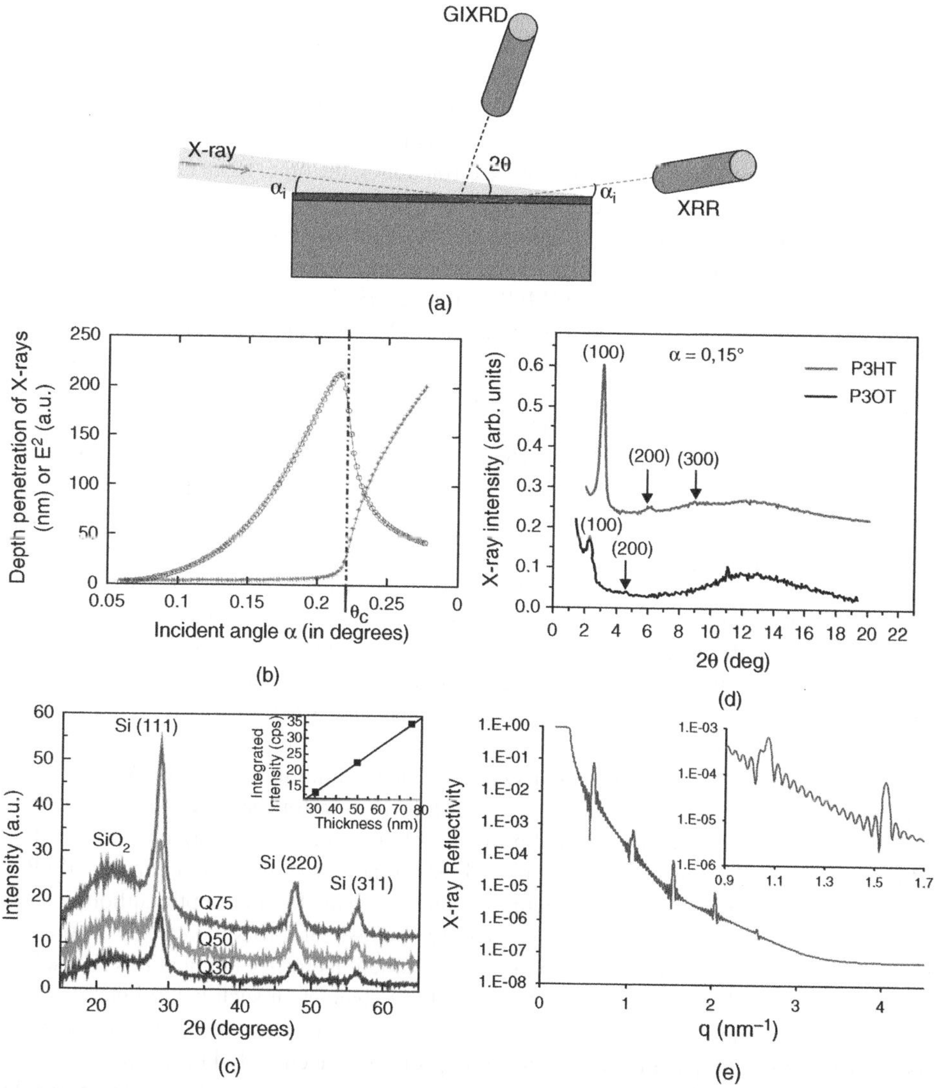

Figure 3.2 (a): Schematic representation of a low-incidence X-ray set-up used for grazing-incidence X-ray diffraction (GIXRD) or X-ray reflectivity (XRR). (b): X-ray penetration depth (+) and square of the electrical field (o) versus the incident angle α_i. (c): Experimental GIXRD curves associated with Si nanoparticles in SiO_2 oxide matrix in thin films of 30, 50 and 75 nm thickness. (Reproduced with permission from Surana, 2012. Copyright © IOP Publishing, 2012.) (d): GIXRD diffraction profiles of P3OT and P3HT obtained with an incident angle of 0.15 degree. (Reproduced with permission from Abad, 2012. Copyright © 2012, Elsevier) (e): XRR of a nonannealed multilayer composed of 15 periods of 6-nm Sn-rich SiO_2/6-nm SiO_2. The inset shows the zoom in the 0.9–1.7 nm^{-1} range of the module of the scattering vector q. (Reproduced with permission from Bellet, 2007. Copyright © 2007, Wirtschaft und Infrastruktur GmbH & Co Planungs-KG.)

π-conjugated polymers [Abad, 2012]. The latter materials have attracted attention due to their potential applications in optoelectronic devices such as organic solar cells [Kim et al., 2007]. Recent investigations have been devoted to better understand the correlation between structural properties and their performance as active layers in organic electronic devices. One concept, called a "bulk heterojunction", uses a blend of two materials that is fabricated to create a distributed p-n junction: one acts as an electron donor, the other as an electron acceptor. The two percolative networks created in the blend should mix at a nanometric scale compatible with the mean free path of the exciton formed after light absorption. Polytiophene exhibit suitable properties for such a concept. But their properties are strongly dependent on thermal postprocessing and blend composition [Muller et al., 2008]. The latter influences deeply the efficiency of the device into which they are integrated. A deep understanding of the dynamics of the pure components of the blend should help to better control and optimise the processing of the mixture. The recent work of Abad [Abad, 2012] is based on this idea and the authors reported a detailed calorimetric and GIXRD investigation of the most widely used poly-alkyl-thiophenes (i.e. P3OT and P3HT). The melting and crystallisation temperatures were identified as a function of temperature scan rates and concentration of the polymers in solution. The incidence angle between the surface film and incident beam was set at 0.15 degrees, corresponding to a penetration depth estimated at 260 nm allowing probing mainly the polymer thin film of about 200 nm thickness. Sharp and broad peaks were observed and assigned, respectively, to lamellar interlayer spacing of the poly-alkyl-thiophenes and to side-chain disorder (see Figure 3.2d). Peak intensity and peak position were followed in situ during thermal annealing as well as the increase of coherent length (deduced from the Scherrer relation). GIXRD results reveal some differences between P3HT and P3OT that can be assigned to the interdigitation of the side chain. The reported improvement in efficiency of organic solar cells upon annealing can be related to this onset of side-chain interdigitation [Abad, 2012]. Such investigation can then provide very useful information for the processing of devices containing an active layer with blends of poly-alkyl-thiophenes such as organic solar cells.

Moreover, since the penetration length is a drastic function of the incident angle, GIXRD can also be used to investigate the structural changes versus the film depth, although such investigation is often not straightforward. A careful analysis of GIXRD data was devoted to $CuInSe_2$, which constitute an essential part of heterojunctions for PV applications [Kotschau and Schock, 2003]. This study showed that structural changes in such heterojunctions can occur in a 5- to 60-nm thick Cu-poor surface layer. This is based on the assumption that the dependency of the lattice parameters versus composition is accurately known. In such case, it is then possible to link the structural depth profile with the compositional depth profile [Kotschau and Schock, 2003].

3.3.3 X-Ray Reflectivity (XRR)

X-ray reflectivity (XRR), being sensitive to the modulation of the material electron density, can provide precise information on the thickness, roughness and density of surfaces, thin films, interfaces and multilayers.

When an X-ray beam impinges on the surface of a sample at a very small angle, the material can be treated as one with a continuous electronic density and one can use similar considerations to those valid for visible light. In fact, the index of refraction n is also a relevant parameter. Let us first focus on specular reflectivity which means that the incident and exit angles (with respect to the sample surface) are identical (as shown in Figure 3.2a).

The reflectivity curve is simply the ratio between reflected and incident X-ray intensity versus the scattering vector defined as:

$$q = 4 \cdot \pi \ \sin\theta / \cdot \lambda \tag{3.3}$$

For X-rays, the real part of n is slightly smaller than unity. This leads to the phenomenon called total external reflection for incident angles smaller than the critical angle α_c, defined by equation (3.2), which is directly related to the material density [Als-Nielsen and McMorrow, 2001], [Daillant and Gibaud, 2009]. Above the critical angle, the reflectivity follows the standard expression as observed in optics, called Fresnel reflectivity which decays asymptotically as q^{-4}.

The remaining features of the specular reflectivity curves reveal interesting structural information along the direction normal to the specimen surface. Unlike XRD, XRR can be equally used for amorphous, polycrystalline or even liquid materials as for crystalline materials, as it essentially probes contrast in the average electron density rather than scattering from individual atoms [Willmott, 2011]. A typical specimen thickness range of 5–200 nm is accessible for a laboratory setup, while synchrotron facilities provide a much larger range spanning approximately 1 to 500 nm.

Let us consider the example of a multilayer structured specimen as those where alternating deposition of two types of layer are considered. Indeed, the XRR signal produced by a sample depends on the scattering length density (SLD) profile along the sample depth, the SLD being proportional to the electron density at a particular point. Although the SLD profile is a continuous varying function it can be often well approximated by a stacking of several slabs, each one of constant SLD. The X-ray waves reflected at different interfaces within the multilayer (or at the external surfaces) give rise to constructive/destructive interference effects and therefore to intensity oscillations in the XRR curve. This can be seen in Figure 3.2e exhibiting the experimental XRR curve [Bellet et al., 2007a] of a Si-rich SiO_2/stoichiometric SiO_2 stacking used to obtain Si nanocristallites after annealing [Zacharias et al., 2002]. These silicon nanostructures have been investigated extensively in recent years for their potential use in photonic and photovoltaic applications; especially with the goal of considering all-Si tandem solar cells [(Conibeer et al., 2006), (Huang et al., 2007), (Surana et al., 2012)]). Often, for silicon quantum dots embedded in SiO_2, control over interdot distance and size is observed in multiple bilayer stacks of silicon-rich oxides and silicon dioxide.

For such systems the total film thickness d gives rise to the so-called periodic Kiessig fringes through: $\Delta q = 2\pi/d$. One single layer would give rise to a modulation inversely proportional to its thickness. In the case of a multilayer system, due to its periodic nature, the thickness of the multilayer period (P) yields Bragg peaks separated by: $\Delta q = 2\pi/P$. Figure 3.2e exhibits both Kiessig fringes (with a high q-frequency) and Bragg peaks (at a lower q-frequency) for a multiple bilayer stacks [Bellet et al., 2007a].

The homogeneity of the multilayer system is an essential condition to observe these oscillations. This gives us a direct way for measuring the periodicity with a good accuracy (typically of the order of a tenth of a nanometre). Moreover, the real specimen surface and interfaces are not flat and smooth on the nanometre scale. Part of the incident beam is then diffused in other directions than the specular one. The XRR intensity damping then allows measurement of the surface and the interface roughnesses.

To obtain quantitative information about the multilayers, it is essential to perform simulations and data fitting. The analysis of the sample requires a minimum knowledge of the deposited layer or multilayer (concerning the chemical composition and thickness, even approximately). One can use refinement procedures that minimise the difference between the calculated and the measured XRR curve. Simulation program such as MOTOFIT [Nelson, 2006] can efficiently be used for this purpose.

3.3.4 Other X-Ray Techniques

For brevity, only a few other X-ray techniques will be described in this chapter. Let us only focus on two types of methods: the first one aims at determining spatial features in the studied material in terms of shapes and sizes either in the direct space (X-ray imaging) or in the reciprocal space (small-angle X-ray scattering: SAXS). The second important group discussed here will concern energy -selective response of a material leading to resonant soft X-ray scattering and resonant soft X-ray reflectivity.

If we consider synchrotron X-ray imaging, major developments have emerged in recent years and concern phenomena such as: absorption, phase contrast, diffraction and fluorescence. Indeed synchrotron radiation enables one to achieve quantitative 3D images of a specimen, with spatial resolution that, in some conditions, can reach the submicrometre range [Salvo et al., 2003]. This allows characterisation of specimens with three-dimensional imaging, high spatial resolution, exploiting X-ray beam coherency in situ and real-time observations (annealing, experiments under pressure, etc.). For instance, synchrotron X-ray imaging during solidification of multicrystalline silicon for photovoltaic applications reveals information on the dynamical evolution of the solid/liquid interface as well as on grain structure, strain and twinning [Tandjaoui et al., 2012]. The generation of a three-dimensional image is performed by analysing several transmission radiographic projections of a specimen taken at different angles. This requires radiation that has an absorption length comparable with the specimen size [Willmott, 2011].

Small-angle X-ray scattering (SAXS) aims at determining the general structural features of systems that have characteristic dimensions between a few nanometres and micrometers. Compared to X-ray diffraction, SAXS has a modest resolution (1–3 nm) that is not sufficient to reveal the atomic structure of materials. However, it gives information related to the shape and size of nano- or submicrometre objects (particles, pores, precipitates, etc.). This makes SAXS an ideal complement to time-consuming analytical techniques such as electron microscopy. Unlike XRD, SAXS is purely a contrast technique, whereby the scattering signal arises from the difference in the average electron density between two phases [Willmott, 2011]. It can therefore be used for noncrystalline or semicrystalline systems. Finally, note that the grazing-incidence SAXS (called GISAXS) can be a powerful nondestructive tool to study nanoscale structures on surfaces, interfaces and thin films [Renaud et al., 2009]. SAXS experiments can be performed either in a synchrotron or in a laboratory. Concerning materials for third-generation solar cells, let us mention the investigation of Si nanocristallites within SiO_2 (for tandem cells) size [Bellet et al., 2007b]. SAXS can provide information relating to the size and shape of nanostructures with a very high statistical average.

Due to the limited electron density contrast between materials when dealing with soft matter, resonant soft X-ray scattering (here soft refers to the relatively long X-ray wavelength) can be a powerful technique for some material problems involved in solar cells. For instance, the bulk nanomorphology of organic bulk heterojunction devices (especially those composed of all-polymer devices) is difficult to characterise. A rather recent review highlights the main challenges of morphology characterisation in organic photovoltaic at multiple length scales [Giridharagopal and Ginger, 2010]. Resonant soft X-ray scattering can be an efficient method for investigating the nanomorphology of such materials [Swaraj et al., 2010]. Indeed, soft X-ray scattering provides greatly enhanced scattering contrast by adding elemental sensitivity, as its low photon energy (C, O, and N *K* edges) matches the energy of core-level electrons of the constituent atoms. One can then deduce information related to domain purity and domain sizes down to a few nanometres. The investigation of bulk heterojunction photovoltaic thin films by using this technique [Swaraj et al., 2010] reveals the behaviour of the domain during annealing. The dominant domain size appears to be much larger than the exciton diffusion length, even after annealing [Swaraj et al., 2010]. This study concludes that the lack of the morphological control would contribute to the relatively poor performance of all-polymer solar cells.

The resonant soft X-ray reflectivity can be used as well. It provides quantitative information on the interface roughness. This method is especially suitable for low-Z material thin films: through the use of specific energies, the sensitivity to the polymer/polymer or polymer/vacuum interface can be selectively enhanced [Wang and Ade, 2005]. As an example, the influence of annealing and interface roughness on the performance of polymer-based solar cells demonstrates that such an experimental method is well suited for studying this type of material. This investigation shows that organic photovoltaic performance can be sensitive to interfacial order, and that heterojunction sharpness should be considered as a requirement for high-performance devices [Yan et al., 2010].

Without any doubt the synchrotron methods will enable scientists to perform more and more useful experiments in the future. The spatial, time and energetic resolutions are improving along with building of new synchrotrons or upgrading existing ones (see for instance the upgrading of the European Synchrotron Radiation Facility [ESRF website]). One of the other great advantages is the use of the very high brilliance opening ways for in situ experiments during deposition or annealing. As an example, one can mention the in situ studies of the recrystallisation process of $CuInS_2$ thin film that were performed using energy-dispersive XRD [Thomas et al., 2011]. Related to X-ray techniques, recent evolutions concern the more frequent use of synchrotron radiation, which allows investigating smaller diffracting volume, tracking structural changes with an information depth selected at will and allowing in situ experiments during a thermal annealing for instance. Concerning laboratory experimental setups the improvement comes mainly from the versatility of the diffractometer, the resolution and configuration of which can be adapted to the sample needs and the use of 1-dimensional detectors of higher dynamics that reduce the acquisition time as well as increase the signal over background ratio.

3.4 ELECTRON MICROSCOPY METHODS

Electron microscopy comprises a range of different methods that use the various signals arising from the interaction of the electron beam with the specimen to obtain information about its surface topography and morphology, its chemical composition and finally its crystalline structure. Electron microscopy is a very useful and versatile technique for characterising material at the micrometre, nanometre and even atomic level allowing us to better understand the synthesis–structure–properties relationship in materials ([Reimer, 1985)]; [Williams and Carter, 2009]; [Yao and Wang, 2005]).

Electronic sources have nowadays a high brilliance, which allows very fast probing of very small volumes. On the other hand, due to the high interaction between electrons and matter it is not possible to sample macroscopic volumes: typical thicknesses of 100 nm are required for transmission electron microscopy and thicknesses of less than 0.5–1 μm contribute to the signal for scanning electron microscopy.

As shown schematically by Figure 3.3a, the interactions between electrons and the specimen may be either elastic (transmitted, diffracted or backscattered beams) or inelastic with energy losses. For the latter case, this occurs when the electron beam strikes an orbiting shell electron of the specimen, causing for instance ionisation with the ejection of secondary electrons of low energy or excitation of the electron towards high energy levels that decays by emission of characteristic X-rays. All these energy losses depend highly on the chemical nature of the sample. Figures 3.3a and b represent an overview of the electron–matter interactions. Clearly this involves a number of different phenomena within a specific interaction volume below the specimen surface. The resulting radiations are recorded through dedicated detectors. The size of the interaction volume is affected by different factors such

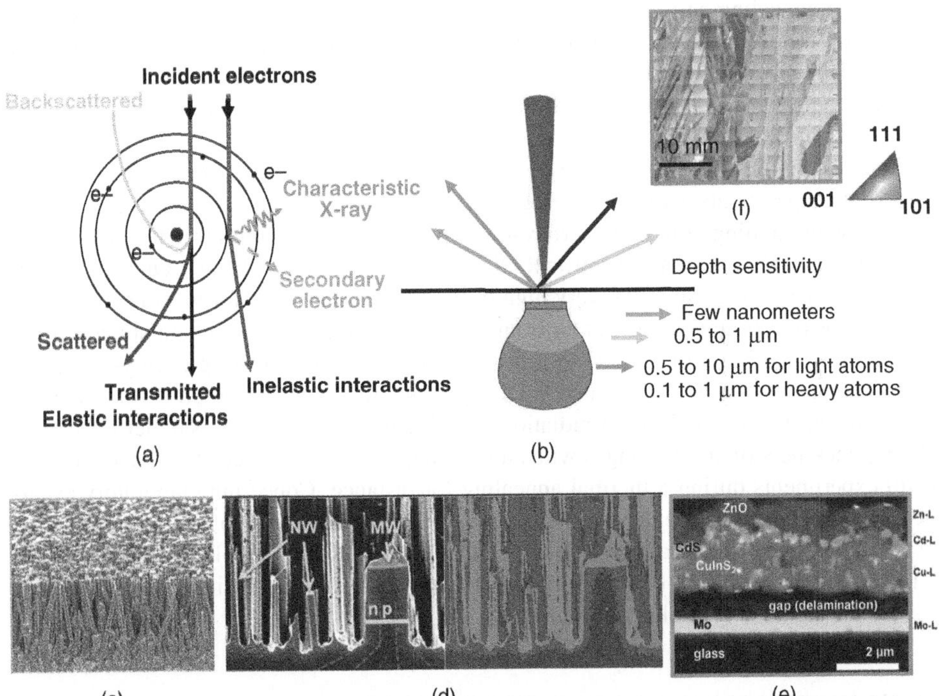

Figure 3.3 Electron–matter interactions. (a): Schematic representation of the main elastic and inelastic interactions. (b): Sketch of the sample depth vs. interaction type. (c): SEM image using secondary electrons of ZnO nanowires giving topological contrast. (Reproduced from Puyoo, 2012. Copyright © 2012, American Chemical Society.) (d): LV-SEM image of doped Si nanowires: a topological and chemical contrast (n-i-p dopants) is obtained. The scale bar is 2 μm. (Reproduced with permission from Jung, 2010. Copyright © IOP Publishing, 2010.) (e): EDX elemental distribution mapping of a CIGS/ interface superimposed to a SEM image reveals the spatial variation of the chemical composition. (Reproduced with permission from Abou-Ras, 2011b. Copyright © 2011, Elsevier.) (f): EBSD mapping of a polycrystalline Si solar cell can give some information related to the grain size and orientation as well as the grain boundaries. (Reproduced with permission from Naerland 2008. Copyright © John Wiley & Sons, Ltd, 2008.)

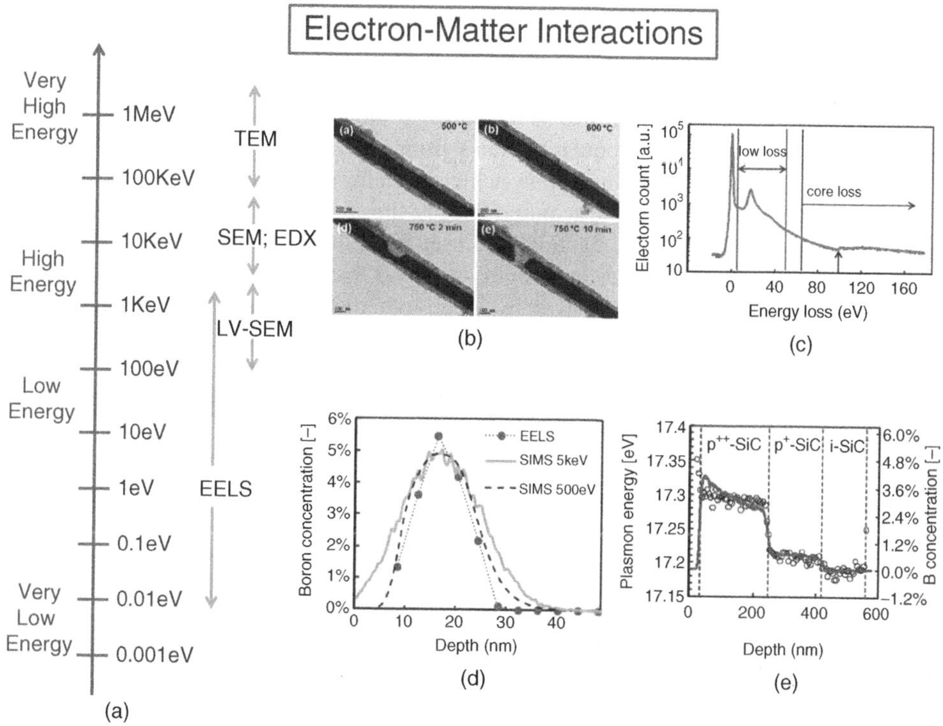

Figure 3.4 Electron–matter interactions. (a) energy scale showing the characteristic energies in play in different electronic characterisation methods (TEM: transmission electron microscopy, SEM: scanning electron microscopy, EDX: energy-dispersive X-ray, LV-SEM: low-voltage scanning electron microscopy, EELS: electron energy loss spectroscopy). (b) in situ TEM observation of Ag nanowires embedded in TiO_2 nanoparticles during thermal annealing showing a morphological instability of the nanowire shape. (Reproduced with permission from Ramaswamy, 2012. Copyright © 2012, Royal Society of Chemistry.) (c) Typical EELS spectrum showing the zero-loss peak, the low-loss region due to Plasmon absorption and the core-loss region. The ionisation energy of Si is also indicated (99 eV). (Reproduced with permission from Van Aken, 2012. Copyright © 2012, Elsevier.) (d) Boron profiles determined by EELS (circles) and SIMS (lines). EELS data were collected with a 5 nm probe beam. SIMS data were measured at two different ion energies. (Reproduced with permission from Van Aken, 2012. Copyright © 2012, Elsevier.) (e) Plasmon energy (open circles) versus depth in a triple stack of SiC associated with different dopant concentrations. The dashed lines indicate the locations of the interfaces. The B concentration, deduced from SIMS measurements, is shown as the blue line. (Reproduced with permission from Van Aken, 2012. Copyright © 2012, Elsevier.)

as specimen chemistry and the energy of the incident electrons, which ranges from 1 keV to more than 300 keV (see Figure 3.4a).

In the first part of this section we will discuss the interaction between the electron beam and matter then focus on scanning electron microscopy (SEM). Then, electron backscattering diffraction (EBSD) will be briefly presented, followed by transmission electron microscopy (TEM) and finally by the electron energy loss spectroscopy (EELS). Figure 3.4a deals with the electron–matter interactions and shows the correspondence between an electronic method cited above and the typical involved energy.

3.4.1 Electron–Specimen Interactions and Scanning Electron Microscopy (SEM)

Scanning electron microscopy (SEM) is without any doubt the most widely used of all electron-beam instruments. Its popularity stems from its versatility (with various imaging modes), the excellent spatial resolution now achievable, almost no sample preparation, rather straightforward interpretation of the acquired images, the accessibility of associated spectroscopic and diffraction techniques, and finally its user-friendliness [Yao and Wang, 2005]. For thin-film solar cells SEM provides an insight into layer thicknesses with cleaved sections, complex 3D surface topographies, as well as various other features with resolutions down to 1 nm. The continuous improvements of the main components of a SEM (field emission gun (FEG), probe forming lenses, detectors, image acquisition, etc.) have contributed to making modern high-resolution FEG-SEM a powerful and high-throughput tool for investigating the properties of materials at the pertinent length scale.

Scanning electron microscopes usually operate with a beam voltage in the range 5–30 keV. Electrons used for imaging are either secondary (energy lower than 50 eV so their escape thickness is reduced) or backscattered electrons (energy close to the incoming beam). An interesting and useful feature arises from the difference of imaging when secondary or backscattered electrons are used. This results in different contrasts in the images and provides information on compositions, microstructures and surface potentials. Images recorded with a secondary electron detector contain information from the near surface (1–2 nm for conductors, over 10 nm for insulating layers). Therefore, these electrons are used to produce topographical images of specimen with high spatial resolution. Backscattered electrons are mainly used for SEM imaging contrasts. They probe a region of 100 to 500 nm thick depending on the incoming beam energy and are affected by the chemical composition (signal intensity increases with the atomic number (Z) of an element) and crystallography.

Figure 3.3c exhibits a secondary-electron SEM image of a ZnO nanowire network [Puyoo et al., 2012]. This figure shows that the ZnO nanowires are well aligned and their length distribution is rather sharp; such networks are being integrated into dye-sensitised solar cells [Puyoo et al., 2012].

The development of low-voltage SEM broadens the application range including for instance nonconductive samples (as it minimises charging effects) or delicate specimens (reducing then the radiation damage). Material types that benefit from low-energy beams are, for example, organic semiconductor films or contact resist. Another consequence of using low-voltage SEM is that it allows ultrahigh-resolution images with subnanometre resolution. Decreasing beam energy provides additional flexibility in optimising image contrast. Indeed, a low acceleration voltage, below 2 keV, is critical in obtaining 2-dimensional dopant distributions with a sensitivity range from 10^{15} to 10^{22} at/cm^3. These aspects are crucial for the development and design of PV technology. A detailed analysis of the mechanisms that give rise to dopant contrast can be found in El-Gomati et al. [El-Gomati et al., 2005]. As an example, Jung et al. [Jung et al., 2010] have shown a distinctive contrast difference in secondary electron brightness between n and p dopants for doping as low as 10^{16} atoms/cm^3 in Si nanowires. Low-voltage SEM also enables the detection of doped regions of semiconductors [El-Gomati et al., 2005].

Figure 3.3d illustrates the use of low-voltage SEM imaging of doped Si nanowires where a topological and chemical contrast (n-i-p dopants) can be clearly obtained [Jung et al., 2010].

The short discussion below will first deal with the use of electrons for compositional analysis of materials. Readers are invited to refer to books or reviews, for more detailed information ([Reimer, 1985)]; [Williams and Carter, 2009]; [Yao and Wang, 2005]). Very frequently, scanning electron microscopes are also equipped with energy-dispersive X-ray (EDX) detectors that are used for analysing local elemental compositions. Indeed during the inelastic scattering of the beam electrons, inner-shell vacancies are produced on the specimen atoms when an electron is removed from the shell and ejected from the atom (see Figure 3.3a). The resulting shell vacancy can be filled by an outer-shell electron leading to a photon creation whose energy corresponds to the difference in energies between the two shells that is characteristic of a specific element. The detection of such X-ray emission reveals the atom's fingerprint and is then an invaluable tool for chemical identification analysis. The EDX spectrum consists of X-ray lines recorded at different energy positions that can be correlated to one specific element. The spatial resolution for this technique is mainly determined by the interaction volume size, which depends on the atomic number and the beam voltage. Typically, for a sample containing Au and C atoms the probe depths will be of 75 nm (750 nm) for Au and 300 nm (7500 nm) for C for a beam energy of 5 keV (30 keV). The effective spatial resolution for EDX mapping is therefore much larger than the SEM resolution.

Obtaining quantitative chemical composition requires certain precautions: an ideal sample should be conductive, homogeneous, flat and thicker than the interaction volume. Moreover, values should be obtained using previously calibrated standards. Carbon coating and sample polishing can improve the accuracy under the conventionally assumed chemical composition uncertainty value of $\pm 5\%$ [Carlton et al., 2004]. When the microstructure of the sample is known then simulations can be performed, but for instance no inhomogeneity can be extracted from a single measurement.

An interesting illustration concerns the analysis of $Cu(In, Ga)(S, Se)_2$ thin-film solar cells [Abou-Ras et al., 2011b]. This contribution indeed gives an overview of the various scanning and transmission electron microscopy techniques applied on $Cu(In, Ga)(S, Se)_2$ thin-film solar cells. Indeed, the authors clearly showed that the combinations of several electron microscopy methods can considerably improve the scientific understanding of the properties of the material.

Figure 3.3e exhibits an EDX mapping of the CIGS/interface that reveals the spatial variation of the chemical composition [Abou-Ras et al., 2011b].

3.4.2 Electron Backscattering Diffraction (EBSD)

Electron backscattering diffraction (EBSD) was first developed for measuring the microtexture (misorientation and boundaries) of samples and later for identification of micrometre or submicrometre crystalline phases. The combination of field emission electron sourced SEM and EBSD has now allowed EBSD to be applied to nanocrystalline materials (grain size smaller than 100 nm) and ultrafine grained materials (grain size < 500 nm) [Schwartz et al., 2009].

EBSD patterns are obtained in the SEM by illuminating a highly tilted specimen (typically at 70 degrees) with a stationary electron beam. Electron beams diffracted at a specific set of atomic planes form, when emerging from the specimen, a diffraction pattern where straight

Kikuchi lines are separated by an angle of 2θ (θ being the Bragg angle). An EBSD pattern consists of a large number of parallel lines of intensity and corresponds to a map of the angular relationships that exist within the crystal. Indeed, the spacing between a pair of lines is inversely related to the spacing of atomic planes within the crystal (as a Fourier transform consequence) from which the local crystallographic orientation can be deduced. From this repeated analysis, an orientation map is eventually generated that is generally superposed to the classical SEM image.

The lateral resolution of EBSD is strongly influenced by the atomic number of the studied material, the accelerating voltage of the SEM, the focused probe size and the sample tilt. The depth resolution of EBSD is also strongly dependent on the specimen atomic number and the operating voltage of the SEM. Since the exit depth of the backscattered electrons contributing to the EBSD patterns are only a few tens of nanometres, EBSD is then a very surface-sensitive technique. Let us note that both spatial resolutions (lateral as well as in depth) are improved by operating at lower voltages or for high atomic numbers.

In contrast to XRD, where texture measurements give the crystallographic orientation distribution of a macroscopic volume, EBSD provides crystallographic orientation with nanometre dimensions and is accessible at different scales from the several micrometres to a few nanometres.

The quality of the EBSD data depends mainly on the quality of the specimen surface preparation as it is affected by surface roughness, defect surface preparation and oxide layers. A very useful tool for the preparation of nanocrystalline and delicate material for EBSD is the focused ion beam (FIB) tool. This technique appears to be the best in comparison with other polishing techniques for investigation of, for instance, the crystalline structure of CdTe thin films [Moutinho et al., 2008]. The advantage of FIB preparation is that the sample can be prepared from a very specific sample location, allowing for instance to analyse material with a very well-defined position dependency. Moreover, combined SEM/FIB machines enables investigation of orientation images versus surface distance by successive ion milling of the surface. New developments also tend toward in situ surface modifications coupled to in situ processing such as: oxidation or annealing tensile deformation [Wright and Nowell, 2009].

Orientation maps can be used to quantitatively determine grain size and shape, as well as crystal orientation and type of grain boundaries with low/high misorientation, twin boundaries, etc., which allow correlation of these features with the electrical properties. Figure 3.3f shows an EBSD mapping of polycrystal Si solar cell where grain and twin boundaries can be clearly seen and investigated [Naerland et al., 2008]. The characterisation of grain boundaries is a key parameter for polycrystalline solar cells. Indeed grains can have both advantageous and adverse effects. The presence of grain boundaries generally reduces the density of defects and impurities. On the other hand, the presence of grains is believed to decrease photovoltaic efficiency by recombination of photogenerated carriers at grain boundaries. This is the case for multicrystalline silicon grain boundaries with a misorientation angle of less than 10 degrees [Tachibana et al., 2009]. Surprisingly in CdTe, grain boundaries participate in the energy-conversion process as efficient collectors and transporters, which leads to the higher performance of thin polycrystalline film with respect to single-crystal device [Visoly-Fisher et al., 2004].

Readers interested in grain-boundary effects on multicrystalline Si materials for PV are invited to read Chapter 5 of the present book written by Ruud Schropp.

In the case of CIGS absorber layers EBSD has been used, in combination with electron beam induced current (EBIC) and scanning spreading resistance microscopy (SSRM), to investigate the grain-boundary properties [Kawamura et al., 2011]. These authors were also able to conclude that twin boundaries of CIGS do not contribute to charge-carrier recombination.

3.4.3 Transmission Electron Microscopy (TEM)

Transmission electron microscopy (TEM) is widely used for investigations such as microstructural analyses, lattice defects, strain mapping and chemical characterisation at the nanometre scale of thin films and interfaces. In a TEM experiment, the elastic interaction between electrons and material dominates the contrast observed in the transmitted electron image, while inelastic events contain information of the specimen chemical composition.

While SEM often does not require any special sample preparation (except deposition of a thin conductive layer at the sample surface to avoid charging effects), a successful TEM experiment depends drastically on sample preparation. Indeed the stringent limitations on sample thickness that are dictated by the restricted penetration and inelastic scattering of the electron beam impose the preparation of very thin specimens (of about 100 nm depending on the absorption coefficient of the considered material). Several preparation methods do exist [Brandon and Kaplan, 2008], among them: mechanical or electrochemical thinning or ion milling. A small-angle wedge cleavage technique is also widely used since this method is fast and requires minimal sample preparation, which avoids sample damage [Walck and McCaffrey, 1997]. It is also worth noting that when a high-energy electron beam (over 200 keV) is impinging a small specimen area, some induced phenomena can occur damaging the sample or modifying artificially the sample characteristics. This is, for instance, rather well known in the investigation of small particles for which the electron beam can induce crystallisation. The corroboration between electron microscopy and other suitable techniques, such as XRD or Raman, is therefore an essential requirement for a successful experimental investigation.

Contrast arises either from mass thickness, diffraction or phase contrast. For an amorphous material the elastic scattering is random and the intensity scattered out of the direct beam (obtained by placing an aperture beneath the specimen) depends on the sample thickness and density. For crystalline specimens diffraction can instead occur associated with crystal planes (*hkl*) satisfying the Bragg conditions. If an aperture is placed in such a way that only the directly transmitted beam is going through the imaging system, a bright-field image is obtained. If only one of the diffracted beams is accepted then a dark-field image is formed. For both cases the contrast is mainly due to the presence of crystal lattice defects (diffraction contrast) [Brandon and Kaplan, 2008]. Finally, a larger aperture can be used to admit several diffracted beams to the imaging system that interfere in the image plane to yield a periodic image reflecting the periodicity in the crystal (perpendicular to the incident beam direction), which is called phase contrast.

TEM is very often used for investigating specimens devoted for solar cells as it allows the study of interface formation and material qualities. For instance, microcrystalline silicon is considered as a complex material composed of Si microcrystalline phase (Si nanocrystals) as well as amorphous Si phase. High-resolution TEM enables the observation of Si nanocrystals (having a diameter between 10 and 20 nm) embedded in amorphous Si [Vallat-Sauvain et al., 2007]. But, as is very often observed, defects such as grain boundaries, cracks, etc. can also be observed inside the nanocristallites [Vallat-Sauvain et al., 2007]. Numerous examples can also be cited where TEM can be considered as a very useful tool. In the case of the amorphous–crystalline heterostructure silicon solar cells [Van Sark et al., 2012], the high TEM spatial resolution is needed to image the amorphous Si:H / crystalline Si. Since the improvement of the intrinsic a-Si:H(i)/c-Si interface passivation under annealing requires the interface to be atomically sharp (i.e. without any epitaxial film at the interface [Van Sark et al., 2012]) high-resolution TEM appears to be the best technique to investigate such interfaces.

It is also worth noting that electron microscopy allows in situ observations leading then to a direct observation of the dynamic properties at the nanoscale through imaging and diffraction. A recent illustration concerns the in situ TEM observation of the Al-induced layer exchange method [Birajdar et al., 2012]. The latter could be a promising tool for preparing large (>10 μm) grained Si thin films on low-cost substrates, with potential applications in solar cells. Birajdar et al. used in situ TEM to investigate the mechanism of Al transport during Al-induced exchange and crystallisation of amorphous Si [Birajdar et al., 2012]. These authors observed significant realignment of grain boundaries and grain coarsening close to the crystallisation front of the Al layer. Another reported in situ TEM concerns Ag nanowires (embedded or not in TiO_2) which are investigated for their use as transparent electrodes in solar cells [Langley et al., 2013]. In situ TEM observation (Figure 3.4b) during thermal annealing reveals a morphological instability of the Ag nanowires while the use of TiO_2 shell drastically improves their thermal stability [Ramasamy et al., 2012].

Finally, the four-dimensional ultrafast microscopy has been recently developed especially by Chemistry Nobel prizewinner A. Zewail and his group at Caltech Institute of Technology [Zewail and Thomas, 2010a; Zewail and Four, 2010b; Flannigan and Zewail, 2012]. The integration in electronic microscopy of the fourth dimension, time, opens up numerous opportunities. The spatial resolution is at the atomic scale, while time resolution is of the order of femtoseconds. Therefore the advent of near-field and convergent beam four-dimensional ultrafast electron microscopy variants can, for instance, be applied in nanoscale plasmonics and photonics [Zewail and Thomas, 2010a; Zewail and Four, 2010b; Flannigan and Zewail, 2012]. One can therefore reasonably anticipate further exploration in solar cell materials within the near future.

3.4.4 Electron Energy Loss Spectroscopy (EELS)

Electron energy loss spectroscopy (EELS) is now a very used tool in TEM since the intimate connection between unoccupied states and core shells allows the measurement of energy distribution above the Fermi level. This technique can provide rather similar information to that deduced from a synchrotron method: X-ray absorption spectrometry. EELS relies on the analysis of the energy spectrum of the inelastically scattered electrons. EELS peaks corresponding to the K or L shell ionisations can be used to identify a very wide variety of atoms including light elements such as boron. The energy loss spectrum is measured in the thin-film transmission geometry. Generally speaking, an optimum specimen thickness exists for EELS analysis at which the counting statistics are adequate, while the probability of multiple scattering is small [Brandon and Kaplan, 2008]. Several processes contribute to the energy losses of the electrons: the electron transitions (both within and between different shells of the atom), plasmon excitation (quantised charge oscillations) and absorption edges (associated with atomic ionisation). Figure 3.4c shows typical EELS spectrum with the zero-loss peak, the low-region associated with plasmon absorption and the core-loss region. The ionisation energy of Si is also indicated (99 eV) [Van Aken et al., 2012].

In fact, the analysis of the chemical composition of microstructural features is mainly performed by using the characteristic signal observed in a SEM: a specimen volume of about 1 μm^3 (with a spatial resolution of 1 μm) can be routinely observed. While this technique is very often used, two disadvantages are associated with it: the low collection efficiency for the X-ray detection and the very low efficiency of the excitation of characteristic radiation for the lighter elements. A much better spatial resolution can be achieved with EELS, which is also better suited for light atoms. Instrumental developments in EELS have improved the energy resolution to 1 eV (compared to 130 eV for EDX systems) with a spatial resolution of about 0.5 nm as well as a detection limit of less than 100 atoms [Brandon and Kaplan, 2008]. The improvement in energy resolution for EELS enables one now to investigate fine structure in the neighbourhood of the absorption edges. Such microanalysis can go much further than a simple chemical composition analysis as it can provide information on the chemical state and the local atomic symmetry or order. An extensive review of recent instrumental developments and possible uses can be found in [Egerton 2009].

An illustration of such techniques concerns, for instance, the TEM and EELS microanalysis of polycrystalline silicon thin-film solar cells [Stöger et al., 2001]. These authors showed that a mixture of both ZnO (which is used as an electrical contact material) and the polycrystalline Si thin films is present at their interface. This interface region was estimated to be about 13 nm thick.

Let us also mention the use of EELS for the detection of dopants with nanometre-scale spatial resolution, which is a key parameter for solar cells. Boron concentrations, as low as $3 \times 10^{20}/cm^3$, corresponding to 0.6% with a 5-nm resolution were detected in an amorphous silicon layer Si [Duchamp et al., 2011]. Figures 3.4d and e deal both with the boron depth concentration using SIMS and EELS investigations in SiC layers containing different boron concentrations [Van Aken et al., 2012]. Plasmon energy (open circles) versus depth in a triple SiC stack is shown to be in good agreement with the boron concentration as deduced from SIMS measurements [Van Aken et al., 2012].

3.5 SPECTROSCOPY METHODS

Current spectroscopy methods offer powerful tools for investigating optical and electronic properties, and play an important role in the understanding of materials for solar cells. A few of them are described below. Spectroscopy techniques all work on the principle that, under certain conditions, matter absorbs or emits energy depending on the incident photons energy. Each spectroscopic technique operates over a limited frequency range depending on the process and magnitude of the energy changes. Figure 3.5a shows the typical energies in relation with the pertinent physical phenomena playing a key role in the interaction between photons and a specimen.

3.5.1 X-Ray Photoelectron Spectroscopy (XPS)

X-ray photoelectron spectroscopy (XPS), also called ESCA (electron spectroscopy for chemical analysis) is a surface (2 to 10 atomic layers depth) analysis technique that permits

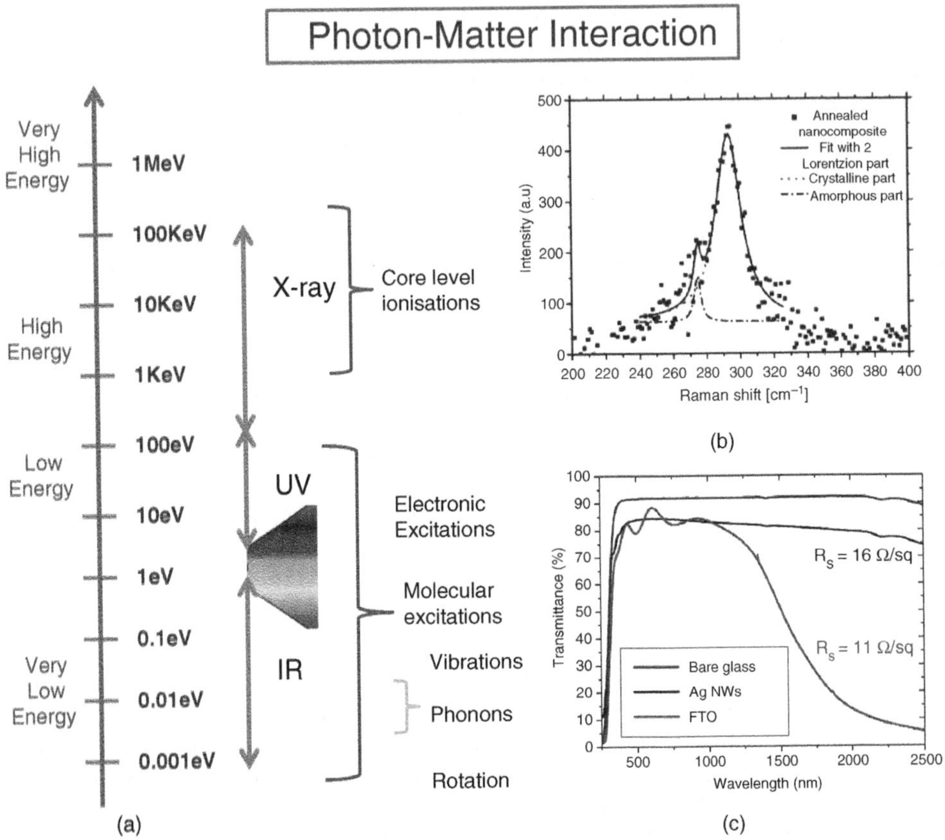

Figure 3.5 Photon–matter interactions. (a) Energy scale showing schematically some of the physical phenomena occurring at different energies. (b): Raman spectra of a nanocomposite composed of Ge nanoparticles embedded in Al_2O_3 matrix annealed at 450 °C. The spectrum was fitted with two Lorentzian functions, associated with crystalline and amorphous Ge. (Reproduced with permission from Parola, 2012. Copyright © 2012, Springer.) (c): Optical total transmittance of a bare substrate (blue) and two transparent conductive materials in the UV-VIS-NIR region. A spray pyrolysis deposited fluorine-doped tin oxide (FTO) and an Ag nanowire network deposited by spin coating are depicted in red and black, respectively. The associated sheet resistance is reported as 16 and 11 ohms per square for Ag nanowire network and FTO, respectively. (Courtesy of Gaël Giusti, from Grenoble INP.)

determination of the chemical composition as well as the chemical environment of atoms (except hydrogen and helium) belonging to a thin layer or to the top surface of a specimen. XPS consists of bombarding a sample with X-rays in a high-vacuum chamber to analyse the kinetic energy of the ejected electrons from an inner-shell orbital of the atoms. XPS is especially important in surface engineering and science, mainly for the investigation of solid-state materials. The principle of XPS is the following: the photons of a monochromatic X-ray beam incident on a specimen are absorbed by the atoms and secondary electrons will be ejected [Brandon and Kaplan, 2008]. The energy of these photoelectrons is related to photon energy ($h.\nu$) and to the electron binding energy (E_b) through the relation:

$$E_k = h\nu - E_b \tag{3.4}$$

The electron binding energy of the electrons depends on the chemical environment of the atoms and may be different (chemical shift) between a pure material and a compound, making XPS a very useful technique to analyse chemical states. Generally speaking, an XPS spectrum is composed of several photoemission peaks plus a background generated by the multiply scattered secondary electrons generated in deeper layers of the sample. Therefore, the observed peaks in the spectrum are associated with photoelectrons generated only by layers close to the surface, from a depth lower than the free mean path of the secondary electrons, which could be as small as 2 to 5 atomic layers. XPS is therefore extremely sensitive for analysing surface composition and chemical binding with a depth resolution close to atomic dimensions.

If coupled with sputtering, depth profiling and interface analysis can also be performed. Indeed, an Ar^+ sputtering is frequently used before measurement to avoid any contamination. One of the drawbacks of XPS is its poor lateral spatial resolution (about 10 μm). One way to avoid this problem is to use Auger analysis by using an electron beam. Auger emission spectroscopy (AES) starts by ejecting an inner-shell atomic electron from an atom. The created vacancy is then filled by a second atomic electron from a higher shell. The excess energy generated by this transition can be used either to eject a third electron (called an Auger electron) in a radiationless process or by emitting a photon (EDX analysis). The Auger process is more probable for low atomic number atoms and photon emission is more probable for high atomic number atoms. Auger analysis can be performed with an incident electron beam or with incident X-rays, however the spatial resolution is better with an electron beam (10–15 nm). As XPS, Auger analysis permits study of the surface chemical states (chemical shift) related to the atom environment and atom identification. Quantitative analysis is more complicated than XPS analysis and the investigation of insulators is difficult with an incident electron beam. However, in some cases, chemical shifts are more pronounced for Auger electron than for the XPS photoelectron, leading to better detection.

Moreover, excitation of a specimen by an incident energetic electron beam scanned over the surface enables a secondary-electron image to be obtained. Focusing and scanning of the incident electron beam can be accomplished with a good lateral resolution (of a few tens of nm).

In photovoltaics, XPS is used to analyse the chemical composition and states of thin-film layers such as antireflection coatings or thin-film solar cell absorbers. Interface information such as electronic properties (energy band alignment) can also be deduced for thin-film solar cells. An interesting illustration of the use of XPS, combined with ultraviolet photoelectron spectroscopy (UPS) concerns the investigation of transparent conductive oxides (TCOs) for photovoltaics [Klein et al., 2010]. By using both these techniques the authors present an extensive data set of technologically relevant electronic properties of TCOs, such as Fermi level, work function and energy-band alignment. They also illustrate the interplay between these key parameters and other material features such as chemical defects [Klein et al., 2010].

3.5.2 Secondary Ion Mass Spectrometry (SIMS)

Secondary ion mass spectrometry (SIMS) is a technique used to analyse the composition of solid surfaces and thin films by sputtering the surface of the specimen with a focused primary ion beam and collecting and analysing ejected secondary ions. These secondary ions are measured with a mass spectrometer to determine the composition of the surface.

SIMS is able to detect elements present in the parts per billion range, i.e. $\sim 10^{13}$ atoms/cm^3. Typically, a secondary ion mass spectrometer works under high vacuum and a primary ion beam of Ar^+, O^-, or Cs^+, ... $O_2{}^+$ is typically used for the detection of electropositive species, Cs^+ for electronegative species, and Ga^+ for improved lateral resolution. The SIMS primary ion beam can be focused to less than 1 mm in diameter. A mass analyser is used to separate the ions according to their mass and charge ratio and an ion-detection unit permits quantification of the atomic fraction. SIMS analysis is mainly used in photovoltaics to analyse dopant profiles, compound composition and metallic contamination. Compared to other techniques SIMS is very sensitive and can give quantitative concentrations (when compared to standards) with a good depth resolution. On the other hand, it has no spatial resolution and is locally destructive. For instance, SIMS depth profiling of CdTe-based solar cells grown on sapphire was reported by Emziane et al. [Emziane et al., 2006]. These authors investigated CdTe/CdS/TCO solar cell structures on sapphire substrates. They investigated the influence of using sapphire or glass as the substrate as well as diffusion of several elements such as Si, In, Cl, Na, O, Sn, S, Te, F and Br. The effects of annealing or any other process have also been analysed by SIMS [Emziane et al., 2006].

SIMS is generally used for characterising inorganic materials, but it is also possible to use a modified SIMS method compatible with soft materials, called gentle SIMS (G-SIMS). A recent study concerns the characterisation of organic solar cell materials by G-SIMS and demonstrates that this technique helps to identify both PCBM and P3HT materials through the enhancement of molecules and characteristic ions of the two molecules [Franquet et al., 2012].

3.5.3 Rutherford Backscattering Spectrometry (RBS)

In a backscattering experiment high-energy particles (He^+ ions of several MeV) are scattered on the studied film and the analysis of the energy of the backscattered particles allows a determination of the elements in the sample. This nondestructive method enables one to obtain a composition profile versus the film depth. RBS is very sensitive for heavy atoms, for example, it can detect less than one monolayer of In at the surface of a GaN layer [Monroy et al., 2003]. RBS has been used often for studying inorganic solar cells; one can, for instance, mention the work of Arafah and Ahmad-Bitar [Arafah and Ahmad-Bitar, 2000] who studied induced defects and structural changes resulting from the processing of CdTe and CdS thin film such as thermal annealing at 400 °C. Such treatment leads to major structural redistribution as well as the detection of new trapping states. This investigation, conducted by combining RBS and thermoluminescence, sheds light on interface defect density and formation of an intermixed interface layer [Arafah and Ahmad-Bitar, 2000], which are known to be determinant for the solar cell efficiency.

3.5.4 Raman Spectroscopy

Inelastic light (Raman) scattering has proven to be a versatile and powerful technique for studying a diverse range of materials [Cooper et al., 2012]. This method is based on the scattering of light as it passes through a transparent media: the light undergoes a change in frequency and a random alteration in phase due to a change in rotational and vibrational energy of the scattering molecules. In other words, a photon of energy $h\nu_P$ is incident upon a ground-state molecule and in a single interaction both excites it to a virtual state of energy $h\nu_P$ and de-excites it with the emission of a photon $h\nu_S$, leaving the molecule in a vibrationally excited state $h\nu_R$. The scattered photon can lose some of its energy to a vibrational mode: such a process produces a positive Stokes shift. If the particular vibrational mode is already in an excited state (a common feature at high temperature) the scattered photon may then gain some energy (anti-Stokes shift). Several characteristics of this technique can be mentioned as followed [Cooper et al., 2012]:

- Inelastic scattering can provide important symmetry, energy and lifetime information regarding a remarkable range of excitations in materials including one- or two- photons excitations, plasmon modes, etc. This method allows, for instance, to deduce information related to the crystalline properties of a specimen (to distinguish between amorphous or crystalline phases) as well as concerning any strains. As an example, Figure 3.5b exhibits a Raman spectrum where one can clearly distinguish the contribution from the amorphous and crystalline phases of a nanocomposite composed of Ge nanoparticles embedded in Al_2O_3 [Parola et al., 2012].
- Since inelastic light scattering is mainly based on the use of visible lasers as excitation sources, it allows the performance of experiments under "extreme" conditions such as high pressure or high magnetic field;
- Surface-enhanced Raman scattering (SERS) can be used for investigating very small samples and even nanometre-scale materials. SERS is a Raman spectroscopic method that provides greatly enhanced Raman signal from Raman-active molecules that have been adsorbed onto certain specially prepared metal surfaces. The importance of SERS is that it is both surface selective and highly sensitive. Such enhancement can correspond to several orders of magnitude increase of the Raman signal.

Let us note that the spectral selection rules (i.e. what vibrational mode are Raman active or inactive) are very important information in determining the structure of a molecule or a solid. The Raman method is an excellent way for "fingerprinting" molecules. It is also worth noting that time-resolved Raman spectroscopy can study the transient phenomena induced by an optical pump pulse through the analysis of the time evolution of the spontaneous Raman spectrum measured with a second optical probe pulse.

Micro-Raman consists of using a specially designed Raman spectrometer integrated with an optical microscope, enabling a high spatial resolution to be reached. Micro-Raman mappings were used, for instance, for characterising layers for crystalline silicon thin-film solar cells [Kunz et al., 2011]. The grain structure of the Si layers was revealed by Raman peak area mapping. The prevailing twin-boundary structure showed no residual stress, while a high density of stacking faults in the growth direction was revealed [Kunz et al., 2011].

An interesting illustration of the combined use of XPS, Raman and EELS can be found in the work of Sato et al. [Sato et al., 2009], which concerned the doping and characterisation of boron atoms in nanocrystalline silicon particles. Indeed these authors showed that most of the heavily doped B atoms in Si nanocristallites were electrically activated in the core of Si particles of 4.3 nm diameter [Sato et al., 2009].

3.5.5 UV-VIS-NIR Spectroscopy

When molecules or a material are irradiated with a UV-VIS-NIR (for ultraviolet-visible-near-infrared) of a continuous wavelength, the wave can be absorbed if its energy is the same as the inherent modes (vibration, rotation, etc.) of the molecular bonds or the solid. An absorption peak will then appear. This technique is complementary to Raman spectroscopy. As a powerful tool, UV-VIS-NIR can be used to: 1) identify a known component, 2) study the formation of new chemical bonds and 3) perform quantitative analysis for a component of interest. The optical transmittance T is defined as the ratio between the transmitted intensity (I) and the incoming one (I_0). The percent transmittance is then often used. The absorbance A is as well very often used and defined as follows:

$$A = \log\left(\frac{I_0}{I}\right) = \log\left(\frac{1}{T}\right) \tag{3.5}$$

The classical Beer–Lambert law states that absorbance A is directly proportional to the path length (L) and the concentration (c) of the absorbing species:$A = \varepsilon \cdot L \cdot c$, where ε is the molar extinction coefficient (mol^{-1} cm^{-1}), at least when assuming that the reflection is negligible.

UV-VIS-NIR spectrum analysis interprets the features of a spectrum in terms of the nature of the sample of interest, using quantum-mechanical models that relate detailed atomic or molecular structure to spectral frequencies, linewidths, and intensities.

One of the main and most obvious uses of UV-VIS-NIR spectroscopy in solar cell materials concerns the dye absorption for dye-sensitised solar cells (also called Gratzel cells or DSSCs) as well as the material used as transparent electrodes. Transparent electrodes are a necessary component in solar cells (among other applications) and have been the subject of much research. Materials that offer simultaneously good transparency and conductivity are transparent conducting oxides (TCOs) [Minami, 2005], [Granqvist, 2007], [Ruske, 2012], [Klein, 2012] as well as upcoming materials like carbon nanotubes, graphene or metallic nanowire networks [Hecht et al., 2011], [Ellmer et al., 2012], [Langley et al., 2013].

For instance, TCOs such as doped tin oxide can be produced on a large scale on float glass and are used for instance for DSSCs. Probably the best-known material is tin-doped indium oxide (ITO) as well as zinc-oxide-based thin films (like aluminium-doped Zn oxide: AZO) [Ellmer et al., 2008]. For Si-based devices such as a-Si/c-Si heterostructure solar cells, ITO or AZO are often considered as transparent electrodes. In the VIS-NIR spectral range, the dielectric function depends upon two major contributions, from the valence electrons and from the free carriers. The bandgap of a semiconductor can be obtained, while its determination is not fully straightforward. Indeed, the Burstein–Moss effect (an effect due to increasing doping that leads to a blocking of the lowest states in the conduction band) should be considered, as well as with a bandgap narrowing.

UV-VIS-NIR provides a direct way to determine the transmission of a TCO with two components: the diffuse and the specular ones. The ratio between the diffuse transmitted intensity divided by the total transmitted intensity is called the haze factor. Attempts are made to maximise the haze factor in a solar cell (to increase the optical path of the photons in the absorbing layer) and to minimise it in a flat-panel display (for a better spatial resolution).

Figure 3.5c exhibits the total transmission for FTO (fluorine-doped tin oxide, another well-known TCO) and a Ag nanowire network in the UV-VIS-NIR spectra. While these two types of transparent conductive materials have similar transmission in the domain [400–1000 nm], they possess different features: the Ag nanowire network exhibits better transmittance in NIR and has a larger haze factor, FTO has a better thermal stability if a process of "high" temperature is required during the solar cell fabrication. The optimisation of TCO layers for a solar cell is almost always accompanied by careful weighting of the optical, electrical and interfacial aspects.

3.6 CONCLUDING REMARKS AND PERSPECTIVES

This chapter briefly presents characterisation techniques useful for the understanding of solar cell materials. The shortness of this contribution has obliged the authors to select only a few techniques, stressing their principles and their most recent evolutions. This chapter can only scratch the surface of all material characterisation involved in solar cell research.

Once more, it is worth noting that although some methods could be very powerful for some analyses, no single method can elucidate any material problem under investigation. This should result from the crossinterpretation of several complementary techniques.

Some more recently developed characterisation methods that will certainly be very useful in the near future have not yet been considered in this chapter. These methods should, however, provide important contribution in the future and will be more widely adopted by the photovoltaic community. We briefly give a few examples of such techniques below.

Among recent characterisation techniques that will help in the future to better understand the subprocesses at play in emerging photovoltaic materials and technologies, scanning probes techniques are very promising. A recent article [O'Dea et al., 2012] reviewed the possible contribution of scanning probe techniques for addressing fundamental processes involved in solar cell operation for both inorganic thin films and organic bulk heterojunctions. These experimental methods allow local investigation of the electrical and optical properties and enable one to perform spatially resolved measurements for surface photovoltage and photocurrent that are particularly useful in understanding charge generation and separation [O'Dea et al., 2012]. These scanning probe microscopy methods include scanning tunnelling microscopy, photoconductive atomic force microscopy, near-field scanning optical microscopy, time-resolved electric force microscopy as well as Kelvin probe force microscopy (KPFM). The latter, for instance, could be used for measuring relative surface electrostatic potential and work-function distribution or for studying grain boundaries in inorganic solar cell materials. Already, numerous contributions using scanning probe techniques have been published and have deepened the understanding of solar cell operation. The increasing availability of commercial instruments should enlarge the use of these methods for making local measurements, enabling the direct correlation of physical (morphology, physical properties) and electronic (photocurrent, photovoltage, charge trapping, energy levels) phenomena [O'Dea et al., 2012].

Moreover, among newly emerging fields, time-resolved soft X-ray spectroscopy could play an interesting role in the near future since it could provide useful information with extreme time resolution (several tens of femtoseconds) as well as its capability to probe the electronic structure of complex systems in very fine detail [Bovensiepen et al., 2010]. This stems from emerging light sources such as the accelerator-based free-electron laser sources that give powerful soft X-ray sources with ultrashort pulses. This enables stroboscopic pump-probe experiments with one or more pulses.

Among others, advanced optical characterisation techniques seem to be still in the relatively early stages of development even if they have been used already to shed light on the physical properties in a large range of solid-state systems. Characterisation techniques without contact (i.e. luminescence for instance) are evolving rather rapidly nowadays and already constitute very useful characterisation tools. Also, the use of ultrashort X-ray and electron pulses in time-resolved structural dynamics in different materials should enable researchers to observe processes in much more detail and therefore improve the intimate understanding of the solar cell phenomena especially with noncontact and nondestructive manner with sub-100 fs and sub-100 nm resolution [Prasankumar, 2012].

The development of all these characterisation methods associated with modelling approaches and the new concepts of solar cells that are emerging [Lopez et al., 2012], should finally lead to increased solar cell efficiency and decreased cost, contributing to make solar energy a clean and efficient energy for the future.

ACKNOWLEDGEMENTS

The authors would like to warmly thank for fruitful discussions: Valentina Cantelli, Stéphane Coindeau, Vincent Consonni, Jean-Luc Deschanvres, Béatrice Doisneau, Gaël Giusti, Carmen Jimenez, Anne Kaminski, Dan Langley, Stéphanie Parola, Stéphanie Pouget, Hervé Roussel, Eirini Sarigiannidou as well as Shujuan Huang and Gavin Conibeer during the authors' sabbatical year at UNSW (Sydney, Australia) and finally Maguy Bellet for all her support.

REFERENCES

Abad, J., *Solar Energy Materials and Solar Cells* **97** (2012) 109.

Abou-Ras, D., T. Kirchartz, U. Rau, eds. "*Advanced Characterization Techniques for Thin Film Solar Cells*" Wiley-VCH, Verlag GmbH&Co kGaA, Weinheim, Germany (2011).

Abou-Ras, D., J. Dietrich, J. Kavalakkatt, M. Nichterwitz, S.S Schmidt, C.T. Koch, R. Caballero, J. Klaer, T. Rissom, *Solar Energy Materials and Solar Cells* **95** (2011) 1452.

Als-Nielsen, J., D. McMorrow, *Elements of Modern X-ray Physics*, John Wiley & Sons, Chichester, England (2001).

Arafah, D.-E., R. Ahmad-Bitar, *Solar Energy Materials and Solar Cells* **64** (2000) 45.

Beaucarne, G., A. Slaoui "*Thin Film Solar Cells: Fabrication, Characterization and Applications*", J. Poortmanns, V. Arkhipov, John Wiley and Sons, Ltd Chichester, England (2006); in chapter 3: "Thin Film Polycrystalline Silicon Solar Cells".

Bellet, D., E. Bellet-Amalric, T. Hanley, S. Huang, E.C. Cho, E. Pink, G. Scardera, G. Conibeer, D. König, and M.A. Green, *Proceedings of the 22th European PV Solar Energy Conference*, Milano, Italy (2007) p. 477.

Bellet, D., E. Bellet-Amalric, T. Hanley, A. Nelson, D. Song, S. Huang, T. Fangsuwannarak, S.W. Park, E. Pink, G. Scardera, E.C. Cho, G. Conibeer, D. König, and M.A. Green, *Proceedings of the 22th European PV Solar Energy Conference, Milano*, Italy (2007) p. 472.

Birajdar, B.I., T. Antesberger, B. Butz, M. Stutzmann, E. Spiecker, *Scripta Materialia* **66** (2012) 550.

Boscherini, F., M. Birkholz, J.-Y. Buffière, D. Chateigner, P.F. Fewster, S. Heun, Proceedings of the EMRS 2009 Spring Meeting, Symposium "X-Ray Techniques for advanced materials, nanostructures and thin films: from laboratory sources to synchrotron radiation", *Nuclear Instruments and Methods in Physics Research B* **268** (2010).

Bovensiepen, U., H. Petek, M. Wolf, eds. "*Dynamics at Solid State Surfaces and Interfaces*" Wiley-CVH Verlag GmbH&Co. KGaA Weinheim, Germany (2010).Vol.1, chapter 14, "Electron Transfer investigated by X-Ray Spectroscopy", W. Wurth and A. Föhlisch.

Brandon, D., W.D. Kaplan, "*Microstructural Characterization of Materials*", John Wiley & Sons, Ltd, Chichester, England (2008).

Carlton, R.A., C.E. Lyman, J.E. Roberts, *Scanning* **26** (2004) 167.

Conibeer, G., M. Green, R. Corkish, Y. Cho, E.-C. Cho et al., *Thin Solid Films* **511** (2006) 654.

Consonni, V., G. Rey, H. Roussel, D. Bellet, *Journal of Applied Physics* **111** (2012) 33523.

Consonni, V., G. Rey, H. Roussel, B. Doisneau, E. Blanquet, D. Bellet, *Acta Materialia* **61** (2013) 22.

Cooper, S.L., et al. "*Optical Techniques for Solid-State Materials Characterization*", edited by R.P. Prasankuma and A.J. Taylor, CRC Press, Taylor 1 Francis Group, LCC, Boca Raton, FL (2012) in chapter 6 "Raman scattering as a tool for studying complex materials".

Cullity, B.D., S.R. Stock, "*Elements of X-Ray Diffraction*", United States Edition, 3rd edition, Prentice Hall, New Jersey (2001).

Daillant, J., A. Gibaud (eds.), "*X-ray and Neutron Reflectivity: Principles and Applications*"; Lecture Notes in Physics, Vol. 770, Springer Berlin, Heidelberg, Germany (2009).

Duchamp, M., C.B. Boothroyd, A. Kovács, S. Kadkhodazadeh, T. Kasama, M.S. Moreno, B.B. Van Aken, J. P. Barnes, M. Veillerot, S.B. Newcomb, R.E. Dunin-Borkowski, *Journal of Physics: Conference Series* **326** (2011) 012052.

Durose, K., S.E. Asher, W. Jaegermann, D. Levi, B.E. McCandless, W. Metzger et al. *Progress in Photovoltaics: Research and Applications* **12** (2004) 177.

Egerton, R.F., *Reports in Progress in Physics* **72** (2009) 16502.

El-Gomati, M., F. Zaggout, H. Jayacody, S. Tear, K. Wilson, *Surface and Interface Analysis* **37** (2005) 901.

Ellmer, K., A. Klein, B. Rech, eds. "*Transparent Conductive Zinc Oxide: Basics and Applications in Thin Films Solar Cells*", Springer, Berlin, Germany (2008).

Ellmer, K., *Nature Photonics* **6** (2012) 809.

Emziane, M., K. Durose, D.P. Halliday, N. Romeo, A. Bosio, *Thin Solid Films* **511–512** (2006) 66.

ESRF website: http//www.esrf.eu.

Flannigan, D.J., A.H. Zewail, *Accounts of Chemical Research*, **45** (2012) 1828.

Franquet, A., T. Conard, E. Voroshazi, C. Poleunis, D. Cheyns, W. Vandervorst, *Surface and Interface Analysis* **45** (2013) 430.

Giridharagopal, R., D.S. Ginger, *Journal of Physical Chemical Letters* **7** (2010) 1160.

Granqvist, C.G., *Solar Energy Materials and Solar Cells* **91** (2007) 1529.

Guo, J., "*X-Rays in Nanoscience*", Wiley-VCH, Weinheim, Germany (2010).

Hao, X.J., E.C. Cho, G. Scardera, Y.S. Shen, E. Bellet-Amalric, D. Bellet, G. Conibeer and M.A. Green, *Solar Energy Materials and Solar Cells* **93** (2009) 1524.

Harris, G.B., *Philosophical Magazine* **43** (1952) 113.

Hecht, D.S., L. Hu, G. Irvin, *Advanced Materials* **23** (2011) 1482.

Huang, S., E.C. Cho, G. Conibeer, M. Green, D. Bellet, E. Bellet, S. Cheng, *Journal of Applied Physics* **102** (2007) 114304.

Jung, J.Y., Z. Guo, S.W. Jee, H.D. Um, K.T. Park, M.S. Hyun, J.M. Yang and J.H. Lee, *Nanotechnology* **21** (2010) 445303.

Kawamura, M., T. Yamada, N. Suyama, A. Yamada, M. Konagai, *Japanese Journal of Applied Physics* **49** (2010) 62301.

Kim, J.Y., K. Lee, N.E. Coates, D. Moses, T.-Q. Nguyen, M. Dante, A.J. Heeger, *Science* **317** (2007) 222.

Klein, A., C. Körber, A. Wachau, F. Säuberlich, Y. Gassenbauer, S.P. Harvey, D.E. Proffit, T.O. Mason, *Materials* **3** (2010) 4892.

Klein, A., *Journal of the American Ceramic Society* **96** (2012) 331.

Kocks, J.F., C. N. Tomé, H. -R. Wenk, "Texture and Anisotropy: Preferred Orientations in Polycrystals and their Effect on Materials Properties", Cambridge University Press, New York USA (2000).

Kotsar, Y., B. Doisneau, E. Bellet-Amalric, A. Das, E. Sarigiannidou, and E. Monroy, *Journal of Applied Physics* **110** (2011) 033501.

Kötschau, I.M., H.W. Schock, *Journal of Physics and Chemistry of Solids* **64** (2003) 1559.

Kunz, T., M.T. Hessmann, B. Meidel, C.J. Brabec, *Journal of Crystal Growth* **314** (2011) 53.

Langley, D., G. Giusti, C. Mayousse, C. Celle, D. Bellet, J.-P. Simonato, *Nanotechnology* **24** (2013) 452001.

Lomov, A.A., D. Bellet, G. Dolino, *Physica Status Solidi B* **190** (1995) 219.

Lopez, A.B.C., A. Marti Vega, A. Luque Lopez "*Next Generation of Photovoltaics, New Concepts*", Springer Series on Optical Science 165, Springer Berlin, Heidelberg, Germany (2012).

Minami, T., *Semiconductor Science and Technology* **20** (2005) S35.

Mittemeijer, E.J., P. Scardi, eds. "*Diffraction Analysis of the Microstructure of Materials*", Springer-Verlag, Berlin, Germany (2004).

Monroy, E., N. Gogneau, F. Enjalbert, F. Fossard, D. Jalabert, E. Bellet-Amalric, Le Si Dang and B. Daudin *Journal of Applied Physics* **94** (2003) 3121.

Moutinho, H.R., R.G. Dhere, M.J. Romero, C.-S. Jiang, B. To, and M.M. Al-Jassim, *Journal of Vacuum Science and Technology A* **26** (2008) 1068.

Müller, C., T. A. M. Ferenczi, M. Campoy-Quiles, J., D.D.C. Bradley, P. Smith, N. Stingelin-Stutzmann, J. Nelson, *Advances in Materials* **20** (2008) 3510.

Naerland, T.U., L. Arnberg and A. Holt, *Progress in Photovoltaics: Research Applications* **17** (2009) 289.

Nelson, A., *Journal of Applied Crystallography* **39** (2006) 273.

O'Dea, J.R., L.M. Brown, N. Hoepker, J.A.A Marohn, S. Sadewasser, *Material Research Society Bulletin* **37** (2012) 642.

Parola, S., E. Quesnel, V. Muffato, L. Guetaz, H. Szambolics, J. Bartringer, A. Slaoui, *Journal of Nanoparticle Research* **14** (2012) 1085.

Pavlinsky, G.V., "*Fundamentals of X-Ray Physics*", Cambridge International Science Publishing Ltd, Cambridge (2008).

Poortmanns, J., V. Arkhipov, "*Thin Films Solar Cells: Fabrication, Characterization and Applications*", John Wiley and Sons, Ltd, Chichester, England (2007).

Prasankumar, R.P., "*Optical Techniques for Solid-State Materials Characterization*", CRC Press, Boca Raton, FL (2012).

Puyoo, E., G. Rey, E. Appert, V. Consonni, D. Bellet, *Journal of Physical Chemistry C* **116** (2012) 18117.

Ramasamy, P., D.-M. Seo, S.-H. Kim, J. Kim, *Journal of Materials Chemistry* **22** (2012) 11651.

Reimer, L., "*Scanning Electron Microscopy, Physics of Image Formation and Microanalysis*", Spring Series in Optical Sciences (ed. P.W. Hawkes) vol. 45, Springer Berlin (1985).

Renaud, G., R. Lazzari, F. Leroy, *Surface Science Reports* **64** (2009) 255.

Robin, I.C., R. André, H. Mariette, S. Tatarenko, Le Si Dang, J. Bleuse, E. Bellet-Amalric, J.M. Gérard, *Nanotechnology* **16** (2005) 1116.

Ruske, F., in "*Physics and Technology of Amorphous-Crystalline Heterostructure Silicon Solar Cells*", edited by W. Van Sark, L. Korte, F. Roca, Springer-Verlag, Berlin, Heidelberg, Germany (2012), Chapter 9: "Deposition and Properties of TCOs".

Salvo, L., P. Cloetens, E. Maire, S. Zabler, J.J. Blandin, J.Y. Buffière, W. Ludwig, E. Boller, D. Bellet, C. Josserond, *Nuclear Instruments and Methods in Physics Research B* **200** (2003) 273.

Sato, K., N. Fukata, K. Hirakuri, *Applied Physics Letters* **94** (2009) 161902.

Schwartz, A.J., M. Kumar, B.L. Adams, D.P. Field, eds. "*Electron Backscatter Diffraction in Material Science*", second edition, Springer Science and Business Media, New York USA (2009).

Stöger, M., P. Schattschneider, V. Schlosser, R. Schneider, H. Kirmse, W. Neumann, *Solar Energy Materials and Solar Cells* **70** (2001) 39.

Surana, K., H. Lepage, J.M. Lebrun, B. Doisneau, D. Bellet, L. Vandroux, G. Le Carval, M. Baudrit, P. Thony, P. Mur, *Nanotechnology* **23** (2012) 105401.

Swaraj, S., C. Wang, H. Yan, B. Watts, J. Luning, C.R. McNeill, H. Ade, *Nanoletters* **10** (2010) 2863.

Tachibana, T., J. Masuda, K. Imai, A. Ogura, Y. Ohshita1, K. Arafune, and M. Tajima *Japanese Journal of Applied Physics* **48** (2009) 121202.

Tandjaoui, A., N. Mangelinck-Noël, G. Reinhart, J.-J. Furter, B. Billia, T. Lafford, J. Baruchel, X. Guichard, *Energy Procedia*, **27** (2012) 82.

Tanner, B.K., T. P. A. Hase, T.A. Lafford, M.S. Goorsky, *Advances in X-Ray Analysis* **47** (2004) 309.

Thomas, D., R. Mainz, H. Rodriguez-Alvarez, B. Marsen, D. Abou-Ras, M. Klaus, Ch. Genzel, H.-W. Schock, *Thin Solid Films* **519** (2011) 7193.

Toyama, T., K. Matsune, H. Oda, M. Ohta, H. Okamoto, *Journal of Physics D: Applied Physics* **39** (2006) 1537.

Vallat-Sauvain, E., A. Shah, J. Bailat, in "*Thin Films Solar Cells: Fabrication, Characterization and Applications*", J. Poortmanns, V. Arkhipov, John Wiley and Sons, Ltd, Chichester, England (2007), chapter 4: "Advances in Microscrystalline Silicon Solar Cell Technologies".

Van Aken, B.B., M. Duchamp, C.B. Boothroyd, R.E. Dunin-Borkowski, W.J. Soppe, *Journal of Non-crystalline Solids* **358** (2012) 2179.

Van Sark, W., L. Korte, F. Roca, eds. "*Physics and Technology of Amorphous-Crystalline Heterostructure Silicon Solar Cells*", Springer-Verlag, Berlin, Heidelberg, Germany (2012), in Chapter 7: "Intrinsic and Doped a-Si:H/c-Si Interface Passivation" by S. De Wolf.

Visoly-Fischer, I., S.R. Cohen, A. Ruzin and D. Cahen *Advanced Materials* **16** (2004) 879.

Walck, S.D., J.P. McCaffrey *Thin Solid Films* **308** (1997) 399.

Wang, C., H. Ade, *Applied Physics Letters* **87** (2005) 214109.

Warren, B.E., *X-Ray Diffraction*, Dover Publications Inc., New York (1990).

Williams, D.B., C.B. Carter, "*Transmission Electron Microscopy: A Textbook for Material Science*", Springer Science, New York (2009).

Willmott, P.R., "*An Introduction to Synchrotron Radiation: Techniques and Applications*", John Wiley & Sons, Chichester, England (2011).

Wright, S.I., and M.M. Nowell "A review of in situ EBSD Studies" in *Electron Backscatter Diffraction in Material Science*, Springer, New York USA (2009).

Yan, H., S. Swaraj, C. Wang, I. Hwang, N.C. Greenham, C. Groves, H. Ade, C.R. McNeill, *Advances in Functional Materials* **20** (2010) 4329.

Yao, N., Z.L. Wang, eds. "*Handbook of Microscopy for Nanotechnology*" Springer, New York (2005).

Zacharias, M., J. Heitmann, R. Scholz, U. Kahler, M. Schmidt, J. Bläsing, *Applied Physics Letters* **80** (2002) 661.

Zewail, A., J.M. Thomas, "*4D Electron Microscopy: Imaging in Space and Time*", Imperial College Press, London (2010).

Zewail, A., A.H. Four, *Science* **328** (2010) 187.

Zhang, S., L. Li, A. Kumar, "*Materials Characterization Techniques*", CRC Press, Taylor & Francis Group, Boca Raton, FL (2009).

Zheng, X., R.-H. Horng, D.-S. Wuu, M.-T. Chu, W.-Y. Liao, M.-H. Wu, R.-M. Lin, Y.-C. Lu, *Applied Physics Letters* **93** (2008) 261108.

4 Developments in Crystalline Silicon Solar Cells

Martin A. Green
School of Photovoltaic and Renewable Energy Engineering, University of New South Wales, Australia

4.1 INTRODUCTION

The photovoltaic industry has grown at over 40%/year over most of the last decade, largely due to promotional programs initiated by a number of governments. Particularly significant has been the guaranteed feed-in tariff scheme in Germany, where system owners have been guaranteed a fixed price for each unit of electricity supplied over a 20-year period.

The majority of solar cells fabricated to date have been based on silicon in monocrystalline or large-grained polycrystalline form, with the latter often referred to as "multicrystalline" silicon. There are two main reasons for this. One is that silicon is an elemental semiconductor with good stability and a well-balanced set of electronic, physical and chemical properties, the same set of strengths that have made silicon the preferred material for microelectronics. The second is that the success of silicon in microelectronics created an enormous industry where the economies of scale have directly benefited the early growth of the then smaller photovoltaics industry.

A significant fraction of silicon cells have been fabricated using thin wafers cut from large cylindrical monocrystalline ingots prepared by the Czochralski (CZ) crystal growth process and doped to about one part per million with boron during the ingot growth. A larger number have used multicrystalline wafers sliced from ingots prepared by a simpler directional solidification technique that produces large-grained polycrystalline ingots. To produce a cell, these generally boron-doped starting wafers have phosphorus diffused at high temperatures a fraction of a micrometer into the surface to form the p-n junction required for photovoltaic action. Metal contacts to both the n- and the p-type side of the junction are generally formed by screen printing a metal paste that is then densified by firing at high temperature. Each cell is typically 15–20 cm either in diameter or along either side if square or rectangular.

Cells are interconnected and packaged into a weatherproof, glass-faced package referred to as a module. Each module typically contains 36–72 cells soldered together in series or in series/parallel arrangement. Since each individual cell gives a maximum output of just over 0.6 V in sunlight, 36-series-connected cells result in a module of over 20 V maximum output voltage, sufficient for fully charging a normal 12 V lead-acid battery. The output current of each cell depends on its size and the sunlight intensity (solar irradiance) but generally would lie in the 6–10 A range in bright sunshine. The packaging consists of a glass/polymer

Solar Cell Materials: Developing Technologies, First Edition. Edited by Gavin Conibeer and Arthur Willoughby.

laminate with the positive and negative leads from the interconnected cells brought out in a junction box attached to the module rear. Such modules have proved extremely reliable in the field with most manufacturers offering a 20–25-year warranty on the module power output, one of the longest warranties on any commercial product.

The efficiency of the cells in the module would typically lie in the 16–19% range, about 60% of the fundamental limit of 29% for silicon (Kerr et al., 2003). Module efficiency is slightly lower than that of the constituent cells due to the area lost by frames and gaps between cells, with module efficiency generally lying in the 14–16% range. Over the last decade, commercial cells and modules of significantly higher performance have become available using more advanced cell processing technology, discussed in more detail in the text. This more advanced technology produces cells up to 24% efficiency and module efficiency as high as 21%. (Unless otherwise noted, all efficiencies quoted are under standard test conditions, namely with a cell temperature of 25 °C under 1000 W m^{-2} intensity of sunlight with the standard global air mass 1.5 spectral distribution.)

This chapter discusses recent developments in such crystalline silicon solar cells that have involved higher efficiency, larger ingots and thinner wafers, and on a better understanding of such prevalent defects as those associated with boron–oxygen complexes.

4.2 PRESENT MARKET OVERVIEW

A recent market survey reports that, of all solar photovoltaic module sales in 2011, 38% were based on monocrystalline silicon wafers, similar to those used in microelectronics (Mints, 2012). A further 48% were based on lower-quality multicrystalline silicon wafers. As previously mentioned, these are large-grained polycrystalline wafers produced by slicing from large ingots of directionally solidified silicon, an approach developed specifically for photovoltaics. Close to 3% of module sales were based on multicrystalline silicon ribbons technology again developed specifically for photovoltaics, but with the advantage of not requiring wafering.

Combined, these "bulk" silicon approaches accounted for a commanding total of 86% of annual production in 2011. About 11% of total production was accounted for by thin films not involving silicon, specifically by cells based on the chalcogenides Te and Se mainly in the form of polycrystalline thin films of CdTe but also a small number of modules based on $CuInSe_2$ (CIS). Most of the remaining production (3%) was made up by thin-film amorphous silicon solar cells, including multijunction stacked "tandem" cells.

Figure 4.1 shows a sampling of the nominal performance of modules on the market from different manufacturers based on bulk silicon wafers and the above-mentioned thin films. As noted, silicon modules accounted for 86% of production in 2011. These modules have a well-established reputation for reliability and durability, with many systems installed over 20 years ago still performing creditably after more than 20 years in the field (Realini et al., 2001).

Most commercial bulk silicon modules have energy-conversion efficiency, the ratio of electrical power output to solar power on the total module area, in the 14–16% range with two manufacturers producing modules in the 18–20% range. Those in the latter range are based on monocrystalline silicon wafers and use unique manufacturing approaches. The remainder are based on monocrystalline or multicrystalline wafers but, irrespective of wafer

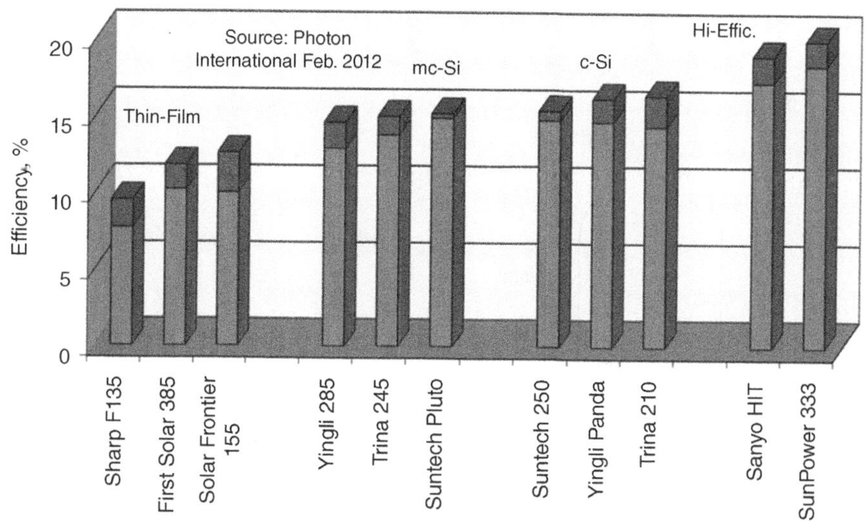

Figure 4.1 Survey of the performance of solar modules from major manufacturers on the European market in 2012. The chart shows the nominal energy-conversion efficiency under standard test conditions for several module types based on the manufacturer's rating and the module's total framed area. The bar at the top shows the likely range for delivered product falling within specifications (mc = multicrystalline; c = crystalline).

type, use a very similar processing sequence based on screen printing the metal contacts to the cells, with only one exception.

4.3 SILICON WAFERS

4.3.1 Standard Process

Silicon solar cell technology originally not only benefited directly from the economies of scale of the silicon microelectronics industry, but it also was able to use scrap material from this industry, since requirements for material quality in photovoltaics are less demanding than in microelectronics. The photovoltaic industry more recently has grown to a size where it needs much more silicon than used in microelectronics. This gives opportunities for silicon-preparation approaches to be commercialised that have been directly customised for photovoltaics.

For microelectronics, the starting point for producing the requisite high-quality "semiconductor grade" silicon is a lower grade of silicon known as "metallurgical grade", produced by the reduction in an arc furnace of quartzite by carbon, the latter generally in the form of wood chips. This metallurgical grade silicon is of about 98% purity and is produced in large quantities for the steel and aluminium industries. A relatively small fraction is refined for microelectronic and solar applications by conversion to a volatile intermediary which can be purified by fractional distillation. The purified intermediate compound is then decomposed to re-extract the silicon in a highly purified form. Generally, the metallurgical-grade silicon is converted by hydrochloric acid to trichlorosilane that is then purified to 99.9999999% (nine

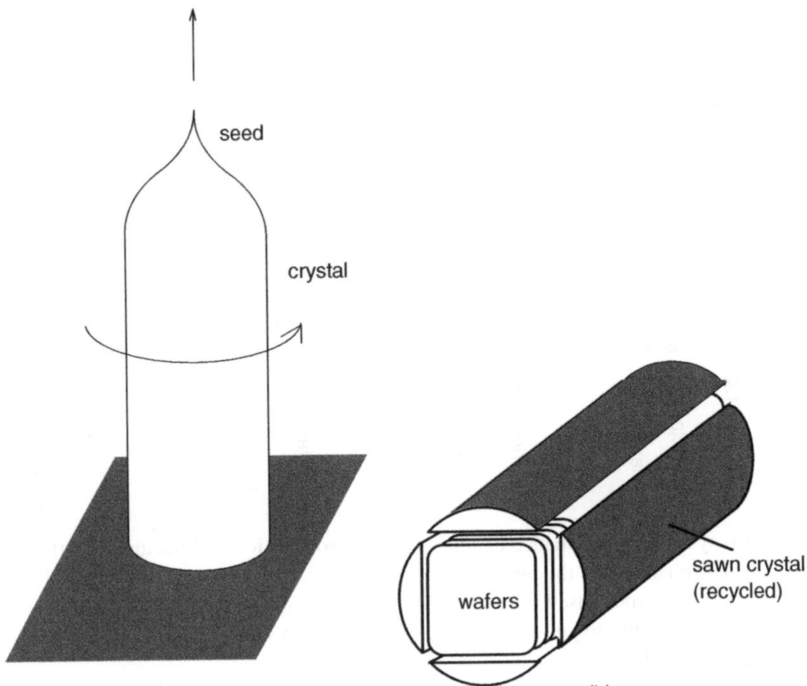

Figure 4.2 (a) Czochralski (CZ) growth; (b) squared-off CZ ingot. (Reproduced with permission from Green and Hansen, 1998.)

"nines") purity by fractional distillation. Silicon is then extracted from the trichlorosilane by reducing the latter by hydrogen at high temperature (Luque and Hegedus, 2011).

In this process, electrically heated silicon rods are exposed to a trichlorosilane/hydrogen mixture that reacts on the surface of the rods, depositing silicon onto them and hence building up their cross section. These rods grow with a fine-grain polycrystalline silicon microstructure. After the rod diameter has increased to the required size, the process is stopped and the rods mechanically broken into smaller chunks, which maintain "nine-nines" purity. These chunks then become the starting point for the growth of ingots of good crystalline quality.

As previously mentioned, crystalline ingots are generally grown by the Czochralski crystal growth process. In this process, the purified silicon chunks are melted in a quartz crucible along with small pieces of silicon heavily doped with boron. This produces a boron-doped melt into which a seed crystal is dipped and slowly withdrawn (Figure 4.2a). For high-quality crystals, good temperature uniformity and slow and steady growth are required. Typically ingots are grown to about 15–20 cm in diameter and 1–2 meters in length, weighing 150–200 kg. The crystallographic orientation of the seed is transferred to the grown crystal. Generally, for photovoltaics, the crystal is grown with a preferred orientation so that the wafers that are sliced from the crystal perpendicular to the growth axis have surfaces parallel to {100} crystallographic planes.

Prior to slicing these ingots into wafers, the ingots are generally subject to a centreless grinding operation to remove the slight fluctuations in diameter along the length of the ingot

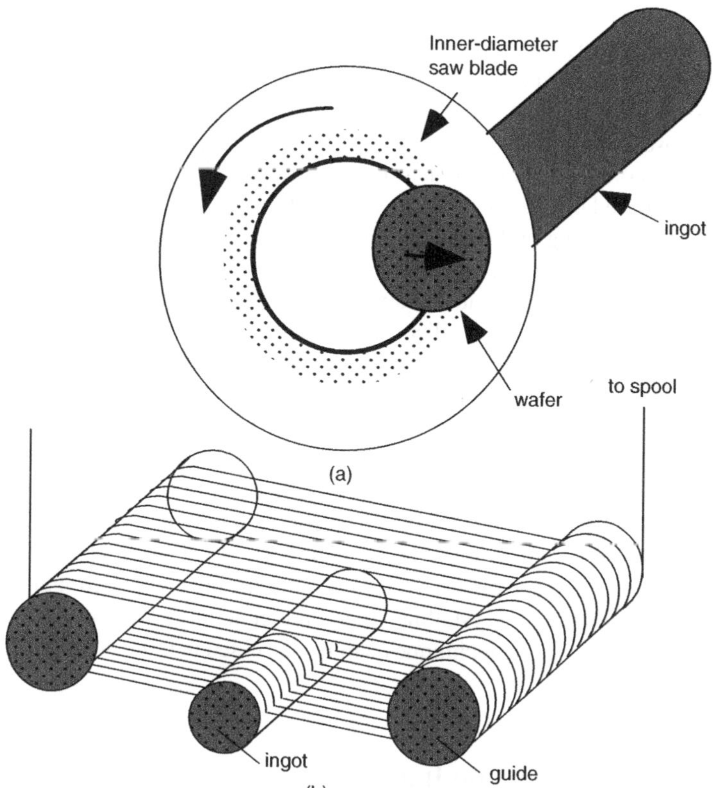

Figure 4.3 (a) Inner diameter wafer sawing; (b) continuous wire sawing. (Reproduced with permission from Green and Hansen, 2002 and Dietl et al., 1978.)

that occur during crystal growth. Alternatively, the ingots can be "squared-off" by sawing off large sections parallel to the growth axis (Figure 4.2b), giving "quasisquare" wafers after wafering. The large pieces of silicon sawn off in this approach are then recycled by remelting as feedstock for the CZ growth.

The technique traditionally used in microelectronics for sawing wafers from ingots was based on the use of inner diameter saws. In this technique (Figure 4.3a), thin metal sheet blades are given dimensional solidity by being held in tension around their outer perimeter. The cutting surface is a diamond impregnated region surrounding a hole within the tensioned metal sheet. This technique gives excellent dimensional tolerance, although there are limitations arising from the thickness of the silicon wafers that is possible to produce while still maintaining high yield. Other limitations arise from the wastage of silicon as "kerf" loss during cutting. Generally, about 10–15 wafers per centimeter of ingot length can be achieved by this process. A more widely used technique in photovoltaics is based on wire sawing (Figure 4.3b). In this case, tensioned wire is used to guide an abrasive slurry through the ingot. Advantages are thinner wafers and less surface damage for these wafers as well as lower kerf or cutting loss, allowing the sawing of over 20 wafers per centimetre, a figure that is steadily improving and may reach 50 by 2020.

In the 1970s and early 1980s, several other options for preparing silicon feedstock were investigated as part of a large US government photovoltaic program encouraged by the Carter administration (Christensen, 1985). During this program, a diversity of alternative routes to producing pure silicon was investigated. These ranged from those involving radically different approaches to those exploring only minor changes from the sequence outlined above, such as the use of different compounds of silicon as the intermediate during the purification process. One such process, based on the use of silane as the intermediate (Christensen, 1985), is now used commercially, resulting in the lowest manufacturing cost, largely due to much reduced energy consumption. Parallel development has been conducted outside this program, notably in Germany (Aulich, 1996) and Japan. In Japan in the 1990s, Kawasaki Steel investigated an alternative route to preparing cheap silicon feedstock from metallurgical-grade precursors (Sakaguchi et al., 1997). Over recent years, commercial quantities of silicon prepared from such metallurgical routes have become available.

An alternative to the standard Czochralski process for producing crystalline ingots is the floatzone (FZ) process. Although some studies have predicted superior economics when compared to the Czochralski process for cell production due to the elimination of consumables such as quartz crucibles, the FZ process, as commercially implemented, is capable of accepting feedstocks only in the form of high-quality cylindrical rods. This makes it unsuitable for using low-cost off-grade material. However, the casting and directional solidification processes used to produce multicrystalline silicon are generally extremely tolerant of poor quality feedstock material. These techniques will be discussed in more detail in the following section.

4.3.2 Multicrystalline Silicon Ingots

In 2011, about 48% of the world's photovoltaic production was based on multicrystalline silicon wafers. Several companies have developed commercial processes for producing the precursor multicrystalline silicon ingots (Ferrazza, 1995; Luque and Hegedus, 2011). Advantages over the Czochralski process are lower capital costs, higher throughput and a higher tolerance to poor feedstock quality.

Basically, the technique involves controllably solidifying molten silicon in a suitable container to give silicon ingots with large columnar grains generally growing from the bottom of the crucible upwards (Figure 4.4a). Pioneers of this approach for modern photovoltaics in the mid-1970s were Wacker Chemitronic of Germany (Authier, 1978) and Solarex of the USA (Lindmayer, 1976). In the 1980s, other manufacturers including Eurosolare/Crystallox, Kyocera, Bayer, Crystal Systems and Sumitomo Sitex developed in-house processes capable of producing good quality multicrystalline material. Techniques differed between these manufacturers in the choice of crucible material, the method of loading the crucible with silicon and the method for controlling the cooling of the melt. A good summary of this earlier work can be found elsewhere (Ferrazza, 1995). More recently, processes have converged with several equipment vendors now supplying the industry with relatively standardised equipment (Luque and Hegedus, 2011). The size of nominally rectilinear ingots can be very large with these ingots now weighing 400–600 kg, with ongoing increase in size expected. The large ingots are sawn into smaller sections as shown in Figure 4.4b, eventually to give wafers generally 156 mm along the sides. These smaller sections are sawn into wafers by continuous wire sawing processes. The resulting multicrystalline wafers are capable of producing

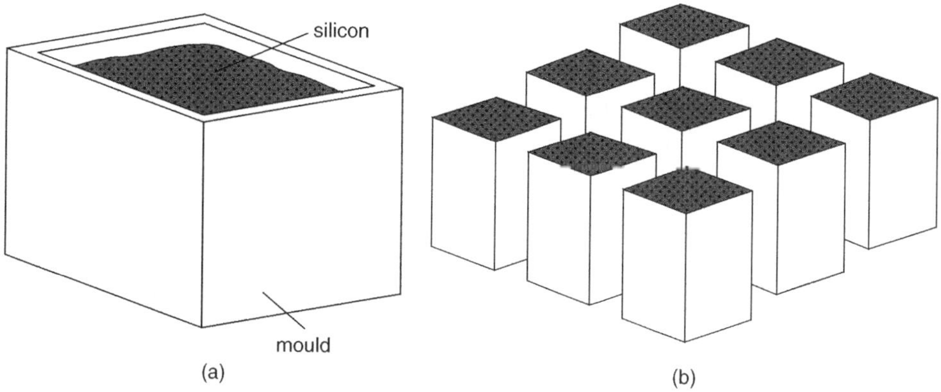

Figure 4.4 (a) Directional solidification of silicon within a mould; (b) sawing of large ingot into smaller bricks shown is a generation-3 ingot where 3 × 3 bricks are sawn from the ingot. Some manufacturers are now using generation-6 ingots, where 6 × 6 bricks are produced from each ingot. (Reproduced with permission from Green and Hansen, 1998.)

cells of about 90% of the performance of a monocrystalline cell fabricated on a CZ wafer. However, because of the higher packing density possible due to their square geometry, this performance difference is largely masked at the module level with multicrystalline module performance lying in the range demonstrated by modules made from monocrystalline cells (Figure 4.1).

4.3.3 Ribbon Silicon

Although there is the potential for substantial cost reductions in both the cost of preparing the silicon feedstock and in forming crystalline or multicrystalline ingots from it, one unavoidable cost with the silicon wafer approach is the cost of sawing the ingot into wafers. This has provided the rationale for investigating approaches that produce silicon directly in the form of self-supporting sheets without the need for sawing (Bergin, 1980; Shulz and Sirtl, 1984; Luque and Hegedus, 2011).

The first ribbon approach to be commercialised was that based on the edge-defined film-fed growth (EFG) method of Figure 4. 5. As originally investigated in the early 1970s, this technique involved the pulling of a thin sheet of silicon ribbon from a strip of molten silicon formed by capillary action at the top of a graphite dye (Figure 4.5a). Substantially higher throughput was obtained with the more symmetrical configuration shown in Figure 4.5b where the ribbon is pulled in the form of a hollow nonagon. Individual wafers are then cut from the sides of the nonagon, normally by laser scribing wafers from each of the sides. The material produced is multicrystalline with elongated grains and is of a similar quality to the standard directionally solidified multicrystalline material. Commercial cells made from this material have been available sporadically since the early 1980s, with steady production beginning in the early 2000s but ceasing circa 2010.

An even older ribbon growth process is the dendritic web approach of Figure 4.6 first described by Westinghouse in the 1960s. In this approach, close thermal control is used to

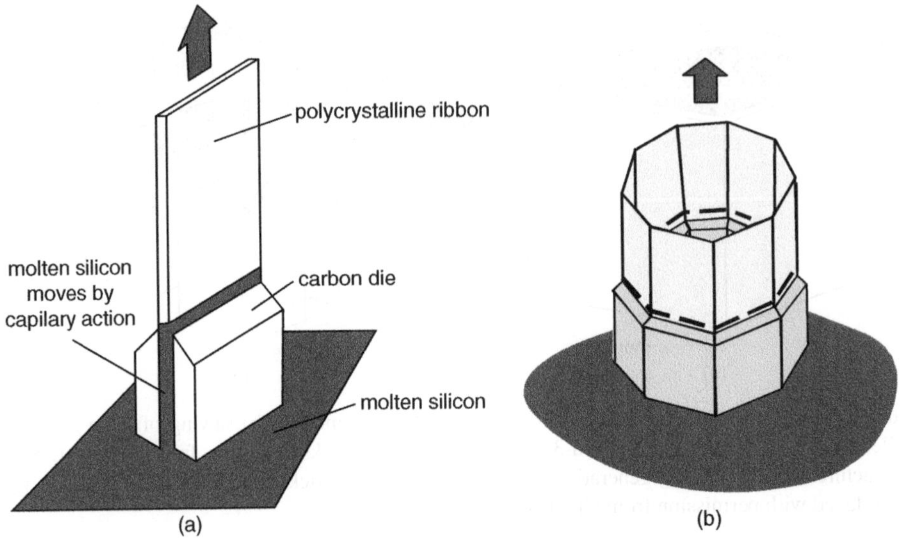

Figure 4.5 (a) Edge-defined, film-fed growth (EFG) method; (b) growth of a nonagonal ribbon of silicon using the EFG method. (Reproduced with permission from Green and Hansen, 1998.)

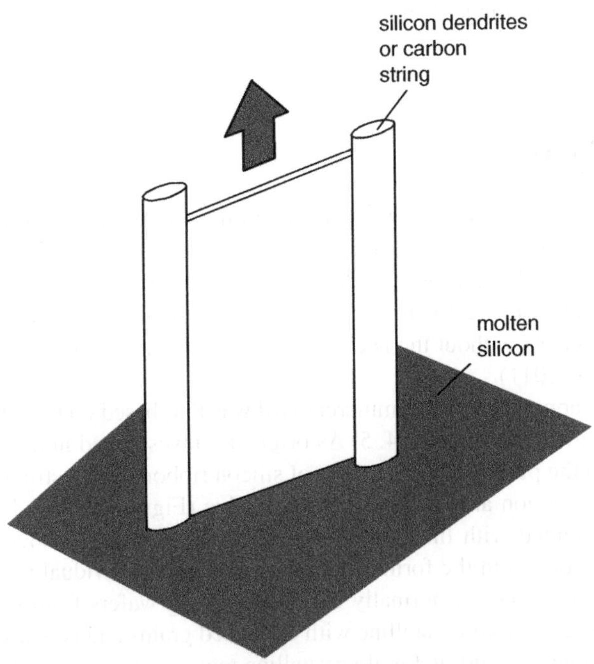

Figure 4.6 Schematic illustrating either the dendritic web growth process or the string ribbon approach. (Reproduced with permission from Green and Hansen, 1998.)

cause two dendrites spaced several centimetres from each other to solidify first during the growth step. When these are drawn from the melt, a thin sheet of molten silicon is trapped between them. This quickly solidifies to form a ribbon.

A related approach is the string ribbon approach. In this case, the molten silicon is trapped between two graphite strings that are drawn from the melt. This relaxes the requirement upon thermal control, compared to the previous dendritic web approach. The string ribbon approach was commercialised by Evergreen Solar (Janoch et al., 1997; Wallace et al., 1997) and was in large-scale production from circa 2005 until 2011. Throughput was increased during this period by pulling multiple "string" ribbons from the one melt.

Despite the promising start with these approaches, no manufacturer is presently using ribbon silicon in commercial production. This is attributed to what the author calls, using terms from bicycle racing, the "breakaway" versus "peloton" effect. Even though unique technology may give a company a head start, the combined efforts of a whole industry all working on the development of mainstream product can often prevail in the end. The only chance for the breakaway to stay ahead is if sufficient riders cross over from the peloton.

4.4 CELL PROCESSING

4.4.1 Screen-Printed Cells

Most of the monocrystalline and multicrystalline cells fabricated during 2012 used the screen-printed solar cell structure shown schematically in Figure 4.7. The associated processing sequence was developed during the early 1970s, along with several alternative solar cell processing approaches. By the early 1980s, this approach had displaced the alternatives to emerge as the commercial standard (Green, 1995; Green, 2005). One advantage was that it was able to use the same screen printers, drying and firing furnaces for applying cell contacts as developed for thick-film, hybrid microelectronics.

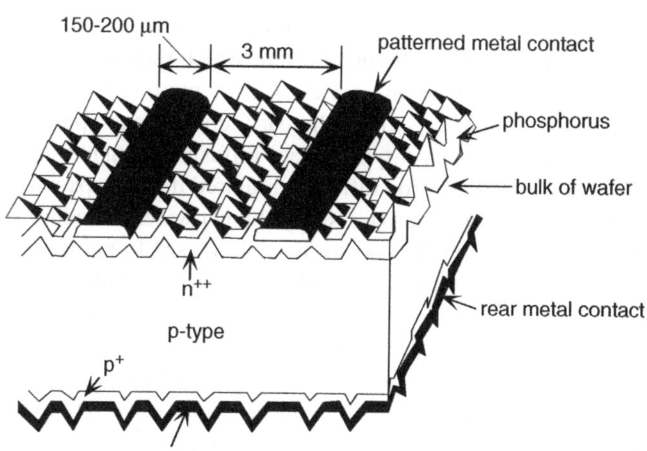

Figure 4.7 Screen-printed solar cell (not to scale). Finger spacing has now been reduced to the 80–120 micrometer range. (Reproduced with permission from Green, 1995.)

For monocrystalline screen-printed cells, the main cell-processing steps (Jester, 2002; Luque and Hegedus, 2011) consist of wafer cleaning and chemical etching, usually anisotropically to form the micrometer-sized, crystallographically defined pyramids apparent on the wafer surface in Figure 4.7, followed by p-n junction formation, either in the same style of dopant diffusion furnaces as used in microelectronics or using simpler approaches based on the spraying or spinning-on of dopant sources followed, in this case, by diffusion in the same type of belt furnaces used for contact firing.

After the diffusion, the oxide left on the surface after diffusion is removed by wet chemical etching. The junction that wraps around the edge of the silicon wafer is also removed, usually by "floating" the wafer on an etch bath so only the rear is contacted. A silicon nitride antireflection coating is then applied generally by plasma enhanced chemical vapour deposition, either in a tube or an inline system.

Contacts are applied in the form of metal pastes, with their pattern on the cell surface defined by printing through an appropriately patterned emulsion mask or screen. A silver paste is generally screened onto the top surface of the cell and both silver and aluminium pastes onto the rear, in two steps.

The strengths of this technology are the simplicity of cell processing and the ready availability of the required processing equipment. The penalty for this simplicity and convenience are the compromises in cell design have been needed to accommodate the less than ideal features arising from the use of screen printing as a method for applying the cell's top contact, in particular.

Both for low contact resistance and to overcome the high shadowing losses otherwise resulting from the relatively coarse metal fingers produced by the screen-printing approach, relatively high concentrations of dopants are required in the top surface region of the cell. These high concentrations reduce the silicon quality in this region, where the strongest light absorption occurs. This reduces the cell's response, particularly to the blue wavelengths of sunlight that are particularly strongly absorbed in this region (Green, 1995). This constrains cell performance to levels below those that are fundamentally possible, although improvements in the formulation of the paste are steadily allowing increasingly lightly doped surface regions to be contacted.

The p-type wafers used in the majority of commercial screen-printed cells are doped with boron during the preparation of the ingots from which they are sliced. This is partly for historical reasons, since regions with such doping are more resistant to damage by the high-energy particles found in space, where the early commercial applications of silicon cells were found. The complementary n-type surface diffusion using phosphorus is also less demanding than would be a p-type boron diffusion required if an n-type wafer were used.

Recent practice has been to push to progressively thinner wafers to reduce silicon material costs (Jester, 2002). Standard wafer thicknesses have decreased from over 0.3 mm to below 0.2mm over recent years. Also over the last decade, a disadvantage arising from the use of p-type wafers has been recognised and has been the subject of much recent research (Rein et al., 2000). Oxygen is also unintentionally incorporated into the wafer during the growth of the original ingot, with this oxygen originating from the crucible holding the molten silicon. This oxygen is mostly inert and, in fact, improves wafer strength. However, under illumination, some is able to interact with the boron dopants to form an electrically active boron–oxygen complex that detracts from device performance. Module output dropped about 3% relative under the first few hours of light exposure as a consequence (Eikelboom and Jansen, 2000; De Wolf et al., 2000). This was originally accommodated within the manufacturer's warranty, which is generally specified as involving less than 10% module output loss over the first

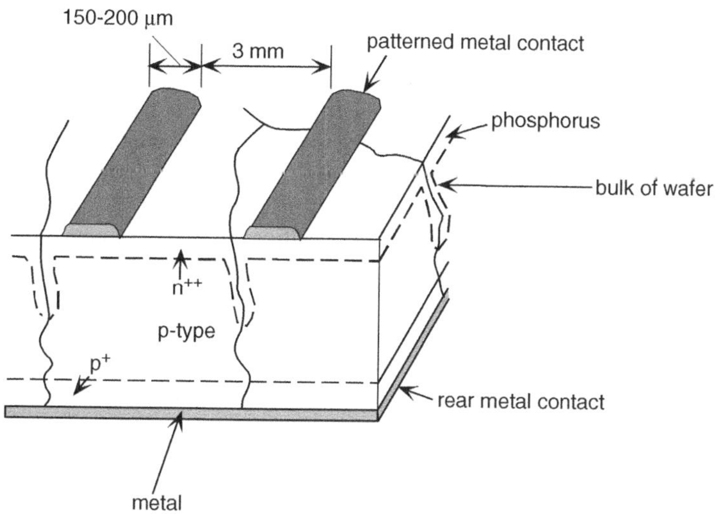

Figure 4.8 Screen-printed multicrystalline silicon solar cell. Finger spacing has now been reduced to the 80–120 micrometer range. (Reproduced with permission from Green and Hansen, 2002.)

10 years, and less than 20% over the first 20 years. More recently, there is evidence that this defect can be permanently deactivated by appropriate processing, which opens up a range of possibilities for further exploiting p-type material.

Most modules produced during recent years used multicrystalline silicon wafers rather than monocrystalline. As previously discussed, these wafers are sliced from large-grained polycrystalline ingots produced more simply than the crystalline ingots from which monocrystalline wafers are sliced. Grains are generally much larger than the wafer thickness (less than 0.2 mm) and hence extend through the wafer as in Figure 4.8.

During 2012, almost all commercially processed multicrystalline wafers were processed with a screen-printing sequence similar to that outlined for monocrystalline wafers above, although with some differences. The lack of control over grain orientation means that the crystallographic surface texturing used for monocrystalline silicon cells as in Figure 4.7 is not particularly effective. Reflection control using quarter-wavelength antireflection coatings therefore is essential for good performance, although surface etching to produce nonpyramidal, nonreflective features is also used in conjunction with these coatings (Luque and Hegedus, 2011). There is also a higher density of crystallographic defects in multicrystalline wafers, not only represented by grain boundaries but also by dislocations and point defects. Often there will be higher levels of impurities as well, since less-pure source material and more rugged manufacturing equipment can be tolerated. There will also be a wide variation in crystallographic quality and impurity levels from wafers cut from the same ingot. Wafers from different suppliers will also often differ widely in terms of the processing conditions that give the best results.

Introducing hydrogen in atomic form into the wafer during processing is a particularly effective way of reducing the scatter in final cell performance resulting from the above variances. A method for incorporating hydrogen demonstrated in the early 1980s (Morita et al., 1982; Kimura, 1984; Wenham et al., 1985) was by the plasma-enhanced chemical vapour deposition (PECVD) of a silicon nitride antireflection coating during cell processing, using silane (SiH_4) and usually ammonia (NH_3) as the source gases. These sources ensure the abundance of atomic hydrogen during deposition, together with its incorporation into the deposited coating and its subsequent diffusion into the underlying wafer.

High-throughput equipment for such deposition has been available for some time (Schmela, 2007), allowing these advantages to be captured by all manufacturers. This development helps bridge the gap in performance between monocrystalline and multicrystalline cell performance. At the module level, the gap is further reduced by the higher packing density possible for the generally square multicrystalline wafers, as opposed to the circular or trimmed 'quasisquare" monocrystalline wafers that are the most economical option, if cut from cylindrical ingots (Figure 4.2).

As in the monocrystalline case, the industry is moving towards progressively thinner multicrystalline wafers. There is also, in principle, more flexibility in controlling oxygen levels in these wafers, which reduces the impact of the boron–oxygen defect issue.

A relatively recent commercial development has been the use of a seed layer along the internal base of the crucible in Figure 4.4a. This allows solidification of the melt with the resulting ingot having predominantly monocrystalline structure. Such "quasimono" material is capable of producing cells with performance ranging from that of multicrystalline cells right up to that from conventional monocrystalline material. Ongoing refinement of this approach may allow ingots ultimately to be produced at close to the relatively low costs associated with multicrystalline silicon, while giving the performance of monocrystalline material. This work has stimulated interest in a range of controlled-crystallisation approaches leading to higher quality multicrystalline wafers.

4.4.2 Buried-Contact and Laser Doped, Selective-Emitter Solar Cells

The buried-contact cell design of Figure 4.9 was developed by the author's group in the early 1980s as a low-cost approach to incorporating some of the gains in laboratory performance of this era into commercial sequences (Green, 1995).

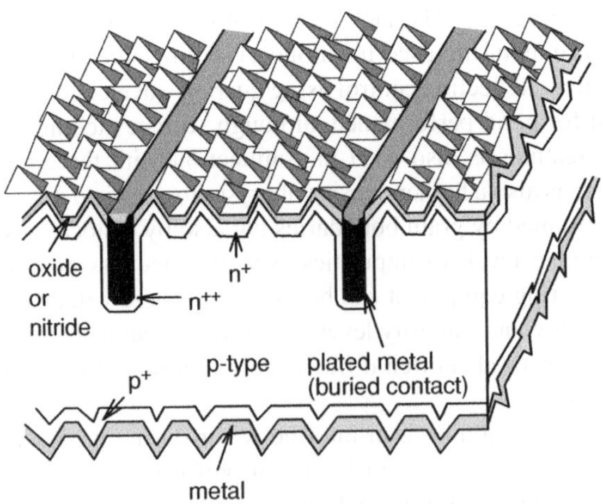

Figure 4.9 Buried-contact solar cell (not to scale). (Reproduced with permission from Green, 1995.)

The key feature of this approach is the use of a laser to form grooves into the top surface of the cell, through a previously lightly diffused layer and dielectric coating. These grooves expose fresh silicon that can be heavily doped during a second diffusion, confined by the dielectric to the grooved region. Similarly, the dielectric confines a subsequently electrolessly plated metal layer to this region. Finally, the dielectric serves as an antireflection coating for the final cell.

The advantage of this approach is that the quality of the silicon in the surface region of the cell need not be sacrificed, allowing full response to blue wavelengths. There are also advantages in reduced shadowing of the top surface of the cell by the narrower fingers resulting from this approach and inherently lower series resistance (Green, 1995).

The resulting 10–20% performance advantage compared to contemporary screen-printing cells translated to a nearly proportionate cost advantage for similar production volumes, due primarily to the shared high material costs when combined with similar processing costs. Since the sequence can extract the full performance benefits from improved wafer quality, control of the boron–oxygen defects mentioned in Section 4.4.1 would give even greater advantages for this technology (Rein et al., 2000).

BP Solar, a top-10 solar cell manufacturer as recently as 2006, commercialised this technology in the early 1990s as its premium "Saturn" product line (Mason et al., 2002). Production ceased in 2006 as the company withdrew from cell manufacturing.

A more recent version of this general approach is the laser doped, selective emitter solar cell that forgoes the use of deep grooves while eliminating the need for an extra diffusion. Sequences where the same approach is applied to both front and rear surfaces have recently given excellent results and may provide a path for transferring efficiencies associated with the PERL cell of Section 4.4.5 into commercial production.

4.4.3 HIT Cell

An alternative approach to a higher efficiency commercial solar cell is the HIT (heterojunction with thin intrinsic layers) cell of Figure 4.10. This cell combines both crystalline and amorphous silicon cell design features in the one structure.

Hydrogenated amorphous silicon, prepared by plasma-enhanced chemical vapour deposition (PECVD), has a higher bandgap than crystalline material. Consequently, this material forms a high-bandgap heterointerface with the underlying silicon wafer, providing a very effective, low recombination cap on this wafer. The uppermost thin heavily doped p-type amorphous silicon layer forms a junction with the underlying n-type crystalline wafer. An intervening, very thin intrinsic amorphous silicon layer plays an important role in obtaining high performance levels (Sakata et al., 2000). A reversed polarity structure on the rear of the wafer provides the equivalent of a "back surface field" (Green, 1995).

Since the conductivity of even heavily doped amorphous silicon is quite low, due to poor carrier mobility, transparent conducting oxides are required on both front and rear surfaces to allow lateral carrier transport to metal contacts screen printed on both surfaces. Since the rear contact is patterned, the cell can respond to light from both directions. This can improve output power in installations where the rear of the module is exposed to light scattered from the surroundings (Chieng and Green, 1993).

There are several other interesting technical features. This quality of surface passivation offered by the amorphous silicon layers is so high that record output voltages for silicon have

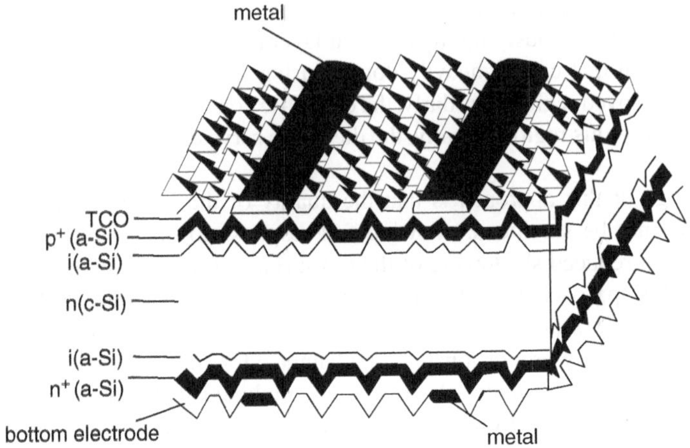

Figure 4.10 HIT solar cell (not to scale).

been confirmed for this approach (Sakata et al., 2000), forming the basis for its high energy-conversion efficiency. Also, the approach uses n-type, phosphorus-doped wafers, almost uniquely within the industry at present. This overcomes the issues with boron–oxygen defects mentioned in Section 4.4.1, since there appears to be no corresponding problems with phosphorus–oxygen defects. Processing temperatures after wafer formation are also those typical of amorphous silicon solar cell processing, much lower than those for normal crystalline silicon.

The main technical weakness of the approach is that the required transparent conducting oxide layers are neither perfectly transparent nor perfectly conducting. This forces a trade-off between light absorption in these layers and lateral resistance losses. Light absorbed in the heavily doped amorphous layers in these devices is also wasted. These absorption losses result in 5–10% current loss, tending to offset the high voltage outputs previously noted.

This is the second highest efficiency commercial cell processing sequence at present, with nominal module efficiency up to 19% (Figure 4.1). Individual cells have reached confirmed efficiencies up to 24.7% (Green et al., 2013).

4.4.4 Rear-Contact Cell

The rear or interdigitated back contact (IBC) cell with the structure shown in Figure 4.11 was suggested some time age as an alternative approach to producing high efficiency cells. Both contacts are on the rear surface of the cell as illustrated (Swartz, 1982). Diffusion is localized to the regions immediately beneath the contact areas shown (the rear surface oxide has been removed in these areas).

In principle, this structure has several advantages over conventional cell approaches. The most obvious are the elimination of contact shading of the top surface and the ability of the rear contacts to cover almost all of the rear surface, reducing the resistance losses due to current flow in these contacts. Since the current flow is perpendicular to the surface in the diffused region in the back contact approach, the resistance losses due to lateral flow in such

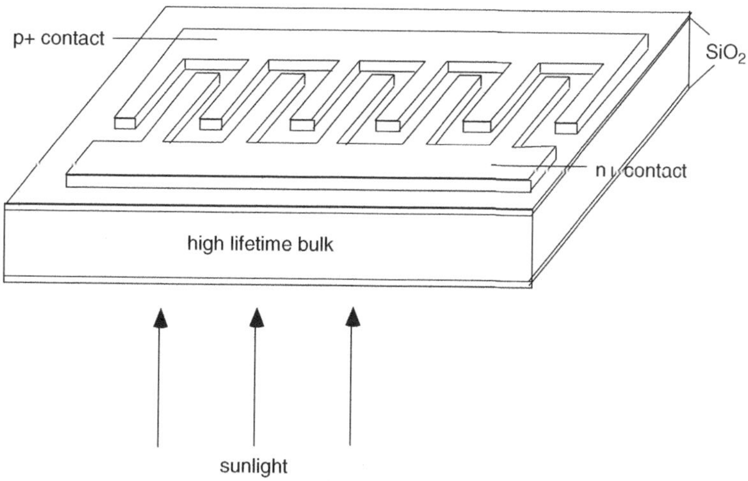

Figure 4.11 The interdigitated back contact (IBC) solar cell. Both n-type and p-type junctions lie on the nonilluminated side of the cell under the contact metallization. (Based on Schwartz, 1982. Copyright © 1982, Elsevier.)

regions in conventional cells are eliminated. The illuminated surface is in a pristine state, making a very low surface recombination velocity possible.

This is fortunate since surface recombination velocities have to be very low for successful cell operation. Another requirement is very good bulk properties, with diffusion lengths of the order of 1 mm being essential. This implies the use of high-quality starting material and high-quality processing.

This rear-contact cell design was refined at Stanford University in the 1980s (Green, 1995). In the 1990s, a new company SunPower was formed to commercialise this technology, mainly for concentrated sunlight applications. In the early 2000s, the fabrication sequence was greatly simplified and the company expanded production of standard nonconcentrating modules with module efficiencies up to 22% offered (Figure 4.1). Cell efficiencies up to 24.2% have been independently confirmed (Green et al., 2012) as had modules efficiencies up to 22.4% (aperture area basis, rather than the total area basis of Figure 4.1).

A sketch of the present commercial cell structure is shown in Figure 4.12 (Swanson, 2006). Note that a "floating" phosphorus diffusion is used along the illuminated surface to control recombination along this surface (King et al., 1988).

4.4.5 PERL Solar Cell

Although the buried-contact, HIT and rear-contact cells have greatly improved the performance of commercially available modules over the last decade, there is still scope for even higher efficiency solar modules based on crystalline silicon. The high-efficiency laboratory silicon cell is the PERL cell (passivated emitter, rear locally diffused cell) shown in Figure 4.13. This cell design took silicon cell efficiency to 23% by the end of the 1980s, an enormous improvement over the figure of 17%, the highest value only 7 years earlier. Since then, further improvements in PERL cells have taken their efficiency to 25.0%. The PERL

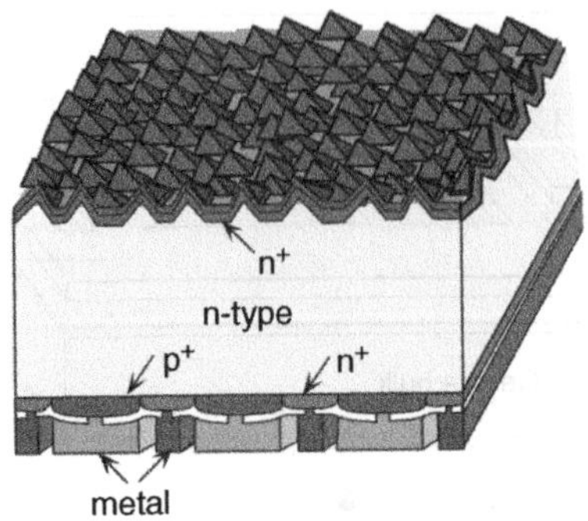

Figure 4.12 SunPower all-back-contact solar cell. (Reproduced with permission from Green and Hansen, 2002.)

cell shares many features in common with the rear point contact cell, including almost complete enshroudment in a passivating oxide layer and small area contacts passified by local heavy diffusions. However, it is a more robust design, being more tolerant of poor surface passivation and poor bulk lifetimes.

Major improvements over recent years include the growth of much thinner oxide for top-surface passivation, which allows the direct application of a double-layer antireflection coating to increase short-circuit current (Zhao and Green, 1991), the use of an "annealing" sequence for this top oxide and localised top contact points to increase open-circuit voltage and improved rear surface passivation and reduced metallisation resistance to improve fill factor.

To maximise cell performance, as much light as possible of useful wavelengths should be coupled into and absorbed by the cell. Modern cell designs such as the PERL cell of Figure 4.13 incorporate several features of a primarily optical nature to achieve this result.

The inverted pyramids along the top surface serve primarily in such an optical role. Most light incident on this structure will hit one of the side walls of the pyramids at the first point of incidence with the majority of this light coupled into the cell. That reflected will be reflected downwards, ensuring that it has at least a second chance of entering into the cell. Some of the light incident near the bottom of the pyramids has three such chances.

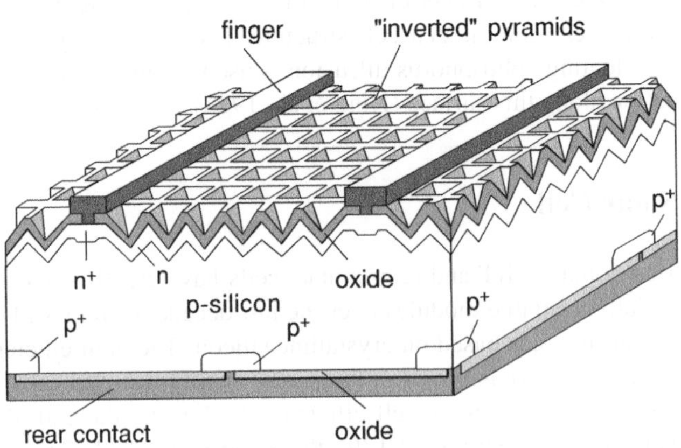

Figure 4.13 24.7% efficiency PERL cell.

The pyramids are covered by an oxide layer of appropriate thickness to act as a quarter-wavelength antireflection coating. In more recent designs, this oxide is grown thin and a double-layer antireflection coating applied (Zhao and Green, 1991).

Light coupled into the cell moves obliquely across the cell towards the rear surface with most absorbed on the way. Weakly absorbed light reaching the rear is reflected by the very efficient reflector formed by the combination of the rear oxide layer covered by an aluminium layer (Green, 1995). The reflectance from this combination depends upon the angle of incidence of the light and the thickness of the oxide layer, but is typically above 95% for angles of incidence close to the normal, decreasing to below 90% as the incidence angle approaches that for total internal reflection at the silicon/oxide interface (24.7°), and increasing again to close to 100% once this angle is exceeded.

Light reflected from the rear then moves towards the top surface. Some reaching this surface strikes a face of a pyramid of opposite orientation to that which coupled it into the cell. Most of this immediately escapes from the cell. Light striking other faces of the pyramid is totally internally reflected. This results in about half the light striking the top surface internally at this stage being reflected back across the cell towards the rear contact. The amount of light escaping after the first double pass depends on the precise geometry involved. It can be reduced by destroying some of the symmetries involved, for example by using tilted inverted pyramids or by using a 'tiler's pattern' (Green, 1995). The latter approach is currently used in PERL cell designs.

The combination of the inverted pyramids and the rear reflector therefore forms a very efficient light-trapping scheme, increasing the pathlength of weakly absorbed light within the cell. Effective pathlength enhancement factors (Green, 1995) above 40 are measured. The light trapping boosts the infrared response of the cell. The external responsivity (amps per watt of incident light) of PERL cells peaks at longer wavelengths at higher values than previous silicon cells with values of 0.75 A/W measured at 1.02 μm wavelength. Energy-conversion efficiency under monochromatic light peaks at the same wavelength with values above 45% measured (Green et al., 1992). Further improvements could push this figure to above 50% at 1.06 μm.

Other optical losses are due to reflection from, and absorption in, the top metal fingers of the cell. This can be minimised by making these lines as fine as possible with, ideally, as large an aspect ratio (height to width ratio) as possible. Alternatively, optical approaches can be used to steer incoming light away from these lines or to ensure that light reflected from them eventually finds its way to the cell surface (Green et al., 1986; Cuevas et al., 1990).

Present PERL cells lose about 5% of incoming light due to absorption or reflection loss associated with these metal fingers, when combined with reflection from the unmetallised top surface of the cell. They also lose 1–2% in performance from the use of a less than optimum light-trapping scheme and from less than 100% reflection of light from the rear surface of the cell. There is therefore some scope for small to moderate gains in performance by further improving the optical properties of these cells.

Although such advanced cell designs have been used for spacecraft and high-value terrestrial applications such as solar car racing (Zhao et al., 1997), the multiple photolithographic steps required in their fabrication have made them too expensive for low cost terrestrial applications. However, recent developments in laser-processed cells (Section 4.4.2) make it likely that PERL cells can be fabricated at low cost at some stage in the future. As noted best cell efficiency is 25.0% and best module efficiency is 22.9%, again on an aperture area basis, both the highest ever for silicon (Green et al., 2013).

4.5 CONCLUSION

Crystalline silicon wafer technology appears likely to be the dominant photovoltaic technology for at least the next decade, judging from recent manufacturing investments and price reductions. Notable recent trends have been an increasing interest in high energy-conversion efficiency monocrystalline silicon wafer sequences, the rapid transition to much thinner silicon wafers and the development of "quasimono" and related "controlled crystallisation" directionally solidified ingot sequences.

Quite revolutionary departures from the standard screen-printing cell processing approach have been demonstrated in production by BP Solar, with its "Saturn" processes, based on laser grooved, buried contacts, as well as by Sanyo, with its crystalline/amorphous silicon hybrid HIT cells and SunPower, with its back-contact cell. These cells have demonstrated a clear advantage over standard production cells, resulting in product with a distinct performance margin. Other companies such as Suntech-Power, JA Solar and Yingli have more recently introduced high efficiency sequences into production under brand names such as "Pluto", "Maple" and "Panda".

The increased competition within the photovoltaic industry over recent years means that there is more opportunity for innovative ideas to be adapted commercially and for the industry to cut costs by this process. The future should deliver significantly higher performance silicon photovoltaics at lower costs.

ACKNOWLEDGEMENTS

This Program has been supported by the Australian Government through the Australian Renewable Energy Agency (ARENA). Responsibility for the views, information or advice expressed herein is not accepted by the Australian Government.

REFERENCES

H.A. Aulich (1996), "Review of Solar-grade Silicon Research Programs Carried Out in Germany During the 80s", *9th Sunshine Workshop on Crystalline Silicon Solar Cells*, Tokyo.

B. Authier (1978), "*Poly-crystalline Silicon with Columnar Structure*", Festkörperprobleme XVIII, Freidr. Vieweg & Son, Wiesbaden.

D.O. Bergin (1980), "Shaped Crystal Growth–a Selected Bibliography", *Journal of Crystal Growth* **50**, 381–396.

C. Chieng and M.A. Green, (1993), "Computer simulation of enhanced output from bifacial photovoltaic modules", *Progress in Photovoltaics*, **1**, 293–299.

E. Christensen (ed.) (1985), Final Report, *Flat Plate Solar Array Project*, Jet Propulsion Laboratory.

A. Cuevas, R.A. Sinton, and R.M. Swanson, "Point- and Planar-Junction P-I-N Silicon Solar Cells for Concentration Applications, Fabrication, Performance and Stability", *Conf. Record, 21st IEEE Photovoltaic Specialists Conf.*, Kissimimee, May, 1990, 327–332.

S. De Wolf., P. Choulat, J. Szlufcik, I. Périchaud, S. Martinuzzi, C. Haßler, and W. Krumbe, (2000), "Light-induced Degradation of Very Low Resistivity Multi-crystalline Silicon Solar Cells". *In Conference Record, 28th IEEE Photovoltaic Specialists Conference*, Anchorage, 53–56, Anchorage, Alaska.

J. Dietl, D. Helmreich, and E. Sirtl (1981), "*Crystals: Growth Properties and Applications*", **5**, Springer-Verlag, Berlin, Heidelberg.

J.A. Eikelboom and M.J. Jansen (2000), "Characteristion of PV modules of new generations; results of tests and simulations, *Report ECN-C-00-067*, p. 18.

F. Ferrazza (1995), in *Polycrystalline Semiconductors IV – Physics, Chemistry and Technology*, S. Pizzini, H. P. Strunk and J. H. Werner (eds.), *Solid State Phenomena*, Trans. Tech. Publ., Zug, Switzerland.

M.A. Green, J. Zhao, A. Wang, and S.R. Wenham (1992), "45% Efficient Silicon Photovoltaic Cell Under Monochromatic Light", *IEEE, Electron Device Letters*, **13**, 317–318.

M.A. Green, (1995), *Silicon Solar Cells: Advanced Principles and Practice*, (Bridge Printery, Sydney).

M.A. Green and J. Hansen (1998), *Catalogue of Photovoltaic Drawings*, Photovoltaics Special Research Centre, University of New South Wales, Sydney.

M.A. Green and J. Hansen (2002), *Catalogue of Photovoltaic Drawings*, Photovoltaics Special Research Centre, University of New South Wales, Sydney.

M.A. Green (2005), "Silicon Photovoltaic Modules: A Brief History over the First 50 Years", *Progress in Photovoltaics*, **13**, 447–455.

M.A. Green and J. Ho (2011), *Catalogue of Photovoltaic Drawings*, ARC Photovoltaics Centre of Excellence, University of New South Wales, Sydney.

M.A. Green, K. Emery, Y. Hishikawa, and W. Warta (2013), "Solar Cell Efficiency Tables (Version 42)", *Progress in Photovoltaics*, **21**, pp. 827–837.

M.A. Green, J. Zhao, A.W. Blakers, M. Taouk, and S. Narayanan (1986), "25-Percent Efficient Low-Resistivity Silicon Concentrator Solar Cells", *IEEE Electron Device Letters*, **EDL-7**, 583–585.

R. Janoch, R. Wallace, and J. I. Hanoka (1997), "Commercialization of Silicon Sheet via the String Ribbon Crystal Growth Technique", *Conf. Record, 26th IEEE Photovoltaic Specialists Conference*, Anaheim, September/October, 95–98.

T.L. Jester (2002), "Crystalline Silicon Manufacturing Progress", *Progress in Photovoltaics*, **10**, 99–106.

M.J. Kerr, A. Cuevas, and P. Campbell (2003), "Limiting Efficiency of Crystalline Silicon Solar Cells Due to Coulomb-Enhanced Auger Recombination" *Progress in Photovoltaics*, **11**, 97–104.

K. Kimura (1984), "Recent Developments in Polycrystalline Silicon Solar Cells", *In: Tech. Digest of the Inst., PVSEC-1*, Kobe, Japan, 37–42.

R.R. King, R.A. Sinton, and R.M. Swanson (1988), "Front and Back Surface Fields for Point-Contact Solar Cells", *Conf. Record, 20th IEEE Photovoltaic Specialists Conf.*, Las Vegas, September, 538–544.

J. Lindmayer (1976), "Semi-crystalline silicon solar cells", *Conf. Record, 12th IEEE Photovoltaic Specialists Conference*, Baton Rouge, November, 82–85.

A. Luque and S. Hegedus (2011), "*Handbook of Photovoltaic Engineering*" (Wiley, Chichester).

N.B. Mason, T.M. Bruton, and M.A. Balbuena (2002), "Laser Grooved Buried Grid Silicon Solar Cells – From Pilot Line to 50 MWp in Ten Years". *In Conference Record, PV in Europe*, Rome, Italy.

P. Mints (2012), "*Solar in the Future Energy Mix*", Navigant Consulting Inc., May 14.

H. Morita et al. (1982), *Japanese Journal of Applied Physics*, **21** (Suppl. 21–2), 47.

A. Realini, E. Bura, N. Cereghetti, D. Chianese, and S. Rezzonico (2001), "Study of a 20 Year Old PV Plant (MTBF Project)", in *Conference Record, 17th European Photovoltaic Solar Energy Conference*, 447–450 Munich, Germany.

S. Rein, W. Warta, and S.W. Glunz (2000), "Investigation of Carrier Lifetime in P-type CZ-Silicon: Specific Limitations and Realistic Prediction of Cell Performance", in *Conference Proceedings, 28th IEEE Photovoltaic Specialists Conference*, 57–60, Anchorage, Alaska.

Y. Sakaguchi Y.N. Yuge, N. Nakamura, H. Baba, K. Hanazawa, M. Abe, and Y. Kato (1997), "Purification of Metallic Grade Silicon Up to Solar Grade by NEDO Melt Purification Process", *Conf. Proceedings, 14th European Photovoltaic Solar Energy Conference*, Barcelona, 157–160.

H. Sakata, T. Nakai, T. Baba, M. Taguchi, S. Tsuge, K. Uchihashi, and S. Kiyama (2000), 20.7% Highest Efficiency Large Area (100.5 cm^2) HIT^{TM} Cell. *Conference Proceedings, 28th IEEE Photovoltaic Specialists Conference*, 7–12, Anchorage, Alaska.

M. Schmela (2007), "High Speed Processing: Market Survey on SiN Deposition Systems", *Photon International*, March, 126–135.

R.J. Schwartz (1982), "Review of Silicon Solar Cells for High Concentrations", *Solar Cells*, **6**, 17–38.

M. Schulz and E. Sirtl (1984), "Silicon Sheet" in *Landolt-Börstein: Numerical Data and Functional Relationships in Science and Technology*, Springer-Verlag, Berlin, **17c**, Section 6.1.2.5.3, 52–54 and 442–444.

R. Swanson (2006), "Photovoltaics: The Path From Niche to Mainstream Supplier of Clean Energy", *Solar Power 2006*, San Jose, October.

R.L. Wallace, J.I. Hanoka, A. Rohatgi, and G. Crotty (1997), "Thin Silicon String Ribbon", *Solar Energy Materials Solar Cells*, **48**, 179–186.

S.R. Wenham, M.R. Willison, S. Narayanan, and M.A. Green (1985), "Efficiency Improvement in Screen Printed Polycrystalline Silicon Solar Cells by Plasma Treatments", *18th IEEE Photovoltaic Specialists Conf.*, Las Vegas, p. 1008.

J. Zhao and M.A. Green (1991), "Optimized Antireflection Coatings for High Efficiency Silicon Solar Cells", *IEEE Transactions on Electron Devices*, **38**, 1925–1934.

J. Zhao, A. Wang, F. Yun, G. Zhang, D.M. Roche, S.R. Wenham, and M.A. Green (1997), "20,000 PERL Silicon Cells for the '1996 World Solar Challenge' Solar Car Race", *Progress in Photovoltaics*, **5**, 269–276.

5 Amorphous and Microcrystalline Silicon Solar Cells

R.E.I. Schropp

Energy research Center of the Netherlands (ECN), Solar Energy, The Netherlands and Department of Applied Physics, Eindhoven University of Technology, The Netherlands

5.1 INTRODUCTION

The first semiconductor-based solar cells with energy-conversion efficiencies larger than 10% were made of silicon, in the years 1950–1960. At present, 80–90% of the solar PV modules produced worldwide are based on crystalline silicon (c-Si) wafers. The global PV demand is likely to surpass 40 GW 2013. Thin-film solar cells, now making up ~10% of the market, are attractive because they can be produced at low cost, in an energy-efficient way, using only a small amount of abundantly available materials. On the rapidly growing PV market, the relative share of thin-film solar cells has recently increased considerably, mainly due to low-cost CdTe thin-film technology. While CdTe leads the way among thin-film PV technologies, there is an opportunity for all thin-film technologies to capture a larger market share as the materials and energy consumption for mass production are inherently low. Among the available thin-film technologies, silicon thin-film technology has great potential as the technology readily allows for the manufacturing of tandem and triple-junction cells.

Whereas c-Si solar cells are commonly made from self-supporting slices sawn from a silicon ingot (with a typical thickness $d \sim 200$–$300\ \mu m$), the active silicon layers in thin-film silicon PV are deposited from the gas phase (using SiH_4 feedstock gas, often diluted in H_2) in a low-temperature process on a cheap substrate such as glass, steel or plastic. Thin-film silicon exists in different phases, ranging from amorphous via microcrystalline to single crystalline (the latter via epitaxy and transfer techniques, which is beyond the scope of this chapter). In contrast to the periodic lattice that characterises the crystalline form, there is only very short-range order in amorphous silicon (a-Si:H). However, most atoms are fourfold coordinated while hydrogen plays an important role in the growth process as well as in the passivation of dangling bonds. Owing to variations from the ideal tetrahedral bond lengths and bond angles, electronic band-tail states are present. They play an important role in the electronic density of states distribution and in the transport of carriers. In addition, due to remaining unpassivated dangling bonds, there is always a significant number of midgap states (defects) in a-Si:H.

In n-type or p-type doped a-Si:H, the dangling-bond density is so high that photo-generated charge carriers in an abrupt p-n junction are lost by recombination. Therefore, a-Si:H-based solar cells are made in the p-i-n or in the n-i-p configuration. The same applies

Solar Cell Materials: Developing Technologies, First Edition. Edited by Gavin Conibeer and Arthur Willoughby.

to microcrystalline silicon (μc-Si:H) solar cells. The vast majority of charge carriers is generated in the intrinsic absorber layer (the i-layer) and drifts to the external contacts due to the field that is set up by the p- and n-type doped layers. Due to the long-range disorder in the amorphous silicon network, the momentum-conservation law for electron excitation by photon absorption is relaxed. This makes a-Si:H behave like a material with a *direct band gap* (no additional phonon is required for the electronic transition). Therefore, the absorption coefficient of a-Si:H is much higher than that of c-Si and thus a-Si:H solar cells can be made very thin. The typical thickness for single-junction cell is $d \sim 0.2$–0.4 μm.

Based on this concept, the first experimental a-Si:H p-i-n solar cell, with an energy-conversion efficiency of 2.4%, was made by Carlson and Wronski [1] in 1976. Presently, the efficiency of such cells is well above 10% [2]. The major disadvantage of a-Si:H, however, which causes the relatively low stabilised efficiencies, is that the dangling-bond density in the i-layer reaches a higher concentration over time upon exposure of the solar cell to light (the so-called Staebler–Wronski effect [3]). It is generally assumed that the presence of H combined with that of weak Si–Si bonds plays a role in the underlying microscopic process. Because of this light-induced defect creation the performance, for relatively stable cells, saturates at a level that is 10–20% lower than the initial efficiency. Therefore, there has been a search for materials with a lower H content, better medium-range order (MRO), and a lower concentration of dihydride bonding configurations (a parameter that frequently appears to be correlated with light-induced defect creation). One form of thin-film silicon offering better resistance to light-induced defect creation is protocrystalline silicon [4]. Protocrystalline silicon is an amorphous material that is characterised by an enhanced MRO. A fingerprint for enhanced MRO is the reduced width of the first scattering peak in X-ray scattering measurements (FWHM $< 5.5°$). In this FWHM the partial pair distribution functions based on next-nearest-neighbor separations play an important role. A higher correlation length obtained from the narrowed FWHM means that the bonds are more ordered. The structure of the material is still amorphous but the bond-angle distortion is reduced, leading to stronger Si–Si bonds as compared to "classical" amorphous silicon. The enhanced stability of protocrystalline silicon against light-induced degradation is thought to be due to the enhanced MRO. It has an optical band gap similar that is slightly higher than amorphous silicon (1.8–1.95 eV). Protocrystalline silicon is usually achieved in plasma-enhanced chemical vapour deposition (PECVD) by strong dilution of the silane (SiH_4) with H_2. Microcrystalline silicon, a mixed-phase material consisting of nanocrystals embedded in amorphous tissue, also appears to be stable when the volume fraction of nanocrystals is sufficiently large (>40%). The optical band gap of microcrystalline silicon is 1.0–1.1 eV.

Since the optical band gap can be altered by creating materials with different crystalline contents, thin-film silicon, unlike any other thin-film material, can relatively easily be used in tandem and triple-junction solar cells. Such multijunction cells, due to their spectrum splitting capability, have true potential for high-conversion efficiency. The two band gaps mentioned above together form an ideal match with the terrestrial solar spectrum. Starting from the first reports on practical microcrystalline cells in 1994, much research effort has been devoted worldwide to the development of both fundamental knowledge and technological skills that are needed to improve thin-film silicon multijunction solar cells.

The challenges of the present research are (i) to enhance the network ordering of amorphous semiconductors, mainly for improving the stability, (ii) to increase the deposition rate, in particular for microcrystalline silicon; (iii) to develop thin doped layers, compatible with the new, fast deposition techniques; and (iv) to design light-trapping configurations, by utilising textured surfaces and dielectric mirrors, or by applying photonic and plasmonic concepts [5].

5.2 DEPOSITION METHODS

The first amorphous silicon layers were deposited in an rf-driven glow discharge using silane [6]. This deposition technique is now usually called plasma-enhanced chemical vapour deposition (PECVD).

PECVD has been the workhorse of the thin-film semiconductor industry and is generally used for the deposition of thin-film silicon solar cells. Some of the energy transferred to silane molecules in the collisions with electrons is radiated as visible light; hence the deposition method is also called glow-discharge deposition. An important advantage of PECVD deposition is that the deposition temperature of device-quality silicon thin films can be kept between 140 °C and 250 °C. This allows the use of a variety of low-cost materials as a substrate, such as glass, stainless steel and flexible plastic foils. The rf glow-discharge technique is based on the dissociation of silicon-containing gasses in an rf plasma, usually at a frequency of 13.56 MHz. The plasma is generated between two electrodes, and the substrate is attached to the grounded electrode. Inelastic collisions of energetic free electrons with the source gas molecules not only result in the production of silicon-containing radicals (Si_xH_y) and hydrogen radicals, but also in the production of ions. These species can react further with other radicals, ions and molecules in the gas phase. Radicals that diffuse to the substrate can contribute to the film growth. For the growth of a compact layer, it is important that the surface mobility of the growth precursors is high. This surface mobility is greatly improved by passivation of the dangling bonds at the growth surface by hydrogen [7, 8]. A growth radical reaching the surface of a growing film attaches to one of the hydrogen-passivated surface silicon dangling bonds. In order to find an energetically favourable position, the radical has to diffuse over the surface by hopping over the hydrogenated surface atoms. Hydrogen can be removed by thermal excitation or by abstraction by a SiH_3 radical, in which the dangling bond and a SiH_4 molecule is formed [7]. Crosslinking with the neighboring silicon atoms finally results in film growth, under the release of molecular hydrogen. For the deposition of device quality a-Si:H, SiH_3 radicals are the preferred growth precursors [7, 9]. Other radicals, such as SiH_2 and higher silane radicals, having higher sticking coefficients than SiH_3, can be directly incorporated into the hydrogen-terminated surface [10]. The contribution of these radicals to the growth, however, results in films with poor quality [7, 11], and therefore the presence of these radicals in the plasma should be avoided. This can be achieved by lowering the electron temperature in the plasma and/or by increasing the deposition temperature. A lower electron temperature in the plasma is achieved by using a higher plasma excitation frequency. A high deposition temperature of up to 350 °C is required at high deposition rates in order to achieve sufficient diffusion of growth radicals on the surface. Such an elevated substrate temperature has the drawback that thermal damage to the previously deposited layers in the solar cell structure may occur.

Ions also contribute to the growth; their contribution is estimated to be 10% [7, 12]. The ions that are produced in the plasma are accelerated under the influence of the electric field between the plasma and the electrodes. The resulting ion bombardment on the growing surface strongly affects the material properties. The energy that is released when an ion is stopped and neutralised at the film surface improves the surface mobility of the growth precursors as well as the crosslinking process, resulting in a more dense silicon network. Therefore, ions are believed to play a beneficial role during growth of device-quality thin-film silicon [13]. On the other hand, energetic ions with a kinetic energy higher than 20 eV are thought to cause defects in the material [14, 15], particularly in μc-Si:H [16].

Summarising, the deposition of thin-film silicon from a silane plasma can be described as a four-step process [14]:

(i) The primary reactions in the gas phase, in which SiH_4 molecules are decomposed by electron-impact excitation, generate various neutral radicals and molecules, positive and negative ions and electrons.
(ii) The secondary reactions in the plasma between molecules and ions and radicals that result in formation of reactive species and eventually in formation of large silicon–hydrogen clusters, which are described in the literature as dust or powder particles. Neutral species diffuse to the substrate, positive ions bombard the growing film and negative ions are trapped within the plasma.
(iii) Interaction of radicals with the surface of the growing film, such as radical diffusion, chemical bonding, hydrogen sticking to the surface or abstraction from the surface.
(iv) The subsurface release of hydrogen and relaxation of the silicon network.

The PECVD technique still delivers the best cell efficiencies using a-Si:H, proto-Si:H or μc-Si:H intrinsic absorber layers. To obtain device-quality a-Si:H in a laboratory rf PECVD deposition system, the typical processing conditions that have generally been used are silane flow 20–50 sccm, process pressure 0.5–0.7 mbar, substrate temperature 200–250 °C, rf power density 20–50 mW/cm^2, electrode distance 1–3 cm. A typical deposition rate is 0.2 nm/s. To obtain protocrystalline or microcrystalline material, additional hydrogen source gas is added to the silane feedstock gas during the plasma process, which generally slows down the deposition rate even further. Typically, H_2/SiH_4 gas flow ratios of 10 or higher [17] have been used.

Such deposition rates are widely considered to be too low for cost-effective mass production of solar cells. For high-volume mass production a deposition rate of 2–3 nm/s is required. A considerable research effort is therefore devoted to increasing the deposition rate of protocrystalline silicon and μc-Si:H while maintaining material quality. A schematic diagram of an rf PECVD deposition configuration is given in Figure 5.1.

5.2.1 Modifications of Direct PECVD Techniques

High-pressure, high-power regime. Using the high-pressure, high-power (hphP) rf PECVD regime, device-quality a-Si:H films have been fabricated at a deposition rate of 1.2 nm/s [18]. The pressure and power density were 5–10 mbar and 270–530 mW/cm^2, respectively. By varying the silane to hydrogen flow ratio or by changing the deposition pressure at a fixed flow ratio, a transition from amorphous to microcrystalline silicon growth was obtained. Single-junction a-Si:H solar cells, having a-Si:H absorbers deposited at 190 °C and $r_d = 1.2$ nm/s, reached a stabilised efficiency of 6.5%. The band gap for this high-rate a-Si:H material is about 50 meV higher than for "classic" a-Si:H i-layers prepared at the same deposition temperature of 190 °C (low hydrogen dilution and low deposition rate (0.15 nm/s). The wide

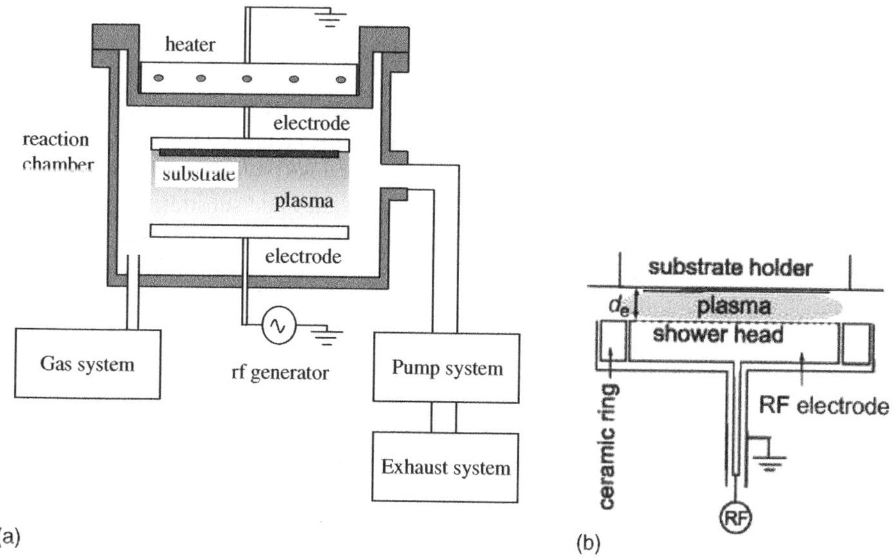

Figure 5.1 Schematic diagram of an rf PECVD chamber. In part (a), a conventional electrode configuration is shown, along with the vacuum chamber, heater, gas supply system, pump system, and exhaust system. In part (b), the electrode configuration is shown for high-pressure conditions at low interelectrode distance. At small interelectrode distances, the use of a shower head is required.

band gap and correspondingly high V_{oc} (typically 0.88–0.9 V) make these high-rate a-Si:H cells interesting candidates for top cells in tandem solar cells.

Dusty regime. Similar widening of the band gap is observed for polymorphous silicon (pm-Si:H) [19]. This material is deposited in a regime that is close to the powder formation. The powder formation is achieved by using plasma conditions that favour secondary reactions including the increase of pressure, power, interelectrode distance, and the decrease of the substrate temperature. Under this regime, small crystalline particles of 3–5 nm can be formed in the plasma, which are embedded in the growing amorphous film. To keep the plasma in the transition regime, high dilutions of SiH_4 in H_2 ($R > 30$) are used.

Very high frequency PECVD. The research group in Neuchâtel investigated the influence of increasing the plasma excitation frequency in PECVD in the VHF range from 13.56 up to 150 MHz [20]. They demonstrated that by increasing the excitation frequency from 13.56 MHz up to 70 MHz at constant plasma power, the deposition rate increases monotonously from 3 to 10 Å/s, while good quality of the a-Si:H films is maintained [21]. The shift to higher excitation frequencies modifies the electron energy distribution function in the plasma, which leads to a faster dissociation rate of source gases, and thus in higher deposition rates. The higher operating frequencies allow higher power densities to be applied in the discharge, while staying in a powder-free operational regime [22].

5.2.2 Remote PECVD Techniques

Expanding thermal plasma CVD (ETP CVD). In the ETP CVD [23] technique the plasma generation and the film deposition are spatially separated. The plasma is generated by a dc discharge in a cascaded arc in argon and/or hydrogen. The power to the plasma is typically 2–8 kW. The pressure in the plasma source is 200–700 mbar. The plasma expands supersonically into the deposition chamber, where the pressure is typically 0.1–0.3 mbar. Pure SiH_4 is injected into the plasma jet from an injection ring. The silane is dissociated by the reactive species from the plasma source. The deposition rate of a-Si:H layers is dependent on a number of factors, such as gas flows and arc current, and can reach 800 Å/s [23]. In order to deposit device-quality a-Si:H at high deposition rates (>70 Å/s), elevated substrate temperatures of around 400 °C have to be used [24]. Applying an external rf bias on the substrate during ETP CVD deposition of a-Si:H has proved to be a tool to provide extra energy to the surface due to a moderate ion bombardment of the growing surface and facilitates a reduction of the deposition temperature of ~ 100 °C [24].

Hot-wire CVD. Since the first patent in 1979, the HWCVD technique has been further developed and presently it is a viable method for the deposition of silicon-based thin films and solar cells [25–28]. This technique can be considered a remote technique, since it is based on the decomposition of silicon-containing gases at a catalytic hot surface, while the substrate itself has no active role in generating the active precursors, unlike in the case of PECVD where it usually has at least a role as the grounded electrode. The absence of the requirement to achieve an equipotential plane at the substrate makes it easier to transport either rigid or foil-type substrate materials during deposition and to scale up to large areas [29]. Because source gases are catalytically decomposed, the method is often referred to as thermocatalytic CVD (TCCVD) [30] or catalytic CVD (Cat-CVD) [31].

Usually, tungsten or tantalum filaments are used as catalyst, with filament temperatures roughly between 1400 °C and 2100 °C. Depending on the filament material and temperature, silane can be entirely dissociated into atomic Si and H at the catalyst surface. The created species react further with unreacted SiH_4 in the gas phase, so that various growth precursors are formed. As no ions are generated there during deposition, the gas-phase species and reactions that occur are different from those in conventional PECVD processes. Furthermore, the absence of any plasma ensures that no particles are trapped, thus eliminating one important source of dust.

Advantages of the HWCVD technique over the PECVD method are the high deposition rate, the scalability to large areas, and the relatively low equipment costs. Ultrahigh deposition rates of over 10 nm/s have been obtained in the case of a-Si:H [32]. In addition, it is possible to deposit amorphous silicon films with a low hydrogen content below 1 at.% that have a low defect density [27]. Solar cells incorporating such layers show less degradation upon light soaking than cells with a conventional PECVD deposited layer [33–35]. This work has initiated considerable interest in the HWCVD deposition method and today many groups study HWCVD thin-film silicon and its alloys for various applications such as solar cells [36], passivation layers [37], and thin-film transistors [38].

The reduced Staebler–Wronski effect has first been attributed to the very low hydrogen content of these hot-wire deposited layers (1–4 at.%) [65], but could also be explained by the presence of enhanced MRO [39, 40]. The absence of ions has the advantage that no defects are created as a result of energetic ion bombardment. However, as there are no low-impact species to densify the layer, it is still unclear how a low hydrogen content, device-quality amorphous silicon is deposited with the HWCVD method. One aspect is, however, that the H flux to the growing substrate surface is five to ten times higher than in PECVD, and that this H is active in subsurface reactions.

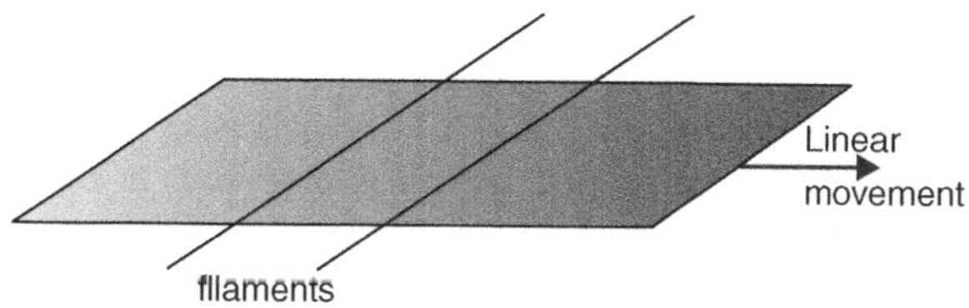

Figure 5.2 The schematic arrangement of two wires with respect to the substrate and the direction of linear motion.

A drawback of the HWCVD method in comparison to PECVD is the lack of independent control of the substrate temperature. Due to the heat radiation from the filaments, it is difficult to use substrate temperatures below 200 °C. Artificial substrate cooling is a possibility, but the effects on material quality have not been fully investigated. Another aspect is that reactions of the source gasses with the filaments may result in the formation of metal silicides on the filaments. These silicides change the catalytic properties and cause aging of the filaments. Eventually the filaments become brittle, particularly at the relatively "cold" ends of the filament. Any early breakage of the filaments can be prevented by physically shielding the filament ends or by flushing them with hydrogen. By choosing appropriate designs for both the deposition chamber geometry as well as the catalyst geometry, and by an appropriate (pre-) treatment of the filaments, filament lifetime issues can be overcome [41–43].

5.2.3 Inline HWCVD Deposition

If long and straight hot catalytic filaments are used, the radical source can be regarded as a linear source. Transport of insulating substrates (such as glass) during deposition can be easily performed as the substrate is not part of the decomposition mechanism, as in plasma-enhanced CVD. These features make HWCVD a very suitable technique for homogeneous deposition of thin films on continuously moving substrates in an inline manufacturing system or in a roll-to-roll system [44]. The dust-free deposition of thin-film silicon greatly simplifies the mounting of the substrates, which can either be rigid or a flexible web, for inline manufacturing at reduced cost. Amorphous (protocrystalline) as well as nanocrystalline silicon thin films with device-quality properties have been achieved on moving substrates. A schematic diagram of a 2-wire laboratory pilot hot-wire arrangement is shown in Figure 5.2. The local deposition rate is relatively high, at 1 nm/s, and a linear speed up to 30 cm/min has been used to produce device-quality i-layers with protocrystalline nature. We have recently further improved our thus-deposited p-i-n solar cells to 8.3% efficiency (despite air breaks after the p-layer and before the n-layer) [29].

5.3 MATERIAL PROPERTIES

Irrespective of the specific plasma process used for deposition of silicon thin films (see Section 5.2), dilution of the silicon-containing (e.g., SiH_4) source gas with hydrogen results into silicon thin films with different ordering, morphologies, and textures from the "classic" amorphous silicon. Using moderately hydrogen-diluted mixtures, a material is deposited that has a higher medium-range structural order than standard amorphous silicon. This material also has a higher stability against light-induced changes than amorphous silicon [4] and is often referred to as protocrystalline silicon [45]. At increasing hydrogen-dilution ratios, a transition from amorphous to crystalline growth is observed. In this transition regime, several material morphologies have been obtained, which are referred to as polymorphous silicon (pm-Si:H) [19], (hydrogenated) microcrystalline silicon (μc-Si:H) [46] or nanocrystalline silicon (nc-Si:H), quasiamorphous [47] or nanoamorphous silicon (na-Si:H) [48] or heterogeneous silicon (het-Si:H) [49], and polycrystalline silicon (poly-Si) [50], depending on whether nanocrystals are present and on the crystalline fraction of the material.

5.3.1 Protocrystalline Silicon

Using conventional PECVD techniques, where hydrogen dilution of the silane (SiH_4) feedstock gas is necessary to produce protocrystalline silicon [4, 45, 51], it has thus far not been possible to reach a high deposition rate for such material. Rates higher than 3 Å/s for the production of protocrystalline silicon using conventional PECVD have not been reported yet and typically, 1 Å/s is used. In addition, the deposition parameters for achieving a protocrystalline regime are dependent on the type of substrate as well as on film thickness [52,53] due to structural evolution of the ordering along the growth direction.

An enhanced stability for thin silicon solar cells has also been reported for material with another structure, which is referred to as polymorphous (pm-Si:H) [19]. This material has high hydrogen content and is deposited in a regime that is close to that where powder formation occurs. Powder formation occurs under plasma conditions close to the transition regime that favour secondary reactions, such as high pressure, high power, large interelectrode distance, and low substrate temperature. To keep the plasma in the transition regime, high dilutions of SiH_4 in H_2 ($R > 30$) are also required. Under this regime, small crystalline particles of 3–5 nm are formed in the plasma that can be embedded in the growing amorphous film.

A technique that is not plasma based, hot-wire chemical vapour deposition (HWCVD), has been shown to be capable of producing stable a-Si:H films in the protocrystalline regime (i.e. with enhanced MRO) [39]) at relatively high rate. In order to give an example of the characterisation of protocrystalline silicon, the hot-wire deposited material is more extensively presented here.

In 1997, the deposition rate for these films was already high (5–8 Å/s), but a possible drawback was the high substrate temperature regime (360–425 °C) in which the materials were obtained. X-ray diffraction measurements indicated that the most stable materials exhibited the narrowest widths of the first scattering peaks. This characteristic parameter can be considered a fingerprint for protocrystalline materials [54].

As the type of substrate has an influence on the degree of ordering in the deposited films, it has further been determined that PECVD conditions leading to *protocrystalline* silicon on n^+-type doped layers lead to *microcrystalline* silicon on plain stainless steel [55]. It was thought that for hot-wire-deposited material either high substrate temperature or low bonded hydrogen content was needed to obtain protocrystalline material. In 2004, however, it was reported [40] for the first time that materials fabricated by HWCVD at relatively low substrate temperature (250 °C) were in fact protocrystalline Si:H materials. Moreover, materials with protocrystalline characteristics were achieved in a high deposition rate regime (>10 Å/s) without using any H_2 dilution [40]. Light soaking for over 1500 h under 100 mW/cm^2, AM 1.5 conditions showed that cells incorporating such material have excellent stability. The changes in the FF were within 10%, while the absolute FF saturated at around 0.65.

5.3.2 Microcrystalline or Nanocrystalline Silicon

In order to obtain good optoelectronic quality (low defect density) μc-Si:H, ion damage to the growing surface should be avoided, while keeping a high atomic H density in the gas phase to interact with the growing surface. In order to reach high deposition rates, new PECVD regimes featuring higher rf frequencies (very high frequency, "VHF" [20–22]), conditions of source gas depletion at high pressure (high-pressure depletion, "HPD" [11] or hphP [18]) have been explored. Further, alternative plasma excitation techniques such as microwave PECVD (MW-PECVD) [56] or the expanding thermal plasma CVD (ETP-CVD) technique [23, 57], and advanced electrode designs are explored. Another approach is to abandon the mechanism of plasma decomposition altogether and use HWCVD or Cat-CVD techniques [31, 58].

There are three interrelated models explaining μc-Si:H growth. Although the role of atomic H in these models is different, there is no controversy about it having an important role.

(i) The surface diffusion model [59], first proposed by Matsuda [17] in 1983, states that the chemically activated H diffusion at the surface as well as the coverage of the surface by hydrogen are important prerequisites for the growth precursors to find favourable sites for microcrystalline growth.

(ii) The chemical transport model by Veprek et al. [60] in 1987 is based on the equilibrium between growth and etching at the surface. If the etching rate is high (comparable to the growth rate) microcrystalline silicon material is deposited. The selective etching model, proposed by several groups [61–63] is based on the chemical transport model and the difference in etching rate between the dense crystalline regions (lower etching rate), and the amorphous tissue (higher etching rate).

(iii) Various growth-zone models [11] describe mechanisms that are responsible for crystal nucleation taking place in the subsurface region of the growing film. The responsible processes are based on interaction with indiffused atomic hydrogen. Layer-by-layer growth of μc-Si:H seems to be an appropriate system to study the hydrogen-induced nucleation in the subsurface.

From optical emission spectroscopy (OES) measurements, it has become clear that there exists a pressure- and temperature-dependent threshold for the flux of atomic hydrogen to the growing surface [17]. Atomic hydrogen can originate from SiH_4 and from H_2. An important loss mechanism for atomic hydrogen is the recombination reaction with SiH_4, $H + SiH_4 \rightarrow H_2 + SiH_3$. Hydrogen dilution thus slows down the H-annihilation reaction.

Another possible method to increase the atomic hydrogen density is to operate under silane depletion conditions, in which the silane is decomposed faster than that supplied by the flow of feedstock gas. High-density plasmas, where the silane depletion can be easily realised, have been applied for depositing microcrystalline silicon [64–67]. This has even succeeded using pure silane [64, 67].

As mentioned above, ion bombardment above a certain ion energy (20 eV) causes damage [15], particularly in μc-Si:H [16]. The magnitude of the ion damage can be expressed by the product of ion energy and ion flux density per deposited monolayer. If the rf power density is increased under depletion conditions, the ion damage per monolayer increases because there is no increase in deposition rate.

The HPD regime has been investigated [68, 69] as an approach to suppress the ion damage, while maintaining a high atomic hydrogen density. The HPD method combined with the VHF technique improves the crystalline volume fraction in the very high deposition rate regime (higher than 2 nm/s). In recent work, in addition, a small interelectrode distance was implemented that was made possible by using a showerhead electrode [70]. The cell results using this combined approach are highlighted in the section on solar cells.

Microcrystalline silicon was first reported by Veprek and Marecek in Europe in 1968 [71], and in Japan by Matsuda et al. [72] and Hamasaki et al. [73] in 1980.

Microcrystalline silicon is a mixed-phase material consisting of a-Si:H, embedded nano-sized crystallites or conglomerates of nanocrystallites, and grain boundaries. During the growth of the layer, the formation of crystallites starts with a nucleation phase after an amorphous incubation phase. During continued layer deposition, clusters of crystallites grow (crystallisation phase) until a saturated crystalline fraction is reached. These processes are very much dependent on the deposition conditions. In general, crystalline growth is enhanced by the presence of atomic hydrogen, which chemically interacts with the growing surface [46, 74, 75].

A schematic representation of the incubation and crystallisation phase as a function of the source gas dilution with hydrogen is shown in Figure 5.3. The incubation phase and crystallisation phase have a large influence on the optoelectronic material properties. The optical absorption in μc-Si:H is because of local absorption in both the amorphous and the crystalline part of the volume.

Compared to a-Si:H, the absorption of high-energy photons ($h\nu$ > 1.7 eV and higher energies) is low, due to the fact that the band gap in the crystalline volume fraction is

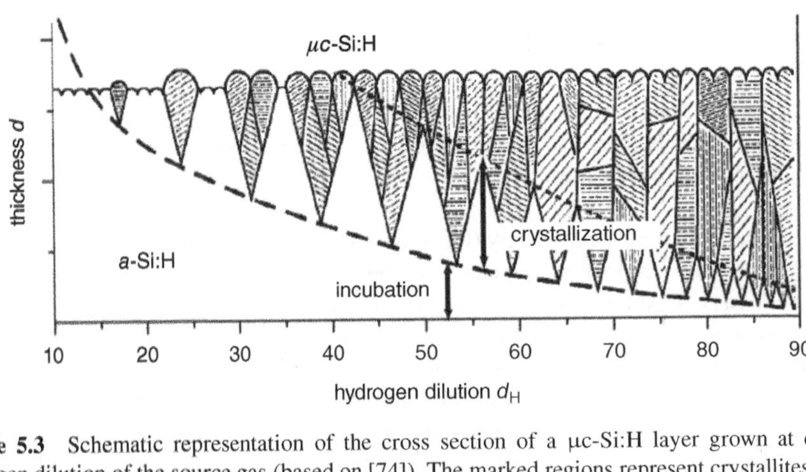

Figure 5.3 Schematic representation of the cross section of a μc-Si:H layer grown at different hydrogen dilution of the source gas (based on [74]). The marked regions represent crystallites and the white region the amorphous fraction. (Reproduced with permission from Collins, 2003. Copyright © 2003, Elsevier.)

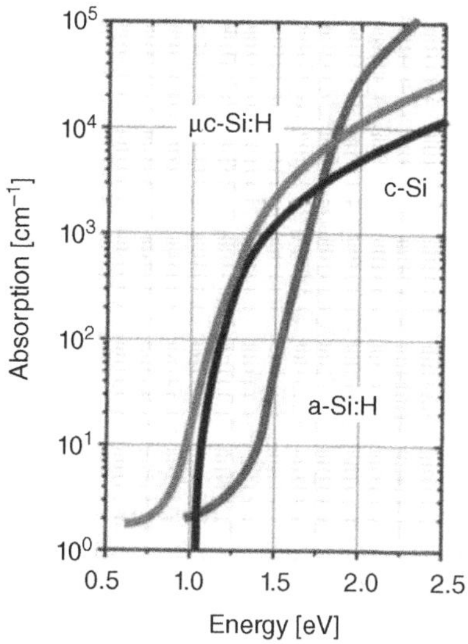

Figure 5.4 The absorption coefficient α of c-Si, a-Si:H, and μc-Si:H.

indirect, whereas in the low-energy part ($h\nu < 1.7$ eV) the absorption of μc-Si:H is higher than in amorphous silicon, due to the smaller band gap of 1.0–1.1 eV. The optical absorption spectra of μc-Si:H, a-Si:H, and single crystal Si (c-Si) are compared in Figure 5.4.

In mixed-phase material (μc-Si:H), the charge-carrier transport mechanism is not yet completely understood. Whether the electrical properties are determined by the amorphous or the crystalline fraction of the volume is dependent on the transport path. If the crystalline fraction is high enough, percolation takes place along interconnected paths through the crystallites. From geometrical considerations, the percolation threshold is ~33%. In n-type and p-type microcrystalline silicon, percolation is observed at even lower crystalline volume fractions. Of these two doping types, n-type material seems to exhibit percolation at the lowest crystalline content, in experiments at Utrecht University at only ~10%, much lower than theoretical predictions [76]. For intrinsic microcrystalline silicon, the threshold has been found at a crystalline content that is instead higher than predicted. For example, Kocka et al. [77] observe this threshold at 65%. Overall, it can be stated that the transport of charge carriers in μc-Si:H is influenced by (1) the crystalline fraction, (2) electronic barriers because of the difference in mobility gaps of crystalline grains and grain boundaries or amorphous tissues between grains, (3) doping due to impurities such as oxygen and (4) grain-boundary defects.

Obviously, grain-boundary properties play an important role in carrier transport in μc-Si:H. Matsui et al. [78] studied electron and hole transport in separate devices and concluded that μc-Si:H becomes more n-type at higher crystalline fractions, leading to low activation energies for the dark conductivity. Elsewhere, it has been found that the grain-boundary defect density increases with the crystalline fraction [79].

The performance of solar cells based on transition-type microcrystalline silicon is based on the principle that a considerable amorphous silicon content passivates the grain-boundary defects, whereas the crystalline fraction is high enough to bring about sufficient absorption in the near-infrared part of the spectrum. This principle is fundamentally different from that in polycrystalline silicon [80,81], which has no a-Si:H content but rather contains large crystals oriented in the [110] direction, with a low grain-boundary defect density, extending from bottom to top in the growth direction.

The optoelectronic properties of μc-Si:H are thus sensitively dependent on the structure of the crystallites and the grain boundaries. Dangling bonds give rise to localised states that are effective recombination centres for charge carriers. Most of the defects are concentrated at the grain boundaries and in the amorphous phase. Dangling-bond densities N_d can be as low as for compact a-Si:H, but in the low-deposition temperature regime (below $T_s = 180\ °C$) N_d is usually reported to be higher [82,83].

The presence of oxygen in concentrations higher than $1 \times 10^{18}\ cm^{-3}$ may deteriorate the electronic properties of the layers considerably [84]. Interconnected voids adjacent to grain boundaries can act as a diffusion path for impurities (such as oxygen, acting as an n-type dopant), after deposition as well as during deposition [11, 85, 86]. The n-type doping effect leads to an unfavourable field distribution in p-i-n or n-i-p type solar cells. It is thus necessary to develop a material with high compactness, so that grain boundaries are fully passivated and impurities, such as oxygen, do not deteriorate the transport mechanism.

5.4 SINGLE-JUNCTION CELL

This section first discusses the basic operation of a basic thin-film silicon solar cell and then presents the thin-film structure and technology. Thin-film silicon solar cells were, in the first instance, merely based on a-Si:H as the single absorber material. Single-junction solar cells of this type are now commercially available and the operation principles of thin-film drift-type cells are best explained with the help of this basic cell structure.

Intrinsic (i.e. undoped) amorphous silicon is a semiconductor with a mobility band gap of approximately $E_{gap} = 1.7\text{–}1.8$ eV at room temperature. Within the band gap, band tail states and midgap defect states are present. In n-type or p-type-doped a-Si:H layers, the defect density is so high that excited carriers have effectively zero mean free path and thus photogenerated electrons and holes recombine locally within the doped layer. Therefore, high-quality (low defect density) intrinsic layers (i-layers) are used as absorber layers. The absorption of photons leads to the formation of mobile electron–hole pairs in the i-layer. The electrons in the conduction band and holes in the valence band move to the external contacts, aided by the electrical field that is set up by the doped layers on either side.

Because of the localised states within the band gap of the i-layer, charge carriers can intermittently be trapped in band-tail states and eventually may be lost due to recombination (through defect states). The diffusion length of charge carriers in a-Si:H is much shorter than in crystalline Si. Therefore, unlike the situation in c-Si, in a-Si:H, the main charge-transport mechanism is not diffusion, but drift by the electrical field. For this reason, a-Si:H-based solar cells are often described as drift-type solar cells.

Figure 5.5 shows a schematic energy-band diagram of a p-i-n solar cell under different external voltage conditions. The electric-field distribution over the i-layer is determined by the mobility band gap of the i-layer, the doping levels of the doped layers, the mobility band gaps of the doped layers, the defect density in the i-layer, its spatial distribution and its occupation, and the thickness of the i-layer. The electric field is not constant over the entire i-layer. Due to charged defects in the regions near the p/i and i/n interface, as indicated in Figure 5.6, strong band bending occurs near the p- and the n-layer, and thus the electric

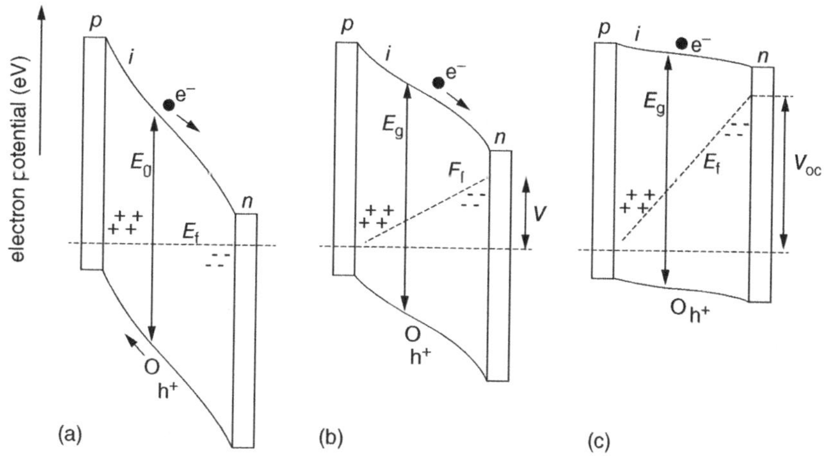

Figure 5.5 Schematic band diagram (with electrons e^- in the valence band and holes h^+ in the conduction band, and an imaginary Fermi level connecting the Fermi levels at the *p*- and n-layer) of a *p-i-n* solar cell in (a) short-circuit condition, (b) operating condition, and (c) open-circuit condition.

field in the middle of the i-layer is lower than that in the regions closer to the doped layers. This is one of the reasons why thin-film silicon solar cells cannot be made as thick as would optically be of interest.

Therefore, light trapping in the absorber layer is commonly applied and is brought about by employing textured or rough substrates, for example, textured transparent conductive oxides (TCOs) [87]. Because of light scattering at the rough substrate, a thin i-layer can absorb the same amount of light as a thicker layer on a flat substrate [88]. In this configuration, the collection of charge carriers is also improved: the recombination rate is lower, due to the larger overall electrical field over the thinner i-layer. Whereas TCOs can have native texture or can be texture-etched by simple immersion in a dilute acid solution, recently research is

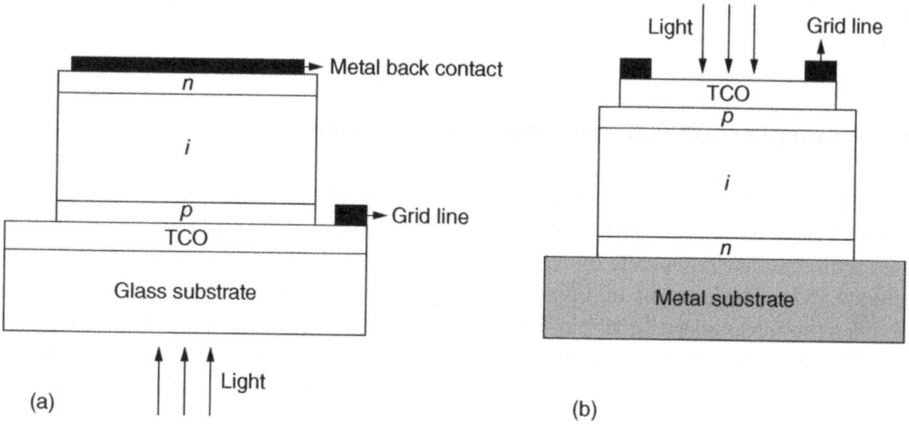

Figure 5.6 Schematic cross section of (a) a p-i-n structured solar cell (superstrate type) and (b) an n-i-p structured solar cell (substrate type). The dimensions are not to scale.

expanding on designed geometric patterns and morphologies that may lead to even higher light absorption in thin layers than the "natural" textures. These two-dimensional patterns can have either photonic (diffraction) properties or plasmonic properties, or both [89–93].

Also in μc-Si:H solar cells, the transport mechanism is predominantly drift driven. Cells need to be kept thin, not only because of the limited charge-carrier collection length, but also for reducing production costs. The charge-carrier transport mechanism is still under discussion; many models and concepts that correlate the V_{oc} to material properties can be found in the literature. Kocka et al. [94] claim that the V_{oc} is decreased when large grain boundaries introduce barriers in the conduction path. Droz et al. [95] experimentally observe a strong correlation between the V_{oc} and the estimated crystalline fraction of the i-layer. Yan et al. [96] describe the V_{oc} with a two-diode model. Werner et al. [97] correlated the V_{oc} to the grain size and they conclude that the grain-boundary recombination velocity for small grain sizes ($<$10 nm) is very low. Apart from the obvious need for further improvement of the charge-transport properties of μc-Si:H, the highest potential for improvement lies in advanced light trapping [98].

At present, the focus in thin-film silicon research has shifted from a-Si:H to different forms of thin-film silicon such as protocrystalline Si, nanocrystalline Si, μc-Si:H (microcrystalline Si) and thin-film poly-Si.

Thin-film solar cells can be made in the substrate or in the superstrate mode (see Figure 5.6). In superstrate type solar cells, the carrier on which the various thin-film materials are deposited serves as a window to the cell. Usually in superstrate cells, glass is used as the carrier. A more technologically challenging version of a superstrate configuration is a cell on a transparent thin polymer foil. This type of cell unavoidably has performance limitations as the limited temperature resistance of most transparent plastics dictates the maximum temperature allowable in further process steps. In substrate-type solar cells, the carrier on which the various thin-film materials are deposited forms the backside of the cell. They are usually made on a stainless-steel carrier that at the same time serves as the back contact. A highly flexible device can be made in the substrate configuration by employing a very thin metal carrier or a metal-coated polymer foil. As the polymer need not be transparent, a temperature-resistant type of polymer can be employed such as polyimid.

The next section discusses the status of the technology of single-junction cells in the laboratory. This summary is not meant to be exhaustive, but gives the highlights of the last decade.

5.4.1 Amorphous (Protocrystalline) Silicon Cells

In 2003, the University of Neuchâtel presented a p-i-n a-Si:H cell with a stabilised efficiency of 9.5% [99] using the VHF PECVD technique. The cell was deposited on a glass substrate coated with textured ZnO made by the low-pressure chemical vapour deposition (LPCVD) technique [82]. The amount of light entering the i-layer was additionally enhanced by applying a multilayer antireflective coating (ARC). This p-i-n cell has an i-layer thickness of only 0.25 μm and the short-circuit current density J_{sc} is very high ($>$17.5 mA/cm^2 stabilised and $>$18 mA/cm^2 initially). The deposition rate for the a-Si:H intrinsic layer was around 5 Å/s.

Six years later, in 2009, Oerlikon [100] presented a 10.03% stabilised efficiency single junction a-Si:H cell. A schematic cross-sectional diagram of the cell structure is given in

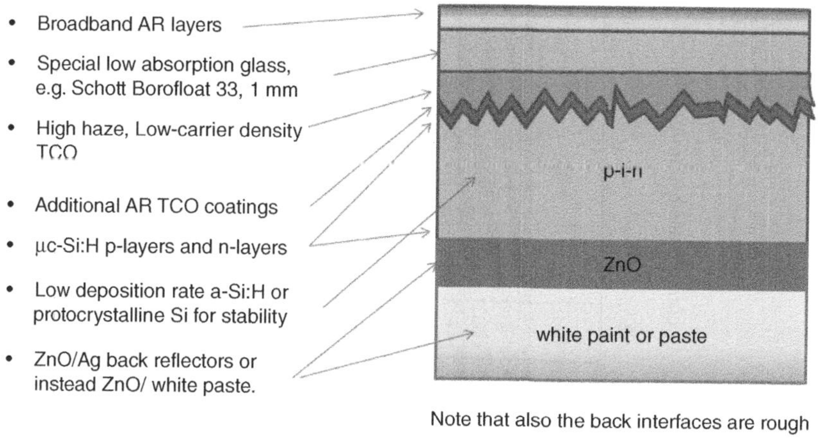

Figure 5.7 Schematic diagram of the present world record stabilised efficiency single-junction a-Si:H solar cell.

Figure 5.7. This is the highest stabilised efficiency for a purely amorphous silicon p-i-n cell structure. The constituents are: 1) an LPCVD ZnO front TCO, so-called "type B", high-haze TCO (70% Haze at 600 nm), 2) Surface texture, pyramidal like (as-grown) plus "new process" for a-Si:H suitable for high-haze pyramidal TCO, 3) a low deposition rate for the i-layer: only 1.75 Å/s, 4) i-layer thickness of 250 nm and a rather small band gap of 1.73 eV, produced in a KAI-M system (41×52 cm^2) at 40.68 MHz, 5) a broadband diffuse type AR layer stack, 6) substrate Schott Borofloat 33, 1 mm thick, 7) LPCVD ZnO back TCO $<$ 200 °C, and 8) a white paint/ white paste back reflector.

Solar cells in the n-i-p configuration incorporating HWCVD protocrystalline layers at a substrate temperature of 250 °C have demonstrated excellent stability against light soaking, where the continuous exposure time was over 1500 h. The change in fill factor and overall performance was within 10% [58]. The stability of these layers against degradation has been attributed to a special void nature that allows enhanced MRO in the material itself.

The record single junction n-i-p a-Si:H solar cells with an HWCVD absorber layer deposited at 16.5 Å/s had an initial efficiency of 9.8% [101]. Also, p-i-n single-junction cells on glass/SnO_2:F substrates have been made using the HWCVD technique. HWCVD is particularly of interest due to its high deposition rates. Cells with HWCVD absorber layers deposited at high rates of 32 and 16 Å/s reached initial efficiencies of 7.5 and 8.5%, respectively [102].

5.4.2 Microcrystalline (μc-Si:H) Silicon Cells

In 1992, Faraji et al. [103] reported a thin-film silicon solar cell with a μc-Si:H:O i-layer. The first solar cell with a μc-Si:H i-layer was reported in 1994 by Meier et al. [104] at IMT Neuchâtel, Switzerland with $\eta = 4.6\%$.

At Utrecht University, the concepts of discharge frequencies in the VHF PECVD regime [104], plasma conditions in the HPD regime [11], and gas distribution through a shower head gas inlet [11, 105] have been combined for the fabrication of high-efficiency, stable,

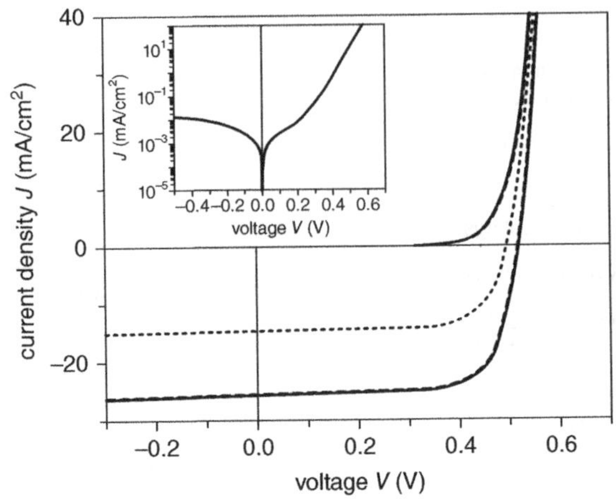

Figure 5.8 Stable 10% μc-Si:H p-i-n cell. The inset shows the *J–V* dark characteristics.

single-junction μc-Si:H solar cells. These concepts have separately been studied by other groups in the following references: VHF PECVD [104, 106–109], HPD [11, 68, 69, 109] and shower head [11, 110].

In order to control the material properties in the growth direction, the hydrogen dilution of silane in the gas phase is graded using various gas ratio profiles with a parabolic shape as a function of deposition time. By optimisation of p-i-n solar cells based on the transition-type μc-Si:H, an efficiency of 10% is obtained for a single-junction solar cell [69] on a texture-etched ZnO:Al front contact (see Figure 5.8). The μc-Si:H i-layer is 1.5 μm thick, deposited at a rate of 0.45 nm/s.

Using a hot-wire-deposited p/i interface layer (buffer layer) in p-i-n μc-Si:H cells with a plasma-deposited bulk i-layer at a conventional rf frequency of 13.56 MHz in the high-pressure, high-power regime, an initial efficiency of 10.3% was obtained at IPV, Jülich [111]. The buffer layers were made at ∼1 Å/s. The bulk layers were made at 11 Å/s. This result is primarily due to the enhancement of V_{oc} (to 568 mV) for the cell with the special HW buffer layer with respect to a cell without it (545 mV). The HW buffer layer also leads to an improvement from 8.7 to 9.4% of regular rf PECVD cells (V_{oc} from 505 to 549 mV). A possible mechanism for this improvement is the protection by the HW buffer of the p-type seed layer from energetic ions in the plasma. The 10.3% μc-Si:H single-junction solar cell shows a small light-induced degradation of 5.8% relative to the initial efficiency, leading to a stabilised efficiency of 9.8%.

Recently, it was shown that oxygenated nanocrystalline p- and n-layers, nc-SiO_x:H, are helpful in reducing the substrate-roughness-induced shunting effects. This is particularly beneficial in μc-Si:H solar cells, as these are sensitive to growth defects. This led to performance gain, in particular due to the shunt resistance and the V_{oc} for μc-Si:H cells [112]. Along with further optimisation of the plasma processes, this has ultimately led to μc-Si:H single-junction cells with a 10.9% *initial*[1] efficiency [113]. Further parameters are $V_{oc} = 0.535$ V, FF = 0.742, $J_{sc} = 27.47$ mA/cm^2. The area of the cell and the i-layer thickness are 0.25 cm^2 and 2 μm, respectively. The deposition rate of the i-layer was 0.3 nm/s.

[1] Early 2013, the record *stabilised* efficiency reported for a single junction μc-Si:H p-i-n solar cell was 10.69%, where the i-layer thickness was 1.8 μm (S. Hänni et al., *Prog. Photovolt: Res. Appl.*, **21** (2013) 821).

5.4.3 Higher Deposition Rate

Since microcrystalline silicon layers have to be made rather thick, especially when they are used in hybrid or micromorph a-Si/μc-Si tandem cells, the deposition rate is a major research challenge. Microcrystalline silicon materials with higher deposition rates are under development at many laboratories worldwide. At Utrecht University, combinations of a high rf discharge power and high total gas flow are chosen, such that the depletion condition is constant. At a deposition rate of 4.5 nm/s, a stabilised conversion efficiency of 6.7% is obtained for a single-junction solar cell with a μc-Si:H i-layer of 1 μm. It is found that the defect density increases one order of magnitude upon an increase in deposition rate from 0.45 to 4.5 nm/s. This increase in defect density is partially attributed to the increased energy of the ion bombardment during plasma deposition. The solar cells with the high-rate μc-Si:H i-layers have been light soaked in AM1.5 light, filtered by an a-Si:H top cell in order to simulate the light exposure of the bottom cell in an a-Si:H/μc-Si:H tandem cell. It appeared that the performance of the solar cells deposited at high rate actually *improved* upon light soaking (mainly due to the improving FF), contrary to the 5–10% degradation with light soaking of the transition materials reported in the literature. The reason for the improvement might be thermal equilibration of the fast-deposited material due to the elevated temperature of 50 °C during illumination, or the improvement of the contacts due to this temperature.

Even higher deposition rates for μc-Si:H have been reported. Matsui et al. [114] and Goya et al. [115] have reported deposition rates of 23 and 22 Å/s, respectively. Matsui et al. used VHFCVD, at a frequency of 100 MHz, and an initial efficiency of 8.8% was obtained. This cell has a 2.3-μm thickness. Goya et al. also used VHFCVD, but with a ladder-shaped electrode, and an initial efficiency of 8.3% was obtained.

Figure 5.9 gives an overview of the efficiencies obtained for μc-Si:H-based single-junction solar cells as a function of the deposition rate of the i-layer and Table 5.1 gives a brief overview of the published results.

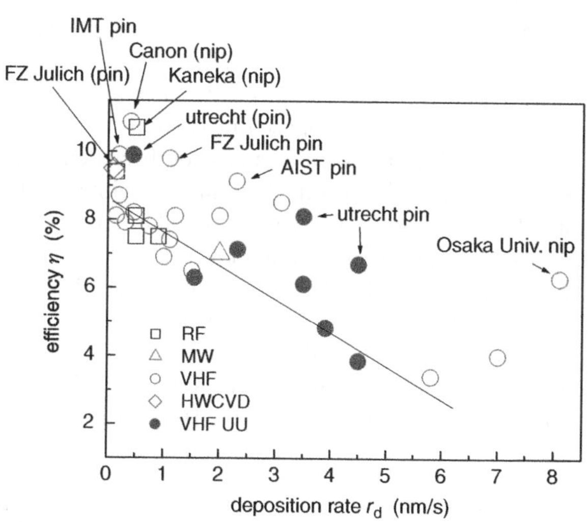

Figure 5.9 Overview of the efficiencies obtained for μc-Si:H-based single-junction solar cells as a function of the deposition rate of the i-layer. (Courtesy J.K. Rath, Utrecht University.)

Table 5.1 Overview of the achievements on single junction μc-Si:H solar cells (small area).

Laboratory	Deposition rate (nm/s)	Initial efficiency (%)	Stabilised efficiency (%)	Reference
EPFL/IMT	0.3	10.9%		[113]
EPFL/IMT	0.16		10.69%	
Utrecht University	0.45	9.9%	10.0%	[70]
IPV Jülich	1.1	10.3%	9.8%	[111]
AIST, Japan	2.2	8.8%		[114]
MHI, Japan	2.3	8.3%		[115]

5.5 MULTIJUNCTION CELLS

Microcrystalline intrinsic silicon (μc-Si:H) is very attractive for application in thin-film silicon solar cells because together with its amorphous counterpart a-Si:H it makes an ideal pair for use in multibandgap tandem solar cells. This combination not only is better matched to the solar spectrum than either a-Si:H or μc-Si:H by itself, but it is also the best possible combination. Theoretical calculations [116] show that the combination of the band gap of a-Si:H (1.7–1.8 eV) and μc-Si:H (1–1.1 eV) is close to the optimal combination for maximum efficiency. Microcrystalline silicon single-junction and μc-Si:H/a-Si:H tandem solar cells are studied by numerous groups worldwide [83, 105, 106, 117, 118].

Microcrystalline silicon doped layers are suitable as doped window layers because they have low optical absorption compared to a-Si:H at photon energies higher than $h\nu = 2$ eV. In this energy range the use of amorphous doped layers in window layers can lead to a significant portion of absorbed photons not resulting in free electron–hole pairs (i.e. parasitic absorption). The electrical properties of μc-Si:H doped layers are superior to those of a-Si:H, because microcrystalline material can be doped more efficiently. The activation energies of the dark conductivity are an order of magnitude smaller than in a-Si:H, because the Fermi level is closer to the valence or the conduction band for p- and n-type-doped layers, respectively. Therefore, the built-in voltage in solar cells is higher when μc-Si:H doped layers are used.

The concept of "micromorph" tandem cells, stacked cells with a *microcrystalline* bottom cell and an *amorphous* top cell, was first proposed in 1996 by the University of Neuchâtel [119]. In 2002, Meier et al. [120] published a micromorph tandem cell with a stabilised efficiency of 10.8% (12.3% initial) of which the bottom cell (with a thickness of $d \sim 2$ μm) was deposited at a rate of $r_d = 0.5$ nm/s. The highest initial efficiency so far, for a thin-film silicon tandem cell, is 14.7%, reported by Yamamoto of Kaneka, Japan [121]. Other high values reported for tandem micromorph *a*-Si:H/μ*c*-Si:H cells are 14.1% (initial) [122] and 11.91% stable [123]. For triple-junction cells, comprising of at least one microcrystalline silicon cell and one amorphous silicon cell, the highest initial efficiencies report to date are 16.3% initial efficiency at Unisolar [124] and 15% at Kaneka [125], respectively. The best stabilised efficiency of a triple junction a-Si:H/a-SiGe:H/μc-Si:H cell is 12.52% [124] and 12.41% for a a-Si:H/μc-Si:H/μc-Si:H triple-junction cell[2] [126].

[2] At the end of 2012, a stabilised efficiency of 13.44% was reported for an a-Si:H/μc-Si:H/μc-Si:H triple junction cell (S.W. Ahn et al., *20th European Photovoltaic Solar Energy Conference* (2012) 3AO.5.1).

To improve the stability behaviour, micromorph or "hybrid" cells are proposed to incorporate a reflecting intermediate layer (IL) at the tunnel recombination junction between the amorphous and the microcrystalline cell. The IL allows one to keep the amorphous cell very thin (<0.3 μm, in tandem cells), while enough high current can be generated. Without an IL, the amorphous top cell would have to be made so thick that an excessive Staebler–Wronski effect cannot be avoided. The optical effect of an IL is stronger in n-i-p type tandem cells than in p-i-n type tandem cells [127], and the required thickness (or rather the $n \cdot d$ product) is dependent on the roughness of the back reflector or the front TCO, respectively. Thicknesses have been varied between 30 and 110 nm. A low refractive index value would be beneficial because of lower reflection loss [125]. A low refractive index is obtained using either p-type or n-type doped nc-SiO_x:H, or both dopant types, in the tunnel-recombination junction [128, 129].

Intrinsic microcrystalline silicon with a sufficiently high crystalline volume fraction (>45% as determined by Raman spectroscopy) does not show light-induced degradation of optoelectronic properties [130, 131], whereas a-Si:H shows an increasing defect density upon light soaking (Staebler–Wronski effect [3]). Therefore, the microcrystalline bottom cell is often chosen to be current limiting, so that the tandem cell as a whole behaves more stably. For outdoor application, however, in certain locations a top-cell-limited design instead offers higher annual and daily yield than a bottom-cell-limited tandem cell [132]. This is because under practical outdoor circumstances the operation temperature and light intensity are correlated. This correlation is advantageous for the daily and annual energy yield of the top cell but adverse for the bottom cell.

In addition to the efficiency and stability, the manufacturing costs of a solar cell determine its feasibility as an industrial product. Therefore, the trend in research is toward higher deposition rates or smaller thicknesses, while minimising the loss in material quality and cell performance.

Further development and optimisation of a-Si:H/μc-Si:H tandems will remain very important because it is expected that in the near future (first half of this century), its market share can be considerable. For example, in the JRC Photovoltaic Status Report of 2010 [133], it was predicted that in 2020, the European market share for thin-film silicon (most probably a-Si:H/μc-Si:H tandems) will be 30%. This shows the importance of thin-film multibandgap cells as second-generation solar cells.

5.6 MODULES AND PRODUCTION

Within the silicon-based PV technologies, thin-film silicon has important advantages over c-Si wafer technology. Among the thin-film PV technologies it is one of the most mature technologies, because commercial modules are available from multiple manufacturers and turn-key production equipment is available from multiple companies as well. Silicon thin-film solar cells can be made using nontoxic constituents only. Moreover, the constituent elements are abundant in the earth's crust. The deposition processes can easily be scaled up to an area larger than several m^2 in a batch process, while in continuous roll-to-roll processes the area can even be in excess of 1000 m^2 (and is cut afterward into submodules of the desired size). The processing temperatures are generally low: for Si deposition, the temperature is less than 250 °C and for the textured TCO layers (depending on the type), the temperatures are also in this range or at most 550 °C. The energy payback times are short (depending on the location of the system, typically between one and three years). Thin-film silicon PV is the only thin-film technology so far that allows effective stacking of cells to form truly spectrum-splitting tandems or triple-junction cells. Monolithic series connection avoids soldering and tabbing (as in c-Si modules) and thus provides modules with the desired output voltage that are manufacturable completely inline, while area losses due to these connections are presently as low as 5%. The efficiency of these thin-film drift-type devices shows a low temperature coefficient, while the fill factor and V_{oc} are only very weakly affected by the illumination intensity. Owing to this, the energy yield of thin film Si PV modules, in particular those consisting of multijunction cells [134], is 10% higher than c-Si PV modules with the same power rating under standard test conditions (STC conditions: 100 mW/cm^2, AM1.5 spectrum and 25 °C) [135].

There are several ongoing methods to obtain a high deposition rate on a large area. The principle challenges in applying VHF PECVD on an industrial scale are (i) the uniformity of the deposited layers on a large substrate, where the effect of standing electrical waves at high frequencies (>60 MHz) may limit its application; (ii) effective coupling of the power to the plasma. The remarkable features of the HWCVD process, such as the high deposition rate, the high gas utilisation, the low-pressure process (avoiding dust formation and therefore avoiding the need for frequent chamber cleaning), the large-area deposition capability (achieved by the use of multiple filaments and/or filament grids [136]), and the decoupling of the gas dissociation from the deposition on substrate, are all very attractive for industrial production [58]. The issues regarding scaling-up of the HWCVD technique concern the uniform deposition over the large area, the filament lifetime and filament ageing. The Japanese company Anelva [136] introduced a HWCVD deposition system for large-area deposition, and Ulvac Inc. developed a large-area deposition apparatus with an effective deposition area of 150×85 cm^2 [137].

A few examples of successful demonstrations of solar cells on a large area and production activities are discussed here. The first four examples deal with deposition on glass, the latter three deal with roll-to-roll deposition on foil.

Oerlikon Solar (previously Unaxis Solar presently TEL Solar) has adapted their KAI PECVD commercial deposition system, originally designed for deposition of the active matrix for liquid crystal displays, to allow deposition of a-Si:H and μc-Si:H silicon at a plasma frequency of 40 MHz [138]. The full Gen5 size is at present very commonly used for thin-film solar panels. Oerlikon Solar, one of the manufacturers of turn-key solar cell manufacturing lines, is fully committed to this panel size. In 2008, Applied Materials introduced panels of size Gen8.5 (2.2×2.6 m^2; or 5.7 m^2), good for 8 LCD panels of 46 inches, or for four solar panels of 1.1×1.3 m^2.

In the 1980s, display technology benefitted from the know-how on plasma deposition and equipment intended for solar cells. In the 2000s, thin-film solar technology benefitted from the scaled-up versions of plasma deposition equipment, and presently Gen5 or Gen5.5 is the most commonly used size for thin-film solar modules. Scaling has always been the major cost driver for displays. For solar cells, the cost per installed watt is another important factor, and therefore primary attention for sustained growth in photovoltaics should be given to cell efficiency.

Applied Materials has introduced the equipment for 5.7 m^2 display panels into the thin-film solar cell manufacturing market and, although the company discontinued the sale of turn-key manufacturing lines, some 10–12 customers worldwide have been operating these production lines. The full size (2.2 m × 2.6 m) modules can be readily installed or be cut into 4 pieces for various applications. Gen10 (>2.8 m × 3.1 m; 9.0 m^2) and Gen11 (>3.0 m × 3.3 m) will be the next sizes in production of LCDs (investments by Sharp, Samsung, Sony, and LG). Gen10 may also be attractive for solar cell deposition. A thin-film plasma-enhanced CVD (PECVD) machine has a $14 \times 14\ m^2$ footprint (Applied Materials AKT-90K).

Mitsubishi Heavy Industries (MHI) have developed a large-area VHF-PECVD production apparatus for depositing on $1.1 \times 1.4\ m^2$ substrate implementing a ladder-shaped electrode and using a phase modulation method [139]. The base frequency in the process is 60 MHz and the modulation frequency is varied to minimise the thickness uniformity. The average deposition rate is 11 Å/s and a thickness uniformity within ± 18% was obtained. Amorphous silicon p-i-n modules with stable aperture efficiency of 8% have been made. Since October 2002, a-Si:H modules have been commercially produced. Also, a-Si:H/nc-Si:H tandem modules are produced with a power output of 130 W for $1.1 \times 1.4\ m^2$ panel areas, and triple-junction cells are under investigation.

Kaneka Corporation is producing modules on glass of 4 mm thickness with an area of $91 \times 45.5\ cm^2$ and $91 \times 91\ cm^2$. In 1993, the goal for mass production was a stable efficiency of 8%, based on a single-junction p-i-n structure made by large-area CVD and precision laser patterning. Later, thin microcrystalline Si technology was introduced. In 1995, a 10% double junction was obtained on a large area. Light-soaking tests of a $91 \times 45.5\ cm^2$ module showed a degradation of less than 8% (light-soaking conditions 0.75 suns, 38 °C) [140]. In 1999, commercial production was at a level of 20 MW/year. In 2008, production of a-Si:H/nc-Si:H modules was started with a thin intermediate interlayer as discussed above. Recently, 11.7% was obtained for a thin a-Si:H/nc-Si:H submodule. In production, the module efficiency is 9% for $1.0 \times 1.2\ m^2$ panels.

United Solar Ovonic's first roll-to-roll facility was built in 1986. In 1991, it was modified to obtain a capacity of 2 MW/year. The production capacity of United Solar has since been increasing gradually. The accumulated installation once amounted to > 400 MWp, but production has recently been discontinued. In the production machine, six rolls of stainless steel are mounted vertically, and the foils run once through a 240-ft long machine. All layers of the triple-junction solar cell structure are deposited automatically. The length of the stainless-steel foil is 1.5 mile (2.3 km) and its thickness is 5 mil (125 μm). This machine has been duplicated in Auburn Hills, Michigan, where manufacturing started in 2006 [141]. The stainless-steel roll is first washed in a roll-to-roll washing machine, then the back Al/ZnO reflector is deposited, then the triple-junction Si-based alloy solar cell, and finally the top contact, which also has antireflection properties, is deposited. After a short passivation step, the electrodes are bonded, cells are cut and interconnected, followed by lamination, framing and finishing. The active area efficiency is 10.4% over the long run [141]. Typical modules have a power of ~40 W indoors, outdoors it is ~38 W. The light-induced degradation is 12% in 800 h, with a generally observed stable efficiency of 8%. Recently, a stable module ($806\ cm^2$) efficiency of 11.8% has been obtained for a triple-junction configuration. Just before closing down, the company presented approaches for Si-based thin-film PV towards 25% efficiency.

Fuji Electric Co. Ltd, after first developing thin-film silicon tandem modules on $30 \times 40\ cm^2$ glass substrates, adopted a "stepping roll" deposition method on rolls of polyimid foil. The products use the concept of the so-called series connection through apertures formed on film (SCAF) series connection. There are two kinds of holes: series connecting holes, and current collection holes. As a part of the process sequence, the basic steps are to punch series connecting holes, followed by a metallisation step (sputtering), then current-collection holes are punched, and then the solar cell structure including the transparent top contact is deposited and laser scribed. One roll is 1.5 km long and in one run 30 kW can be produced. In R&D the stabilised efficiency of $40 \times 80\ cm^2$ a-Si/a-SiGe modules is 9%

(initial aperture area efficiency 10.1%). The completed modules are less than 1 mm thick and weigh approximately only 1 kg/m^2. Examples of the first applications launched in the market, since October 2004, are power sources for mobile use and on the rooftop of water reservoir ponds at water-purification plants. Another type of module is a 4 m long steel plate integrated module, which can be used for installation as metal rooftops of factories or public facilities. The low weight of such roof plates reduces the requirements with respect to the construction of the building. The annual capacity is presently 24 MW (Kumamoto factory). The typical tandem modules are based on a-Si/a-SiGe cells, 40 × 80 cm^2, and have a power output of 24 W(DC 80 V) [142].

Hyet Solar b.v. has bought the assets of Helianthos b.v. from Nuon in 2012. Helianthos introduced a manufacturing concept in which a-Si:H solar cells are produced in a roll-to-roll manner on a temporary metal substrate (temporary superstrate process) [143, 144]. The cell structure is first deposited on a SnO_2:F-coated metal foil. At one point in the process, the solar cell is transferred to a permanent plastic substrate on which cells are monolithically series connected to produce lightweight PV modules. In the pilot-line modules with efficiency above 7% are obtained on 1 ft wide foil. In order to increase efficiency, a-Si:H/nc-Si devices are incorporated in the temporary superstrate process. This allows for cost reduction in both the module production (high-throughput roll-to-roll). As a result of research, a-Si:H/nc-Si:H modules on 10 × 10 cm^2 foils with an aperture area efficiency of 8.6% (initial 9%) have been presented. Single-junction encapsulated modules from the roll-to-roll pilot line have an efficiency of 7% (stable 5.5–6%) [145].

In conclusion, thin-film Si PV has reached a mature stage with commercially attractive module efficiencies. A few of the highest values for module total area stabilised efficiencies are presently 10.0% (Oerlikon, 1.4 m^2, tandem module 143 W), 10.0% (Sanyo, 1.54 m^2), 9.5% (Sharp, 1.42 m^2), 9.0% (Kaneka, 1.22 m^2), 8.9% (Nexpower, 1.58 m^2), and 8.4% on the largest area modules (Sunfilm, 5.72 m^2). The latest record stabilised efficiency for Unisolar on 400 cm^2 modules is 11.2%, indicating the immediate potential of triple-junction, large-area module technology. The intermediate term potential can be deduced from the initial efficiencies of close to 15% and in the long-term several roadmaps predict 18% module efficiency.

ACKNOWLEDGMENTS

I would to thank all former and present colleagues worldwide, for their constant dedication to advance the thin-film silicon photovoltaic field and for providing the right atmosphere to do so. Much of this overview has been made possible through studies performed by my Ph.D. students, for which I am very grateful.

REFERENCES

[1] D.E. Carlson and C.R. Wronski, *Appl. Phys. Lett.*, **28** (1976) 671.
[2] S. Benagli, D. Borrello, E. Vallat-Sauvain, J. Meier, U. Kroll, J. Hötzel, J. Bailat, J. Steinhauser, M. Marmelo, G. Monteduro, L. Castens, *Proc. 24th European Photovoltaic Solar Energy Conference*, Hamburg, Germany (2009) 2293.
[3] D.L. Staebler and C.R. Wronski, *Appl. Phys. Lett.*, **31** (1977) 292.

[4] S. Guha, K.L. Narasimhan and S.M. Pietruszko, *J. Appl. Phys.*, **52** (1981) 859.
[5] A. Polman and H.A. Atwater, *Nature Mater.*, **11**, 174–177 (2012).
[6] H.F. Sterling, R.C.G. Swann, *Solid-State Electron.*, **8** (1965) 653.
[7] A. Matsuda, K. Nomoto, Y. Takeuchi, A. Suzuki, A. Yuuki and J. Perrin, *Surf. Sci.*, **227** (1990) 50.
[8] Y. Toyoshima, K. Arai, A. Matsuda and K. Tanaka, *J. Non-Cryst. Solids*, **137–138** (1991) 765–770.
[9] G. Bruno, P. Capezzuto and A. Madan, Plasma *Deposition of Amorphous Silicon-Based Materials*, Academic Press, San Diego, CA, USA, 1995.
[10] R. A. Street, *Hydrogenated Amorphous Silicon*, Cambridge University Press, Cambridge, 1991.
[11] M. Kondo and A. Matsuda, *Curr. Opin. Solid State Mater.*, **6** (2002) 445.
[12] E.A.G. Hamers, J. Bezemer and W. F. van der Weg, *Appl. Phys. Lett.*, **75** (5) (1999) 609–611.
[13] A. Matsuda, *Thin Solid Films*, **337** (1996) 1.
[14] R. E. I. Schropp and M. Zeman, *Amorphous and Microcrystalline Silicon Solar Cells: Modeling, Materials and Device Technology*, Springer, ISBN 0-7923-8317-6, London, 1998.
[15] W. Shindoh and T. Ohmi, *J. Appl. Phys.*, **79** (1996) 2347.
[16] M. Kondo, M. Fukawa, L. Guo and A. Matsuda, *J. Non-Cryst. Solids*, **84** (2000) 266–269.
[17] A. Matsuda, *J. Non-Cryst. Solids*, **767** (1983) 59–60.
[18] B. Rech, T. Roschek, J. Müller, S. Wieder and H. Wagner, *Sol. Energy Mater. Solar Cells*, **66** (2001) 267–273.
[19] P. Roca i Cabarrocas, *J. Non-Cryst. Solids*, **31** (2000) 266–269.
[20] H. Curtins, N. Wyrsch and A.V. Shah, *Electron. Lett.*, **23** (5) (1987) 228–230.
[21] A. Shah, J. Dutta, N. Wyrsch, K. Prasad, H. Curtins, F. Finger, A. Howling and Ch. Hollenstein, *Mater. Res. Soc. Symp. Proc.*, **258** (1992) 15–26.
[22] J.L. Dorier, C. Hollenstein, A.A. Howling and U. Kroll, *J. Vac. Sci. Technol. A.*, **10** (4) (1992) 1048–1052.
[23] W.M.M. Kessels, R.J. Severens, A.H.M. Smets, B.A. Korevaar, G.J. Adriaenssens, D.C. Schram and M.C.M. van de Sanden, *J. Appl. Phys.*, **89** (2001) 2404.
[24] A.H.M. Smets, W.M.M. Kessels and M.C.M. van de Sanden, *Mater. Res. Soc. Symp. Proc.*, **808** (2004) 383.
[25] H. Wiesmann, A.K. Gosh, T. McMahon and M. Strongin, *J. Appl. Phys.*, **50** (1979) 3752.
[26] H. Matsumura, *J. Appl. Phys.*, **65** (1989) 4396.
[27] A.H. Mahan, J. Carapella, B.P. Nelson, R.S. Crandall and I. Balberg, *J. Appl. Phys.*, **69** (1991) 6728.
[28] K.F. Feenstra, R.E.I. Schropp and W.F. van der Weg, *J. Appl. Phys.*, **85** (1999) 6843.
[29] R.E.I. Schropp, C.O. van Bommel, C.H.M. van der Werf, Y. Liu, J. Xin, *35th IEEE Photovoltaics Specialist Conf.*, Honolulu, HI, USA (2010) p.3739.
[30] B. Schröder, U. Weber, H. Seitz, A. Ledermann and C. Mukherjee, *Thin Solid Films*, **395** (2001) 298.
[31] H. Matsumura, *Jpn. J. Appl. Phys.*, **25** (12) (1986) L949.
[32] B.P. Nelson, E. Iwaniczko, A.H. Mahan, Q. Wang, Y. Xu, R.S. Crandall and H.M. Branz, *Thin Solid Films*, **395** (2001) 292.
[33] M. Vanecek, A.H. Mahan, B.P. Nelson and R.S. Crandall, *Proceedings of the 1 1th European Photovoltaic Solar Energy Conference* (L. Guimaraes, W. Palz, C. De Reyff, H. Kiess and P. Helm, eds.), p. 96, Montreux, Switzerland, October 12–16, 1992.
[34] M. Vanecek, Z. Remes, J. Fric, R.S. Crandall and A.H. Mahan, *Proceedings of the 12th European Photovoltaic Solar Energy Conference* (R. Hill, W. Palz and P. Helm, eds.), p. 354, Amsterdam, The Netherlands, April 11–15, 1994.
[35] A.H. Mahan, E. Iwaniczko, B.P. Nelson, R.C. Reedy Jr., R.S. Crandall, S. Guha and J. Yang, *Proceedings of the 25th IEEE Photovoltaic Specialists Conference*, p. 1065, Washington, D.C., USA, May 13–17, 1996.

[36] B. Schröder, *Thin Solid Films*, **430** (2003) 1–6.
[37] V. Verlaan, C.H.M. van der Werf, W.M. Arnoldbik, H.C. Rieffe, I.G. Romijn, W.J. Soppe, A.W. Weeber, H.D. Goldbach and R.E.I. Schropp, *20th European Photovoltaic Solar Energy Conference*, p. 1434, Barcelona, Spain, June 6–10, 2005.
[38] H. Meiling and R.E.I. Schropp, *Appl. Phys. Lett.*, **70** (20) (1997) 2681–2683.
[39] A.H. Mahan, D.L. Williamson and T.E. Furtak, *Mater. Res. Soc. Symp. Proc.*, **467** (1997) 657.
[40] R.E.I. Schropp, M.K. van Veen, C.H.M. van der Werf, D.L. Williamson and A.H. Mahan, *Mater. Res. Soc. Symp. Proc.*, **808** (2004) A8.4.1.
[41] P. Alpuim, V. Chu and J.P. Conde, *J. Non-Cryst. Solids*, **110** (2000) 266–269.
[42] H. Matsumura, *Thin Solid Films*, **395** (2001) 1.
[43] M.K. van Veen and R.E.I. Schropp, *Appl. Phys. Lett.*, **82** (2003) 287.
[44] H. Li, R.H. Franken, R.L. Stolk, C.H.M. van der Werf, J.K. Rath and R.E.I. Schropp, *Thin Solid Films*, **516** (2008) 755.
[45] R. J. Koval, J. Koh, Z. Lu, L. Jiao, R. W. Collins and C. R. Wronski, *Appl. Phys. Lett.*, **75** (1999) 1553.
[46] E. Vallat-Sauvain, U. Kroll, J. Meier, A. Shah and J. Pohl, *J. Appl. Phys.*, **87** (2000) 3137.
[47] A.R. Middya, S. Hamma, S. Hazra, S. Ray and C. Longeaud, *Mater. Res. Soc. Proc.*, **664** (2001) A9.5.
[48] H. Yu, R. Cui, F. Meng, C. Zhao, H. Wang, H. Yang, H. Chen, P. Luo, D. Tang, S. Lin, Y. He, T. Sun and Q. Ye, *15th Int. PVSEC*, p. 1048, Shanghai, China AD-9, 2005.
[49] R. E. I. Schropp, B. Stannowski, A. M. Brockhoff, P. A. T. T. van Veenendaal and J.K. Rath, *Mater. Phys. Mech.*, **1** (2000) 73.
[50] R.E.I. Schropp and J.K. Rath, *IEEE Trans. Electron Devices*, **46** (1999) 2069.
[51] D.V. Tsu, B.S. Chao, S.R. Ovshinsky, S. Guha and J. Yang, *Appl. Phys. Lett.*, **71** (1997) 1317.
[52] J.H. Koh, Y. Lee, H. Fujiwara, C.R. Wronski and R.W. Collins, *Appl. Phys. Lett.*, **73** (1998) 1526.
[53] R.W. Collins and A.S. Ferlauto, *Curr. Opin. Solid State Mater. Sci.*, **6/5** (2002) 425.
[54] D.L. Williamson, *Mater. Res. Soc. Symp. Proc.*, **557** (1999) 251.
[55] S. Guha, J. Yang, D.L. Williamson, Y. Lubianiker, J.D. Cohen, A.H. Mahan, *Appl. Phys. Lett.*, **74** (1999) 1860.
[56] K. Saito, M. Sano, K. Ogawa and I. Kajita, *J. Non-Cryst. Solids*, **164–166** (1993) 689–692.
[57] A.H.M. Smets, W.M.M. Kessels and M.C.M. van de Sanden, *Appl. Phys. Lett.*, **82** (6) (2003) 865–867.
[58] R.E.I. Schropp, *Thin Solid Films*, **451−452** (2004) 455.
[59] T. Akasaka and I. Shimizu, *Appl. Phys. Lett.*, **66** (1995) 3441.
[60] S. Veprek, F.A. Sarrott and Z. Iqbal, *Phys. Rev.*, **B 36** (1987) 3344.
[61] I. Solomon, B. Drevillon and N. Layadi, *J. Non-Cryst. Solids*, 164–166 (1991) 989.
[62] C.C. Tsai, G.B. Anderson and R. Thompson, *Mater. Res. Soc. Symp. Proc.*, **192** (1990) 475.
[63] S. Kumar, B. Drevillon, C. Godet, *J. Appl. Phys.*, **60** (1986) 1542.
[64] M. Scheib, B. Schröder and H. Oechsner, *J. Non-Cryst. Solids*, **895** (1996) 198–200.
[65] R. Nozawa, K. Murata, M. Ito, M. Hori and T. Goto, *J. Vac. Sci. Technol.*, **A 17** (1999) 2542.
[66] R. Knox, V. Dalal, B. Moradi and G. Chumanov, *J. Vac. Sci. Technol.*, **A 11** (1993) 1896.
[67] K. Endo, M. Isomura, M. Taguchi, H. Tarui and S. Kiyama, *Sol. Energy Mater. Solar Cells*, **66** (2001) 283.
[68] L. Guo, M. Kondo, M. Fukawa, K. Saitoh and A. Matsuda, *Jpn. J. Appl. Phys.*, **37** (1998) L1 116.
[69] M. Fukawa, S. Suzuki, L. Guo, M. Kondo and A. Matsuda, *Sol. Energy Mater. Solar Cells*, **66** (2001) 217.
[70] A. Gordijn, J.K. Rath and R.E.I. Schropp, *Prog. Photovolt.: Res. Appl.*, **13** (2005) 1–7.
[71] S. Veprek and V. Marecek, *Solid-State Electron.*, **11** (1968) 683.
[72] A. Matsuda, S. Yamasaki and H. Yamamoto, *Jpn. J. Appl. Phys.*, **19** (1980) L305.

[73] T. Hamasaki, H. Kurata and Y. Osaka, *Appl. Phys. Lett.*, **37** (1980) 1084.
[74] R.W. Collins, A.S. Ferlauto and C.R. Wronski, *Sol. Energy Mater. Solar Cells*, **78** (2003) 143.
[75] G. M. Ferreira, A.S. Ferlauto, Chi Chen, R.J. Koval, J.M. Pearce, C. Ross, C.R. Wronski and R. W. Collins, *J. Non-Cryst. Solids*, **338–340** (2004) 13–18.
[76] A. Gordijn, J.K. Rath and R.E.I. Schropp, *Mater. Res. Soc. Symp. Proc.*, **762** (2003) 637.
[77] J. Kocka, A. Fejfar and I. Pelant, *Sol. Energy Mater. Solar Cells*, **78** (2003) 493.
[78] T. Matsui, R. Muhida and Y. Hamakawa, *Appl. Phys. Lett.*, **81** (25) (2002) 4751.
[79] A.L. Baia Neto, A. Lambertz and F. Finger, *J. Non-Cryst. Solids*, **274** (2000) 299–302.
[80] J.K. Rath, H. Meiling and R.E.I. Schropp, *Jpn. J. Appl. Phys.*, **36** (1997) 5436.
[81] P.A.T.T. van Veenendaal, T.J. Savenije, J.K. Rath and R.E.I. Schropp, *Thin Solid Films*, **175** (2002) 403–404.
[82] O. Kluth, B. Rech, L. Houben, S. Wieder, G. Schoepe, C. Beneking and H. Wagner, *Thin Solid Films*, **351** (1999) 247.
[83] A. Matsuda, *J. Non-Cryst. Solids*, **1** (2004) 338–340.
[84] T. Kamei and T. Wada, *J. Appl. Phys.*, **96** (4) (2004) 2087.
[85] J. Kocka, H. StuchlikovaandA. Fejfar, *J. Non-Cryst. Solids*, **299–302**, (2002) 355–359.
[86] M. Konagai, T.Tsushima and R. Asomoza, *Thin Solid Films*, **395** (2001) 152.
[87] J. Müller, B. Rech, J. Springer, and M. Vanecek, *Sol. Energy*, **77**, (2004) 917–930.
[88] S. Fahr, C. Rockstuhl, and F. Lederer, *Appl. Phys. Lett.*, **92** (2008) 171114.
[89] H. A. Atwater and A. Polman, *Nature Mater.*, **9** (2010) 205–213.
[90] F. J. Beck, A. Polman, and K. R. Catchpole, *J. Appl. Phys.*, **105** (2009) 114310.
[91] C. Haase and H. Stiebig, *Appl. Phys. Lett.*, **91** (2007) 061116.
[92] V. E. Ferry, M. A. Verschuuren, H. B. T. Li, R. E. I. Schropp, H. A. Atwater, and A. Polman, *Appl. Phys. Lett.*, **95** (2009) 183503.
[93] V. E. Ferry, M. A. Verschuuren, H. B. T. Li, E. Verhagen, R.J. Walters, R. E. I. Schropp, H. A. Atwater, and A. Polman, *Opt. Exp.*, **18** (2010) A237.
[94] J. Kocka, J. Stuchlik and A. Fejfar, *Appl. Phys. Lett.*, **79** (2001) 2540.
[95] C. Droz, E. Vallat-Sauvain and A. Shah, *Sol. Energy Mater. Solar Cells*, **8** (2004) 61.
[96] B. Yan, J. Yang, G. Yue, K. Lord and S. Guha, *Proc. of 3rd World Conference on Photovoltaic Energy Conversion*, p. 1627, Osaka, May 11–18, 2003.
[97] J.H. Werner, K. Taretto and U. Rau, *Solid State Phenomena*, **80–81** (2001) 299.
[98] A.V. Shah, J. Meier and J. Bailat, *J. Non-Cryst. Solids*, **338–340** (2004) 639.
[99] J. Meier, J. Spitznagel, U. Kroll, C. Bucher, S. Faÿ, T. Moriarty and A. Shah, *Proc. of the 3rd World Conference on Photovoltaic Energy Conversion*, paper S20-B9-06, Osaka, 2003.
[100] S. Benagli, J. Hoetzel, U. Kroll, J. Meier, E. Vallat-Sauvain, D. Borrello, J. Spitznagel, O. Kluth, M. Keller, R. Kravets, M. Kupich, D. Koch-Ospelt, *24th EUPVSEC, Hamburg* (2009).
[101] A.H. Mahan, R.C. Reedy, E. Iwaniczko, Q. Wang, B.P. Nelson, Y. Xu, A.C. Gallagher, H.M. Branz, R.S. Crandall, J. Yang and S. Guha, *Mater. Res. Soc. Symp. Proc.*, **507** (1998) 119.
[102] R.H. Franken, C.H.M. van der Werf, J. Loeffler, J.K. Rath and R.E.I. Schropp, *Thin Solid Films*, **501** (2006) 47–50.
[103] M. Faraji, S. Gokhale and S.V. Ghaisas, *Appl. Phys. Lett.*, **60** (26) (1992) 3289.
[104] J. Meier, R. Flückiger, H. Keppner and A. Shah, *Appl. Phys. Lett.*, **65** (7) (1994) 860.
[105] B. Rech, T. Roschek, T. Repmann, J. Müller, R. Schmitz and W. Appenzeller, *Thin Solid Films*, **427** (1–2) (2003)157–165.
[106] A.V. Shah, J. Meier and U. Graf, *Sol. Energy Mater. Solar Cells*, **78** (2003) 469.
[107] A. Matsuda, *J. Non-Cryst. Solids*, **1** (2004) 338–340.
[108] M. Fukawa, S. Suzuki and A. Matsuda, *Sol. Energy Mater. Sol. Cells*, **66** (2001) 217.
[109] Y. Mai, S. Klein and F. Finger, *Appl. Phys. Lett.*, **85** (14) (2004) 2839.
[110] B. Rech, T. Repmann, S. Wieder, M. Ruske, and U. Stephan, *Thin Solid Films*, **502** (2006) 300–305.

[111] S. Klein, Y. Mai, F. Finger, M.N. Donker and R. Carius, *15th International Photovoltaic Science and Engineering Conference (PVSEC-15)*, pp. 736–737, Shanghai, China, 2005.
[112] M. Despeisse, C. Battaglia, M. Boccard, G. Bugnon, M. Charrière, P. Cuony, S. Hänni, L. Löfgren, F. Meillaud, G. Parascandolo, Th. Söderström, and C. Ballif, *Phys. Status Solidi*, **A 208** (2011) 1863.
[113] G. Bugnon, G. Parascandolo, Th. Söderström, P. Cuony, M. Despeisse, S. Hänni, J. Holovsky, F. Meillaud, and C. Ballif, *Adv. Funct. Mater.*, **22** (2012) 3665.
[114] T. Matsui, A. Matsuda and M. Kondo, *Proc. of 19th European PVSEC*, 1407, Paris, 2004.
[115] S. Goya, Y. Nakano, T. Watanabe, N. Yamashita and Y. Yonekura, *Proc. of 19th European Photovoltaic Solar Energy Conference*, 1407, Paris, 2004.
[116] K. Zweibel, *Chem. Eng. News*, **64** (1986) 34.
[117] K. Yamamoto, A. Nakajima and Y. Tawada, *Sol. Energy*, **77** (6) (2004) 939.
[118] O. Vetterl, F. Finger and H. Wagner, *Sol. Energy Mater. Solar Cells*, **62** (2000) 97.
[119] J. Meier et al., *Mater. Res. Soc. Proc. Ser.*, **420** (1996) 3.
[120] J. Meier, S. Dubail and A. Shah, *Sol. Energy Mater. Solar Cells*, **74** (2002) 457.
[121] K. Yamamoto, *Proc. of 3rd World Conference on Photovoltaic Energy Conversion*, S2O-B9-03, Osaka, 2003.
[122] K. Yamamoto, M. Yoshimi, Y. Tawada, S. Fukuda, T. Sawada, T. Meguro, H. Takata, T. Suezaki, Y. Koi, K. Hayashi, T. Suzuki, M. Ichikawa, A. Nakajima, *Sol. Energy Mater. Solar Cells*, **74** (1–4) (2002) 449–455.
[123] J. Bailat, L. Fesquet, J.-B Orhan, Y. Djeridane, B. Wolf, P. Madliger, J. Steinhauser, S. Benagli, D. Borrello, L. Castens, G. Monteduro, M. Marmelo, B. Dehbozorgi, E. Vallat-Sauvain, X. Multone, D. romang, J.-F. Boucher, J. Meier, U. Kroll, M. Despeisse, G. Bugnon, C. Ballif, S. Marjanovic, G. Kohnke, N. Borrelli, K. Koch, J. Liu, R. Modavis, D. Thelen, S. Vallon, A. Zakharian, D. Weidman, *Proc. 25th European Photovoltaic Solar Energy Conference*, Valencia, Spain (2010) 2720.
[124] S. Guha, D. Cohen, E. Schiff, P. Stradins, P.C. Taylor, J. Yang, *Photovolt. Int.*, **13** (2011) 144.
[125] K. Yamamoto et al., *15th International Photovoltaic Science & Engineering Conference (PVSEC-15)*, pp. 529–530, Shanghai, China, 2005.
[126] Banerjee A, Su T, Beglau D, et al., *Proceedings of the 37th IEEE PVSC* (2011) Seattle, June 2011.
[127] D. Dominé, J. Steinhauser, L. Feitknecht, A. Shah and C. Ballif, *20th European Photovoltaic Solar Energy Conference*, pp. 1600–1604, Barcelona, Spain. June 6–10, 2005.
[128] A. Lambertz, F. Finger, B. Holländer, J.K. Rath, and R.E.I. Schropp, *J. Non-Cryst. Solids*, **358** (2012) 1962.
[129] A. Lambertz, T. Grundler, and F. Finger, *J. Appl. Phys.*, **109** (2011) 113109.
[130] H. Keppner, J. Meier and A. Shah, *Appl. Phys.*, **A 69** (1999) 169.
[131] M. Fonrodona, D. Soler, J. Escarre, J.M. Asensi, J. Bertomeu and J. Andreu, *J. Non-Cryst. Solids*, **659** (2004) 338–340.
[132] A. Nakajima, M. Ichikawa, T. Sawada, M. Yoshimi and K. Yamamoto, *Jpn. J. Appl. Phys.*, **43**, 9A/B (2004) L1162–L1165.
[133] *JRC Photovoltaic Status Report* 2010, http://re.jrc.ec.europa.eu/refsys/.
[134] S. Guha, *Mater. Res. Soc. Symp. Proc.*, **808** (2004) A6.4.1.
[135] R. Gottschalg, C.N. Jardine, R. Ruether, T.R. Betts, G.J. Conibeer, J. Close, D.G. Infield, M.J. Kearney, K.H. Lam, K. Lane, H.P. Pang and R. Tscharner, *Proc. of the 29th IEEE PV Specialists Conference*, New Orleans, LA, USA. pp. 1699–1702, 2002.
[136] K. Ishibashi, M. Karasawa, G. Xu, N. Yokokawa, M. Ikemoto, A. Masuda, H. Matsumura, *Thin Solid Films*, **430** (2003) 58–62.
[137] S. Osono, M. Kitazoe, H. Tsuboi, S. Asari and K. Saito, *Thin Solid Films*, **501** (2006) 61.

[138] J. Meier, U. Kroll, J. Spitznagel, S. Benagli, T. Roschek, G. Pfanner, Ch. Ellert, G. Androutsopoulos, A. Hügli, G. Büchel, D. Plesa, A. Büchel and A. Shah, *Proc. of the 19th EU PVSEC*, pp. 1328–1333, Paris, June 2004.

[139] H. Takatsuka, M. Noda, Y. Yonekura, Y. Takeuchi, Y. Yamauchi, *Sol. Energy*, **77** (2004) 951–960.

[140] Y. Tawada, *3rd World Conference on Photovoltaic Energy Conversion*, 5PLD103, Osaka, May 11–18, 2003.

[141] S. Guha, *Proc. of the 31st IEEE Photovoltaic Specialists Conference*, pp. 12–16, Lake Buena Vista, FL, USA (2005).

[142] M. Shimosawa, K. Tabuchi, M. Uno, S. Kato, M. Tanda, Y. Takeda, S. Iwasaki, Y. Yokoyama, T. Wada, Y. Sakakibara, H. Nishihara, H.E. Nomoto, A. Takano and T. Kamoshita, *20th European Photovoltaic Solar Energy Conference*, paper 3CO.7.1, p. 1533 Barcelona, Spain, June 6–10, 2005.

[143] R.E.I. Schropp, H. Meiling, C.H.M. van der Werf, E. Middelman, E. van Andel, P.M.G.M. Peters, L.V. de Jonge-Meschaninova, J. Winkeler, R.J. Severens, G.J. Jongerden, M. Zeman, M.C.M. van de Sanden, A. Kuipers and C.I.M.A. Spee, *2nd World Conference and Exhibition on Photovoltaic Solar Energy Conversion*, pp. 820–822, Vienna, Austria, July 6–10, 1998.

[144] E. van Andel, E. Middelman and R.E.I. Schropp, *Method of Manufacturing a Photovoltaic Foil*, Priority date 26/09/1996, Filing date 24/09/1997, Publication number, EP1066642, PCT publication number WO9949483.

[145] R. Schlatmann, B. Stannowski, E.A.G. Hamers, J.M. Lenssen, A.G. Talma, G.C. Dubbeldam and G.J. Jongerden, *15th International PV Science & Engineering Conference (PVSEC-15)*, 744, Shanghai, China, 2005.

6 III-V Solar Cells

N.J. Ekins-Daukes
Department of Physics, Imperial College, London, UK

6.1 INTRODUCTION

The III-V semiconductor family possesses a number of attractive properties for fabricating solar cells. The bandgaps of the binary III-V semiconductors are conveniently scattered around the solar spectrum with further bandgap tuning possible through alloys of group III and V elements. Most III-V semiconductors share the zincblende crystal structure, lending the family to heterostructure solar cells and multijunction configurations. The majority of III-V materials have direct bandgaps, leading simultaneously to large absorption coefficients for photons with energies in excess of the bandgap and short radiative lifetimes (ns); the absorption coefficients for a series of III-V semiconductors are shown in Figure 6.1. In comparison to silicon solar cells, the active region of III-V solar cells are thin, no more than a couple of micrometers and epitaxial crystal growth techniques exist that are capable of depositing extremely high-quality layers of III-V semiconductor, leading to almost ideal semiconductor junctions. Many high-performance III-V solar cells have been demonstrated to date, namely the highest efficiency single-junction solar cell (GaAs), multijunction solar cell (InGaP/GaAs) and concentrator solar cell. Single-junction cell technology will be discussed in Section 6.2, multijunction cells in Section 6.3. Applications for these solar cells are discussed in Section 6.4.

Despite their high performance and potentially low use of semiconductor material, there are disadvantages to III-V-based solar cells. The materials are relatively expensive and some rare elements, such as indium are used and the metal-organic precursors commonly used during cell fabrication are highly toxic and often pyrophoric. Finally, the lack of a chemically inert and electrically passivating oxide, equivalent to SiO_2, complicates the cell design and places the solar cell at risk of corrosion. Nevertheless, the superior efficiency and radiation tolerance of III-V multijunction solar cells initially led to their use in space and satellite applications and subsequently in terrestrial concentrator systems.

The III-V semiconductor family spans the ultraviolet, visible and infrared wavelengths offering the possibility to efficiently absorb the broad solar spectrum and also finding application for thermophotovoltaics, the conversion of lower temperature radiation to electricity [1]. A map of semiconductor bandgap and lattice parameter for most relevant binary and ternary zincblende III-V materials [2,3], as well as for silicon and germanium [4] is shown in Figure 6.2. A key attribute of the III-V material system is the ability to grow heterostructures composed of different semiconductor alloys, enabling binary, ternary, quaternary and even quinternary alloys to be combined. Traditionally a high-efficiency III-V solar cell would be grown lattice matched to a convenient substrate, often GaAs, Ge or InP placing some

Solar Cell Materials: Developing Technologies, First Edition. Edited by Gavin Conibeer and Arthur Willoughby.

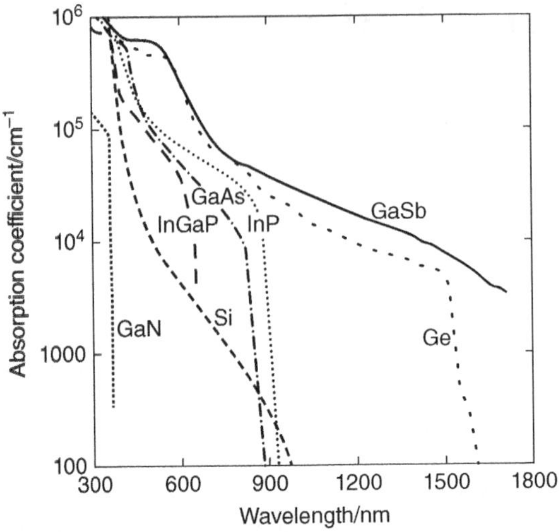

Figure 6.1 Absorption coefficients of a selection of III-V semiconductors (GaN, InGaP, GaAs, InP and GaSb) together with two indirect gap semiconductors (Si and Ge)

restriction on the bandgap combinations that could be attained. However, recent progress in controlling defect introduction during lattice relaxation has enabled highly efficient lattice-mismatched solar cells to be grown, where the lattice parameter varies within the cell.

The very first p/n III-V solar cells were fabricated by diffusing dopants into a GaAs [5], but efficient devices required abrupt junction heterostructres [6] to be formed via liquid phase epitaxy (LPE). Extremely high quality material can be grown using liquid phase epitaxy

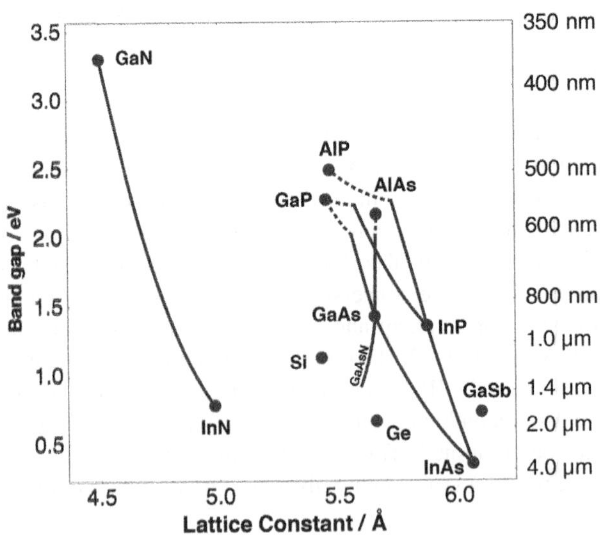

Figure 6.2 Bandgap vs. lattice parameter map for binary and ternary III-V semiconductors. The location of silicon and germanium are shown as possible substrate materials.

and high growth rates of up to 60 μm/h can be attained [7], but being a phase equilibrium approach, there are constraints to the abruptness of interfaces that can be grown as well as limitations on control of alloy composition and layer thickness. As a result, metal-organic vapor phase epitaxy (MOVPE) is now the standard means for producing III-V solar cells, giving a growth rate of 10 μm/h, better interfacial control and reproducible growth over larger-area substrates as well as allowing a wide range of III-V alloys to be grown [8]. The principle disadvantage is the hazardous nature of the metal-organic precursors, many of which are pyrophoric and highly toxic [9, 10]. For some solar cell structures, molecular beam epitaxy (MBE) has been used, giving extremely abrupt interfaces on the monolayer scale and with modern machines, similar versatility to MOVPE can be achieved with regard to alloys that can be grown [11]. While MBE is used for industrial-scale production of high-mobility transistors, the growth rate is relatively slow compared to MOVPE, typically 1 μm/h making it less attractive for III-V solar cell manufacturing. Nevertheless, some effort has been made to to develop MBE techniques for solar cell production [12, 13] and the present world record is held by a 44% efficient solar cell grown by MBE [14]. A hybrid approach, chemical beam epitaxy (CBE) may be able to combine some of the high throughput of MOVPE with the interfacial control of MBE [15, 16] and some promising prototype cells have been grown using this technique [17, 18].

6.2 HOMO- AND HETEROJUNCTION III-V SOLAR CELLS

The development of single-junction III-V solar cells followed that of silicon [19], with the first report of a GaAs photovoltaic device in 1956 [5] but early cells were only a few percent efficient [20] and struggled to match the efficiencies attained using silicon, which was a better controlled semiconductor at that time. Nevertheless, early GaAs solar cells were used in the 1960s on Soviet space missions due to their superior radiation resistance, discussed in more detail in Section 6.4.1. Much progress was made in understanding the III-V semiconductor alloy system over the subsequent two decades, enabling III-V-based light-emitting diodes and lasers to be developed. Simultaneously, the high surface recombination velocity (typically 10^6 cm/s) was identified as a key limiting factor in early device designs [21] leading to much more efficient heterostructre solar cells with possible applications both in space and in high-efficiency concentrator systems [22].

Since high-quality III-V semiconductor junctions can be made with different bandgaps, it is instructive to consider the limiting efficiency that solar cells would have under different illumination conditions. Figure 6.3 shows limiting efficiency calculations following the approach of Shockley and Queisser [23] for the air-mass zero (AM0) [24] spectrum characteristic of the space environment, air mass 1.5 global (AM1.5G) [25] characteristic of conditions for terrestrial flat plate collectors and air-mass 1.5 direct low-AOD (AM1.5D) [26] characteristic of sunlight concentrated by 500 times.

The bandgap of GaAs at 1.42 eV and InP at 1.34 eV are conveniently matched to the peak of these limiting efficiency curves, and in the 1980s, solar cells made from both materials matured considerably. GaAs cell efficiencies consistently achieved conversion efficiencies in excess of 21% AM0 [27] and 24% AM1.5G [12]. InP also received considerable interest as its excellent radiation resistance lends it to space applications. GaSb has a relatively low bandgap and has found application in thermophotovoltaic devices [28]. The developments

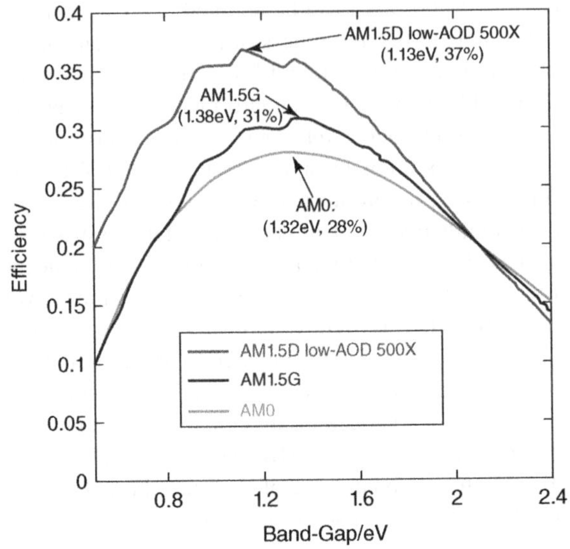

Figure 6.3 Limiting efficiency curves for single-junction solar cells under AM0, AM1.5G and 500× AM1.5D low-AOD spectra.

of these single-junction cells were all important milestones leading to the demonstration of multijunction solar cell described in Section 6.3.

Most III-V solar cells follow the typical homojunction solar cell geometry illustrated in Figure 6.4. Light is incident through a thin emitter layer and absorbed completely in the base. The doping of the emitter layer is usually an order of magnitude higher than the base and correspondingly has a shorter diffusion length; the optimum thicknesses of the layers

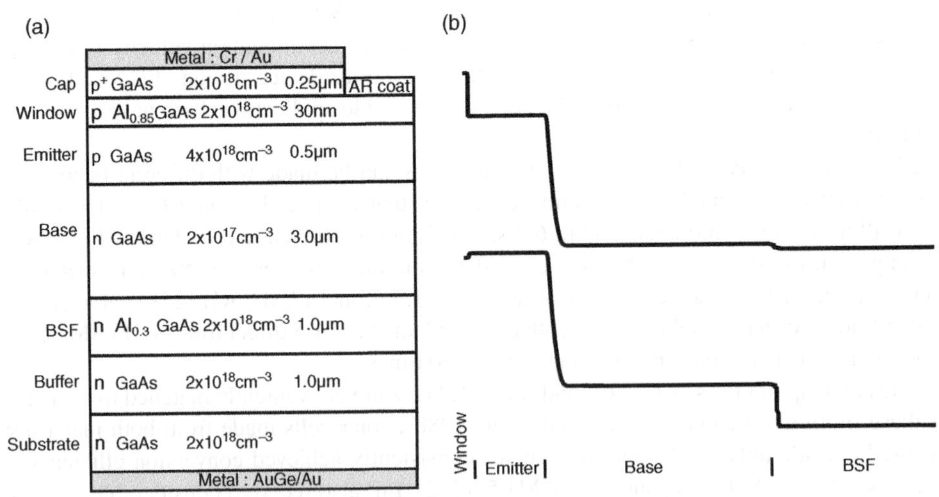

Figure 6.4 (a) Layer structure for a p/n GaAs solar cell with AlGaAs window and (b) its associated band diagram drawn under short-circuit conditions ($V = 0$).

Table 6.1 Parameters a series of GaAs solar cells measured under 1-sun AM1.5G.

Cell Type	Window	Efficiency	V_{oc}/V	J_{sc}/mA/cm^{-2}	FF	Origin and Date
GaAs (thin-film)	??	27.6%	1.107	29.6	84.1	Alta Devices 2010 [49]
GaAs (thin-film)	AlInP window	26.1%	1.045	29.5	84.6	U.Nijmegen [50]
GaAs	AlInP	26.1%	1.038	29.7	84.7	U.Nijmegen [50]
GaAs n/p	InGaP	25.7%	1.039	28.54	86.8	NREL 1990 [39]
GaAs p/n	AlGaAs	24.8%	1.029	27.9	86.4	Spire 1990 [35]

depends upon the diffusion length and photogeneration in each layer. The absence of any suitable passivating material for most III-V solar cells necessitates the use of a window layer, preventing minority carriers from reaching the physical semiconductor surface where significant surface recombination would otherwise take place. The choice of material for the window layer is often limited, leading to trade-offs in the cell design. Minority carrier diffusion in the base layer should not proceed towards the substrate, so a backsurface field (BSF) is often employed, usually via a doping gradient or sometimes using a wide-bandgap heterostucture. The metal grid is defined using standard photolithography [29] and an antireflection coated added [30]. The design of the metal grid is intimately linked to the concentration level that the solar cell operates under as well as the sheet resistivity of the emitter, a problem that is now tacked through numerical computer-aided design [31–33] although analytical methods also exist [34].

The direct bandgap of III-V solar cells and high material quality means than some devices operate close to the radiative limit. In this limit, the dominant solar cell loss is unavoidable radiative recombination.

6.2.1 GaAs Solar Cells

GaAs solar cells hold the record for the highest efficiency single-junction solar cells fabricated to date. Table 6.1 lists some notable cells with different efficiencies showing a remarkable recent result of 27.6% under 1-sun AM1.5G.

A typical p/n GaAs solar cell is shown in Figure 6.4 showing both the layer structure and the associated band diagram. Typically the junction is formed 0.5 μm from the front window layer and a backsurface field (BSF) is formed by the $Al_{0.3}GaAs$ layer to ensure the diffusion gradient for minority holes is directed towards the junction [29]. This structure leads to almost ideal carrier collection in the base and emitter regions, as shown by the simulated internal quantum efficiency (IQE) plot based on data from [35] and shown in Figure 6.5.

The components of the IQE are calculated from analytical solutions to the drift-diffusion equations for a p-n junction [36] and appropriate values for the absorption coefficient for GaAs and $Al_{0.8}GaAs$ [37]. From this calculation, it is clear that significant photogeneration takes place in the emitter and in a high-quality cell such as the present device, most of the photogenerated carriers are collected and contribute to the photocurrent of the device. The loss in collection efficiency at short wavelengths ($\lambda < 600$ nm) stems almost entirely from absorption in the $Al_{0.8}GaAs$ window layer that is not collected as photocurrent. However, omitting the window layer and exposing the minority electrons in the emitter layer to a bare GaAs surface leads to high surface recombination velocity and to QE$<$ 25% for wavelengths below 500 nm. Since the absorption coefficient and diffusion lengths are high enough in GaAs

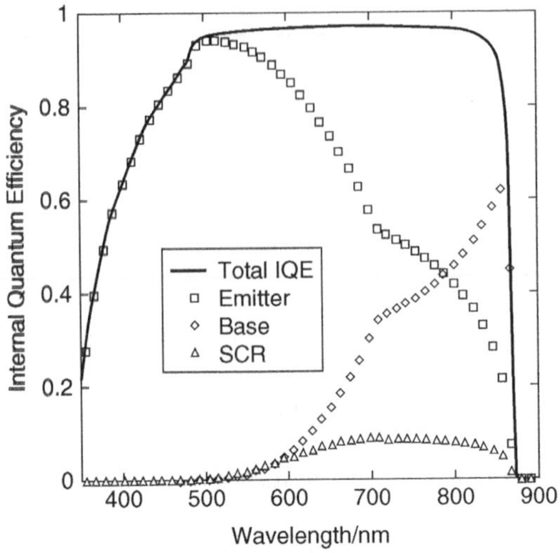

Figure 6.5 Quantum efficiency for a GaAs p/n solar cell, showing the components from the p-type emitter, n-type base and the space-charge-region (SCR).

to yield almost ideal absorption and carrier collection, the principle route for improving the photocurrent from a GaAs solar cell lies in developing a fully transparent passivating surface for GaAs.

InGaP has also been used very successfully as a window layer material on account of extremely low InGaP/GaAs interface recombination velocity [38], resulting in a marginally higher power-conversion efficiency in an n/p configuration [39]. A comparison between cells grown with InGaP and AlGaAs windows is shown in Table 6.1, indicating that the short-circuit current is higher for the InGaP window cell than that using AlGaAs. This result is remarkable since the bandgap of lattice-matched $In_{0.48}GaP$ is lower than that of $Al_{0.8}GaAs$, but the higher J_{sc} is likely due to higher levels of interface recombination at the AlGaAs/GaAs interface [40]. Raising the bandgap of the window further, using an AlInP window results in yet higher short-circuit currents.

In the space environment, radiation induced degradation of diffusion length takes place. Under these circumstances light trapping schemes can be considered, allowing a physically thin solar cell to be fabricated but using a surface texture to increase the pathlength of light inside the solar cell, as commonly employed in silicon solar cells [41]. The simplest is to incorporate a Bragg reflector to double the optical pathlength [42], but in principle an enhancement of $4n^2$ is possible, where n is the semiconductor refractive index, if internally propagating light is scattered randomly [43]. This has been attempted in a V-groove configuration, formed using anisotropic etchants that preferentially expose (111) planes [44,45] and resulted in a 13% enhancement in short-circuit current over a planar device [46] although the increase in junction area will reduce the open-circuit voltage. More recently, plasmonic [47] and nanoscale dielectric textures [48] have been attempted, showing relative improvements in absorption.

In high-quality n-GaAs films the photoluminescence lifetime exceeds the radiative lifetime [51], an effect that can be attributed to photon recycling; the reabsorption of radiative recombination. In the conventional solar cell structure shown in Figure 6.4, absorption in the

GaAs substrate is sufficiently high to quench any beneficial effect due to photon recycling, but if the substrate is removed, the confinement of light becomes much stronger and the beneficial effects of is expected to become apparent [52, 53]. Epitaxial lift-off techniques [54–56] provide a means of separating an epitaxially grown solar cell from the substrate and have recently enabled record-breaking GaAs solar cells to be fabricated. In Table 6.1, comparing the entries with AlInP windows illustrates the effect of optical confinement in the solar cell. The thin-film device has been lifted off its substrate and shows a marginally higher voltage than the otherwise identical device grown on a substrate. While the differences between these solar cells is small, the most recent achievement of 27.6% from a thin-film device shows a remarkable increase in voltage. At the time of writing no details of the cell design are available but this highly efficient solar cell is almost certain to be exploiting the beneficial effects of photon recycling.

In principle, epitaxial lift-off is a nondestructive process, leaving the wafer on which the solar cell was grown intact after the device has been removed. This offers the possibility of reusing the substrate and hence reducing the usage of III-V material [57]. Inevitably some roughing of the wafer occurs during lift-off and polishing is required to regain an epi-ready surface, so the economic advantage of wafer reuse depends on the respective costs of the wafer and polishing steps. At present the cost of reusing a 4-inch GaAs wafer is roughly half that of a new wafer [58]. An alternative route for reducing the quantity of valuable III-V semiconductor is to grow the material on an inexpensive foreign substrate.

For use in space, GaAs is often grown on a Ge substrate since the materials are almost lattice matched and Ge is physically stronger. Since the specific power (W/kg) of the solar cell is of importance, it is possible to thin the Ge substrate giving a solar cell with the same mechanical strength, but with a lower weight.

Growth of GaAs on silicon is attractive since silicon is much less expensive, is significantly stronger than GaAs and has a higher thermal conductivity; data are presented in Table 6.2. However growth on silicon is challenging, not only due to the 4% lattice mismatch evident in Figure 6.2 but also due to the difference in thermal expansion coefficient. Nevertheless, GaAs on silicon 18% efficient solar cells have been achieved either via direct growth of GaAs on silicon [59,60] or the use of graded Si-SiGe-GaAs buffer layers [61]. Alternatively, antiphase domain-free surfaces of GaP have been achieved on a silicon substrate [62] opening up a possible route to grade the lattice constant using III-V materials. This is discussed in more detail with regard to multijunction solar cells in Section 6.3.1.2

Growth of polycrystalline GaAs has also been attempted, with very early attempts achieving 4% on molybdenum [63] and 13% on polycrystalline GaAs substrates with passivation of grain boundaries [64, 65]. Higher efficiencies have since been achieved upon

Table 6.2 Common substrate material properties: thermal conductivity σ, thermal expansion coefficent ρ and density ρ. (Data extracted from Landolt and Bornstein, Springer materials: The landolt-bornstein database.)

Material	$\sigma/\text{W cm}^{-1}\,\text{K}^{-1}$	$\alpha/10^{-6}\text{K}^{-1}$	$\rho/\text{g cm}^{-3}$
Silicon	1.56	2.58	2.33
Ge	0.58	5.90	5.32
GaAs	0.455	6.86	5.32
InP	0.68	4.75	4.81

polycrystalline Ge substrates resulting in efficient GaAs solar cells of 20% AM1.5 achieved on an optical grade polycrystalline (sub-mm grain) Ge substrate [66]. Since this landmark result, emphasis has shifted to depositing Ge polycrystalline thin films on a foreign substrate, with some promising > 1 mm grainsizes reported on ceramics [67] and recent attempts at growth on molybdenum [68,69].

6.2.2 InP Solar Cells

InP solar cells are highly desirable as space solar cells on account of their exceptional tolerance to radiation [70]. Typical device designs closely follow that of GaAs except that there is no suitable window material to passivate the front surface. Instead, shallow homojunction designs with appropriate metal contacts [71] have reached efficiencies of 19% AM0 [72]. A typical device structure and band diagram are shown in Figure 6.6. Using a stepped doping profile at the emitter and thinning it as much as possible, photogeneration in the emitter is minimised, while minority holes are prevented from reaching the bare InP n^{++} surface.

Unlike the GaAs solar cell were the emitter plays the dominant role in photoabsorption, the calculated IQE for the photocurrent from an InP shallow heterojunction is almost entirely due to the base and space charge region. The loss in short-wavelength IQE is due to photogeneration in the emitter that is effectively lost through recombination in the heavily doped layers or via the surface. Another consequence of such aggressive step doping of the emitter is that the bandgap narrows in the more heavily doped layer, thereby lowering the voltage of the solar cell. If a suitable window layer were found for InP, these losses would be eliminated and efficiencies similar to that achieved with GaAs could be expected. An approach using ITO has been attempted [71] and V-groove light-trapping structures have also been formed on InP solar cells showing reduced surface reflectivity and a relative increase in short-circuit current and efficiency [73].

Other than the window problem, other disadvantages of working with InP is the high cost of the substrate and low mechanical strength. For both these reasons it is desirable to consider growth of InP on foreign substrates or epitaxial lift-off routes, especially if the

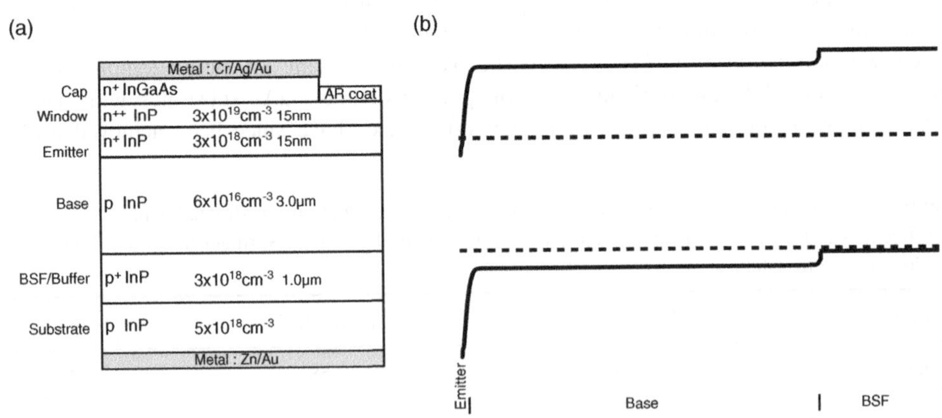

Figure 6.6 (a) Layer structure for a $n^{++}n/p$ shallow homojunction and (b) its associated band diagram drawn at a typical operating voltage ($V = 0.752$)

substrate can be reused. The mismatch between InP and Si is large (8%), yet as shown in Table 6.2 the difference in thermal expansion coefficient is lower, resulting in InP/Si films with lower residual strain than GaAs/Si films. Epitaxial lift-off and substrate reuse has also been demonstrated on InP and efficiencies consistently above 14% have been demonstrated on thin cells grown on reused substrates [74].

6.2.3 InGaAsP

Under solar concentration the limiting efficiency rises and the optimal bandgap energy shifts to lower energy. The reason for this difference has very little to do with the solar spectrum, but arises from the increased voltage of the solar cell under concentrated sunlight. Since the optimal bandgap energy for power-conversion is a balance between photogeneration and recombination, increasing the solar concentration shifts the balance to favour higher levels of recombination and hence a lower optimal band-gap energy [75,76]. For a single-junction solar cell, it is shown in Figure 6.3 a maximum efficiency of 37% is attained at an absorption threshold of 1.13 eV. The InGaAsP material system can be grown across a wide energy range on the InP substrate [77] and has resulted in a 1.14 eV concentrator solar cell that attained an efficiency of 27.5% at 171 suns [78].

6.2.4 GaN

The GaN family of materials are unusual in possessing very high energy direct bandgaps. GaN itself has a bandgap that is too high for efficient solar power conversion, 3.6 eV or 344 nm, however as indicated in Figure 6.2 alloying with the InN material system yields

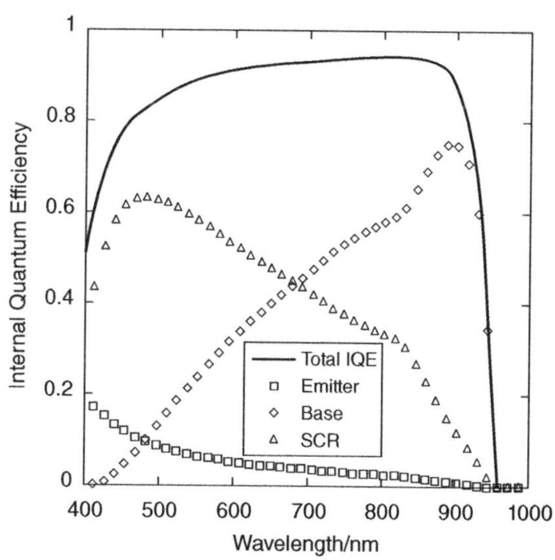

Figure 6.7 Quantum efficiency for a InP $n^{++}n/p$ solar cell, showing the components from the p-type emitter, n-type base and the space-charge-region (SCR).

much lower energy gaps, offering the possibility to span the entire solar spectrum with one ternary alloy [79]. The GaN semiconductor family have high absorption coefficients, enabling thin junction devices to be made [80] and to date InGaN devices with absorption extending to 430 nm with internal quantum efficiencies of up to 97% have been demonstrated [81].

6.3 MULTIJUNCTION SOLAR CELLS

Jackson proposed the multijunction concept in 1955 [82], whereby the solar spectrum is absorbed selectively by solar cells of different bandgaps, thereby preserving a greater fraction of the free-energy received from high energy photons. Figure 6.8 shows how the efficiency rises rapidly with increasing number of junctions. Since the limit for an infinite number of junctions is 67% under one sun, it is clear that much of the gain to be made through this approach can be realised with only a couple of junctions. The key requirement is to partition the solar spectrum at the appropriate point and direct the sunlight to semiconductors with different bandgap energies chosen to match the bandwidth of sunlight incident upon them.

The partitioning of the solar spectrum can take place internally within the solar cell simply by stacking semiconductors of decreasing bandgap energy upon one another. This can be achieved most elegantly by integrating all the component junctions and semiconductors into a monolithic crystal structure, or most simply by mechanically stacking one solar cell on top of another. Alternatively, the solar spectrum can be partitioned externally using a dispersive optical system, directing sunlight onto separate solar cells with different bandgaps.

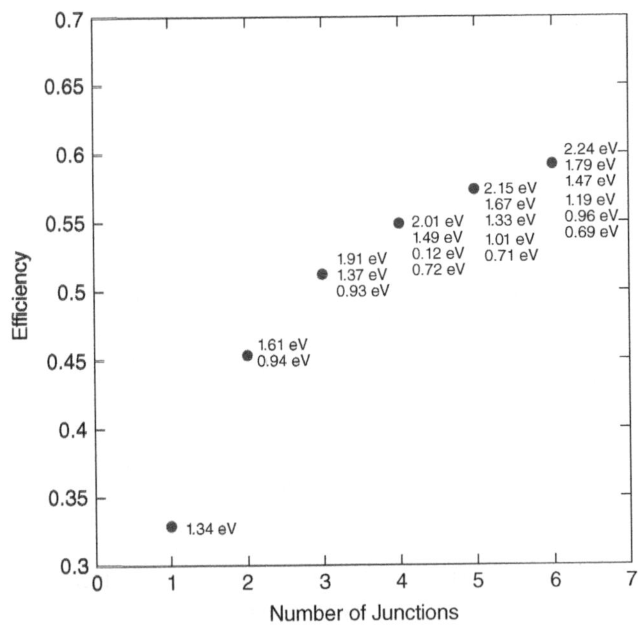

Figure 6.8 Efficiency for a series constrained multijunction solar cell under AM1.5G illumination. (Data extracted with permission from Brown, 2002. [83] Copyright © 2002, Elsevier.)

6.3.1 Monolithic Multijunction Solar Cells

Integrating all the components for a multijunction solar cell into one monolithic structure places some constraints upon the multijunction solar cell design. In general, the highest-quality semiconductor junctions are achieved using lattice-matched semiconductor material. However, since the subcells are connected together in series internally, the current passing the solar cell will be limited by the junction producing the lowest current. It is therefore important to engineer subcells producing similar current at maximum power and this can be difficult to achieve using only lattice-matched materials, so a number of monolithic lattice-mismatched (or metamorphic) multijunction solar cells have been attempted where the lattice parameter is graded from one semiconductor junction to another.

The typical band structure for a double-junction solar cell is shown in Figure 6.9, showing the physical stacking of semiconductor layers and the band structure. Many of the components from the single-junction solar cell are present, notably the presence of a window layer and backsurface field. The interconnection of the junctions is critically important in a monolithic cell and is achieved using a tunnel junction. Typically, tunnel junctions are formed by growing a heavily doped n^{++}, layer adjacent to a heavily doped p^{++} layer, thereby bringing the conduction band of the n^{++} semiconductor to almost overlap energetically with the valence band of the p^{++} semiconductor. In principle, the remaining barrier is sufficiently thin to enable a substantial tunnelling current to pass, but simulations show that the high tunnel currents measured experimentally are likely to be enhanced through trap-assisted tunelling [84, 85]. In practice, tunnel junctions can be difficult to achieve. Firstly the junction must be transparent, meaning relatively high-bandgap semiconductors must be used, increasing the doping density required to achieve the same peak tunnel current [86]. Secondly, actually

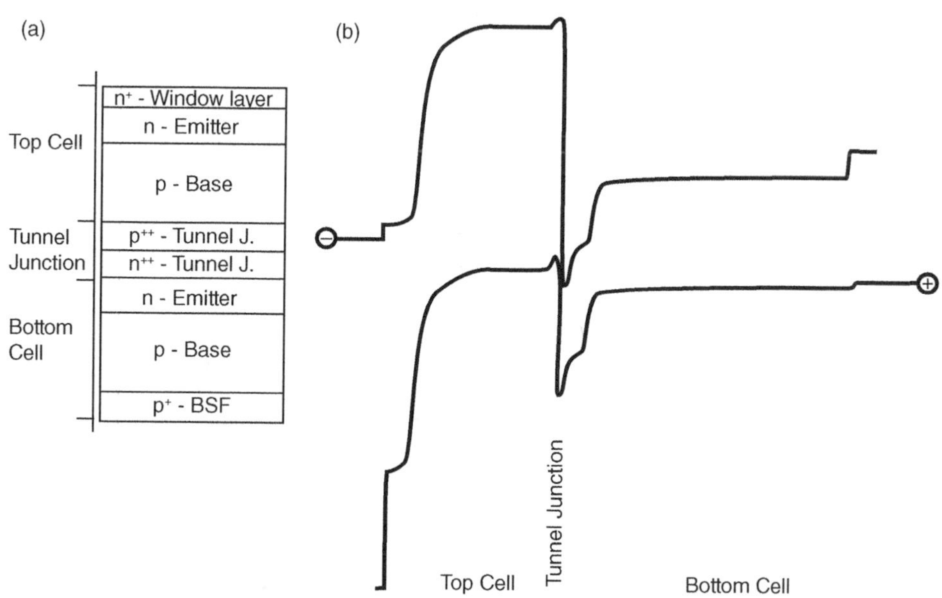

Figure 6.9 A monolithic double-junction solar cell with a tunnel junction interconnect: (a) schematic layer diagram and (b) band diagram showing top cell, tunnel junction, lower cell and rear BSF.

achieving very abrupt p^{++}/n^{++} layers and preventing dopant diffusion during the remainder of the epitaxial growth is also challenging. One important means for preventing dopant diffusion has been the use high-bandgap materials (typically AlGaAs or AlInGaP) that act as diffusion barriers around the tunnel junction [87–89] as well as serving as useful window/BSF layers; for clarity these barriers are not shown in Figure 6.9.

6.3.1.1 Lattice-Matched Multijunction Solar Cells

A limiting efficiency map for series constrained double-junction solar cells is shown in Figure 6.10 for both the AM0 and 500× AM1.5D spectra. Much effort was invested in developing AlGaAs/GaAs double-junction solar cells ultimately resulting in an efficiency of 27.6% AM1.5G and 23.0% AM0 [90]; interestingly this solar cell did not use a tunnel junction, but rather a front metal interconnection scheme with shading losses mitigated by a prismatic lens.

While AlGaAs yields a reasonable top cell, good-quality junctions require very clean precursors and careful control of the growth conditions [12,91]. The development of good-quality InGaP alloys during the late 1980s marked a turning point in the technological development of III-V multijunction solar cells [92], since high-quality, high-bandgap (1.9 eV) junctions could now be reliably grown on a GaAs substrate without the contamination issues encountered with AlGaAs. In 1994 an InGaP/GaAs solar cell surpassed the best AlGaAs/GaAs solar cell, achieving 29.5% under 1-sun AM1.5G [93] using a GaAs tunnel junction and a thin $Al_{0.53}In_{0.47}P$ top surface window layer. More recently, a 32.6% efficient InGaP/GaAs solar cell has been demonstrated operating at 1026 suns [94].

The growth of InGaP is complicated by the existence of lateral compositional ordering [95] that results in alternating phases of Ga- and In-rich phases and a concomitant reduction in band-gap energy from 1.9 eV for fully disordered InGaP to 1.8 eV. During MOVPE growth, ordering of InGaP can be controlled by substrate orientation [96] or the use of antimony, that acts as a surfactant and stabilises the growth surface [97,98]. The successful

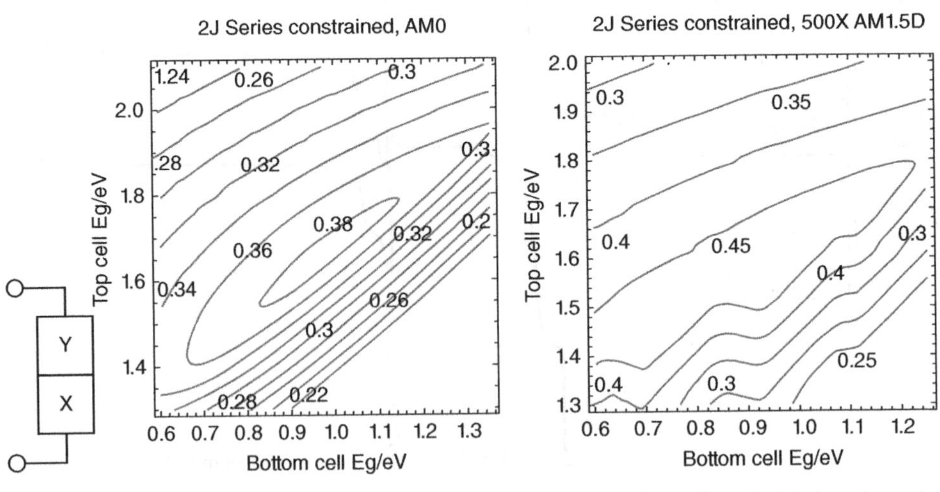

Figure 6.10 Iso efficiency contour maps showing the limiting efficiency for a double-junction solar cell under AM0 and 500× AM1.5D.

control of ordering has been key to the improvements made to the InGaP/GaAs double-junction solar cell, resulting in a 30.3% efficiency under 1-sun AM1.5G [99] in 1997. Under most illumination conditions, this solar cell could be improved by either raising the top-cell bandgap or lowering the bottom-cell bandgap. The current-matching condition is achieved by growing a thin InGaP top cell, allowing some light with energy above the cell band-gap to leak into the junction below [39]. Although this approach of thinning subcells to attain current matching leads to a small loss in voltage, the impact on the efficiency is relatively small and enables the highest-quality semiconductors to be used. An alternative is to raise the bandgap of the top cell using AlInGaP and although promising junctions have been demonstrated [100] the need to achieve complete carrier collection with a thick AlInGaP junction has proven difficult resulting in no net improvement in efficiency in the full multijunction structure. In general, the InGaP subcell is grown in an n/p configuration owing to difficulties in attaining high p-type doping [101], in addition to achieving higher electron mobility and hence lower sheet resistivity in the emitter.

Monolithic double-junction solar cells have also been attempted on the InP substrate [102]. $In_{0.53}GaAs$ is lattice matched to the InP with a bandgap energy of 0.72 eV. Since the bandgap of InP is 1.3 eV, the bandgap combination is not optimal under one sun illumination [78], but becomes better matched under concentrated sunlight. This InP/InGaAs monolithic double-junction solar cell achieved a 1-sun AM0 efficiency of 22.2% [103], but was limited by series resistance and surface recombination; the authors consider that 26% AM0 should be attainable with this approach.

As discussed earlier, GaAs-based solar cells can be grown on Ge substrates. If a junction is formed at the GaAs/Ge interface, a double-junction cell results, albeit with suboptimal bandgap combination [104]. The concept can be extended by growing an InGaP/GaAs solar cell on a Ge substrate to form a triple-junction InGaP/GaAs/Ge solar cell. The 0.66 eV Ge junction overproduces in current under all realistic solar spectra, so while the configuration is far from optimal, the Ge substrate can be employed both for its mechanical strength and to provide a small increase in cell voltage; Figure 6.11 shows the limiting efficiency of a

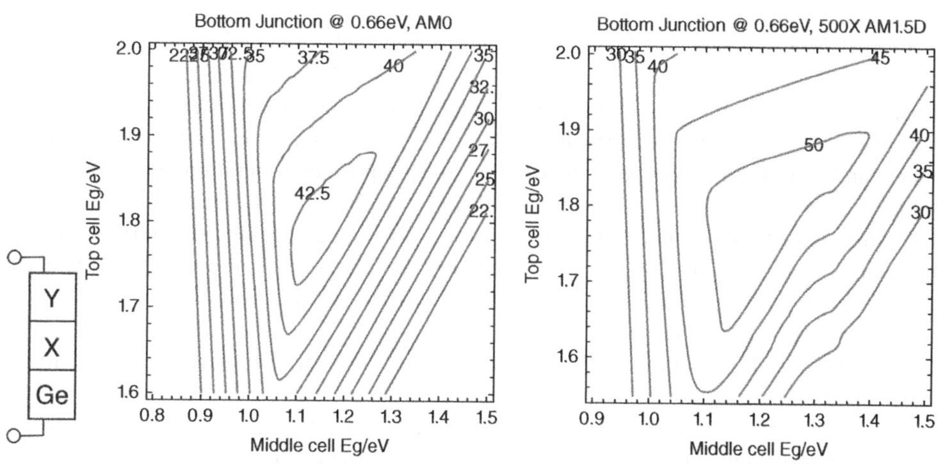

Figure 6.11 Isoefficiency contour maps showing the limiting efficiency for a series connected triple-junction solar cell grown on an active 0.66 eV Ge substrate under AM0 and 500× AM1.5D.

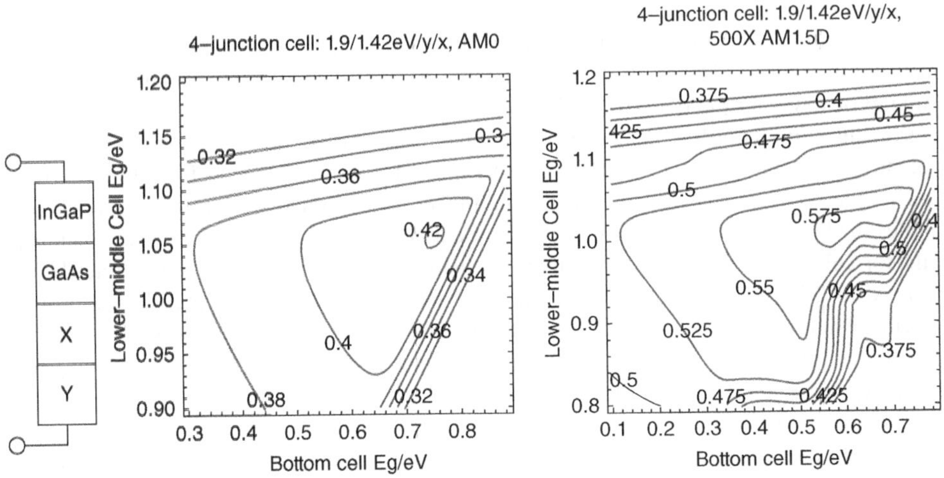

Figure 6.12 Isoefficiency contour maps showing the limiting efficiency for a series connected four-junction solar cell under AM0 and 500× AM1.5D. The top two junctions are assumed to be 1.9 eV InGaP and 1.42 eV GaAs with the following two junctions varied as the *y* and *x* parameters, respectively.

Ge-substrate-based triple-junction solar cell. Since Ge has a marginally larger lattice constant than GaAs, the entire solar cell structure can be grown at slightly closer to the optimum bandgap configuration, replacing GaAs with $In_{0.01}GaAs$ [105]. This configuration is now the standard space solar cell technology and is also widely employed in solar concentrator systems.

Some further improvements can be made to the InGaP/GaAs/Ge solar cell, for example increasing the current of the middle cell by lowering its bandgap using strain-balanced quantum wells [106], but ultimately the next step in the development of the III-V multijunction solar cell is to introduce a fourth junction [107]. The isoefficiency map shown in Figure 6.12 shows the limiting efficiency of a 4-junction solar cell assuming that the top two junctions are formed from the standard InGaP/GaAs lattice-matched combination. If a 0.66 eV Ge substrate is used as the lower junction, a 1-eV lattice-matched alloy would be required. Much effort has been invested in dilute nitride InGaAsN alloys to achieve this goal, but the poor diffusion length in this material has frustrated efforts to match the currents of the other subcells in the multijunction stack either in p/n [108] or p-i-n configuration [109]. However, two innovations during growth have enabled high-quality 1-eV material to be grown by MBE; the use of biased deflection plates to prevent N ion damage and the introduction of Sb into the InGaAsN alloy [110]. These have led to the demonstration of a record efficiency of 43.5% under concentration, but at the time of writing it is unclear what precise device structure has been used [14].

Alternative candidate alloys such as GaAsB [111], GaAsBi [112] have been attempted but to date the performance of junctions formed from these alloys have been insufficient to result in efficiency gain. Some attempts have been made to adjust the solar cell structure to compensate for the current-poor 1-eV cell, for example growing a low current, high-voltage 4-junction solar cell consisting of two InGaP junctions, followed by a $In_{0.01}GaAs$ junction, a InGaAsN junction has been grown on inactive Ge, resulting in an open-circuit voltage of

3.9 V but an efficiency of 19.55% under AM0 [113]. Attaining a lattice-matched four-junction solar cell on a Ge substrate remains a long-term research challenge.

6.3.1.2 Lattice-Mismatched Multijunction Solar Cells

Much greater freedom of choice in the bandgap energies is obtained if the lattice constant of subcells is allowed to vary. In a monolithic solar cell, the lattice parameter can be adjusted by growing graded buffer layers of strained semiconductor material, deliberately introducing arrays of misfit dislocations to relax the lattice. The rate at which strain is introduced into the buffer layer and the growth temperature are critical to ensure that the dislocations introduced during relaxation are able to bend into 90° misfit dislocations and do not spawn excessive threading dislocations [114]. With careful control of these parameters, threading dislocation densities as low as 9.5×10^4 cm^{-2} have been achieved in lattice-mismatched InGaAs [115], while the boundary for acceptable device performance is estimated to be 1×10^5 cm^{-2} [60]. Under conditions where excessive dislocation density has accumulated during growth, strained layer superlattices and thermal cycle annealing has been shown to have remedial value [116, 117]. Since the substrate of choice should host at least one lattice-matched alloy, in the discussion that follows the technologies are grouped around the substrate upon which mismatched layers are grown.

GaAs substrates: Recognising that AlGaAs or InGaP alloys can be used to make lattice-matched top cells on a GaAs substrate, there has been a long-standing effort to develop good-quality relaxed InGaAs subcells and early cells delivered promising results [118–120]. However, integration of the sub-cells into a monolithic stack proved challenging since the preferred AlGaAs/GaAs/InGaAs architecture in the late 1980s required two graded mismatched layers. Starting with a GaAs substrate, the lattice parameter was graded to 1.15 eV InGaAs to form the bottom junction, then the lattice parameter was graded back to GaAs to enable the GaAs middle junction and a lattice-matched AlGaAs top junction to be grown [121].

One of the fundamental limitations of growing the lattice-mismatched layer first is that any threading dislocations that are nucleated in the buffer layer will introduce additional recombination not only in the mismatched layer, but also in all subsequent junctions. One means of overcoming this problem is to perform epitaxy on both sides of the GaAs substrate, growing the lattice-matched layer on one side and the lattice-mismatched layer on the other [122], this approach has recently resulted in a 42.3% efficiency [123]. Alternatively, the problem is solved using the inverted metamorphic (IMM) solar cell, in which the high-bandgap lattice-matched junctions are grown first on a GaAs substrate. The last layer to be grown is then the relaxed InGaAs layer, confining defects associated with the relaxed buffer layer growth to that junction. Since the cell is effectively grown upside down, the substrate must be removed and the top metal contacts fabricated upon a thin solar cell [124]. While this adds complexity to the cell manufacturing, it brings exceptional performance; efficiencies of 35.8% AM1.5G [125], 32.8% AM0 [126] and 40.8% at 326 suns [127] have been demonstrated with this approach. The inverted metamorphic approach can also accommodate further relaxed junctions, a four-junction device has been demonstrated with an AM0 efficiency of 33.6% [128].

Ge substrates: As with the lattice matched configuration, Ge provides a convenient low-bandgap bottom cell, thereby moving the relaxed layer into the middle of the solar cell. Starting with the Ge substrate to form the bottom cell, the lattice parameter is graded to

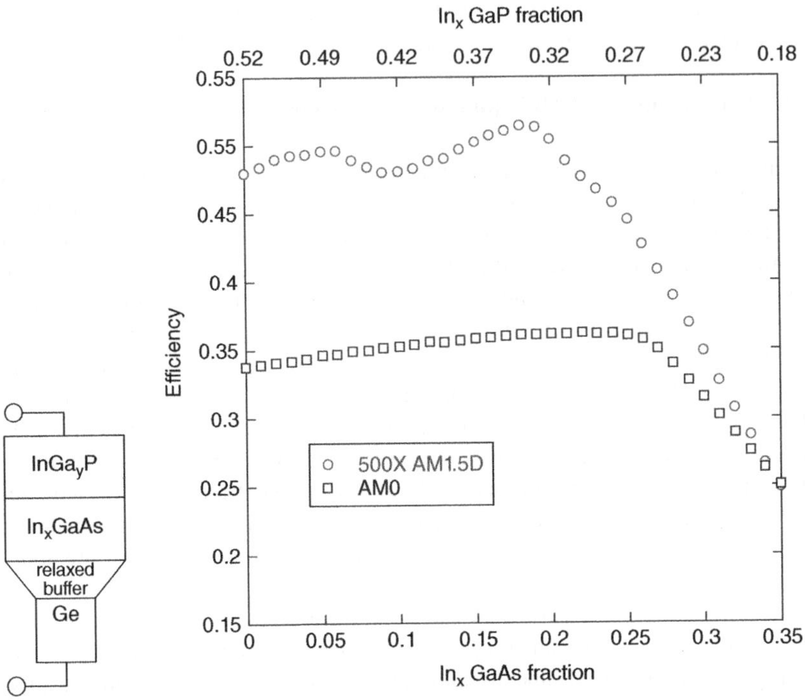

Figure 6.13 Limiting efficiency for a lattice-mismatched $InGa_yP/In_xGaAs/Ge$ where one relaxation step is introduced between the In_xGaAs and Ge layer. The $InGa_yP/In_xGaAs$ junction share the same lattice parameter.

$In_{0.17}GaAs$ to enable a 1.18-eV middle cell to be grown and then a 1.77-eV $In_{0.65}GaP$ top cell is grown lattice matched to the InGaAs layer. As shown in Figure 6.13 this approach does not greatly enhance the efficiency of 1-sun solar cells, but does lead to a good match for the top two subcells under 500× AM1.5D concentrated sunlight; an $InGa_{0.35}P/In_{0.17}GaAs/Ge$ material combination has achieved 41.1% under 454× concentrated AM1.5D [129].

Si substrates: One route for achieving good multijunction solar cells on silicon substrates has been to grade from Si to Ge, effectively creating a Ge virtual substrate and growing a conventional InGaP/GaAs cell on top. Efficiencies are marginally below control devices grown on GaAs [130] at 17.0% AM0 (20.6% for GaAs control), but the approach appears to be promising. In addition to providing a substrate, the silicon can provide a valuable 1.1-eV junction if it is actually incorporated into the multijunction solar cell stack. The buffer layer must then be transparent and a direct approach is to grow an AlGaAs layer on a Si substrate; while control of dislocations with this method is problematic an efficiency of 21.2% was attained under AM0 [131]. A more controlled approach is to start with a GaP nucleation layer grown on the Si substrate. Careful preparation of double-stepped Si substrates and high-temperate growth of GaP to encourage annihilation of antiphase boundaries has led to antiphase domain-free GaP grown on silicon [62].

From the GaP surface, grading with either compressively strained GaAsP or InGaP is possible. A comparison between GaAsP and InGaP buffer layers [132] shows clear advantages in using compressively strained GaAsP buffer layers since InGaP buffer layers are believed to phase separate, leading to instability in the relaxation rate in the buffer layer [133].

In the mid-1990s a 12.5% efficient GaAsP/Si double-junction solar cell was demonstrated [134] with an etch pit density of 1×10^7 cm^{-2} indicating a very high threading dislocation density. More recently, GaAsP/Si buffer layers have been grown with threading dislocation densities below the detection limit of TEM (1×10^6 cm^{-2}) suggesting that the efficiency of the GaAsP/Si double-junction solar cell could be substantially improved [135].

Since the target bandgaps for a multijunction solar cell with a 1.1-eV substrate are high, there has been interest in growing GaN-based solar cells on silicon; an activity that is also of interest for light-emitting diodes [136]. A convenient band alignment between InGaN and Si enables the valence band of $In_{0.46}GaN$ to align directly with the conduction band of silicon, connecting the two subcells without the need for a tunnel junction [137]. In practice this advantage is somewhat lessened by the fact that this interface must also accommodate the graded buffer layer between silicon, but a proof of concept GaN/Si device has been demonstrated [138].

Since the GaN/InN material system spans the solar spectrum, it potentially enables a wide range of lattice-mismatched junctions from different compositions of the same ternary alloy [79]. This approach has also been demonstrated in GaAsN, where amorphous layers have been grown on glass spanning from 2.7–1.2 eV, offering a possible approach to flat-plate thin-film photovoltaic panels [139].

6.3.2 Mechanically Stacked Multijunction Solar Cells

The difficulty of integrating disparate semiconductor junctions can be exchanged with the challenge of fabricating a bifacial solar cell and mechanically stacking each cell one upon another. Typically, the bifacial top cell is mounted in a holder above the lower cell, enabling four electrical terminals to be made to the cells. Not only does this approach allow bandgap configurations that are more optimally matched to the solar spectrum, the independent electrical connection to each subcell lifts the need for current matching making the stack much more tolerant to variations in the solar spectrum. To illustrate this effect, the 500× AM1.5D double- and triple-junction solar cell contour plot shown in Figure 6.14 is recalculated lifting the constraint of current matching, enabling each subcell to operate at its maximum power point.

The combination of GaAs/Ge (1.42 eV, 0.66 eV) while not optimum gives a reasonable double-junction limiting efficiency under AM0 and concentrated light, and a 26.1% device has been demonstrated under 285 times solar concentration under the AM1.5D spectrum [140]. The structure can be improved by raising the lower-cell bandgap, so replacing the Ge junction with GaSb with a bandgap of 0.72 eV resulted in a mechanical stack with an efficiency of 35.6% under a concentration of 100 suns (AM0) [141]. Growing $In_{0.53}GaAs$ lattice matched to InP yields a 0.75 eV junction, so a GaAs/InGaAs stack has been attempted achieving an efficiency of 28.1% under one sun AM1.5G. Recognising that even silicon could provide a useful boost in a mechanical stack, a 31% GaAs/Si mechanical stack has been demonstrated at 350× AM1.5D [142].

Mechanical stacks have also been made using a combination of monolithic multijunction solar cells, for example a triple-junction AlGaAs/GaAs stacked on InGaAsP achieved an efficiency of 25.2% AM0 [143]. With the development of lattice-matched InGaP/GaAs

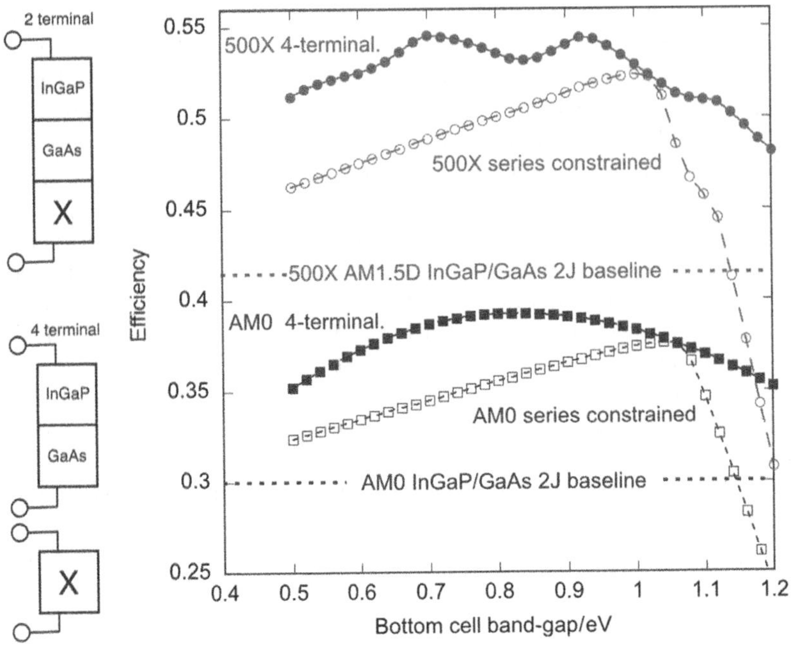

Figure 6.14 Limiting efficiency a InGaP/GaAs/*x* triple-junction solar cell, where the x-axis indicates the bandgap of the bottom junction. The calculation is performed under AM0 and 500× AM1.5D for a series-constrained cell (comparable to the inverted metamorphic solar cell) and a four-terminal mechanically stacked solar cell, contacts made to the InGaP/GaAs lattice-matched cell and independently to the bottom cell.

monolithic double-junction solar cells, the natural extension of this concept is to mechanically stack the monolithic InGaP/GaAs cell with an optimal bottom cell. The limiting efficiency of this arrangement is shown in Figure 6.14 under AM0 and 500× AM1.5D and for series constrained and a four-terminal device (two connections to the bottom cell, two connections to the InGaP/GaAs top cell). For reference, the horizontal lines showing the limiting efficiency of a bare double-junction InGaP/GaAs monolithic cell are shown. At the point of current matching the series constrained and four-terminal device coincide, but a significant efficiency gain can be attained by allowing the bottom cell to operate independently. In practice, an InGaP/GaAs cell stacked on 0.72 eV GaSb has achieved 34% AM0 [144], while a InGaP/GaAs cell stacked on InGaAs lattice matched to InP attained 33% under one sun AM1.5G [145]. While not a mechanical stack, a three-terminal monolithic InGaP/GaAs/Ge solar cell has been proposed, allowing the Ge junction to be biased separately via a rear contacting scheme [146].

A more modern approach to mechanical stacking of solar cells is to grow separate junctions on separate substrates, perform epitaxial lift-off and bond the two layers together. While technically a mechanical stack, these devices are two-terminal devices and are therefore subject to the much narrower bandgap constraints presented in Figures 6.10–6.12. The most immediate technological goal is to bond a low-bandgap combination of InGaAsP/InGaAs grown on InP with the high-gap combination of InGaP/GaAs, resulting in a four-junctions solar cell [147]. To date, low-band-gap InGaAs solar cells grown on InP have been successfully transferred to Si substrates [148] and Ge layers have been bonded onto Si substrates followed by the growth of InGaP/GaAs multijunction solar cells on the Ge template [149].

6.4 APPLICATIONS

6.4.1 III-V Space Photovoltaic Systems

III-V solar cells have a long track record in space, starting in the 1960s with GaAs solar cells used in the Russian spacecraft Venera 2 and 3 to Venus and later in the 1970s to power Russian moon vehicles. In 1986, 70 m^2 of GaAs solar cells were deployed on the MIR space station that operated for 15 years [150]. Most commercial satellites were powered by silicon space cells until the late 1990s when the Iridium communication satellite constellation was launched using single-junction GaAs on Ge solar cells. Since then, the advent of multijunction solar cells has led to the triple-junction III-V solar cells becoming the standard technology for use on commercial satellites.

The principle advantage offered by III-V solar cells is derived from superior radiation resistance over silicon and in multijunction form, a superior efficiency. Depending on the orbit or nature of the space mission, the solar cells will be exposed to significant fluences of high-energy electrons and protons that can significantly damage the operation of solar cells and other electronic components [151]. It is usually the efficiency of the solar cell at its end of life (EOL) that is most important, ensuring that the photovoltaic power system can meet the needs of the spacecraft for the duration of its expected operation. The cell design can be optimised to give high radiation hardness, but this is at the expense of the beginning-of-life (BOL) efficiency and always constrained by the underlying radiation damage rate of the semiconductor material. Traditionally, the response of III-V solar cells to the radiation environment was determined from extensive ground testing where the effects of radiation of both electrons and protons are measured at many different energies and combined to give an equivalent 1-MeV electron fluence [152]. The method is accurate, but requires extensive testing of new solar cell materials under many proton and electron fluences. The process can be simplified if instead of an equivalent 1-MeV electron fluence, the displacement damage dose is used as a measure of the radiation damage to the solar cell. It has been shown that solar cell performance after proton and electron damage at different energies and fluences collapses onto a single displacement damage dose curve for single-junction GaAs solar cells [153] and multijunction solar cells [154]. Since the displacement damage dose is the product of the fluence and the nonionising radiation energy loss rate, which can be calculated, it is only necessary to test the solar cell response under one or two proton and electron fluences. Not only does this make the process of radiation testing less labour intensive, it also gives a deeper insight into the physical mechanisms behind the degradation enabling predictive models for cell degradation to be developed [155].

In general, the specific power of a space solar cell at EOL is of principle interest. Table 6.3 compares a silicon solar cell against a standard 26% III-V multijunction solar cell. Despite the significantly lower EOL efficiency, the silicon cell results in a much higher specific power at the cell level on account of its low weight. However, the weight of the panel structure required to give the space array the necessary rigidity to survive the launch changes the figure of merit firmly in favour of the high-efficiency solar cell [156].

Table 6.3 Specific power figures of merit for space photovoltaic arrays at both the cell and panel level. EOL is measured at 28 °C after a 1-MeV equivalent electron fluence of $5 \times 10^14\ cm^{-2}$ corresponding to geostationary orbit. (Data extracted from Fatemi, 2000. Copyright © 2000, IEEE.)

Cell type	η BOL	η EOL	EOL Cell W/kg	EOL Panel W/kg
Si cell	17.0	14.1	676	75
III-V cell	26.0	23.9	360	108

The next stage in the evolution of space photovoltaics is likely to be the addition of a fourth junction, possibly in inverted metamorphic configuration [157]. The removal of the cell's substrate provides an interesting opportunity to fabricate a truly lightweight 'Space Solar Sheet' with a specific power of 5 kW/kg at the cell level [158]. However, the absence of a substrate is expected to make the cell susceptible from radiation damage from both sides, making the cell less radiation hard overall [159].

6.4.2 III-V Concentrator Photovoltaic Systems

One of the constraints of any flat-plate terrestrial photovoltaic system is the need to cover large areas with active semiconductor material. Solar concentrators change this paradigm by using an optical system, usually lenses or mirrors, to collect sunlight and direct it onto small but highly efficient solar cells [160]. This shifts the emphasis away from manufacturing vast areas of semiconductor material, to the more established industries of metal and glass production. When concentration levels rise above a couple of hundred suns, the solar cell forms only a small part of the overall system cost so the efficiency of the cell becomes of paramount importance. At the time of writing, triple-junction III-V concentrator solar cells are being manufactured commercially with efficiencies of 40%.

A simple expression for estimating the cost of electricity from a concentrator system [161] is shown below. The module costs cover the cost of the optical concentration system and module assembly, the tracking costs cover the support structure and mechanical drive that orientates the concentrator module towards the sun. The balance-of-system (BOS) costs cover site preparation, installation, cabling and DC to AC inverters and are typically separated into area- and power-related BOS costs. The energy produced accounts for the insolation of the location as well as the cell efficiency. For simplicity, the model does not account for any economic discount rate and simply considers the number of years of operation of the system.

$$\text{Cost of electricity}[\$/\text{kwh}] = \frac{\text{module}[\$/\text{m}^2] + \text{tracking}[\$/\text{m}^2] + \text{BOS}_{\text{area}}[\$/\text{m}^2] + \text{BOS}_{\text{power}}[\$/\text{W}] \times \text{peak power}[W/\text{m}^2] + \text{cell}[\$/\text{m}^2]}{\text{Energy yield}[\text{kWh}/\text{m}^2/\text{year}] \times \text{operation time}[\text{years}]}$$

For typical c-Si flat-plate systems at the time of writing, the cell cost makes up about 60% of the total system cost, with the balance of system costs occupying about one third and the remainder as module cost. For a concentrator system, the module and tracking costs become much more significant. Figure 6.15 shows the electricity cost for a 500× concentrator system operating over ten years in a location with high direct irradiance (2555 kWh/m^2/year) as a

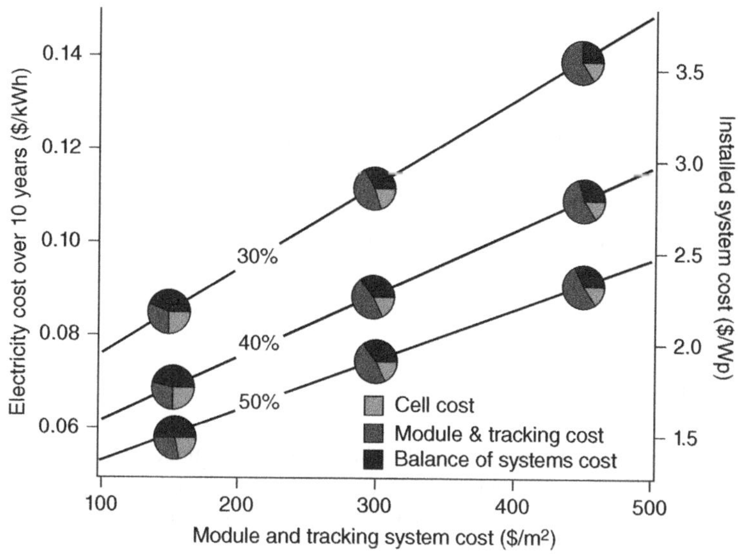

Figure 6.15 Concentrator photovoltaic system cost over 10 years plotted as a function of module and tracking cost and solar cell efficiency. The pie-charts indicate the breakdown of cell cost, module+tracker cost and balance of system costs.

function of combined module and tracking cost and cell efficiency. A cell cost of $6/cm^2 is assumed, in line with the present consensus for automated III-V cell manufacturing [161]. The calculation further assumes an optical efficiency of 85% and a similar electrical derating for connection and inverter losses. The area and power related BOS costs are estimated at $100/m^2 and $0.4/W, respectively. Lines are plotted corresponding to cell efficiencies of 30%, 40% and 50%, indicating that electricity from concentrator systems can become competitive with retail electricity grid prices and can approach the wholesale electricity prices at the extremes [162, 163]. The pie charts illustrate the breakdown of cost between the combined module and tracking costs and the balance of system costs. It is evident that the cell cost is the tertiary cost in the system, illustrating the importance of cost effective, high-performance and durable optical and mechanical systems as well as highly efficient photovoltaic devices.

The relative merit of a CPV system over other technologies depends strongly on the environment in which it is installed. High direct irradiance is a clear requirement and tracking the panel will deliver higher energy yield than from a fixed flat-plate installation. Since a series connected multijunction concentrator solar cell is sensitive to the spectrum of the sun, it is important to correctly account for variations in the solar spectrum when estimating the energy yield [164]. The BOS_{area} term can be a significant discriminating factor, since installing a CPV system typically involves sinking support posts into the ground, while for flat-plate technologies the land may need to be levelled and cleared to accommodate a regular parquet of flat-plate panels orientated in the correct direction. It is also interesting to briefly compare thermal concentrating solar power (CSP) and CPV plants. Although the system efficiency for CPV is higher than CSP, the principle advantage of CSP is the ability to store thermal energy and therefore deliver power on demand. However, the minimum size for a cost-effective CSP plant is 50 MW and present technology requires cooling water. CPV plants with highly efficient III-V solar cells are therefore well suited for medium- to large-scale plants in arid regions with high direct irradiance.

The next stage in the development of concentrator photovoltaics is likely to be both an increase in efficiency, and shift towards eliminating the substrate, either via epitaxial lift-off and reuse or growth on silicon or other low-cost substrates. In the longer term, a push towards concentrations over 1000 suns and smaller cell sizes should aid the economics of III-V concentrator photovoltaics, but this requires high-performance yet inexpensive optical systems.

6.5 CONCLUSION

III-V semiconductors enable almost ideal solar cells to be made, owing to their strong absorption and excellent carrier transport. The wide choice of bandgaps and the ability to integrate different alloys into a single device has enabled a wide variety of multi-junction solar cells to be made with efficiencies in excess of 35% at one sun and 40% under concentration. The InGaP/GaAs/Ge-based triple-junction solar cell is now commonly used for commercial space applications and beginning to find applications in terrestrial power generation in solar concentrator systems. Future increases in performance are likely; adjusting the absorption thresholds of the present triple-junction solar cell will help to push concentrator efficiencies towards 45% while adding a fourth junction provides a pathway to efficiencies of 50%. Present manufacturing technologies are capable of meeting the demand for space and concentrator applications.

REFERENCES

[1] T. J. Coutts and M. Fitzgerald. Thermophotovoltaics. *Scientific American*, pages 90–95, Mar 1998.

[2] I. Vurgaftman, J. R. Meyer, and L. R. Ram-Mohan. Band parameters for III–V compound semiconductors and their alloys. *Journal of Applied Physics*, **89**(11):5815, 2001.

[3] I. Vurgaftman and J. R. Meyer. Band parameters for nitrogen-containing semiconductors. *Journal of Applied Physics*, **94**(6):3675, Aug 2003.

[4] Landolt and Bornstein. Springer Materials : The Landolt-Bornstein Database. http://www.springermaterials.com/docs/index.html

[5] D. A. Jenny, J. J. Loferski, and P. Rappaport. Photovoltaic effect in GaAs p-n junctions and solar energy conversion. *Physical Review*, **101**(3):1208–1209, 1956.

[6] J. M. Woodall and H. J. Hovel. High-efficiency $Ga_{1-X}Al_XAs$-GaAs solar cells. *Applied Physics Letters*, **21**(8):379, 1972.

[7] P. Capper and M. Mauk (eds.) *Liquid Phase Epitaxy of Electronic, Optical and Optoelectronic Materials*. John Wiley & Sons, 2007.

[8] Gerald B. Stringfellow. *Organometallic Vapor-Phase Epitaxy*. Academic Press, 2nd edition, 1998.

[9] D. V. Shenai-Khatkhate, R. J. Goyette, R. L. DiCarlo, and G. Dripps. Environment, health and safety issues for sources used in MOVPE growth of compound semiconductors. *Journal of Crystal Growth*, **272**(1-4):816–821, 2004.

[10] A. Moto, S. Tanaka, T. Tanabe, and S. Takagishi. GaInP/GaAs and mechanically stacked GaInAs solar cells grown by MOCVD using TBAs and TBP as V-precursors. *Solar Energy Materials and Solar Cells*, **66**(1-4):585–592, 2001.

[11] Mohamed Henini. *Molecular Beam Epitaxy: From Research to Mass Production*. Elsevier, 2012.

[12] M. R. Melloch, S. P. Tobin, C. Bajgar, A. Keshavarzi, T. B. Stellwag, G. B. Lush, M. S. Lundstrom, and K. Emery. High-efficiency $Al_{0.22}Ga_{0.78}As$ solar-cells grown by molecular-beam epitaxy. *Applied Physics Letters*, **57**(1):52–54, 1990.

[13] P. Leinonen, M. Pessa, J. Haapamaa, and K. Rajennus. Advances in production MBE grown GaInP/GaAs cascade solar cells. In *Proc. IEEE Photovoltaic Specialists Conference*, page 1177, 2000.

[14] Solar Junction. Solar Junction Breaks World Record with 44% Efficient CPV Production Cell. Press release, 2012.

[15] L. M. Fraas, P. S. McLeod, L. D. Partain, R. E. Weiss, and J. A. Cape. GaP and GaAs films grown by vacuum chemical epitaxy using TEGa and TMGa sources. *Journal Of Crystal Growth*, **77**(1-3):386–391, 1986.

[16] J. L. Benchimol, F. Alexandre, B. Lamare, and P. Legay. Benefits of chemical beam epitaxy for micro and optoelectronic applications. *Progress In Crystal Growth And Characterization Of Materials*, **33**(4):473–495, 1996.

[17] A. Freundlich, F. Newman, C. Monier, S. Street, P. Dargan, and M. Levy. High growth rate metal-organic molecular beam epitaxy for the fabrication of GaAs space solar cells. *Progress In Photovoltaics*, **8**(3):333–338, 2000.

[18] Hidetoshi Suzuki, Makoto Inagaki, Takahiko Honda, Yoshio Ohshita, Nobuaki Kojima, and Masafumi Yamaguchi. Improvements in Optoelectrical Properties of GaAsN by Controlling Step Density during Chemical Beam Epitaxy Growth. *Japanese Journal of Applied Physics*, **49**(4):04DP08, April 2010.

[19] D. M. Chapin, C. S. Fuller, and G. L. Pearson. A new silicon pn junction photocell for converting solar radiation into electrical power. *Journal of Applied Physics*, **25**:676, 1954.

[20] D. M. Perkins and E. F. Pasierb. Photovoltaic properties of GaAs thin films. In *Proc. IEEE Photovoltaic Specialists Conference*, pages A–2, 1964.

[21] B. Ellis and T.S. Moss. Calculated efficiencies of practical GaAs and Si solar cells including effect of built-in electric fields. *Solid State Electron*, **13**(1):1–24, 1970.

[22] J.M. Woodall and H.J. Hovel. Outlooks for GaAs terrestrial photovoltaics. *J Vac Sci Technol*, **12**(5):1000–1009, 1975.

[23] W. Shockley and H. J. Queisser. Detailed balance limit of efficiency of p-n junction solar cells. *J Appl Phys*, **32**(3):510, 1961.

[24] ASTM. Standard solar constant and air mass zero spectral irradiance tables. *E490-73a*, Aug 1999.

[25] ASTM. Standard tables for terrestrial solar spectral irradiance at air mass 1.5 for a 37 tilted surface. *E892-87*, Nov 1992.

[26] ASTM. Standard tables for references solar spectral irradiance at air mass 1.**5**: direct normal and hemispherical for a 37degree tilted surface. *G159-96*, Jun 1999.

[27] J. G. Werthen, G. F. Virshup, C. W. Ford, C. R. Lewis, and H. C. Hamaker. 21-percent (one sun, air-mass zero) 4cm2 GaAs space solar-cells. *Applied Physics Letters*, **48**(1):74–75, 1986.

[28] C. W. Hitchcock, R. J. Gutmann, J.M. Borrego, I.B. Bhat, and G.W. Charache. Antimonide-based devices for thermophotovoltaic applications. *IEEE Transactions On Electron Devices*, **46**(10):2154–2161, 1999.

[29] S. P. Tobin, S. M. Vernon, C. Bajgar, L. M. Geoffroy, C. J. Keavney, M. M. Sanfacon, and V. E. Haven. Device processing and analysis of high-efficiency GaAs Solar Cells. *Solar Cells*, **24**(1-2):103–115, 1988.

[30] D. J. Aiken. High performance anti-reflection coatings for broadband multi-junction solar cells. *Solar Energy Materials and Solar Cells*, **64**(4):393–404, 2000.

[31] T. A. Gessert and T. J. Coutts. Grid metallization and antireflection coating optimization for concentrator and one-sun photovoltaic solar-cells. *Journal of Vacuum Science & Technology A*, **10**(4):2013–2024, 1992.

[32] Beatriz Galiana, Carlos Algora, and Ignacio Rey-Stolle. Comparison of 1D and 3D analysis of the front contact influence on GaAs concentrator solar cell performance. *Solar Energy Materials and Solar Cells*, **90**:2589–2604, 2006.

[33] Marc Steiner, Simon P. Philipps, Martin Hermle, Andreas W. Bett, and Frank Dimroth. Validated front contact grid simulation for GaAs solar cells under concentrated sunlight. *Progress in Photovoltaics*, **19**(1):73–83, 2011.

[34] V. M. Andreev, V. A. Grilikhes, and V. D. Rumyantsev. *Photovoltaic Conversion of Concentrated Sunlight*. Wiley, 1997.

[35] S. P. Tobin, S. M. Vernon, C. Bajgar, S. J. Wojtczuk, M. R. Melloch, A. Keshavarzi, T. B. Stellwag, S. Venkatensan, M. S. Lundstrom, and K. A. Emery. Assessment of MOCVD-grown and MBE-grown GaAs for high-efficiency solar-cell applications. *IEEE Transactions on Electron Devices*, **37**(2):469–477, 1990.

[36] J. Nelson. *The Physics of Solar Cells*. Imperial College Press, 2003.

[37] D. E. Aspnes, S. M. Kelso, R. A. Logan, and R. Bhat. Optical-properties of AlxGa1-xAs. *Journal of Applied Physics*, **60**(2):754–767, 1986.

[38] R. K. Ahrenkiel, J. M. Olson, D. J. Dunlavy, B. M. Keyes, and A. E. Kibbler. Recombination velocity of the $Ga_{0.5}In_{0.5}P$/GaAs interface. *Journal of Vacuum Science & Technology A*, **8**(3):3002–3005, 1990.

[39] S. R. Kurtz, J. M. Olson, and A. Kibbler. High efficiency GaAs solar cells using GaInP window layers. In *Proc. IEEE Photovoltaic Specialists Conference*, page 138, 1990.

[40] R. J. Nelson and R. G. Sobers. Interfacial recombination velocity in GaAlAs-GaAs heterostructures. *Applied Physics Letters*, **32**(11):761–763, 1978.

[41] P. Campbell and M. A. Green. Light trapping properties of pyramidally textured surfaces. *Journal of Applied Physics*, **62**(1):243–249, 1987.

[42] M. Z. Shvarts, O. I. Chosta, I. V. Kochnev, V. M. Lantratov, and V. M. Andreev. Radiation resistant AlGaAs/GaAs concentrator solar cells with internal bragg reflector. *Solar Energy Materials and Solar Cells*, **68**(1):105–122, 2001.

[43] T. Tiedje, E. Yablonovitch, G. D. Cody, and B. G. Brooks. Limiting efficiency of silicon solar-cells. *IEEE Transactions on Electron Devices*, **31**:711–716, 1984.

[44] R. J. Roedel and P. M. Holm. The design of anisotropically etched-III-V solar-cells. *Solar Cells*, **11**(3):221–239, 1984.

[45] R. P. Leon, S. G. Bailey, G. A. Mazaris, and W. D. Williams. Formation of a pn junction on an anisotropically etched GaAs surface using metalorganic chemical vapor-deposition. *Applied Physics Letters*, **49**(15):945–947, 1986.

[46] S. G. Bailey, N. Fatemi, D. M. Wilt, G. A. Landis, and R. D. Thomas. GaAs solar cells with V-grooved emitters. In *Proc. European Space Power Conference*, 1989.

[47] Keisuke Nakayama, Katsuaki Tanabe, and Harry A. Atwater. Plasmonic nanoparticle enhanced light absorption in GaAs solar cells. *Applied Physics Letters*, **93**(12):121904, 2008.

[48] J. Tommila, V. Polojarvi, A. Aho, A. Tukiainen, J. Viheriala, J. Salmi, A. Schramm, J. M. Kontio, A. Turtiainen, T. Niemi, and M. Guina. Nanostructured broadband antireflection coatings on AlInP fabricated by nanoimprint lithography. *Solar Energy Materials and Solar Cells*, **94**(10):1845–1848, 2010.

[49] M. A. Green, K. Emery, Y. Hishikawa, and W. Warta. Solar cell efficiency tables (version 37). *Progress in Photovoltaics Research and Applications*, **19**:84–92, 2011.

[50] G. J. Bauhuis, P. Mulder, E. J. Haverkamp, J. C. C. M Huijben, and J. J. Schermer. 26.1% thin-film GaAs solar cell using epitaxial lift-off. *Solar Energy Materials And Solar Cells*, **93**(9):1488–1491, 2009.

[51] R. K. Ahrenkiel, B. M. Keyes, G. B. Lush, M. R. Melloch, M. S. Lundstrom, and H. F. Macmillan. Minority-carrier lifetime and photon recycling in n-GaAs. *Journal of Vacuum Science & Technology A*, **10**(4):990–995, 1992.

[52] G. Lush and M. Lundstrom. Thin-film approaches for high-efficiency III-V-cells. *Solar Cells*, **30**(1-4):337–344, 1991.

[53] A. Marti, J. L. Balenzategui, and R. F. Reyna. Photon recycling and shockley's diode equation. *Journal of Applied Physics*, **82**(8):4067–4075, 1997.

[54] C. O. Bozler, R. W. McClelland, and J. C. C. Fan. Ultrathin, high-efficiency solar-cells made from GaAs films prepared by the CLEFT process. *Electron Device Lett*, **2**(8):203–205, 1981.

[55] E. Yablonovitch, T. Gmitter, J. P. Harbison, and R. Bhat. Extreme selectivity in the lift-off of epitaxial GaAs films. *Applied Physics Letters*, **51**(26):2222–2224, 1987.

[56] J. J. Schermer, P. Mulder, G. J. Bauhuis, M. M. A. J. Voncken, J. van Deelen, E. Haverkamp, and P. K. Larsen. Epitaxial lift-off for large area thin film III/V devices. *Phys Status Solidi A*, **202**(4):501–508, 2005.

[57] C. O. Bozler, R. W. McClelland, J. P. Salerno, and J. C. C. Fan. Single-crystal GaAs films on amorphous substrates by the CLEFT process. *Journal of Vacuum Science & Technology*, **20**(3):720–725, 1982.

[58] G. J. Bauhuis, P. Mulder, E. J. Haverkamp, J. J. Schermer, E. Bongers, G. Oomen, W. Koestler, and G. Strobl. Wafer reuse for repeated growth of III V solar cells. *Progress in Photovoltaics*, **18**(3):155–159, 2010.

[59] Y. Itoh, T. Nishioka, A. Yamamoto, and M. Yamaguchi. GaAs heteroepitaxial growth on Si for solar-cells. *Applied Physics Letters*, **52**(19):1617–1618, 1988.

[60] M. Yamaguchi. Dislocation density reduction in heteroepitaxial iii-v compound films on si substrates for optical-devices. *Journal of Material Research*, **6**(2):376–384, 1991.

[61] C. L. Andre, J. A. Carlin, J. J. Boeckl, D. M. Wilt, M. A. Smith, A. J. Pitera, M. L. Lee, E. A. Fitzgerald, and S. A. Ringel. Investigations of high-performance GaAs solar cells grown on Ge-Si_{1-x}Gex-Si substrates. *IEEE Transactions on Electron Devices*, **52**(6):1055–1060, 2005.

[62] I. Nemeth, B. Kunert, W. Stolz, and K. Volz. Heteroepitaxy of gap on si: Correlation of morphology, anti-phase-domain structure and movpe growth conditions. *Journal of Crystal Growth*, **310**(7-9):1595–1601, 2008.

[63] P. Vohl. GaAs thin-film solar cells. *IEEE Transactions on Electron Devices*, **14**(1):26–30, 1967.

[64] G. W. Turner, J. C. C. Fan, R. P. Gale, and O. Hurtado. Efficient large grained GaAs homojunction solar cells. In *Proc. IEEE Photovoltaic Specialists Conference*, page 1330, 1980.

[65] J. C. Zolper, M. Bohm, and A. M. Barnett. The influence of grain boundaries on the performance of polycrystalline gallium arsenide solar cells. In *Proc. IEEE Photovoltaic Specialists Conference*, 1985.

[66] R. Venkatasubramanian, B. C. O'Quinn, E. Silvola, B. Keyes, and R. Ahrenkiel. 20% (AM1.5) efficiency GaAs solar cells on sub-mm grainsize poly-Ge and its transition to low cost substrates. In *Proc. IEEE Photovoltaic Specialists Conference*, page 811, 1997.

[67] M. G. Mauk, J. R. Balliet, and B. W. Feyock. Large-grain ($> 1-$ mm), recrystallized germanium films on alumina, fused silica, oxide-coated silicon substrates for III-V solar cell applications. *Journal of Crystal Growth*, **250**(1-2):50–56, 2003.

[68] K. M. Jones, M. M. Al-Jassim, F. S. Hasoon, and R. Venkatasubramanian. The morphology and microstructure of thin-film GaAs on Mo substrates. *NCPV Photovoltaics Program Review*, pages 531–536, 1999.

[69] S. J. Polly, C. R. Plourde, S. G. Bailey, C. Leitz, C. Vineis, M. P. Brindak, D. V. Forbes, J. S. McNatt, and R. P. Hubbard, S. M. and Raffaelle. Thin film III-V solar cells on Mo foil. In *Proc. IEEE Photovoltaic Specialists Conference*, page 1377, 2009.

[70] M. Yamaguchi. Radiation-Resistance of Compound Semiconductor Solar-Cells. *Journal Of Applied Physics*, **78**(3):1476–1480, 1995.

[71] T. A. Gessert, M. W. Wanlass, T. J. Coutts, X. Li, and G. S. Horner. Aspects of processing InP-related solar-cells. *Solar Cells*, **27**(1-4):307–320, 1989.

[72] C. J. Keavney, V. E. Haven, and S. M. Vernon. Emitter structures in MOCVD InP Solar Cells. In *Proc. IEEE Photovoltaic Specialists Conference*, page 141, 1990.

[73] P. Jenkins, G. Landis, N. Fatemi, D. Scheiman, X. Li, and S. Bailey. An InP solar cell with a light-trapping front surface. *Solar Energy Materials and Solar Cells*, **33**(2):125–133, Jun 1994.

[74] Kyusang Lee, Kuen-Ting Shiu, Jeramy D. Zimmerman, Christopher K. Renshaw, and Stephen R. Forrest. Multiple growths of epitaxial lift-off solar cells from a single inp substrate. *Appl Phys Lett*, **97**(10):101107, 2010.

[75] Louise C. Hirst and Nicholas J. Ekins-Daukes. Fundamental losses in solar cells. *Progress in Photovoltaics: Research and Applications*, **19**(3):286–293, Aug 2010.

[76] T. Markvart. The thermodynamics of optical etendue. *Journal of Optics A - Pure & Applied optics*, **10**:015008, 2008.

[77] G. E. Stillman, L. W. Cook, N. Tabatabaie, G. E. Bulman, and V. M. Robbins. InGaAsP photodiodes. *IEEE Transactions On Electron Devices*, **30**(4):364–381, April 1983.

[78] J. S. Ward, M. W. Wanlass, K. A. Emery, and T. J. Coutts. GaInAsP solar cells with the ideal bandgap for terrestrial concentrator applications. In *Proc. IEEE Photovoltaic Specialists Conference*, pages 650–654, 1993.

[79] J. Wu, W. Walukiewicz, K. M. Yu, W. Shan, J. W. Ager, E. E. Haller, H. Lu, W. J. Schaff, W. K. Metzger, and S. Kurtz. Superior radiation resistance of In1-xGaxN alloys: Full-solar-spectrum photovoltaic material system. *Journal of Applied Physics*, **94**(10):6477–6482, 2003.

[80] Omkar Jani, Ian Ferguson, Christiana Honsberg, and Sarah Kurtz. Design and characterization of GaN/InGaN solar cells. *Applied Physics Letters*, **91**(13):132117, 2007.

[81] Elison Matioli, Carl Neufeld, Michael Iza, Samantha C. Cruz, Ali A. Al-Heji, Xu Chen, Robert M. Farrell, Stacia Keller, Steven DenBaars, Umesh Mishra, Shuji Nakamura, James Speck, and Claude Weisbuch. High internal and external quantum efficiency ingan/gan solar cells. *Applied Physics Letters*, **98**(2):021102, 2011.

[82] E. D. Jackson. In *Transaction of Conference on use of Solar Energy, Tucson 1955*, volume 5, pages 122–125. U. of Arizona Press, 1958.

[83] A. S. Brown and M. A. Green. Detailed balance limit for the series constrained two terminal tandem solar cell. *Physica E*, **14**:96–100, 2002.

[84] A. Hermle, G. Letay, S. P. Philipps, and A. W. Bett. Numerical simulation of tunnel diodes for multi-junction solar cells. *Progress in Photovoltaics*, **16**(5):409–418, 2008.

[85] K. Jandieri, S. D. Baranovskii, W. Stolz, and F. Gebhard. Analytical theory for favorable defects in tunnel diodes. *Journal of Applied Physics*, **104**(11):114511, 2008.

[86] Jeffrey F. Wheeldon, Christopher E. Valdivia, Alexandre W. Walker, Gitanjali Kolhatkar, Abdelatif Jaouad, Artur Turala, Bruno Riel, Denis Masson, Norbert Puetz, Simon Fafard, Richard Arès, Vincent Aimez, Trevor J. Hall, and Karin Hinzer. Performance comparison of AlGaAs, GaAs and InGaP tunnel junctions for concentrated multijunction solar cells. *Progress in Photovoltaics*, pages 442–452, Nov 2010.

[87] H. Sugiura, C. Amano, A. Yamamoto, and M. Yamaguchi. Double heterostructure GaAs tunnel junction for a AlGaAs/GaAs tandem solar-cell. *Japanese Journal Of Applied Physics Part 1-Regular Papers Brief Communications & Review Papers*, **27**(2):269–272, 1988.

[88] D. Jung, C. A. Parker, J. Ramdani, and S. M. Bedair. AlGaAs/GaInP heterojunction tunnel-diode for cascade solar-cell application. *Journal of Applied Physics*, **74**(3):2090–2093, 1993.

[89] T. Takamoto, M. Yamaguchi, E. Ikeda, T. Agui, H. Kurita, and M. Al-Jassim. Mechanism of Zn and Si diffusion from a highly doped tunnel junction for InGaP/GaAs tandem solar cells. *Journal of Applied Physics*, **85**(3):1481–1486, 1999.

[90] B. C. Chung, G. F. Virshup, S. Hikido, and N. R. Kaminar. 27.6-percent efficiency (1 sun, Air-Mass 1.5) monolithic $Al_{0.37}Ga_{0.63}As$/GaAs 2-junction cascade solar-cell with prismatic cover glass. *Applied Physics Letters*, **55**(17):1741–1743, 1989.

[91] R. K. Ahrenkiel and D. J. Dunlavy. Minority-carrier lifetime in $Al_XGa_{1-X}As$. *Journal of Vacuum Science & Technology A*, **7**(3):822–826, 1989.

[92] J. M. Olson, S. R. Kurtz, and A. E. Kibbler. Purity and purification of source materials for III-V MOCVD. *Journal of Crystal Growth*, **89**:131–136, 1988.

[93] K. A. Bertness, S. R. Kurtz, D. J. Friedman, A. E. Kibbler, C. Kramer, and J.M. Olson. 29.5-percent-efficient GaInP/GaAs tandem solar-cells. *Applied Physics Letters*, **65**:989–991, 1994.

[94] I. Garcia, I. Rey-Stolle, B. Galiana, and C. Algora. A 32.6% efficient lattice-matched dual-junction solar cell working at 1000 suns. *Applied Physics Letters*, **94**(5):053509, 2009.

[95] A. Mascarenhas, S. Kurtz, A. Kibbler, and J. M. Olson. Polarized Band-Edge Photoluminescence and Ordering in $Ga_{0.52}In_{0.48}P$. *Physical Review Letters*, **63**(19):2108–2111, 1989.

[96] S. R. Kurtz, J. M. Olson, D. J. Arnet, M. H. Bode, and K. A. Bertness. Low-band-gap gainp grown on (511)B GaAs substrates. *Applied Physics Letters*, **75**:5110, 1994.

[97] J. K. Shurtleff, R. T. Lee, C. M. Fetzer, and G. B. Stringfellow. Band-gap control of GaInP using Sb as a surfactant. *Applied Physics Letters*, **75**:1914–1916, 1999.

[98] C. M. Fetzer, R. T. Lee, G. B. Stringfellow, X. Q. Liu, A. Sasaki, and N. Ohno. Effect of surfactant Sb on carrier lifetime in GaInP epilayers. *Journal of Applied Physics*, **91**:199–203, 2001.

[99] Tatsuya Takamoto, Eiji Ikeda, Hiroshi Kurita, Masamichi Ohmori, Masafumi Yamaguchi, and Ming-Ju Yang. Two-Terminal Monolithic $In_{0.5}Ga_{0.5}P$/GaAs Tandem Solar Cells with a High Conversion Efficiency of Over 30. *Japanese Journal of Applied Physics*, **36**(10):6215–6220, October 1997.

[100] T. Agui, T. Takamoto, and M. Kaneiwa. Investigation on AlInGaP solar cells for current matched multijunction cells. In *Proc. Third World Conference on Photovoltaic Energy Conversion*, 2003.

[101] S. R. Kurtz, J. M. Olson, K. A. Bertness, K. Sinha, B. McMahon, and S. Asher. Hidden but important parameters in GaInP cell growth. In *Proc. IEEE Photovoltaic Specialists Conference*, page 37, 1996.

[102] M. W. Wanlass, J. S. Ward, K. A. Emery, T. A. Gessert, C. R. Osterwald, and T. J. Coutts. High-performance concentrator tandem solar-cells based on IR-sensitive bottom cells. *Solar Cells*, **30**(1-4):363–371, 1991.

[103] M. W. Wanlass, J. S. Ward, K. A. Emery, A. Duda, and T. J. Coutts. Improved large-area two-terminal InP/GaInAs tandem solar cells. In *Proc. First World Conference on Photovoltaic Energy Conversion*, page 1717, 1994.

[104] S. J. Wojtczuk, S. P. Tobin, C. J. Keavney, C. Bajgar, M. M. Sanfacon, L. M. Geoffroy, T. M. Dixon, S. M. Vernon, J. D. Scofield, and D. S. Ruby. GaAs Ge tandem-cell space concentrator development. *IEEE Transactions on Electron Devices*, **37**(2):455–463, 1990.

[105] T. Takamoto, T. Agui, E. Ikeda, and H. Kurita. High efficiency InGaP/InGaAs tandem solar cells on Ge substrates. In *Proc. IEEE Photovoltaic Specialists Conference*, pages 976–981, 2000.

[106] K. W. J. Barnham. *Nanotechology for photovoltaics*. CRC Press, 2010.

[107] S. R. Kurtz, D. Myers, and J. M. Olson. Projected performance of three and four junction devices using GaAs and GaInP. In *Proc. IEEE Photovoltaic Specialists Conference*, pages 875–878, 1997.

[108] D. J. Friedman, J. F. Geisz, S. R. Kurtz, and J. M. Olson. 1-eV solar cells with GaInNAs active layer. *Journal of Crystal Growth*, **195**:409–415, 1998.

[109] A. J. Ptak, D. J. Friedman, and Sarah Kurtz. Effects of temperature, nitrogen ions, and antimony on wide depletion width GaInNAs. *Journal of Vacuum Science & Technology B*, **25**(3):955–959, 2007.

[110] David B. Jackrel, Seth R. Bank, Homan B. Yuen, Mark A. Wistey, and James S. Harris. Dilute nitride GaInNAs and GaInNAsSb solar cells by molecular beam epitaxy. *Journal Of Applied Physics*, **101**(11):114916, 2007.

[111] J. F. Geisz, D. J. Friedman, S. Kurtz, J. M. Olson, A. B. Swartzlander, R. C. Reedy, and A. G. Norman. Epitaxial growth of BGaAs and BGaInAs by MOCVD. *Journal Of Crystal Growth*, **225**:372–376, 2001.

[112] Angelo Mascarenhas, Rajeev Kini, Yong Zhang, Ryan France, and Aaron Ptak. Comparison of the dilute bismide and nitride alloys GaAsBi and GaAsN. *Physica Status Solidi B-Basic Solid State Physics*, **246**(3):504–507, 2009.

[113] R. R. King, P. C. Colter, D. E. Joslin, K. M. Edmondson, D. D. Krut, N. H. Karam, and S. R. Kurtz. High-voltage, low-current GaInP/GaInP/GaAs/GaInNAs/Ge solar cells. In *Proc. IEEE Photovoltaic Specialists Conference*, page 852, 2002.

[114] E. A. Fitzgerald, S. B. Samavedam, Y. H. Xie, and L. M. Giovane. Influence of strain on semiconductor thin film epitaxy. *Journal Of Vacuum Science & Technology A-Vacuum Surfaces And Films*, **15**(3):1048–1056, 1997.

[115] Kenneth E. Lee and Eugene A. Fitzgerald. High-quality metamorphic compositionally graded InGaAs buffers. *Journal Of Crystal Growth*, **312**(2):250–257, 2010.

[116] T. Sasaki, K. Arafune, W. Metzger, M. J. Romero, K. Jones, M. Al-Jassim, Y. Ohshita, and M. Yamaguchi. Characterization of carrier recombination in lattice-mismatched InGaAs solar cells on GaAs substrates. *Solar Energy Materials And Solar Cells*, **93**(6-7):936–940, 2009.

[117] M. Yamaguchi, M. Tachikawa, Y. Itoh, M. Sugo, and S. Kondo. Thermal annealing effects of defect reduction in GaAs ON Si substrates. *Journal Of Applied Physics*, **68**(9):4518–4522, 1990.

[118] M. J. Ludowise, W. T. Dietze, R. Boettcher, and N. Kaminar. High-efficiency (21.4-percent) $Ga_{0.75}In_{0.25}As/GaAs(E_g = 1.15\ eV)$ concentrator solar-cells and the influence of lattice mismatch on performance. *Applied Physics Letters*, **43**(5):468–470, 1983.

[119] J. G. Werthen, B. A. Arau, C. W. Ford, N. R. Kaminar, J. M. Gee, M. S. Kuryla, M. Ladle Ristow, C. R. Lewis, H. F. MacMillan, and G. F. Virshup. Recent advances in high-efficiency InGaAs concentrator cells. In *Proc. IEEE Photovoltaic Specialists Conference*, pages 640–643, 1988.

[120] F. Dimroth, P. Lanyi, U. Schubert, and A. W. Bett. Movpe grown ga1-xinxas solar cells for gainp/gainas tandem applications. *Journal of Electronic Materials*, **29**(1):42–46, 2000.

[121] C. R. Lewis, H. F. Macmillan, B. C. Chung, G. F. Virshup, D. D. Liu, L. D. Partain, and J. G. Werthen. Recent developments in multijunction solar-cell research. *Solar Cells*, **24**(1-2):171–183, 1988.

[122] H. F. Macmillan, B. C. Chung, H. C. Hamaker, N. R. Kaminar, M. S. Kuryla, M. L. Ristow, D. D. Liu, L. D. Partain, J. C. Schultz, G. F. Virshup, and J. G. Werthen. Recent advances in multi-junction III-V solar-cell development. *Solar Cells*, **27**(1-4):205–217, 1989.

[123] Spire Corporation. Spire semiconductor produces a world-record 42.3% efficient cpv cell. Press release, October 2010.

[124] J. F. Geisz, Sarah Kurtz, M. W. Wanlass, J. S. Ward, A. Duda, D. J. Friedman, J. M. Olson, W. E. McMahon, T. E. Moriarty, and J. T. Kiehl. High-efficiency GaInP/GaAs/InGaAs triple-junction solar cells grown inverted with a metamorphic bottom junction. *Applied Physics Letters*, **91**(2):023502, 2007.

[125] T. Takamoto, T. Agui, A. Yoshida, K. Nakaido, H. Juso, K. Sasaki, K. Nakamora, H. Yamaguchi, T. Kodama, H. Washio, M. Imaizumi, and M. Takahashi. World's highest efficiency triple-junction solar cells fabricated by inverted layers transfer process. In *Proc. IEEE Photovoltaic Specialists Conference*, 2010.

[126] J. Boisvert, D. Law, R. R. King, D. Bhusari, X. Liu, A. Zakaria, W. Hong, S. Mesropian, D. Larrabee, R. Woo, A. Boca, K. M. Edmondson, D. D. Krut, D. Peterson, K. Rouhani, B. Benedikt, and N. H. Karam. Development of advanced space solar cells and Spectrolab. In *Proc. IEEE Photovoltaic Specialists Conference*, 2010.

[127] J. F. Geisz, D. J. Friedman, J. S. Ward, A. Duda, W. J. Olavarria, T. E. Moriarty, J. T. Kiehl, M. J. Romero, A. G. Norman, and K. M. Jones. 40.8% efficient inverted triple-junction solar

cell with two independently metamorphic junctions. *Applied Physics Letters*, **93**(12):123505, 2008.

[128] M. Stan, D. Aiken, B. Cho, A. Cornfeld, V. Ley, P. Patel, P. Sharps, and T. Varghese. High-efficiency quadruple junction solar cells using omvpe with inverted metamorphic device structures. *Journal of Crystal Growth*, **312**(8):1370–1374, 2010.

[129] Wolfgang Guter, Jan Schoene, Simon P. Philipps, Marc Steiner, Gerald Siefer, Alexander Wekkeli, Elke Welser, Eduard Oliva, Andreas W. Bett, and Frank Dimroth. Current-matched triple-junction solar cell reaching 41.1 efficiency under concentrated sunlight. *Applied Physics Letters*, **94**(22):223504, 2009.

[130] M. R. Lueck, C. L. Andre, A. J. Pitera, M. L. Lee, E. A. Fitzgerald, and S. A. Ringel. Dual junction GaInP/GaAs solar cells grown on metamorphic SiGe/Si substrates with high open circuit voltage. *IEEE Electron Device Letters*, **27**(3):142–144, 2006.

[131] T. Soga, K. Baskar, T. Kato, T. Jimbo, and M. Umeno. MOCVD growth of high efficiency current-matched AlGaAs/Si tandem solar cell. *Journal of Crystal Growth*, **174**(1-4):579–584, 1997.

[132] M. J. Mori, S. T. Boles, and E. A. Fitzgerald. Comparison of compressive and tensile relaxed composition-graded GaAsP and (Al)InGaP substrates. *Journal of Vacuum Science & Technology A*, **28**(2):182–188, 2010.

[133] Nathaniel J. Quitoriano and Eugene A. Fitzgerald. Relaxed, high-quality InP on GaAs by using InGaAs and InGaP graded buffers to avoid phase separation. *Journal Of Applied Physics*, **102**(3):–, 2007.

[134] H. Hayashi, T. Soga, H. Nishikawa, T. Jimbo, and M. Umeno. MOCVD growth of GaAsP on Si for tandem solar cell application. In *Proc. First World Conference on Photovoltaic Energy Conversion*, page 1890, 1994.

[135] Tyler J. Grassman, Mark R. Brenner, Maria Gonzalez, Andrew M. Carlin, Raymond R. Unocic, Ryan R. Dehoff, Michael J. Mills, and Steven A. Ringel. Characterization of Metamorphic GaAsP/Si Materials and Devices for Photovoltaic Applications. *IEEE Transactions On Electron Devices*, **57**(12):3361–3369, 2010.

[136] Dandan Zhu, Clifford McAleese, Maik Haeberlen, Carmen Salcianu, Ted Thrush, Menno Kappers, Andrew Phillips, Penelope Lane, Michael Kane, David Wallis, Trevor Martin, Mike Astles, Nicolas Hylton, Phil Dawson, and Colin Humphreys. Efficiency measurement of GaN-based quantum well and light-emitting diode structures grown on silicon substrates. *Journal of Applied Physics*, **109**(1):014502, 2011.

[137] L. Hsu and W. Walukiewicz. Modeling of InGaN/Si tandem solar cells. *Journal of Applied Physics*, **104**(2):024507, 2008.

[138] Lothar A. Reichertz, Iulian Gherasoiu, Kin Man Yu, Vincent M. Kao, Wladek Walukiewicz, and Joel W. Ager. Demonstration of a III-Nitride/Silicon tandem solar cell. *Applied Physics Express*, **2**(12):122202, 2009.

[139] K. M. Yu, S. V. Novikov, R. Broesler, Z. Liliental-Weber, A. X. Levander, V. M. Kao, O. D. Dubon, J. Wu, J. Wu, J.W. Ager III, W. Shan, M. A. Scrapulla, O. D. Dubon Wladek, W. Walukiewicz, K. M. Yu, P. Becla, and C. T. Foxon. Low gap amorphous GaN1-xAsx alloys grown on glass substrate. *Applied Physics Letters*, **97**(10):101906, 2010.

[140] L. D. Partain, M. S. Kuryla, R. E. Weiss, R. A. Ransom, P. S. McLeod, L. M. Fraas, and J. A. Cape. 26.1-percent solar-cell efficiency for Ge mechanically stacked under GaAs. *Journal of Applied Physics*, **62**(7):3010–3015, 1987.

[141] L. Frass, J. Avery, V. S. Sundaram, V. T. Dinh, T. M. Davenport, J. W. Yerkes, J. M. Gee, and K. A. Emery. Over 35% efficient GaAs/GaSb stacked concentrator cell assembilies for terrestrial applications. In *Proc. IEEE Photovoltaic Specialists Conference*, page 190, 1990.

[142] J. M. Gee and G. F. Virshup. A. 31% efficient GaAs/Silicon mechanically stacked multi-junction concentrator solar cell. In *Proc. IEEE Photovoltaic Specialists Conference*, page 754, 1988.

[143] B.C. Chung, G.F. Virshup, M. Klausmeierbrown, M.L. Ristow, and M.W. Wanlass. 25.2-percent efficiency (1-sun, Air Mass-0) AlGaAs/GaAs/InGaAsP 3-junction, 2-terminal solar-cell. *Applied Physics Letters*, **60**(14):1691–1693, 1992.

[144] L. Frass, J. Avery, and D. Scheiman. AM0 calibration of 34% efficient mechanically stacked GaInP/GaAs-GaSb circuits. In *Proc. IEEE Photovoltaic Specialists Conference*, page 912, 2002.

[145] T Takamoto, E. Ikeda, T. Agui, H. Kurita, T. Tanabe, S. Tanaka, H. Matsubara, Y. Mine, S. Takagishi, and M. Yamaguchi. InGaP/GaAs and InGaAs mechanically stacked triple junction solar cells. In *Proc. IEEE Photovoltaic Specialists Conference*, page 1031, 1997.

[146] T. Nagashima, K. Okumura, and K. Murata. Carrier recombination of germanium back-contact type bottom cells for three-terminal tandem solar cells. In *Proc. European Photovoltaic Solar Energy Conference*, page 2203, 2001.

[147] Daniel C. Law, R. R. King, H. Yoon, M. J. Archer, A. Boca, C. M. Fetzer, S. Mesropian, T. Isshiki, M. Haddad, K. M. Edmondson, D. Bhusari, J. Yen, R. A. Sherif, H. A. Atwater, and N. H. Karam. Future technology pathways of terrestrial III-V multijunction solar cells for concentrator photovoltaic systems. *Sol Energ Mat Sol C*, **94**(8):1314–1318, 2010.

[148] James M. Zahler, Katsuaki Tanabe, Corinne Ladous, Tom Pinnington, Frederick D. Newman, and Harry A. Atwater. High efficiency InGaAs solar cells on si by inp layer transfer. *Applied Physics Letters*, **91**(1):012108, 2007.

[149] Melissa J. Archer, Daniel C. Law, Shoghig Mesropian, Moran Haddad, Christopher M. Fetzer, Arthur C. Ackerman, Corinne Ladous, Richard R. King, and Harry A. Atwater. GaInP/GaAs dual junction solar cells on Ge/Si epitaxial templates. *Applied Physics Letters*, **92**(10):103503, 2008.

[150] V.M.Andreev. *GaAs and High-Efficiency Space Cells*. Elsevier, 2003.

[151] C. Claeys and E. Simoen. *Radiation Effects in Advanced Semiconductor Materials and Devices*. Springer Verlag, 2002.

[152] B.E. Anspaugh. *GaAs solar cell radiation handbook*. JPL Publication, 1996.

[153] S.R. Messenger, G.P. Summers, E.A. Burke, R.J. Walters, and M.A. Xapsos. Modeling solar cell degradation in space: A. comparison of the NRL displacement damage dose and the JPL equivalent fluence approaches. *Progress In Photovoltaics*, **9**(2):103–121, 2001.

[154] R.J. Walters, J.H. Warner, G.P. Summers, S.R. Messenger, and J.R. Lorentzen. Radiation response mechanisms in multijunction III-V space solar cells. In *Proc. IEEE Photovoltaic Specialists Conference*, page 542, 2005.

[155] S.R. Messenger, E.M. Jackson, J.H. Warner, and R.J. Walters. SCREAM: A. new code for solar cell degradation prediction using the displacement damage dose approach. In *Proc. IEEE Photovoltaic Specialists Conference*, page 1106, 2010.

[156] N. Fatemi, H.E. Pollard, H.Q. Hou, and P.R. Sharps. Solar array trades between very high-efficiency multi-junction and si space solar cells. In *Proc. IEEE Photovoltaic Specialists Conference*, page 1083, 2000.

[157] D. Chumney, D. Aiken, B. Cho, A. Cornfeld, J. Diaz, V. Ley, J. Mittman, F. Newman, P. Sharps, M. Stan, and T. Varghese. Path to a drop-in replacement for current technologies with the 33% large area imm cell. In *Proc. IEEE Photovoltaic Specialists Conference*, page 113, 2010.

[158] M. Imaizumi, M. Takahashi, and T. Takamoto. JAXA's strategy for development of high-performance space photovoltaics. In *Proc. IEEE Photovoltaic Specialists Conference*, page 128, 2010.

[159] D.M. Wilt, S.R. Messenger, and A. Howard. Technology opportunities to enable high mass specific power. In *Proc. IEEE Photovoltaic Specialists Conference*, page 1500, 2010.

[160] A.L Luque and V.M. Andreev, editors. *Concentrator Photovoltaics*. Springer Verlag.

[161] R.R. King, D. Law, K.M. Edmondson, C.M. Fetzer, G.S. Kinsey, H. Yoon, R.A. Sherif, D.D. Krut, J.H. Ermer, P. Hebert, P. Pien, and N.H. Karam. Multijunction solar cells with over 40%

efficiency and future directions in concentrator PV. In *Proc. European Photovoltaic Solar Energy Conference*, page 11, 2007.

[162] R.M. Swanson. The promise of concentrators. *Progress In Photovoltaics*, **8**(1):93–111, 2000.

[163] M Yamaguchi and A. Luque. High efficiency and high concentration in photovoltaics. *IEEE Transactions On Electron Devices*, **46**:2139–2144, 1999.

[164] N.L.A. Chan, T.B. Young, H.E. Brindley, N.J. Ekins-Daukes, K. Araki, Y. Kemmoku, and M. Yamaguchi. Validation of energy prediction method for a concentrator photovoltaic module in Toyohashi Japan. *Progress in Photovoltaics*, page DOI: 10.1002/pip.2241, 2012.

7 Chalcogenide Thin-Film Solar Cells

M. Paire, S. Delbos, J. Vidal, N. Naghavi and J.F. Guillemoles
Institut de Recherche et Développement sur l'Energie Photovoltaïque (IRDEP), France

7.1 INTRODUCTION

In the years to come, electricity production is bound to increase (from 17 PWh in 2009 to 28–34 PWh in 2035 [1]) and as a consequence the increased demand on fossil fuels as well as possible regulations on CO_2 emissions are bound to increase the wholesale electricity costs. On the other hand, the price-experience factor of 22% that was observed for the last 4 decades should stay at the same level or slightly below [2, 3], leading to a decrease of photovoltaic (PV) electricity costs [2, 4]. A first change of paradigm is now happening, because in some places like South Italy and other sunny/high electricity price regions [5], the price of retail electricity is higher than the cost of PV electricity (grid parity). A second parity, the "fuel parity", will happen when the cost of PV electricity is lower than the marginal costs of operating fossil-fuel-based power plants [6]. According to deployment scenarios, this fuel parity could happen between 2020 and 2040.

As shown in Figure 7.1, from 5% in 2005, to 12% in 2010 of the yearly PV production, thin-film technologies (based on amorphous and microcrystalline silicon, cadmium telluride and CIGS) are growing faster than wafer-based silicon technologies [7]. This is mostly related to the disruptive industrial endeavor of cadmium telluride solar cells, related to a cost reduction breakthrough reaching a record low product price (for the first time less than 1 dollar per watt in 2009). It is anticipated that the share of thin-film technologies will reach about 30% of the total production capacity in 2015 (25 GW/year) [8]. In 2011, a cadmium telluride (CdTe) company, First Solar, was in the first rank for production capacity (2 GW), ahead of Suntech with the crystalline silicon technology (1.9 GW) [9]. New comers are microcrystalline thin-film silicon and copper indium diselenide (CIGS). With about 1 GW production capacity in 2011 (mainly from Solar Frontier in Japan) CIGS technology follows the same accelerated growth as CdTe. It may overtake the CdTe technology in the near future, thanks to better efficiency figures (20.3% record versus 17.3%), better acceptability (almost or completely cadmium-free technology) and more open industrial perspectives (tens of players instead of a few). One can add that CIGS technologies are strong in Europe (Solibro, Manz, Avancis, Solartecture, Nexcis, etc.), where they benefit historically from high levels of R&D experience. CIGS technologies are still not present in China, with no turnkey factories available (proprietary technologies only).

As with the CdTe technology, CIGS technology fits very well with large-scale applications, in centralised plants. The possibility to reach high efficiencies (15% modules [11] from 20.3%

Solar Cell Materials: Developing Technologies, First Edition. Edited by Gavin Conibeer and Arthur Willoughby.

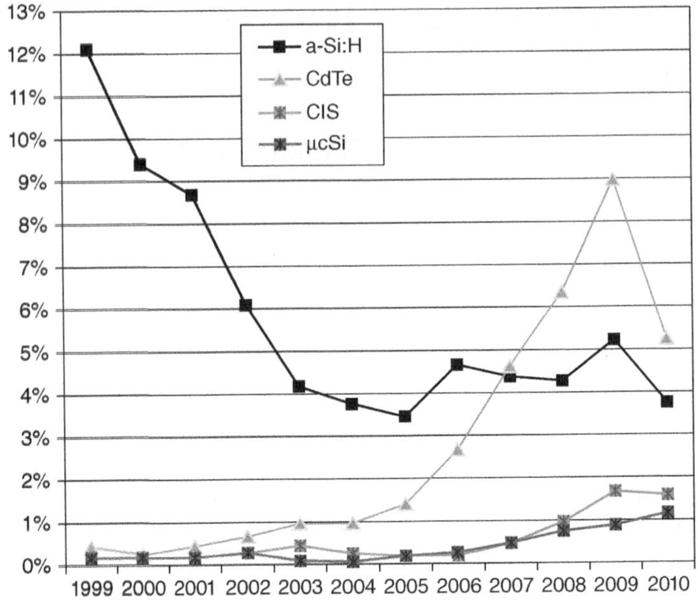

Figure 7.1 Evolution of shares of thin-film technologies within PV global production [7, 10]. (Reproduced with permission from European Photovoltaic Industry Association, 2011. Copyright © 2011, European Photovoltaic Industry Association.)

cells [12]), makes CIGS a good alternative to crystalline silicon for building integration as roofs or facades. Recent efficiency improvements in CdTe technology (18.7% cells [13]) let also expect an important role of this technology. The aesthetic of thin-film modules could be another advantage on the BIPV market. An emerging application, which can develop rapidly, concerns flexible and lightweight CIGS modules. These modules have high efficiencies as compared to state-of-the-art thin-film silicon modules, and could be deployed on industrial roofs. This could be an important market if lightweight modules could be produced.

The basic structure of a CIGS solar cell is a thin-film glass/Mo/CIGS/(CdS or ZnS)/ZnO heterostructure, where the CIGS p-type absorber layer is deposited with a thickness of about 2 micrometres. The Mo back contact layer is about 0.5–1 μm thick, as for ZnO (n-type junction partner), whereas the cadmium/zinc sulfide layer is an interfacial buffer layer with a thickness of 10–50 nm. Prospects to increase further the record efficiency value towards 25%, i.e. closer to the single-crystalline silicon technology, are good. The CIGS technology has made impressive progresses as the module efficiencies are reaching 15.7% on 900 cm^2 [11, 14] and 17.8% [15]. Other companies like Solopower and Nexcis are also reporting record values in the range 13–15% for cells with the electrodeposition technology [16, 17], alternative to previous vacuum-based ones. As a consequence of its performance level, the CIGS technology is reaching efficiencies comparable to polycrystalline-silicon technologies and may compete with single-crystalline silicon technologies in the near future.

The $Cu(In,Ga)Se_2$ solar cells are the product of a long development. In 1975, the Bell laboratory demonstrated a 12% efficient $CuInSe_2$ / CdS solar cell, with CdS evaporated on a single crystal [18]. Since then different breakthroughs have enable efficiencies above 20% to be reached [12]. This value is the same as the record value reached by polycrystalline-wafer-based silicon solar cells. A chemical-bath deposition for the CdS layer was developed.

The control of composition was improved by the introduction of separate sources for Cu, In and Se. Bandgap engineering with the addition of Ga was introduced. The role of sodium, coming from soda-lime glass substrate was discovered. Ingenious deposition process, the three-stage process, was developed to vary stoichiometry and composition profile at will.

The massive deployment of PV foreseen in the years 2020–2040 could lead to 1.5 to 2.5 TWp of worldwide installed PV capacity in 2030, with a demand of 20–150 GWp/y and around 30% of the market taken by thin films [2, 4]. As observed a few years ago (2005–2009) in the case of the disruptive CdTe thin-film technology, the CIGS technology and production has quickly increased since 2010. Large-scale production has started at the GW per year level in 2011, with Solar Frontier (1 GW in 2011), and other companies are also ramping up their production at levels >100 MW/y (i.e. Avancis, Solibro, Honda, Manz). It is now clear that CIGS technology is on the verge of an industrial production breakthrough. Cost competitiveness, which is related to inherent aspects of thin-film solar cell production (reduced number of steps), together with high efficiencies, will now benefit from large manufacturing scale effects. This has proved to be extremely efficient during previous years in the case of silicon technologies. As a consequence, the CIGS technology is in an excellent position to play an important role in the large-scale, multi-GW, industrial deployment of photovoltaics that is occurring now.

The question of the sustainability of the yearly production of 20–150 GWp of PV was raised recently [19]. In particular, the worldwide mining and refining capacity of a few critical elements could be the main problem for the photovoltaic industry in the years to come. Each PV technology has one or more limiting elements, such as silver for crystalline silicon, indium and gallium for CIGS, Tellurium for CdTe, Ruthenium for dye-sensitised solar cells, silver and indium for thin-film silicon, germanium, gold, indium for III-V photovoltaics [20, 21]. The chalcogenide technologies are particularly vulnerable to In, Ga, and Te supply, which has been noted as critical or near critical by the U.S. Department of Energy [9] or the European Commission [10].

Crystalline-silicon PV seems suitable for supplying the demand, but one should keep in mind that the basic assumption used for the third and fourth column of this table is that all worldwide production of silver is used for crystalline PV. The fifth column shows the forecast 2030 demand for Ag, Te, In with respect to the 2008 production. It appears that the

Table 7.1 Production limitations of the PV technologies. Data extracted from [7, 8, 10–12, 14].

PV technology	Limiting element	Max installed capacity based on the limiting element reserves (1.5–2.5 TWp demand)	Max yearly production based on annual production of limiting material (20–150 GWp demand)	Ratio (demand 2030) / (production 2008) of limiting element
c-Si	Ag	5–15 TWp	305 GWp	20
CdTe	Te	0.1–0.8 TWp	2.3–40 GWp	50
CIGS	In	0.1–0.6 TWp	13–26 GWp	3
DSSC*	Ru, In (TCO)	0.06–0.9 TWp	2.2 GWp	n.a.

*DSSC: Dye-sensitised solar cells.

demand will grow enormously. If we focus on In, in 2030 the total demand (including PV and other uses) could be 3 to 7 times the 2008 production level [19, 22], whereas, as In is a byproduct of Zn extraction, the elasticity of the production with respect to demand is low. One author suggests that CIGS and CdTe technologies will not be impeded by resources problems, but he does not take the increase of other sources of demand for In and Te into account [12].

There is, thus, a threat to the brilliant perspectives of the CIGS industry, which is the scarcity of indium (and gallium). The indium production per year is about 500 tons, resulting from the extraction of indium from zinc mining. About 30 tons of indium are presently needed for producing 1 GW of CIGS photovoltaics. With the present technology and considering also the competition with other key indium utilisations in electronics, the limit would be somewhere between 10 and 20 GW per year, with also an increased pressure on costs [19]. This can hamper the production level in the future and thus hinder drastically, or even destroy the chances for further development of the GIGS technology for large-scale production.

The sustainability of PV production is therefore a real question, and the development of novel PV technologies based on abundant and preferably nontoxic elements would alleviate the pressure on all PV technologies in terms of resources.

In the course of this chapter, we will first consider CIGS technology and then we will turn to the new kesterite compounds.

7.2 CIGS

From early investigation [18] to the present large-scale development of CIGS, many insights were gained on the complex materials and devices properties. The present review, while giving an overview of the field's development cannot offer an exhaustive description of the enormous body of work that has been done. For details on specific aspects, the reader could also consult several reference papers such as, for example [23], especially on optoelectronic analysis of the devices, and the book [24] for a very comprehensive review. A good presentation of the processes involved in making the devices can be found in [25, 26].

7.2.1 Device Fabrication

7.2.1.1 Cell and Module Technology

CIGS based cells consist of a stack of thin-film layers deposited on a rigid (such as soda-lime glass (SLG)) or flexible substrate. The complete layer sequence of a Cu-chalcopyrite-based solar cell device is shown in Figure 7.2.

The cell preparation starts with the deposition of a typically 500–1000 nm thick Mo serving as the back contact for the solar cell. The p-n junction of the solar cell is formed by the p-type CIGS absorber and an n-type semiconductor window layer. The absorber is deposited on top of the Mo back contact electrode. This layer has a thickness of 2–2.5 μm. The heterojunction is then completed by chemical-bath deposition (CBD) of CdS (typically 50 nm) and by the sputter deposition of a nominally undoped (intrinsic) i-ZnO layer (usually of thickness of 100 nm) and then heavily doped ZnO layer, commonly by Al (ZnO:Al). The back (e.g., Mo) and front contact layer (e.g., a i-ZnO/ZnO:Al bilayer) facilitate the charge

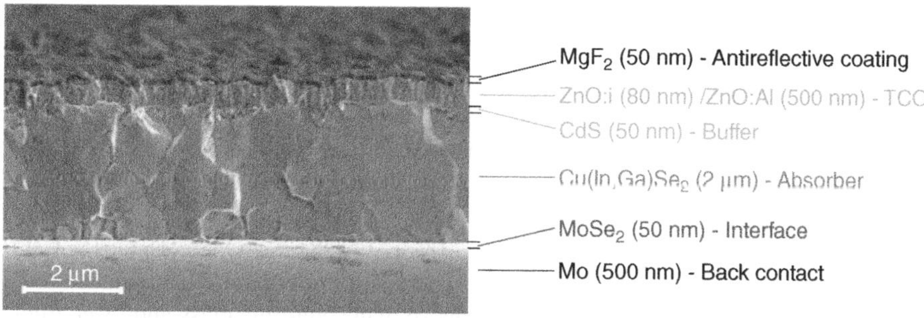

Figure 7.2 Scanning electron microscope image of the cross section of a coevaporated $Cu(In,Ga)Se_2$ solar cell with antireflection coating. (Reproduced with permission from Chirilă, 2011. Copyright © 2011, Nature Publishing Group.)

transport from the p-n junction. As ZnO:Al has a bandgap energy of 3.2 eV it is transparent for the main part of the solar spectrum and therefore is denoted as the window layer of the solar cell. This CIGS solar cell in the so-called "substrate configuration" is illuminated from the ZnO:Al side. For laboratory cells, a Ni-Al grid is deposited on top of the i-ZnO/ZnO:Al window layer to decrease resistive losses. Following the deposition and processing of the active layers of CIGS cells and modules, an antireflecting coating is deposited. Finally, encapsulation is done for mechanical and corrosion protection.

7.2.1.2 Substrate

The preparation of CIGS-based solar cells starts with the deposition of Mo on a glass or flexible substrate [27–29]. The properties of the Mo film and the choice of the substrate are of prime importance for the final device quality. The substrate material for CIGS-based solar cells has to fulfill several requirements:

- resistance to temperatures up to 550 °C;
- no reaction with Mo and absorber layer [30];
- moisture barrier and substantial stability on aggressive environments;
- smooth surface;
- mechanical strength;
- similar thermal expansion coefficient that Mo and CIGS ($8–11 \times 10^{-6}$/K) [31].

7.2.1.2.1 Glass Substrate

Glass is the most common substrate material used by manufacturers of CIGS modules. It is available in large quantities at low cost and presents all the above requirements. However, the choice of the glass is of primary importance for the final device quality. Indeed Na diffuses from the glass through the Mo film into the growing absorber, and plays a crucial role in the device performance. Two types of glass are mainly used:

- Soda-lime glass: This has a thermal expansion coefficient of 9×10^{-6}/K [32], which provides a good match to the CIGS and Mo films. It contains 73% silica, 14% Na_2O, CaO 9% (by mass) as well as other elements in smaller quantities (<1%). These provide sources of alkali impurities such as Na, which diffuses into the Mo and CIGS films during processing. Sodium is a critical component for cell performance. It is reported that the Na diffusion from the soda lime glass into the CIGS film enhances the grain growth of the absorber layer and is also effective in increasing the hole concentration. However, a process that provides a more controllable supply of Na than diffusion from the glass substrate could be preferred for stability. Thus, many manufacturers have chosen to deposit a blocking barrier layer between the glass and molybdenum such as SiO_2 or Al_2O_3, and provide a control source of Na.

- Borosilicate glass: This contains 70% silica, 10% B_2O_3, 8% Na_2O, 8% K_2O, 1% CaO. It has the advantage of containing less sodium, which helps to control the amount of sodium in CIGS absorbers. However, it is more expensive than soda-lime glass and its coefficient of thermal expansion is very different from that of CIGS (4.6×10^{-6}/K).

7.2.1.2.2 Flexible Substrate

The glass substrate has the disadvantage of being heavy and fairly expensive. There is interest in the development of CIGS solar cells on flexible low-mass substrates for novel applications such as space, buildings, etc. Moreover, flexible substrates are suitable for roll-to-roll processes, which could lower production costs. Such substrates can be metal foils or plastic.

7.2.1.2.2.1 Metallic Substrate Metal films have a heat-resistance advantage at the temperatures of synthesis of CIGS films (400–500 °C). They have the disadvantage of requiring a barrier layer to prevent diffusion of metal atoms in the absorber [33]. Martínez et al. [33] proposed the use of Al, Ti, stainless steel or nickel alloys (Kovar) as substrates. They are then covered with SiO_2, acting as a barrier layer, and molybdenum. Other types of barrier layer can be used, such as Al_2O_3 and Cr, but the inconvenience is that they can also diffuse into the $Cu(In,Ga)Se_2$ [34].

Only titanium and Kovar can lead to a molybdenum layer without any fissures after annealing. Efficiencies of 17.5% [35] on stainless steel and 17.7–17.9% on titanium [36, 37] have been achieved. ZSW [38], Flisom, Solopower, and others have already realised completed modules on flexible metallic substrates.

7.2.1.2.2.2 Polymers Plastic substrates (polyimide, Upilex, Capton) cannot resist temperatures above 500 °C, but they have the advantage of allowing a process comprising fewer steps (no barrier layer) and have well-defined surfaces [34]. Efficiencies up to 20.4% [39, 40] have been achieved on polyimide substrates. The deposition of Mo on this type of substrate tends to bend it, but this problem can be circumvented by depositing a layer of material (insulation, titanium or molybdenum) on the back [41]. In these solar cells the Na is added either via evaporation of a NaF layer prior to the absorber deposition [42] or by postdeposition and annealing of a NaF layer on the absorber layer [43].

7.2.1.3 Mo Back Contact

$Cu(In,Ga)Se_2$ solar cells are mainly grown on molybdenum coated substrates. Mo is sputtered on soda-lime glass or flexible substrates (polyimide or stainless steel). The intensive usage of Mo as a back contact, compared to other alternatives [44,45] is due to the in situ formation at the Mo/$Cu(In,Ga)Se_2$ interface of a $MoSe_2$ layer that favours ohmic contact [41, 46–48]. Mo is inertness to the high temperature and the reactive atmosphere used during the

CIGS deposition. Mo deposition is also crucial to control the incorporation of sodium, diffusing from soda-lime glass, to the absorber [49]. Sodium is an important element that diffuses to the grain boundaries, catalyzes oxygenation and passivation of Se vacancies [50], favours the formation of $MoSe_2$ (or MoS_2) [30, 34, 51] and plays a role in the growth of $Cu(In,Ga)Se_2$.

The thickness of the Mo layer is determined by the resistance requirements that depend on the specific cell or module configuration. Typically, the Mo back contact presents a sheet resistance of about 0.2 ohm/sq, and provides good adhesion of the semiconductor to the substrate [52]. Most importantly, in cells without a diffusion barrier, it regulates the diffusion of Na from the SLG to the $Cu(In,Ga)Se_2$.

The two modes of sputter deposition of Mo most commonly used are: deposition by radio-frequency (RF sputtering [46]) or magnetron (DC magnetron [42]) under an argon atmosphere. The deposition by radio frequency avoids the accumulation of charges at the surface layer to be deposited, while the magnetron deposition reduces the pressure of the working gas, which increases the quality of the deposited coatings [53].

When the pressure is varied during the DC magnetron deposition [54]

- The deposition rate varies slightly and is around 9.5 Å.
- There is a low variation of the resistivity of the molybdenum layer (between 0.20 to 0.25 Ω cm) for deposition pressures between 0.2 and 2 mTorr, which can reach 4.1 Ω cm for a deposition pressure of 20 mTorr;
- The adhesion of the films is acceptable for a pressure higher than 2 mTorr (successful test of the tape);
- Compressive stress decreases from 2 mTorr.

Moreover, when produced by DC sputtering, the layer properties are varied primarily by changing the sputter power, resulting in different kinetic energies of the sputtered Mo atoms arriving at the substrate. For a high sputter power (e.g., 2 kW), the adhesion of the Mo film to the SLG substrate is enhanced compared with low-power (e.g., 0.3 kW) films. However, Mo films at high power exhibit a larger resistivity compared with low-power films. Therefore, to achieve both good adhesion and good conductivity, commonly, a Mo multilayer is applied as back contact. First, a thinner adhesive layer is deposited at high argon pressure (6 Pa, 10 min), and/or high sputter power (e.g., 100 nm at 2 kW), then a second thicker and low-resistance layer (1.12–2.4 Pa, 120–150 min) is deposited at low sputter power (e.g., 850 nm at 0.3 kW). Resistivities of about 0.08 Ω cm can be obtained.

7.2.1.3.1 $MoSe_2$ Layer

During selenised absorber deposition a $MoSe_2$ film forms at the Mo surface [55]. $MoSe_2$ may be considered as a buffer layer between Mo and CIGS, which leads to a better energy-band alignment between these two layers and thus reduces recombination at the interface [48, 54–56]. Moreover it permits formation of an ohmic contact [48], which makes the $MoSe_2$ formation an important issue. $MoSe_2$ is a layered semiconductor with p-type conduction, a bandgap of 1.3 eV and weak van der Waals bonding along the *c*-axis. In fact, $MoSe_2$ is regularly found in a hexagonal phase (Figure 7.3), which consists of Se-Mo-Se sheaths oriented perpendicular to the c-axis. From this anisotropic crystal structure, possible applications arise for $MoSe_2$, e.g., as a lubricant [57].

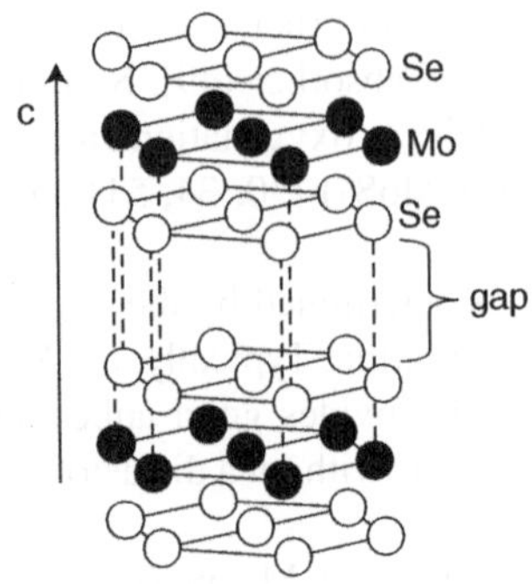

Figure 7.3 Anisotropic crystal structure of hexagonal 2H-$MoSe_2$. It consists of Se-Mo-Se sheaths that are separated from each other by a gap. The *c*-axis of this crystal structure is perpendicular to the Se-Mo-Se sheaths [58].

The crystal orientation of the $MoSe_2$ grains is important for the fabrication of solar cells. Due to the layer structure of the $MoSe_2$ compound, if the *c*-axis is oriented perpendicular to the surface of the Mo layer, the CIGS film would tend to delaminate from the Mo layer due to ease of interlayer cleaving as well as leading to unfavourable electronic transport. Fortunately, the *c*-axis is found to be parallel with, and the van der Waals planes thus perpendicular to, the interface.

Abous-Ras et al. [59, 60] showed that the *c*-axis direction is controlled by the annealing temperature. If the temperature is below 550 °C, the *c*-axis is perpendicular to the substrate and the thickness of $MoSe_2$ is less than 20 nm. When the temperature exceeds 550 °C, the *c*-axis is parallel to the substrate plane and the thickness of $MoSe_2$ becomes important (micrometre). In this case, its formation rate seems to change almost linearly with time, but with a higher speed for lower selenisation time (less than an hour when the temperature is 580 °C).

7.2.1.3.2 Alternative Back Contacts

Various alternative metal contacts to p-type CIGS were examined by Matson et al. [61] (Au, Ni, Al, Ag) concluding that only Au and Ni ensure an ohmic contact. Orgassa et al. [56] fabricated CIGS solar cells with different back-contact materials (W, Mo, Cr, Ta, Nb, V, Ti, Mn) and found that W and Mo contacts provide the best CIGS/back-contact interface passivation.

In the case where a transparent back contact is necessary, such as superstrate or bifacial solar cells, transparent conductive oxides such as ZnO:Al or ITO were used instead of molybdenum [62, 63]. However, in these cases it is necessary to deposit a thin layer of Mo or $MoSe_2$ on top to ensure the good ohmic contact [64]. Efficiencies up to 15.2% were obtained.

In the case of very thin layers of absorber (<1 micrometre), it is possible to use an additional back reflector, such as silver, gold or white paint, leading to significantly increased short-circuit current [65, 66]. Recently Jehl et al., showed that the replacement of Mo by a Au back contact does not degrade the other parameters of the cell, as good ohmicity on CIGS is achieved [67]. An important photocurrent increase compared with regular Mo back contact

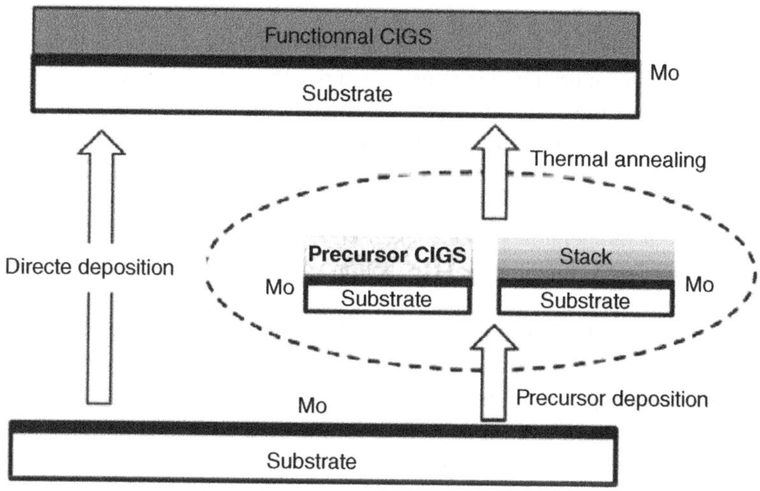

Figure 7.4 Flow chart for chalcopyrite thin-film solar cells deposition. On the left side is the classical one-step coevaporation process. On the right side is the two-step strategy, starting with a precursor preparation by sputtering, electrodeposition, screen printing, of a mixed chalcopyrite layer or stacked layers.

solar cells is achieved by the enhanced light-trapping effect due to the back reflector, leading to an absolute efficiency increase of +2.5% for a CIGS thickness of 0.4 μm.

7.2.1.4 CIGS Absorber

A wide variety of thin-film deposition methods has been used to deposit CIGS thin films. However, the performance of cells varies greatly depending on the processes used to synthesise the CIGS absorber. These processes can be divided into two categories as illustrated in Figure 7.4.

1. The first is a one-step deposit of a hot multinary: growth can be controlled during the process. The coevaporation technique is the most common process that was originally developed by Boeing [68] and then modified and advanced by NREL [69]. It is currently used by Würth Solar and Global Solar.
2. The second approach is a two-step process where first a precursor layer is prepared using low-cost and/or low-temperature methods that facilitate uniform composition. This precursor layer can be an already intermixed CIGS matrix or consist of stacked layers either elemental or binary. Then, the films are annealed in a S or Se atmosphere, at 400 to 600 °C. The optimisation of both properties (composition, structure) and the thermal annealing step (temperature, atmosphere, duration) should lead to final electronic p-type quality of the materials comparable to those achieved with the one-step process. The two-step based process was originally developed by ARCO Solar Inc. in 1981 [70] based on a selenised process.

7.2.1.4.1 Coevaporation Processes

The highest efficiency devices have been deposited by thermal coevaporation from elemental sources [12]. The coevaporation technique requires the simultaneous and homogeneous thermal evaporation of the elements copper, indium, gallium, and selenium in the correct proportions for forming high-quality CIGS. All the constituents, Cu, In, Ga, and Se, can be simultaneously delivered to a substrate heated to 400 to 600 °C and the $Cu(InGa)Se_2$ film is formed in a single growth process. This is usually achieved by thermal evaporation from elemental sources at temperatures greater than 1000 °C for Cu, In, and Ga. The advantage of the evaporation route is that material deposition and film formation are performed during the same processing step.

While a variation of the In/Ga ratios during the deposition process leads to only minor changes in the growth kinetics, variation of the Cu content strongly affects the film growth.

A reproducible coevaporation process requires good control of the elemental fluxes from each evaporation source. While the evaporation rates from each source can be controlled simply by the source temperature, this may not give good reproducibility, especially for the Cu source that is at the highest temperature. Consequently, direct *in situ* measurement of the fluxes is often used to control the evaporation sources [32].

7.2.1.4.1.1 The Single- and Bilayer Process The single-layer process is the simplest technique to produce CIGS layers, since the elements are evaporated simultaneously at constant evaporation rates and substrate temperatures. The resulting CIGS films are always Cu poor. Such a process was applied by Würth Solar.

Advanced preparation sequences called "the bilayer of the Boeing process" starts with excess Cu flux for improved crystallite size and quality and finishes with excess In flux to react with the Cu_xSe surface phase [69]. This bilayer process yields larger grain sizes compared to the constant rate (single-stage) process.

7.2.1.4.1.2 The 3-Stage Process The record $Cu(In,Ga)Se_2$ devices have absorbers that are made by three-stage coevaporation [12, 71]. This is the state-of-the-art deposition technique at the laboratory scale, first introduced in 1994 [72] as an evolution from the previous "Boeing" two-stage process [68]. In the industry, slightly simpler deposition routines are followed [52, 73]. Coevaporation is a physical vapor deposition technique, where Cu, In, Ga and Se are evaporated in a high-vacuum chamber from elemental sources onto a heated substrate (Figure 7.5). The deposition begins with the formation of $(In,Ga)_2Se_3$. Then, Cu is evaporated until a copper-rich compound is reached. The temperature of the substrate has to be brought to 500–600 °C during this phase to enable a good crystallisation with formation of large grains, through the mediation of a liquid Cu_xSe phase that helps mass transport and thus grain growth. The changes in $Cu(In,Ga)Se_2$ stoichiometry can be detected by the changes in thermal emissivity, and thus thermal power needed to maintain the substrate at high temperature. In order to avoid the formation of the highly conductive Cu_xSe binaries on the surface, a third stage is added where In and Ga are evaporated in the presence of a Se overpressure. Se overpressure is favourable during the whole absorber growth in order to prevent the formation of selenium vacancies, which are active electronic defects [74]. The three-stage process naturally leads to a slightly Ga-rich compound near the back substrate and front surface. This Ga gradient is due to the low mobility of Ga compared to In and Cu, which prevents Ga from outdiffusing from the bottom copper-poor phase to the surface copper-rich phase during phase two [75,76]. This gradient naturally creates a front and back

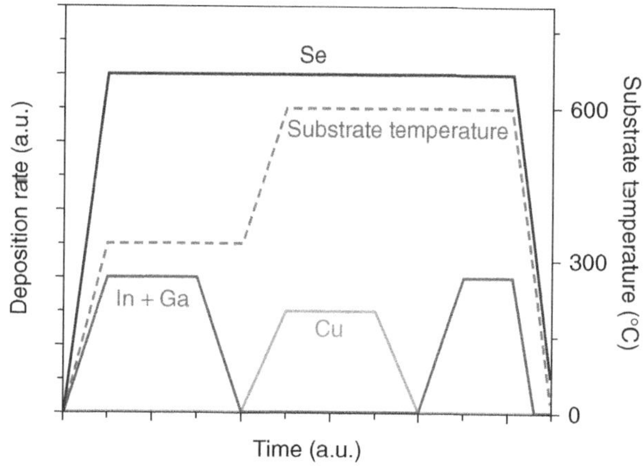

Figure 7.5 Sketch of the three-stage process. Values are given for information.

surface field that are favourable for the device electronic properties. Figure 7.5 displays schematically the sequences of the three-stage process with indications on metal fluxes and substrate temperature.

The three-stage process is a stationary process that is difficult to control, and thus a modified inline process is usually preferred at the industrial scale. The substrate is moved through deposition chambers and is subjected to elemental fluxes of various compositions, basically Cu-rich in a first part and then Cu-poor at the end of the growth [73]. Thus, a Ga gradient is also seen in industrial samples.

7.2.1.4.2 Other Deposition Methods Based on the Two-Step Process

If coevaporation is the state-of-the-art deposition technique for high-efficiency devices, other techniques are in development at the laboratory as well as at the industrial scale. All these methods are based on the two-step process. The two-step process has many variations in both the precursor deposition and the Se reaction step [52]. This approach permits the fabrication of a wide variety of absorbers. Currently, the CIS-based alloy system as a p-type absorber material contains a selenide like $CuInSe_2$ and $Cu(InGa)Se_2$ and a sulfide like $CuInS_2$ and the mixed one like $Cu(InGa)(SeS)_2$ and $Cu(InGa)Se_2$ with a thin $Cu(In\text{-}Ga)(SeS)_2$ surface layer). Moreover, the two-step process generally splits CIGS formation into two stages, one in which the precursor is deposited and one in which the precursor is converted into CIGS. Table 7.2 sums up the different industrial processes and corresponding conversion efficiencies.

Such processes consist of the deposition of a precursor material, followed by thermal annealing in a controlled reactive or inert atmosphere for optimum compound formation via the chalcogenisation reaction. This is commonly referred to as selenisation of or sulfurisation of stacked metal or alloy layers. The metals and alloys can be deposited by a variety of methods that involve vacuum or no vacuum. Decoupling of deposit precursor (stoichiometry control) and the annealing step (grain-growth control), allows a better control of stoichiometry and homogeneity of this technology. However, the use of gallium makes it difficult to control film growth. During selenisation, it is well known from studies on sputtered Cu–In–Ga layers that Se reacts preferentially with In, leading to segregation of Ga [77, 78].

Table 7.2 Overview of the efficiencies of different Cu(In,Ga)Se_2 fabrication processes in the laboratory and industry. Cell and submodules differ from module due to the smaller area. In the absence of an indication the efficiencies are given as total module efficiency. If the abbreviation AA is used, the efficiencies are relative to the module aperture area, otherwise the efficiencies are relative to the module total area. When the sign (?) is present, no precision could be found on the efficiency definition.

Company / Laboratory	Deposition method	Record cell	Record module	Reference
Flisom	Coevaporation	18.7% flexible		[79]
NREL	Coevaporation	19.9%		[71]
Solibro	Coevaporation	17.4% (AA) (16 cm^2)	13.4% (14.7% (AA))	[80]
Würth Solar/Manz	Coevaporation		15.1% (AA)	[81]
ZSW	Coevaporation	20.3%		[12]
Nexcis	Electrodeposition	14.9%	11.8%	[17]
SoloPower	Electrodeposition	13.8%	13.4% (AA) flexible	[16, 82]
Nanosolar	Inkjet printing	17.1% (AA)		[83]
Avancis	Sputtering	15.8% (AA) (30 × 30 cm^2)	12.6% (14.2% AA)	[14, 84]
Heliovolt	Sputtering	14% (?) 11.7% (?) (30 × 30 cm^2)		[85]
MiaSolé	Sputtering		15.7% (AA)15.5 (AA) flexible	[11]
Solar Frontier	Sputtering	17.8% (AA) (30 × 30 cm^2)	13.4% (14.5% AA)	[15]
Sulfurcell	Sputtering		13.1% (AA)	[86]

7.2.1.4.2.1 Vacuum Depositions The most common vacuum processes are sputtering and thermal evaporation (as in Table 7.2). Sputtering is an attractive process because it is easily scalable using commercially available deposition equipment and can provide good uniformity over large areas with high deposition rates. In this case sputtering of sequential copper, indium, gallium, followed by a rapid thermal annealing (RTP) in the presence of selenium or sulfur is achieved at temperatures between 400 to 500 °C for 30 to 60 min for the best device quality material. This method is used at the industrial scale by Solar Frontier, Avancis and Sulfurcell [52] and has resulted in cell efficiencies of 17.8% [15].

7.2.1.4.2.2 Nonvacuum Techniques Nonvacuum methods have also been developed to deposit the precursors [87]. The stack of metallic layers of In, Ga and Cu can be deposited by electrodeposition, preferably on a conductive stainless steel substrate, such as for Nexcis or Solopower. Inkjet-printing approaches of Cu(In,Ga)Se_2 nanoparticles inks are also developed by Nanosolar for example. Pure solution deposition also lead to high-efficiency devices [88]. Nonvacuum techniques for CIGS deposition offer potential reductions in capital cost and many such techniques have been investigated [150]. A large variety of nonvacuum deposition approaches have been developed and reviewed [44, 87] and they can be divided into the following categories depending on the deposition method and the scale of mixing of the precursor materials:

1. electrochemical process;
2. particulate and solution-based process.

1. ***Electrodeposition process***

Electrodeposition of semiconducting materials presents interesting characteristics for large-area, low-cost and generally low-temperature and soft processing of materials. Extensive work on chalcopyrite electrodeposition has yielded fundamental understanding of the process and high-performance devices. Several review papers [45, 89–91], have been devoted to the electrochemical preparation of CIGS and making electrodeposition the most intensely studied nonvacuum deposition method for CIGS.

The electrodeposition and electroless deposition of CIGS from a single solution is difficult due to the wide differences between the standard potentials of Se ($E^{\circ} = +0.75$ V vs. NHE), Cu ($E^{\circ} = +0.34$ V vs. NHE), In ($E^{\circ} = -0.34$ V vs. NHE) and Ga ($E^{\circ} = -0.53$ V vs. NHE). In addition, the electrochemical behaviour of Se is complex; it exhibits several oxidation states ($+6$, $+4$, 0, -2), leading to complex reactions.

Due to solubility issues, most electrodeposition baths have a low pH, though some studies use strongly alkaline solution. In most cases, the metallic elements are introduced as sulfate or chloride salts and more rarely as nitrate salts, while the Se precursor is introduced as selenium oxide or selenous acid. The addition of buffering agents, such as ethylene diamine, is sometimes mentioned in the literature. However, their role is not clearly established and they may act as complexing agents rather than pH buffers [92]. The incorporation of Ga in a one-step process seems to be more difficult because the redox potential of Ga is low and the formation of Ga oxide is very difficult to avoid.

To overcome the thermodynamic and chemical difficulties involved with electrodeposition of CIGS, several strategies have been developed:

1. coelectrodeposition of Cu-(In, Ga)-(Se, S) from a single electrolyte;
2. electrodeposition of stacked layers of metals, alloys or binary selenides;
3. electroless and chemical-bath deposition.

As with other nonvacuum techniques, these layers are normally considered to be precursors and are subjected to a high-temperature annealing step to form CIGS after deposition, often in the presence of a chalcogen source.

Coelectrodeposition of Cu-(In, Ga)-(Se, S) from a single electrolyte This approach intends to try to form directly CIS precursor films by using a unique bath containing the three elements, as initiated by Bhattacharya [93]. This is by far the most investigated case due to the fact that such a process, often called one-step electrodeposition is the simplest in principle, involving only one electrochemical process. The counterpart is that the electrochemical aspects are becoming even more complex, with the possibility of forming either the elements in their elemental form, or as binary compounds in addition to the desired ternary CIS phase [87].

The best coelectrodeposited cells are obtained for the ternary compound Cu-In-Se where Se is exchanged to S after annealing under a S atmosphere with the best efficiencies of 11.4%

[91]. The incorporation of Ga in a one-step process seems to be more difficult because the redox potential of Ga is low and the formation of Ga oxide is very difficult to avoid.

Electrodeposition of stacked layers of metals, alloys or binary selenides To overcome the intrinsic difficulty of In and Ga incorporation, alternative strategies to coelectrodeposition have been developed, often involving the deposition of stacked elemental layers or deposition of alloys, followed by a selenisation or sulfurisation treatment to provide all of the chalcogen [90].The simplest way is thus to deal with the preparation of elemental layers from single element solutions for preparing elemental layers (Cu,In,Ga,Se) and to change from one bath to the other for the preparation of precursor films with stacked layer structure. The thickness of each layer being easily controlled by coulometry, the overall composition of the precursor films is thus also controlled. Such an approach has been used for a long time for preparing copper/indium bilayers that are then treated thermally with a selenium atmosphere to form CISe [92].

Other processing employed during the last decades include; i) electrodeposition of thin Cu and In layers forming a Cu/In precursor stack and reacting this metallic stack with gaseous Se species to form the compound, ii) electrodeposition of a Cu/In/Se stack on a substrate and rapid thermal annealing of the stack to form CIS, iii) electrodeposition of In-Ga, Cu-Ga or Cu-In-Ga metal alloys to form precursor layers and reacting these precursor layers to form the compound, iv) electrodeposition of In-Se and Cu-Se on a substrate forming a stacked precursor such as a Cu/In-Se/Cu-Se structure and annealing the structure in an inert atmosphere to form CIS [91].

Actually, the best efficiencies are obtained by a sequential electrochemical-based process followed by rapid thermal heat treatment enabled synthesis of 13.8% CIGS-based solar cells [92].

2. ***Other methods***

Direct liquid coating describes a variety of techniques characterised by distributing a liquid or a paste to the surface of a substrate, followed by the necessary thermal/chemical treatments to achieve the desired phase. The deposition media used are solutions or particle (usually submicrometer size) suspensions of metal oxide, organic and inorganic compounds, including metal chalcogenide species. The deposition techniques used are mainly printing and spin coating, although any standard process such as spraying, dip coating, doctor blading, ink-jet printing or slit casting can be applied. The precursor film is then transformed by a thermal treatment, generally in a chalcogen-containing atmosphere, to the final crystalline layer. This approach permits the use of low-cost and high-throughput equipment and the deployment of large-scale production facilities with lower capital investment. Although many of the methods discussed are under laboratory development, there are already industrial start-ups employing these promising methods for future large-scale photovoltaic production [94]. Solar cells with efficiencies up to 14% have been reached by these methods [95,96].

7.2.1.5 Buffer Layer

7.2.1.5.1 CdS Buffer Layers

The highest and most reproducible solar-cell efficiencies, independent of the absorber used, are obtained by use of the classical chemical-bath deposited (CBD) CdS buffer layer. Obviously, the material CdS does not lead to higher efficiencies alone, the chemical-bath deposition also has a very important part. In fact, CIGS cells with CdS buffers deposited by physical vapor deposition (PVD) have always shown significantly lower efficiencies than cells with CBD-CdS buffers.

CBD CdS involves the precipitation from the solution of CdS on the absorber. With this method, a thin and conformal layer of buffer layer (30–60 nm) is usually obtained after 5–10 min at temperature ranging from 60 to 70 °C, using an aqueous ammonia solution in the presence of thiourea and cadmium salt (such as $Cd(C_2H_3O_2)_2)_4$, or $CdSO_4$).

The overall reaction of CdS formation can be described as [97]

$$Cd(NH_3)_2^{+4} + SC(NH_2)_2 + 2OH^- \rightarrow CdS + CN_2H_2 +_4 NH_3 + 2H_2O \qquad (7.1)$$

It has been suggested not only that buffer layers deposited by the CBD process function as a protection of the p-n junction from plasma damage during subsequent ZnO sputtering and prevent undesirable shunt paths, but also that chemically deposited buffer layers permit a conformal and pinhole-free coating of the absorber by the CdS layer. Moreover, when the glass/Mo/absorbers stack is immersed into the chemical bath for the buffer deposition, the CIGS layer surface is probably also subjected to beneficial chemical etching of the surface, e.g., the ammonia used in most CBD recipes is thought to be crucial for the cleaning of the absorber surface by removal of oxides and other impurities [98]. In addition, n-type doping of the CIGS region close to the surface by Cd from the CBD solution has been reported [99, 100].

7.2.1.5.2 Cd-Free Buffer Layers

Because of both environmental reasons and the fact that the common CdS layer with a bandgap energy of about 2.4–2.5 eV limits the level of optimum performance of the cells, especially in the short-wavelength range, one of the major objectives in the field of CIGS technology remains the development and implementation in the production line lines of Cd-free alternative buffer layers.

The current understanding is that candidates for alternative buffer material should hold five common properties: (a) The material should be n-type in order to form a pn junction with the absorber layer. (b) The bandgap should be wide enough for limited light absorption. (c) The process and material choice of the buffer layer should provide an alignment of the conduction band with the $Cu(In,Ga)(S,Se)_2$-based absorber and to the undoped ZnO. (d) The process for deposition must have a capability to passivate the surface states of the absorber layer and (e) the deposition should be highly conformal.

The development of Cd-free devices started in 1992 and intensively continued to the current efficiency level of 19%, as shown in different review papers [101, 103, 104]. A general conclusion of the investigations during the last decade suggests that the most relevant materials are films based on In_2S_3, ZnS, $Zn_{1-x}Mg_xO$, $Zn_{1-x}Sn_xO_2$ and their derivatives using the following deposition techniques: chemical-bath deposition (CBD), atomic-layer deposition (ALD), ion-layer gas reaction (ILGAR), sputtering and evaporation (PVD), and ultrasonic spray pyrolysis (USP).

Likewise, a comparison of the efficiencies of Cd-free-based solar cells shows that for the most relevant materials mentioned above, the ability to reach efficiencies equivalent to or higher than the corresponding CdS reference cells is highly dependent on the deposition technique used (Figure 7.6). Most often, a clearly poorer photovoltaic performance of solar

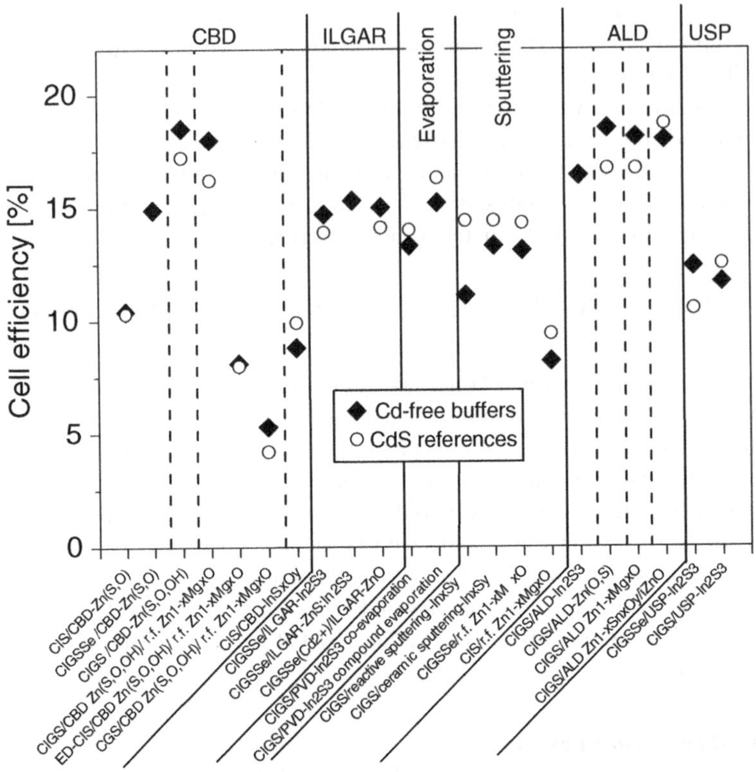

Figure 7.6 Summary of the best chalcopyrite-based solar-cells efficiencies (1 cm^2) realised with different Cd-free buffer-layer materials and deposited by different methods compared to their reference devices containing CBD-CdS [101, 102].

cells with vacuum-deposited buffer layers is found compared with solar cells with chemical-based buffer layer such as CBD, atomic-layer deposition (ALD) or ion layer gas reaction (ILGAR). However, whereas for CBD-CdS buffers high and reproducible efficiencies are obtained regardless of the absorber used, Cd-free cells often reach their highest efficiencies after light soaking and/or heat treatments.

7.2.1.6 i-ZnO/ZnO:Al Window and Ni-Al Grid

In order to complete the device, window layers are deposited on top of solar cells. In general, the window has a much wider bandgap than the absorber. This moves the absorption maximum into the cell and away from the surface, thereby minimising the influence of surface recombination. The window must thus be highly transparent, create an interface to the absorber with low recombination losses, and limit the influence of locally distributed shunts and performance fluctuations. In addition, it must have a low sheet resistance to laterally transport the current over macroscopic distances to the nearest metal contact finger or interconnect.

The most common window applied in CIGS solar cells consists of two layers. The first layer, a thin i-ZnO film, is deposited by rf magnetron sputtering and has normally a thickness of about 50–100 nm. The second layer is the actual front contact and consists of ZnO:Al, also deposited by rf magnetron sputtering. Typical ZnO:Al thicknesses are 300–500 nm.

The current collection by the window layer is commonly supported by a Ni-Al metal grid deposited on the front contact. Ni is applied to reduce the formation of a resistive Al_2O_3 barrier. The Ni and Al layers are deposited by electron-beam evaporation with thicknesses of about 50 nm and 1 μm. Although the application of a grid is beneficial for the solar cell performance, it also contributes to a reduction of the active cell area.

Depositing a film with low lateral resistance directly onto the buffer increases the negative influence of local defects such as pin holes and local fluctuations of absorber properties (e.g., bandgap) [105, 106]. This can be avoided by first depositing a thin (100 nm) ZnO film with lower conductivity, designated as i-ZnO. In the classical chalcopyrite cell, the buffer establishes the interface properties, the undoped ZnO (i-ZnO) decouples poorly performing locations, and the highly doped ZnO is responsible for lateral current transport. The requirements for the window layer can be divided into physical properties and technological requirements. Key issues for the industrial success are high deposition rates, homogeneous large-area coating, and high process stability with easy control leading to high yield. However, the deposition materials and equipment cost should be low and guarantee a high uptime. Because the window layer is deposited after the p-n junction formation the substrate temperatures during deposition are limited to about 200 °C in order to avoid detrimental diffusion.

The desired high resistivity of i-ZnO film is obtained using an undoped ZnO target, if necessary by oxygen admixture to the Ar sputter gas. Because the pure target is not conducting, r.f. sputtering is usually applied. The film thickness of i-ZnO turned out to have a minor influence on cell efficiency when a CdS buffer layer is used and may vary from a few to several hundred nanometres. However, optimisation concerning both resistivity and thickness is mandatory for monolithic module integration when i-ZnO is deposited in situ onto Mo before the subsequent TCO, thereby inhibiting the Mo-TCO interconnect.

In the case of Cd-free buffer layers the undoped buffer layer can strongly influence the performance of cells [107]. For example for CBD-ZnS buffer layers the replacement of i-ZnO by $Zn_{1-x}Mg_xO$ or direct deposition of MOCVD-ZnO:B can improve the efficiency and stability of the cells.

The window layer of the chalcopyrite-based solar cells is usually a degenerately doped semiconductor serving both as n^+-type partner in the p-n heterojunction and as a transparent contact layer (TCO). The most common TCO materials used in CIGS solar cells are highly doped ZnO, namely with gallium (ZGO), aluminum (ZAO), or boron (ZBO) deposited by sputtering or MOCVD.

For economic and ecological reasons magnetron sputtering of ZnO doped with Al or Ga is the industrial standard today, allowing for high efficiencies at moderate cost. Chalcopyrite-based cells on small areas with the highest efficiencies obtained so far are normally coated by ZnO:Al (ZAO) using r.f. sputtering of high-purity ZnO:Al_2O_3 targets. According to the needs of an industrial production r.f. sputtering was replaced successfully by DC sputtering on larger areas for (sub-)module production using low resistive hot isostatic pressed (HIP) targets [108, 109]. The breakthrough for the production was achieved by advances in (pulsed) DC power supplies and target manufacturing when much cheaper sintered ZAO targets become available. Provided there is good cooling of the targets, high power densities far above 10 W/cm^2 led to dynamic deposition rates exceeding 100 nm/min. Today's development is directed toward the implementation of (dual) cylindrical magnetrons with target tubes that increase the target utility fraction above 80% and extend the maintenance intervals dramatically.

7.2.2 Material Properties

7.2.2.1 Description of the Chalcopyrite $Cu(In,Ga)Se_2$

It is remarkable that with a few exceptions, all semiconductors of technological importance share the same structure based on the 4 electrons per atom rule, characteristic of the adamantines. From Si to GaAs and from CdTe to kesterites, the same 4 e- rule applies, yielding a tetrahedral coordination of the atoms in the structure (diamond for column IV, zincblende for II-VI and III-V, Chalcopyrite, stannite or kesterite for ternaries and higher-order multinaries).

$Cu(In,Ga)Se_2$ is a quaternary I-III-VI compound. It can be seen as the alloys of $CuInSe_2$ and $CuGaSe_2$ that have the tetragonal chalcopyrite structure. The chalcopyrite structure is derived from the cubic zincblende, with Se atoms in a face-centred cubic structure and Cu and In (or Ga) atoms occupying half of the tetrahedral holes. A $Cu(In,Ga)Se_2$ absorber is not composed of a perfect quaternary compound but most often contains ternary or binary compounds, such as $InSe_2$, Ga_2Se_3 or Cu_2Se. For more details on the crystallography of $Cu(In,Ga)Se_2$ the reader is advised to report to reference [24].

7.2.2.2 Structure and Composition

The success of CIGS in solar cells is a challenge to common sense. Making a good ohmic contact on p-type semiconductors is notoriously difficult, but it did not appear to be so between Mo and CIGS. Polycrystalline semiconductors tend to have reduced electronic performances, as compared to single crystals, due to carrier trapping or recombination at grain boundaries. This does not seem to be a problem in CIGS, where grain boundaries are essentially inactive. Semiconductors are generally very sensitive to impurities, these being generally lifetime killers. CIGS seems to be relatively immune to most impurities. Ironically, the one found to diffuse from the glass substrate (Na), was also found to improve the quality of the CIGS films structurally and electronically. Moreover, the device itself contains many elements, compounds and interfaces, all potential focus for evolution or reaction. Last but not least, record efficiencies are obtained using a material that is chemically the most complex of the materials used in the various thin-film solar cells (a/Si, c-Si, CdTe, etc.), a compound that has also metastable states and shows significant ionic conductivity [110–113].

In the list above, one will recognise factors that plague or have plagued the development of other thin-film solar cells.

Before discussing in detail this apparent paradox, we will start with the intrinsic stability of CIGS and the global chemical stability of the device (for a more detailed review, see [50]). As far as terrestrial applications are concerned, $Cu(In,Ga)Se_2$-based solar modules have proven their stability in long-term outdoor tests as well as under accelerated lifetime test conditions [114]. Actually, it is not uncommon that cells and modules show some improvement during testing [115], implying that the device as prepared is not optimised. This further suggests a positive evolution of the interfaces with time. The sensitivity of the cell to humidity either via the ZnO window [116] or via the bare CIGS absorber material [117] is mainly a concern at the production level: properly encapsulated, the modules are stable for years. This is remarkable considering the chemical complexity of a system incorporating about ten different elements in layers of a few tens of nm.

More striking is the exceptional tolerance of CIGS to defects of various origins. First, one should keep in mind that CIGS is a nonstoichiometric compound, with deviations from stoichiometry in the percent range. PV-grade material is generally obtained with a Cu content between 22 and 24%. Hence, most defects are intrinsic, with a concentration largely above that of free carriers. The material is strongly self-compensated, but in a way that does not harm its electronic properties. This is quite surprising as generally, large deviations from stoichiometry are a problem for electronic applications (e.g. in WO_x, Cu_xSe, $In2O_{3-x}$, ZnO_{1-x} etc.). The valence electron imbalance is usually compensated by free carriers, so that relatively small deviations (in the 10^{18} cm^{-3} range) already result in electronic degeneracy. In chalcopyrites though, electron imbalance from the 4 electron per atom rule, which are larger by as much as three orders of magnitude, i.e. in the 10^{21} cm^{-3} range) results in nondegenerate semiconductors. This implies that the formation of the compensating defects is energetically favoured as compared to formation of electronic carriers [118]. When this is the case, the Fermi-level position is self-stabilised between a lower and upper value, as has indeed been found in many compound semiconductors [119, 120].

Evaluation of the concentration of point defects in CIGS led to the conclusion that many of the cationic defects have concentration above 10^{18} cm^{-3} [110]. At these concentrations, they most likely form complexes, a fact also supported by ab initio calculations [121]. Parallel to that ability to accommodate intrinsic defects, and certainly not unrelated, the electronic properties of the material appear less sensitive to impurities than is usually found in other semiconductors [110, 122]. The excellent radiation hardness of this type of solar cells, as compared to traditional space cells (Si, InP, InGaP) [123, 124] was also reported. Other studies concluded that a very efficient, low temperature, defect recombination mechanism is needed to explain the exceptional radiation hardness [125, 126]. Finally, in spite of the high density of grain boundaries and other crystallographic defects, the transport properties of the free carrier did not seem to be affected, at least in the transversal direction. Accordingly, very high efficiencies could be obtained. An explanation for this feature was found involving passivation of grain boundaries in air for the p-type material [127, 128].

Surely, all this fortunate tolerance of the compound must be grounded in some of its particular properties. The most prominent one, from a solid-state chemistry point of view, is its nonstoichiometry. Though the actual homogeneity domain of $Cu(In,Ga)Se_2$ is not well known (see the discussion below), it is much larger than that of the other semiconductors, successfully used for solar energy conversion (e.g. CdTe, $Cu_{2-x}S$). The existence of an electronic material with excellent electronic properties, compatible with high-efficiency solar energy conversion, is quite puzzling in this context as most of the defects or defect complexes present must be inactive with respect to carrier recombination. Thus, their corresponding energy levels must either be shallow or altogether outside the bandgap. Again, ab initio calculations on some defect complexes have shown that trend: defect levels of the considered complexes are shallower than that of isolated point defects [121].

The second most important characteristic of CIGS is its mixed ionic–electronic conductivity. Migration of Cu is well documented in crystals and in thin films [111, 112, 129, 130]. The room temperature Cu diffusion coefficient in $CuInSe_2$ (CIS) is in the range of 10^{-13} $\pm$ 10^{-10} cm^2/s [113]. Regardless of the exact values, this implies that Cu can diffuse across the space-charge region (SCR) of the CIGS absorber in a few minutes to several days at most [50]. Such a rapidly migrating species could potentially raise serious concerns on the long-term stability of CIGS-based devices.

Overall, these facts outline the picture of a compound with a 'soft lattice' certainly related to the nonstoichiometry of the material and to its ionic conductivity. It is not surprising then that metastable electronic centres were found in this material. The most studied metastable levels in CIGS are very different from those encountered in a:Si, for instance, on two aspects at least: they lead to an improvement of the device characteristics under operating conditions and they are reversible [131–133]. The 'soft lattice' picture is central to the model proposed to account for the resilience and self-stabilisation of $Cu(In,Ga)Se_2$ [50, 110, 133]. It is because the material is highly disordered, but in a strange way: so that most intrinsic defects complexes are not lifetime killers, and because Cu is mobile, that most defects (impurities, or radiation induced) that may form could be self-passivated. This model uses the similarity between the point defects in a solid and solutes in an electrolyte, an analogy developed earlier [118]. In that picture, the buffering action of the defect pool is essentially similar to the role of a buffer solution in aqueous electrolytes, where complexes in large concentrations (e.g. acids with a concentration such that the pH is close to their pK_A) are used to fix the ionic concentration, e.g. the pH. In other words, the defect complex pool acts as a reservoir, capable of receiving or delivering electronic and ionic charge carriers so that their net amount remains fixed. Such defect reactions therefore act as both electronic and chemical buffers, by controlling both the electron concentration and the Cu concentration, respectively. In this model, a central role is given to mobile Cu as a vector of the buffering equilibrium. The detailed exposition of that mechanism being beyond the scope of this review, the readers are referred to the original work [50, 110].

7.2.2.2.1 Crystal Structure

$CuInSe_2$ crystallises in the chalcopyrite phase, with space group I42d, which forms an isoelectrical analog to the III-V binary semiconductor phase, zincblende. In the chalcopyrite phase, each anion is coordinated to two indium and two copper atoms, while each cation is tetrahedrally coordinated to four anions. The unit cell contains two chemical formula, i.e. 8 atoms. The corresponding Wyckoff positions are Cu (0,0,0), In (0,0,1/2) and (S,Se) (u,1/4,1/8). The chalcopyrite structure is depicted in Figure 7.7.

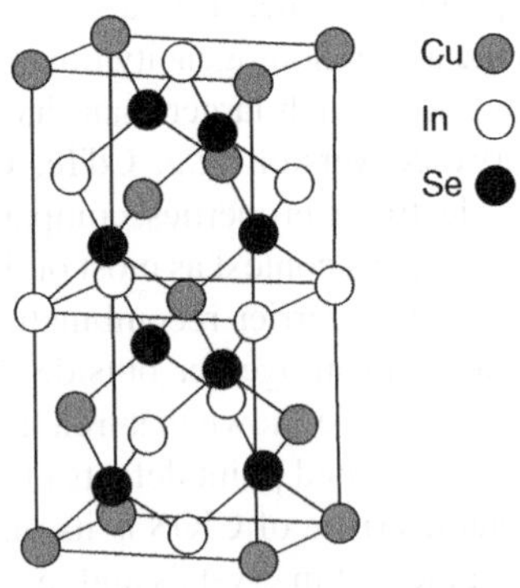

Figure 7.7 Atomic structure of $CuInSe_2$. grey spheres represent copper atoms, white spheres represent indium atoms and black spheres represent selenium atoms. (Reproduced with permission from Shafarman, 2003. Copyright © John Wiley & Sons, Ltd, 2011.)

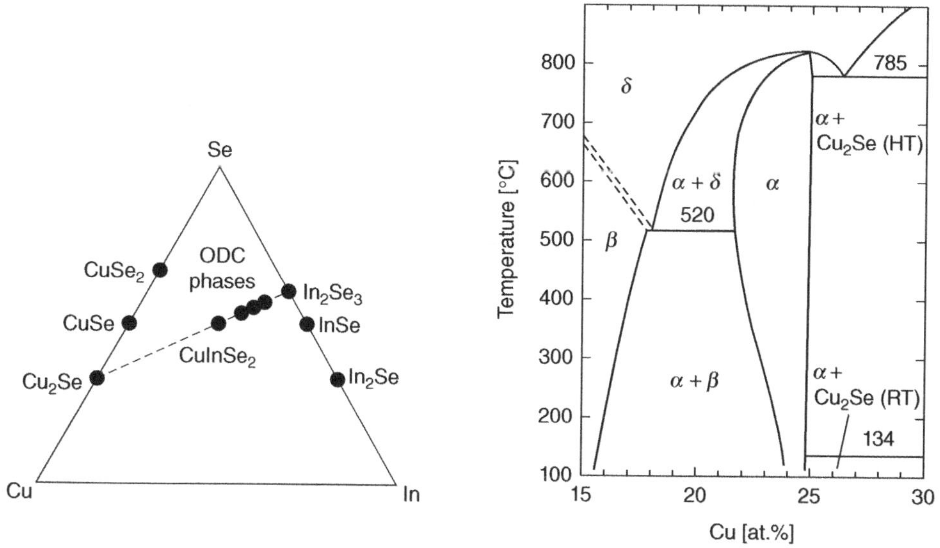

Figure 7.8 Phase diagrams of $CuInSe_2$, ternary triangular diagram showing stable phases at room temperature (left), phase diagram along the quasibinary cut In_2Se_3-Cu_2Se in the range of 10–32% of Cu and in the temperature range 100–1000 °C (right). (Reproduced with permission from Shafarman, 2003. Copyright © John Wiley & Sons, Ltd, 2011.)

If we replace indium atoms by copper atoms, we recover the zincblende phase. However, there are significant differences between the chalcopyrite phase and the zincblende. The existence of two different cations (In and Cu) results in two different bonding lengths R_{In-Se} and R_{Cu-Se} distorting the crystallographic structure in two ways. First the cell is tetragonally distorted, i.e. $\eta = c/2a \neq 1$. For In-based chalcopyrite, the tetragonal distortion is very small $\eta \approx 1{:}006$. Second, the ideal zincblende site of the anion is disturbed leading to a parameter u called the anion displacement. If the two zincblende sublattices forming the chalcopyrite lattice were identical, $u = 1/4$ whereas in the case of chalcopyrites

$$u = \frac{1}{4} + \frac{R^2_{Cu-(S,Se)} - R^2_{In-(S,Se)}}{a^2}$$

where a is the lattice parameter. The value of u lies between 0.21 and 0.235 for CIS.

7.2.2.2.2 Phase Diagram, Nonstoichiometry and Point Defects

Now let us turn to the composition of CIS thin films: they present significant deviations from stoichiometry depending on the conditions of growth. It is not surprising as CIS materials display a very rich phase diagram. Owing to its ternary nature, many possible binary or ternary phases can form depending on temperature and the compositional ranges. Figure 7.8 shows the phase diagram of $CuInSe_2$ as a function of the content of Cu and the temperature. First, the pure chalcopyrite phase α is restricted to a rather small compositional range. Its domain of stability decreases significantly at low temperature. Besides, it is bounded for high fractions of Cu by the segregation of $Cu_{2-y}Se$ phase and for lower fraction of Cu by the appearance of secondary phase β and δ. The β-phase is an ordered defect phase corresponding

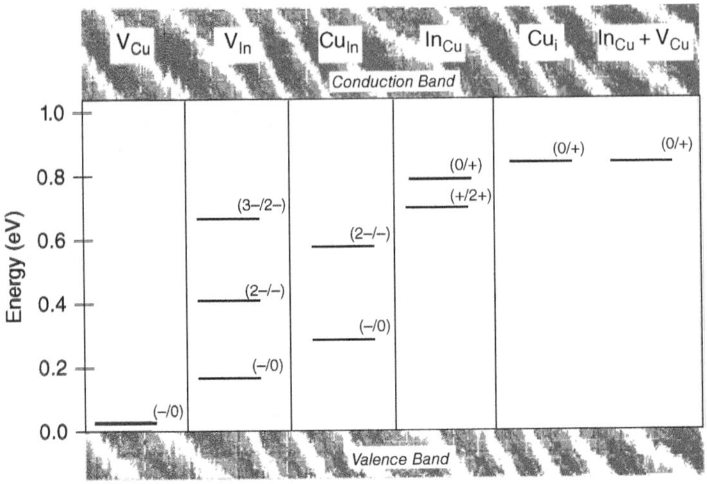

Figure 7.9 Defect transition energy for $CuInSe_2$ calculated by means of ab initio methods. (Reproduced with permission of Zhang, 1998. Copyright © American Physical Society, 2011.)

to $CuIn_3Se_5$, while δ is the high-temperature sphalerite phase. More surprising is the fact that the chalcopyrite phase is not stable for perfect stoichiometry.

Usually, PV-grade CIGS shows a significant Cu-poor character and therefore lies within the stability domain of the α phase. All such deviations from the perfect stoichiometry can therefore be viewed as the fingerprint of the presence of intrinsic defects. Understanding the physics of the defects in CIS allows us to better understand the phase diagram and also the mechanism of doping at work in CIS. Zhang et al. [121] published in 1998 a seminal work on the physics of defects in $CuInSe_2$ using ab initio methods. They drew five important conclusions:

- Defect formation energy in CIS depends strongly on the growth condition through the chemical potential of the atomic species.
- The copper vacancy V_{Cu} has extremely low formation energy.
- V_{Cu} forms shallow acceptor level (see Figure 7.9) which explains the self-doping of CIS.
- Neutral defect pairs with charge transfer involving two copper vacancies V_{Cu} and one indium copper antisite In_{Cu}^{2+} are abundant, and electrically benign due to the neutralisation process of the midgap In_{Cu}^{2+} states by two V_{Cu} (see Figure 7.9).

The first conclusion is a rather general consideration of the ab initio calculation of the defect formation energies. However, it allows us to analyse the benefit of the intrinsic ZnO layer on the device performance: V_{Cu} and In_{Cu} are believed to stabilise the polar surface (112) of CIS and CIS grain boundaries. It is experimentally observed that CIS surfaces usually present a strong Cu depletion that can be associated with doping inversion: while the bulk is p-type, the surface turns out to be n-doped. The fact that the formation energies of V_{Cu} and In_{Cu} are E_f dependent allows one to design the band structure in order to control the position of the Fermi level at the surface. For instance, by introducing a layer of intrinsic ZnO, the Fermi level is pushed away from the conduction band. Consequently, the VCu formation energy significantly increases and counterbalances the type inversion that occurs at the heterojunctions [134].

The two previous conclusions agree well with experimental observations. The p-type character of CIS under Cu-poor growth condition originates from V_{Cu}, seen experimentally as a depletion of Cu atoms. The existence of an ordered defect compound (ODC) phase β results from the repetition of the neutral complex defect $\{2V_{Cu}^{-} + In_{Cu}^{2+}\}$. The two latter defects explain the usual depletion of Cu atom experimentally observed in thin films. In thin films, the β phase is rarely observed due to: (i) diffusion of Na from the glass substrate into CIS [135] and replacement of indium by gallium preventing the formation of such a phase, (ii) high growth temperature extends the stability domain of the α phase as seen in Figure 7.7, (iii) nucleation of the β phase is forbidden by the important interfacial energy originating from electrostatic [76] or mismatch effects [136]. Following this study on defects involving Cu and In, Lany and Zunger [137, 138] have introduced two defects that can explain the experimentally observed metastable electrical behaviour in bulk CIGS [139, 140]. While V_{Se} can explain the p-type persistent photoconductivity, V_{Se}–V_{Cu} may account for the appearance of a deep hole trap that appears at 260 meV above the valence band [139].

The discovery of the morphology enhancement action of Na is one of the main reasons for the second technological step observed in the evolution of the record efficiency. However, the mechanism behind this improvement is still not known exactly and may proceed by multiple effects, two of the many being the effect of Na on thin-film growth and the incorporation of Na into the lattice. During growth, incorporation of Na leads to the formation of $NaSe_x$ that can mediate the inclusion of Se into the film, preventing formation of the V_{Se} donor [136]. Moreover, Na can delay the growth of $CuInSe_2$ at temperature below 380 °C, requiring the growth temperature to be increased and consequently increasing the stability domain of the α phase. The incorporation of Na is, however, believed to occur at the film surface and grain boundary [141]. Another interpretation has been proposed as Na–Se bonds were observed experimentally [142]: Na may replace In or Ga so that $Na_{In;Ga}$ is formed and acts as a shallow acceptor, improving the p-type character of CIS. Replacement of Cu by Na has been proposed and may prevent the formation of the deep defect In_{Cu} [143]. Both hypotheses can be valid, but the large amount of Na required to grow device-quality CIGS is of the order of 0.1% and such amounts of Na are not detected in the sample [144]. Thus, one may favour the assumption that the action of Na is more predominant during film growth than by an actual incorporation into the lattice.

The second hypothesis about the predominance of the α phase was the beneficial effect of Ga. Similarly to Na, Ga addition acts on the growth by slowing down the process [136] and also on the film itself due to its incorporation into the lattice. However, there is absolutely no doubt about the presence of Ga in the lattice. Wei et al. [145] explained the improvement of the film properties by the slightly different defect physics between $CuGaSe_2$ and $CuInSe_2$: GaCu formation energy is higher than InCu so this compensating donor will be much less likely to occur in the film. Moreover, the complex defect $\{2V_{Cu}^{-} + In_{Cu}^{2+}\}$ is less likely in $CuGaSe_2$ than its counterpart in $CuInSe_2$. Therefore, the β phase is less probable in $CuGaSe_2$ and addition of Ga in $CuInSe_2$ makes the domain of stability of β-phase shrink. It was noticed empirically that there exists an optimum of Ga content around Ga/(Ga + In) $\approx$ 0.2–0.3 that makes the solar cells the most efficient [146]. Record cells [71] together with commercial products [147] use such a ratio for their $Cu(In,Ga)Se_2$ absorbers. Other factors are known to impact favourably the growth of CIGS or the properties of the film-like addition of sulfur or oxygen [132]. Grain boundaries play an intriguing role in the performance of CIGS solar cells. Indeed, while Si- and GaAs-based polycrystalline solar cells suffer from large recombination at grain boundaries [148]. CIS polycrystalline solar cells outperform significantly their single-crystalline counterparts: today, record efficiency of 20% [12, 71] is achieved for polycrystalline to be compared with the lower efficiency of the record single-crystal solar cell 13% [149, 150]. Usually, the grain size in CIS is rather small and does not exceed the thickness of thin films i.e. 1–2 μm. As a comparison, the grain size in polycrystalline Si can be larger than 5 mm. Once a grain boundary is formed, states may appear in the bandgap and they can act as recombination centres. Such states result in a charge that is cancelled by the formation of a depletion layer around the grain boundary. Such a space-charge region induces a band bending of the valence and conduction bands: consequently, the grain is thought to act as a hole barrier whose barrier height is defined as Φ_b. Grain boundaries in CIS are believed to originate from the polar (112) free surface passivated with V_{Cu}, In_{Cu} and $\{2V_{Cu}{}^{-} + In_{Cu}{}^{2+}\}$ defects. As in that case the p–d repulsion that governs the valence-band maximum behaviour is relaxed, removing Cu atoms will result in the downshift of the valence band maximum: such internal valence-band offset ΔE_v is often called the neutral barrier as it does not depend on the space-charge region. The total barrier is then $\Phi_b{}^* = \Phi_b + \Delta E_v$ and is typically of the order of 0.2–0.3 eV. Under conditions of illumination, the majority carriers might get trapped in the grain and therefore diminish the grain charge and consequently Φ_b. Therefore, ΔE_v must be sufficient to prevent majority carriers from reaching the recombination centre at the grain boundaries: $\Delta E_v > 0{:}2$ eV is believed to be enough to eclipse such recombination process. Grain boundaries improve the performance of the polycrystalline CIGS solar cell by extending the space-charge region of the p-n junction [151]. They can also serve as a guide for minority-carrier current through the junction. Such a view can only be possible if the grains are orientated perpendicularly to the p-n junction. When grain boundaries are oriented parallel to the heterojunction, a space-charge region results in a dead layer that prevents the current flow [152].

7.2.2.3 Optoelectronic Properties

$CuInSe_2$ has a bandgap of 1.02 eV and $CuGaSe_2$ 1.68 eV [24]. By varying the stoichiometry of the $CuIn_{1-x}Ga_xSe_2$ compound, i.e. adjusting $x = Ga/(In + Ga)$, the bandgap can be engineered. The bandgap dependence on x is of the form $E_g(x) = 1.02 + 0.67x + 0.11x(x - 1)$ [153], but difference between sample can occur and this formula has to be adjusted. The most efficient devices have a Ga content around 0.2–0.3 and thus bandgap energy around 1.1–1.2 eV [12, 71, 154]. The composition is not homogeneous in the absorber and bandgap engineering is introduced via the variation of the Ga content. Composition gradings have a strong impact on the device performance and will be discussed in detail hereafter.

$Cu(In,Ga)Se_2$ presents p-type conduction. The doping level of $Cu(In,Ga)Se_2$ is around 10^{16} cm^{-3} [156]. The shallow acceptor $V_{Cu}{}^{-}$ is assumed to be the principal dopant, as calculated by ab initio [121], but In_{Cu} antisite is a donor impurity that compensates $Cu(In,Ga)Se_2$. As discussed before, it was calculated that several neutral defect complexes exhibit low formation enthalpies, and in particular $(2V_{Cu} + In_{Cu})$ that has an enthalpy close to 0 or even negative. This is beneficial to $Cu(In,Ga)Se_2$ as it limits two defects, V_{Cu} and In_{Cu}, in the bandgap region. Off-stoichiometry can thus be accommodated without formation of high numbers of electronics states in the bandgap that would be harmful for the device performance. However, several donor deep defects that compensate $Cu(In,Ga)Se_2$ have been reported (see Table 7.3). These defects are mainly studied by admittance or modulated

Table 7.3 Electronic parameters of deep traps in $Cu(In,Ga)Se_2$. (Data extracted from Igalson, 2011. Copyright © Springer, 2011.)

Electrical activity	E_c–E_t (eV)	Hole capture cross section (cm^2)	Electron capture cross section (cm^2)	Assignment	Theory [21]
Donor	0.19		6×10^{-17}	$(In_{Cu} + V_{Cu})^+$	$E_c - 0.2$ eV
Donor	0.26		4×10^{-16}	$In_{Cu}{}^+$	$E_c - 0.26$ eV
Donor	0.34		4×10^{-15}	$In_{Cu}{}^{2+}$	$E_c - 0.34$ eV
Donor	0.47	$< 10^{-18}$	5×10^{-16}	O_{Se}	
Donor	> 0.6	5×10^{-14}	$> 5 \times 10^{-14}$	$V_{Se}{}^0$	

photocurrent spectroscopies. Despite intensive study their position and exact origin in the $Cu(In,Ga)Se_2$ layer are still under debate [155, 157–159].

$Cu(In,Ga)Se_2$ is a direct semiconductor thus the absorption coefficient is high (Figure 7.10), as confirmed by spectrophotometry or ellipsometry measurements.

7.2.2.4 $Cu(In,Ga)Se_2$ Grain Boundaries

$Cu(In,Ga)Se_2$ solar cells are usually made of polycrystalline $Cu(In,Ga)Se_2$ absorbers. The typical grain size is around 0.5–1 μm [161]. Surprisingly, the devices do not suffer from the polycrystalline character of the absorber, as polycrystalline devices are more efficient than crystalline ones [162]. There are numerous studies on the $Cu(In,Ga)Se_2$ grain boundaries that can be controversial, showing at least variations between samples. Independent of the composition of the film, it has been found that grain boundaries are both twin boundaries

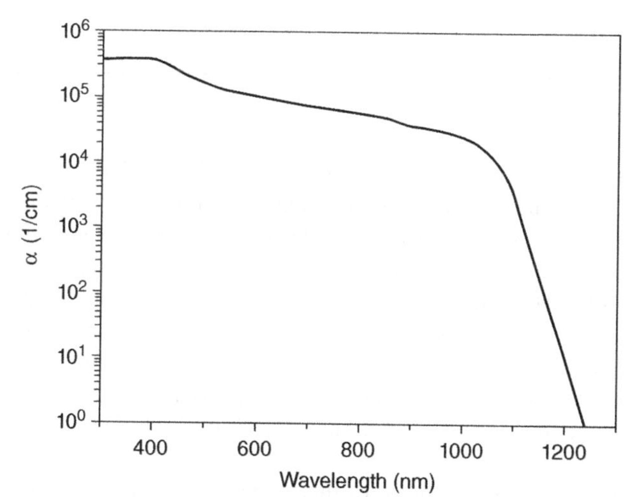

Figure 7.10 Absorption coefficient of $Cu(In,Ga)Se_2$ absorber of bandgap 1.15 eV according to [26]. (Reproduced with permission from Klenk, 2006. Copyright © John Wiley & Sons, Ltd, 2006.)

Table 7.4 Material properties of the $CuInSe_2$ compound. (Data extracted from Guillemoles, 1994.)

Property		Value	Unit
Lattice constant	a	5.78	Å
	c	11.62	Å
Density		5.75	g/cm^3
Molar mass		336	g
Melting Temperature		986	C
Thermal expansion coefficient at 273 K	(a-axis)	8.32×10^{-6}	1/K
	(c-axis)	7.89×10^{-6}	1/K
Thermal conductivity at 273 K		0.086	W/cm/K
Dielectric constant	Low frequency	13.6 ± 2.4	
	High frequency	8.1 ± 1.4	
Mobility	Electrons	100–1000	$cm^2/V/s$
	Holes	20–200	$cm^2/V/s$
Effective mass	Holes (heavy)	0.71	
	Holes (light)	0.092	
	Electrons	0.09	
Intrinsic carrier concentration (300 K)		$2. \times 10^{14}$	cm^{-3}
Minority carrier lifetime	Electrons	0.1–100	ns
Energy gap (300 K)		1.02	eV
Energy gap (0 K)		1.06	eV
Electron affinity		4.3–4.6	eV
Energy-gap temperature coefficient		-2×10^{-4}	eV/K
Energy-gap pressure coefficient		0.028	eV/GPa
Hardness		3.2×10^9	N/m^2
Compressibility		1.4×10^{-11}	m^2/N
Sound speed		2.2×10^5	cm/s
Debye temperature		222	K

(no dangling bonds or strain), which is a favourable situation, and nontwin, "random", grain boundaries [161, 163]. The effective grain-boundary recombination velocity is smaller than 10^3 cm/s at twin boundaries in highly efficient devices [164, 165]. Nontwin boundaries have higher recombination velocities, around 10^4 cm/s, which is the sign of a higher density of defects [163, 166]. The low recombination velocity in twin boundaries can stem from low defect density, passivation and/or an appropriate band diagram. $Cu(In,Ga)Se_2$ grain boundaries present a low band bending (< 200 mV), or even no bending [167], which reflects a low concentration of charged defects. Oxygenation passivates defects at grain boundaries, such as positive charges due to selenium vacancies [128, 168]. Thus, postdeposition annealing can be desirable. It was shown that the presence of sodium, which diffuses from soda-lime glass or is voluntary introduced, catalyzes the oxidation of selenium vacancies [169]. As mentionned earlier, it was also proposed that the twin grain boundaries present negative valence-band offsets. This originates from the Cu-poor composition of twin grain boundaries

[36, 170], due to Cu vacancies and In_{Cu} antisites. This valence-band offset prevents holes from recombining at grain boundaries. The Cu-poor composition of grain boundaries may be due to the diffusion of Cd during chemical-bath deposition and consequent Cu outdiffusion that would also lead to a strong n-type doping of the grain boundaries and thus provoke inversion. For random grain boundaries Cu depletion is not systematic, but an anticorrelation between Cu and In contents is seen [163]. Due to the negative band bending or even inversion of grain boundaries, it was proposed that $Cu(In,Ga)Se_2$ polycrystalline solar cells are self-assembled nanostructured radial junction solar cells, which explains the excellent collection efficiency [167]. In this respect, the multistage deposition process is particularly favourable, as it leads to grain boundaries with more voids than single-step preparations, due to the coalescence of Cu vacancies during the Cu diffusion between Cu-poor and Cu-rich phases. These voids help CdS to cover grain boundaries deep in the absorber (up to 500 nm under the free surface), and enhanced the radial junction behaviour [167].

7.2.2.5 $Cu(In,Ga)Se_2$ Free Surface

The free surface of $Cu(In,Ga)Se_2$ presents similarities with grain-boundary surfaces. In Cu-rich $Cu(In,Ga)Se_2$ films the Fermi level at the surface is nearly merged with the valence band, due to the presence of Cu-rich binaries on the surface. For Cu-poor compositions of the absorber, the surface presents a type inversion, with the Fermi level closer to the conduction band than to valence band (and the E_f $E_f - E_v$ E_v difference can be as large as 1.1 eV). The surface composition of Cu-poor absorbers suggests the presence of an ordered defect compound (ODC), i.e. a $Cu(In,Ga)_3Se_5$ phase. This compound has a higher bandgap than bulk material (the bandgap of $CuIn_3Se_5$ is 1.23 eV and that of $CuGa_3Se_5$ is 1.8 eV) and tends to exhibit n-type conduction. However, the doping of the ODC is too shallow to be the source of a buried homojunction. It is more likely that defects such as Se vacancies and In_{Cu} or Ga_{Cu} antisites at the free surface that can provide more than 10^{14} charges/cm^2 are the cause of the type inversion [128]. It seems that the presence of such surface defects is thermodynamically favourable for the formation of the ODC, explaining their simultaneous presence on numerous samples [24]. The defects on the free surface of $Cu(In,Ga)Se_2$ can be passivated by oxygenation and sodium, as for grain boundaries [128].

7.2.3 Device Properties

7.2.3.1 Device Operation and Performance

$Cu(In,Ga)Se_2$ devices have reached 20% efficiencies [12], with impressive parameters for a polycrystalline thin-film solar cells, given in Figure 7.11.

As explained above, the CIGS bandgap can be modulated easily by tuning the Ga/(In + Ga) ratio of the absorber. As expected, this changes both V_{oc} and I_{sc} (Figure 7.12), while the J_{sc} is decreased by the expected amount from the bandgap change (with a minor decrease in collection efficiency), the V_{oc} does not follow the bandgap increase beyond Ga/(In + Ga) > 0.3, yielding a maximum in the efficiency versus gap curve around 1.15 eV, and a global shape of the curve in Figure 7.13, different from that expected from a Schockley–Queisser argumentation.

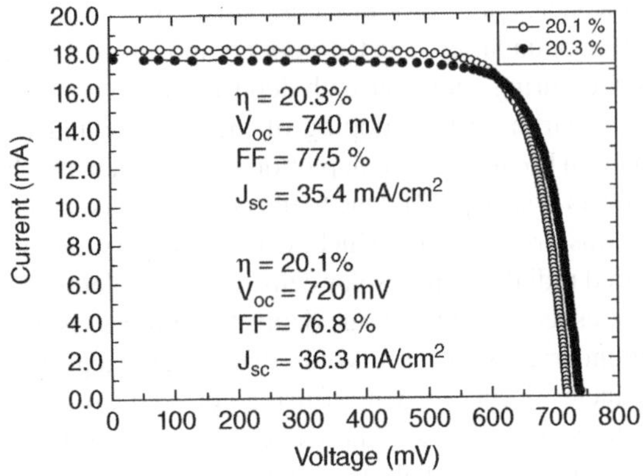

Figure 7.11 Current–voltage characteristic of state-of-the-art devices. (Reproduced with permission from Jackson, 2011. Copyright © John Wiley & Sons, Ltd, 2011.)

There have been explanations proposed on why there is such a sharp optimal composition, and why the V_{oc} increases with E_g much less than expected beyond the optimal composition. Two models seem to prevail. In the first one, the increase of the bandgap causes a decrease of the electron affinity of CIGS, to the point that the band line up with the buffer layer changes from a spike (as in Figure 7.15 and Figure 7.18) to cliff-like, as evidenced by IPES studies, and so the dominant recombination mechanism changes from bulk dominated (quasineutral region in best devices or Schockley–Read–Hall in poorer absorbers) to heterojunction dominated [171].

In a second model, it is the intrinsic quality of the absorber that degrades when the Ga content is too large, and indeed, it seems that the growth mechanism is affected (as well as the doping level) when Ga is increased [23, 146]. Of course, both possibilities could be true.

In the best devices, minority-carrier lifetimes down to 100 ns have been measured [172], demonstrating the very high quality that has been achieved with UHV-grown samples. This

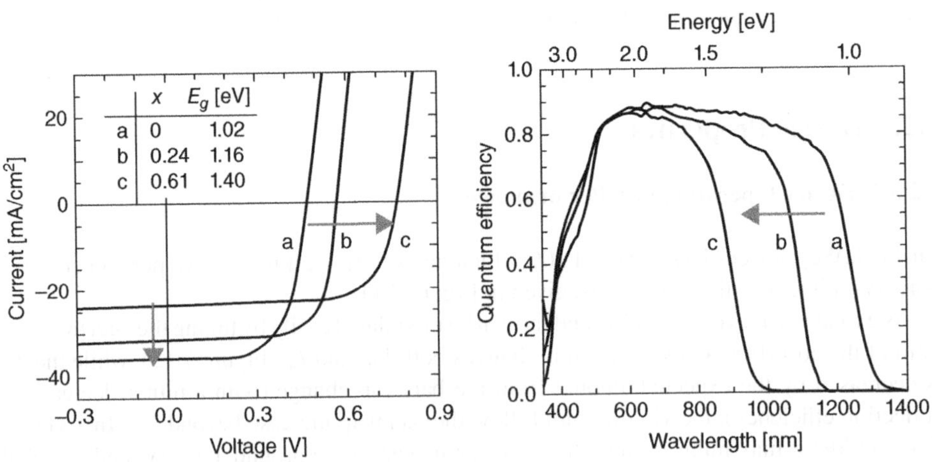

Figure 7.12 (left) Current–voltage curve with varying Ga content. (right) External quantum efficiency as a function of Ga content and thus bandgap. (Reproduced with permission from Shafarman, 2003. Copyright © John Wiley & Sons, Ltd, 2011.)

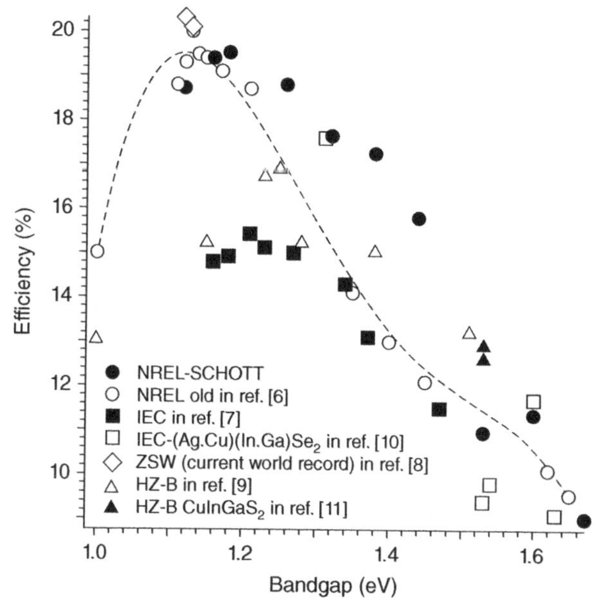

Figure 7.13 Efficiency vs. bandgap for Cu(In,Ga)Se_2 solar cells. (Reproduced with permission from Contreras, 2012. Copyright © John Wiley & Sons, Ltd, 2012.)

values translates into V_{oc} that are similar to those achieved with best single-crystal Si cells (corrected for E_g differences, both having similar gaps). Mainly, it seems that the current is still significantly below what could be achieved, compared again with c-Si solar cells.

When analyzing a typical quantum efficiency curve of a CIGS solar cell (Figure 7.14), one finds important reflection losses, but also significant losses from the window layers (optical absorption by free carriers in ZnO and by CdS), as well as a suboptimal absorption in CIGS.

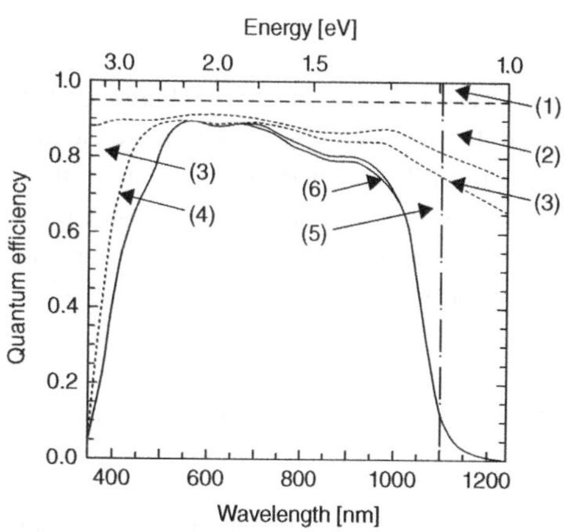

Region in Figure 13.14	Optical loss mechanism	ΔJ [mA/cm²]
(1)	Shading from grid with 4% area coverage	1.7
(2)	Reflection from Cu(InGa)Se_2/CdS/ZnO	3.8
(3)	Absorption in ZnO	1.8
(4)	Absorption in CdS	0.8
(5)	Incomplete generation in Cu(InGa)Se_2	1.9
(6)	Incomplete collection in Cu(InGa)Se_2	0.4

Figure 7.14 Quantum efficiency of a CIGSe device, showing typical current losses. (Reproduced with permission from Shafarman, 2003. Copyright © John Wiley & Sons, Ltd, 2011.)

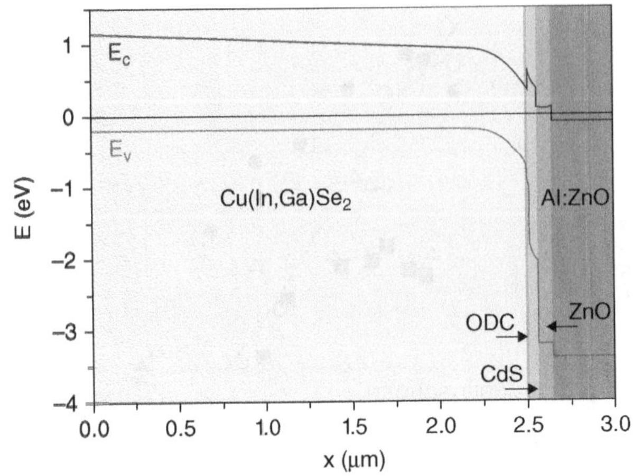

Figure 7.15 Energy-band diagram of the $Cu(In,Ga)Se_2$ solar cell according to reference [156] at zero applied voltage in the dark. The $Cu(In,Ga)Se_2$ bandgap shows a gradient from 1.34 eV at the back contact to 1.08 eV at the front contact. An ordered defect compound layer is included at the absorber surface.

7.2.3.2 Description of the $Cu(In,Ga)Se_2$ Heterojunction: Different Models

The $Cu(In,Ga)Se_2$ solar cell is a heterojunction. Determining the band diagram is not an easy task, and depends on the material properties and specific deposition processes. We will give a standard description of the cell and discuss the most interesting variants.

7.2.3.2.1 General Band Diagram (SCAPS)

A general description of the $Cu(In,Ga)Se_2$ solar cell band diagram can be calculated with SCAPS software [173] and the materials parameters given in ref. [156]. Figure 7.15 shows the energy-band diagram for AM1.5 illumination under zero applied voltage. The absorber presents a bandgap gradient, from 1.34 eV at the back contact to 1.08 eV at the front contact (see Section 7.7.2.3.2.4).

7.2.3.2.2 $Cu(In,Ga)Se_2$ Absorber

The $Cu(In,Ga)Se_2$ absorber can be separated into two regions: majority charge carrier depleted space-charge region and quasineutral region. In the space-charge region the electric field due to the heterojunction is high, the conduction and valence bands are bent, and minority carriers drift towards the interface. In the quasineutral region the carries are subjected to diffusion. Depending on the doping level of the absorber and window layer the width of the space-charge region varies but is usually of the order of 500 nm at zero applied voltage [174].

7.2.3.2.3 $CdS/Cu(In,Ga)Se_2$ Interface

The interface between CdS and $Cu(In,Ga)Se_2$ is crucial for the device performance. The physics of this interface is intricate, and we will shed light on the principal mechanisms at stake.

The heterointerface is never defect free, due to lattice mismatch between $Cu(In,Ga)Se_2$ and CdS for example. These defects can be centres of recombination. According to the

Shockley–Read–Hall model, if we note E_d the energy level of the defects, we can define the recombination rate as [175]:

$$U = S_p S_n \frac{n^{IF} p_a^{IF} - n_i^2}{S_n\left(n^{IF} + n_1\right) + S_p\left(p_a^{IF} + p_1\right)} \tag{7.2}$$

where $n_1 = N_c \exp(-(E_c - E_d)/kT)$, $p_1 = N_v \exp(-(E_d - E_v)/kT)$, with N_c and N_v the density of states in the conduction and valence band, n^{IF} the density of electrons at the interface and p_a^{IF} the density of holes at the interface, on the absorber side, S_p and S_n the recombination velocity for holes and electrons. In the case where the density of defects at the interface N_d is important, the electron density at the interface (in the absorber or the buffer) is higher than the hole concentration, such that $n^{IF}/S_p \gg p_a^{IF}/S_n$, and the defects level is deep (near midgap), then $n^{IF}/S_p \gg n_1/S_p$ and $\gg p_1/S_n$. In this case equation (7.2) becomes [176]:

$$U \approx S_p p_a^{IF} \tag{7.3}$$

Thus, the recombination at the $Cu(In,Ga)Se_2/CdS$ interface is controlled by the hole concentration p_a^{IF}. Designs that tend to decrease p_a^{IF} are thus beneficial for the device performance as they will reduce interface recombination.

When the concentration of interface states N_i is sufficiently large, the Fermi level at the interface can be pinned. The distance between the $Cu(In,Ga)Se_2$ conduction band and the Fermi level at the interface ΔE_{Fn} is equal to [128, 177]:

$$\Delta E_{Fn} = \frac{\Delta\chi - \dfrac{qN_d d_b^2}{2\varepsilon_b} + \dfrac{qN_d d_b}{\varepsilon_b}\left(E_{g,a} - \phi_n\right) + \dfrac{qN_{A,a} d_b w_a}{\varepsilon_b}}{1 + \dfrac{qN_d d_b}{\varepsilon_b}} \tag{7.4}$$

where $\Delta\chi = \Delta E_c^{wb} - \Delta E_c$ and ΔE_c^{wb} is the conduction-band offset between the window (subscript or superscript w) layer and buffer (subscript or superscript b), and ΔE_c is the conduction-band offset between the absorber (subscript or superscript a) and the buffer. ε is the dielectric constant, $E_{g,a}$ the absorber bandgap, ϕ_n the neutrality level of the interface states with respect to the valence band edge, d_b the thickness of the buffer layer, that is supposed to be completely depleted, $N_{A,a}$ the doping level of the absorber and w_a the width of the space-charge region in the absorber. It was calculated that for interface state concentration N_d above 10^{13} cm^{-2} eV^{-1}, the Fermi level can be pinned, so that $\Delta E_{Fn} = E_{g,a} - \phi_n$ (doping levels of CdS and $Cu(In,Ga)Se_2$ are set to 10^{16} cm^{-3}). As a consequence of Fermi-level pinning, the surface of the absorber can be inverted, and a buried homojunction is created. This homojunction is desirable as it makes electrons become majority carriers at the interface, i.e. decreases the density of holes at the interface, and thus interface recombination (equation (7.2)).

Depending on the bandgap of $Cu(In,Ga)Se_2$ the conduction-band offsetΔE_c between buffer and absorber can be positive or negative (In Figure 7.15, the band offset is positive). Positive conduction-band offset occurs for bandgaps at the interface smaller than 1.3 eV, and negative conduction-band offset for bandgaps higher than 1.3 eV. Positive band offsets are desirable designs as the hole concentration at the interface becomes small, due to the high

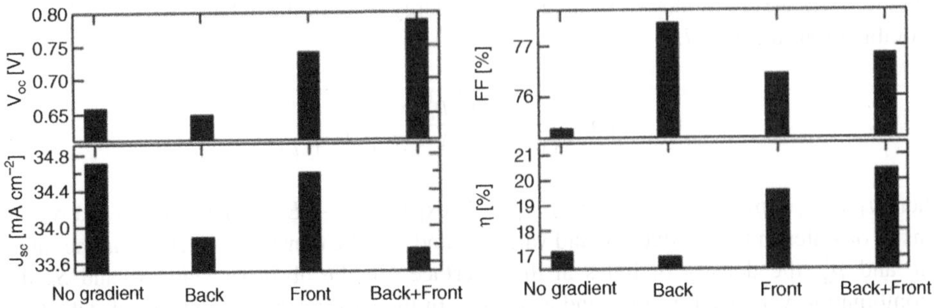

Figure 7.16 Effect of back and front surface gradient on solar cell parameters, for an absorber of 500 nm thickness with 0.2 eV of gradient (front surface gradient extends of 10 nm, back surface gradient extends on 100 nm). (Data extracted with permission from Scheer, 2011. Copyright © 2011, Wiley-VCH Verlag GmbH & Co. KGaA.)

hole barrier ϕ_b^p (difference between valence band and Fermi level at the interface) (see Figure 7.18). For a negative band offset ΔE_c the open-circuit voltage decreases proportionally to ΔE_c [16], due to a reduced ϕ_b^p and thus increased recombination velocity at the interface. Thus, most cells present a positive conduction-band offset at the $Cu(In,Ga)Se_2/CdS$ interface.

The surface of $Cu(In,Ga)Se_2$ can present a composition different from the bulk, forming an ordered defect compound that tends to be Cu poor. This compound, that presents a high concentration of defects, is often represented by a layer of the same material properties as the bulk but with a positive valence-band offset [156]. Thus, this layer can improve the device efficiency by an increase of the hole barrier.

$Cu(In,Ga)Se_2$ absorbers often present front bandgap grading. This grading is set to have a high open-circuit voltage, combined with a good short-circuit current density. The high open-circuit voltage is due to the decrease of interface (or space-charge region) recombination due to a higher bandgap at the $Cu(In,Ga)Se_2/CdS$ interface, and thus higher hole barrier ϕ_b^p. The good short-circuit current density is maintained due to a smaller bandgap region in the absorber. Front grading is thus highly desirable (Figure 7.16).

7.2.3.2.4 Back Contact

$Cu(In,Ga)Se_2$ solar cell also generally present a bandgap gradient at the back contact. This gradient is widely used in industrial processes, and is set to help carrier collection in the absorber by repelling electrons from the recombining back contact [39, 178, 179], and thus improve fill factor (Figure 7.16). This bandgap grading is especially crucial for the thinnest devices or absorbers with small electron diffusion lengths. However, due to the grading the average bandgap is increased and thus the absorption and short-circuit current are reduced. The total effect on the solar cell performance is thus a balance between increased collection and decreased absorption. The association of a front and back contact grading (16) is the most efficient configuration, often referred to as V-shape. However, the bandgap grading has to be controlled in order to avoid the formation of a notch, i.e. difference between minimum and maximum bandgap that is too pronounced (Figure 7.17). This situation leads indeed to increased recombination, by the capture of carriers in a potential well [39].

The back contact is also the location of recombination. Similarly to what we defined for the $Cu(In,Ga)Se_2/CdS$ interface, the back contact can have a high recombination velocity.

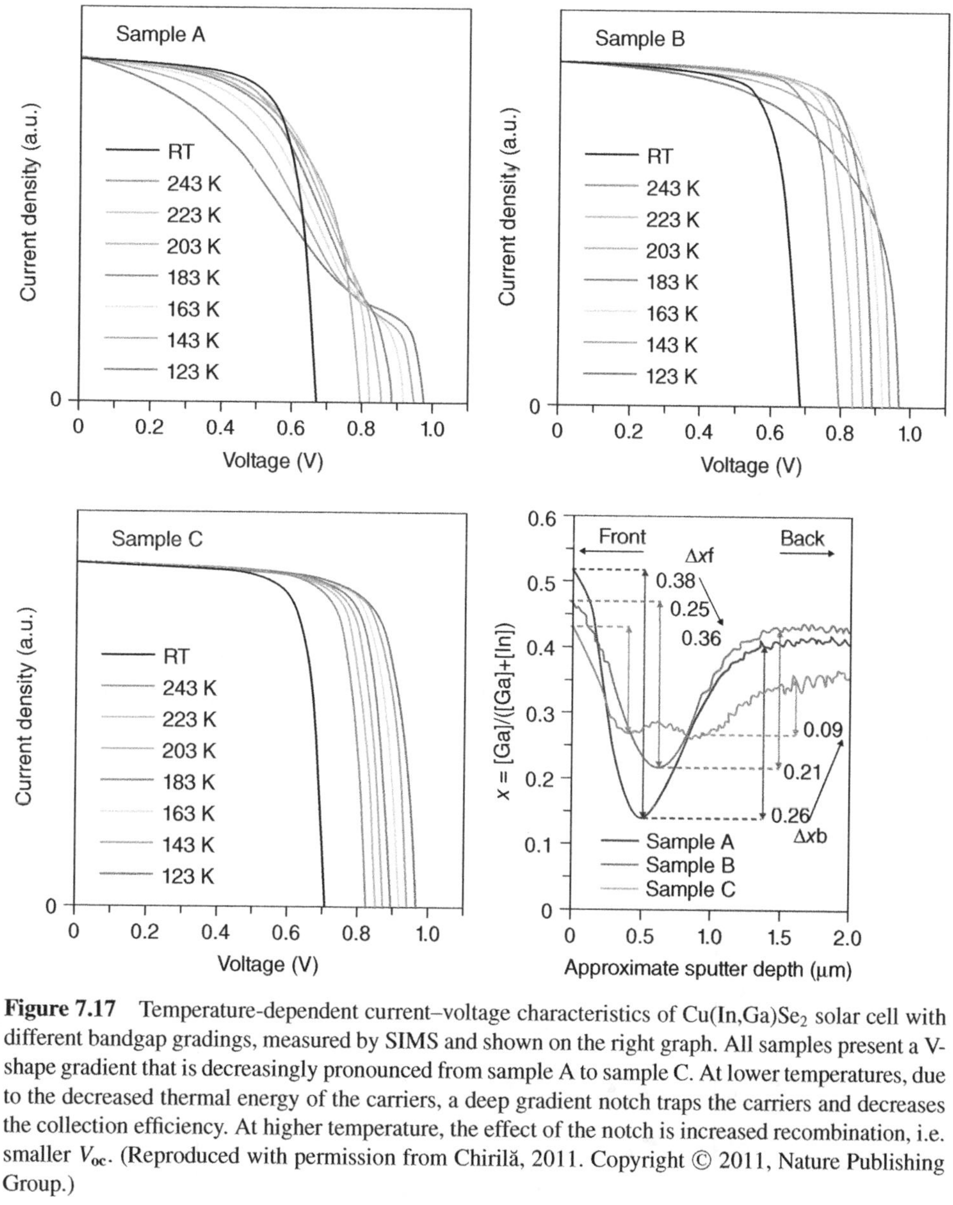

Figure 7.17 Temperature-dependent current–voltage characteristics of $Cu(In,Ga)Se_2$ solar cell with different bandgap gradings, measured by SIMS and shown on the right graph. All samples present a V-shape gradient that is decreasingly pronounced from sample A to sample C. At lower temperatures, due to the decreased thermal energy of the carriers, a deep gradient notch traps the carriers and decreases the collection efficiency. At higher temperature, the effect of the notch is increased recombination, i.e. smaller V_{oc}. (Reproduced with permission from Chirilă, 2011. Copyright © 2011, Nature Publishing Group.)

7.2.3.2.5 Recombination Mechanisms

In a solar cell, the diode current is due to several recombination mechanisms.

As stated before there are several recombination paths in a $Cu(In,Ga)Se_2$ solar cell (Figure 7.18). The back contact (1), quasineutral region (2), space-charge region (3) and $Cu(In,Ga)Se_2/CdS$ absorber interfaces (4) are the principal recombination paths. Due to the high bandgap of the window layers, recombination is neglected there. The tunneling-enhanced space-charge region and interface recombination are also depicted on the inset of Figure 7.18. The tunneling of holes can indeed increase both recombination rates.

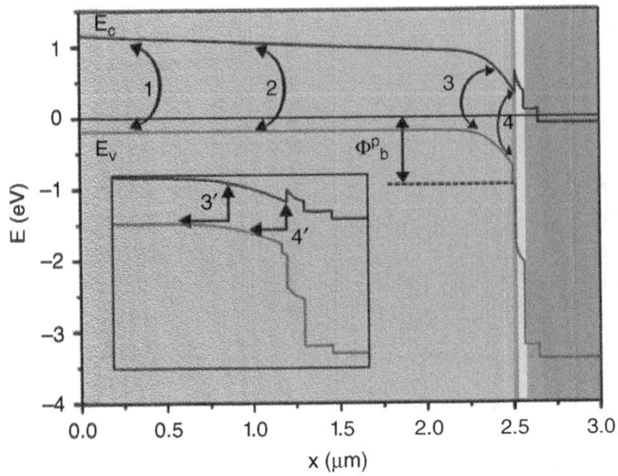

Figure 7.18 Recombination paths: 1 back surface, 2 quasineutral region, 3 space-charge region, 4 interface. In the inset we detail the tunneling-enhanced space-charge region recombination 3′ and the tunneling-enhanced interface recombination 4′. Tunneling increases the density of holes that can participate in the recombination. The hole barrier ϕ_b^P at the interface is also depicted.

The two recombination mechanisms that cannot be avoided in a solar cell are radiative recombination and Auger recombination. Radiative recombination is the inverse of optical absorption: the direct recombination of an electron and a hole that emits a photon. The radiative recombination rate is:

$$R_{rad} = B\left(np - n_i^2\right) \tag{7.5}$$

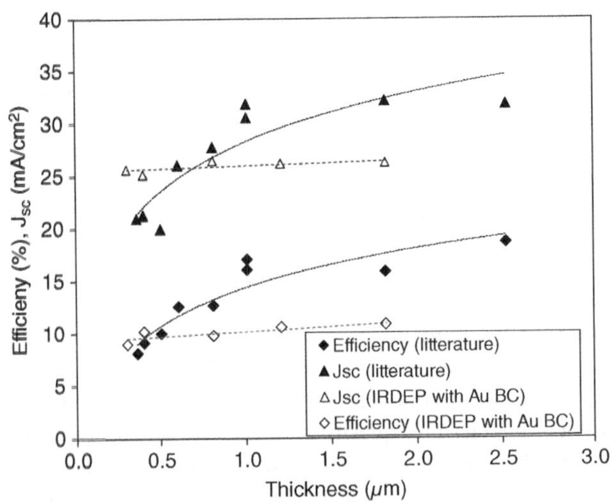

Figure 7.19 Experimental results for thin CIGS layers at Uppsala (2003) [186], NREL (2006) [187] and from IRDEP with gold back contact (Au BC) [67].

where B is the radiative recombination coefficient. This coefficient is of the order of 10^{-10} $cm^3\ s^{-1}$ in a $Cu(In,Ga)Se_2$ solar cell [180]. The product np in the space-charge region or quasineutral region is equal to $n_i^2 \exp(qV/kT)$, due to the constant separation of quasi-Fermi levels, and is equal to the applied voltage. Thus, the diode current corresponding to radiative recombination has a voltage dependence in the term$\exp(qV/kT)$, an ideality factor of 1. The activation energy of the saturation current is equal to the absorber bandgap. More generally the diode ideality factor A can be defined through the voltage dependence of carriers [24] as:

$$(np)^{1/A} = n_i^{2/A} \exp(qV/AkT) = (N_c N_v)^{2/A} \exp\left(-E_g/AkT\right) \exp(qV/AkT) \quad (7.6)$$

The saturation current density is then:

$$J_{0A} = J_{00A} \exp(-E_a/AkT) \quad (7.7)$$

with J_{00A} the reference saturation current density for the ideality factor A, and E_a the activation energy of the saturation current that in this case equals the bandgap energy E_g. In standard $Cu(In,Ga)Se_2$ the radiative recombination is never limiting.

The Auger recombination is also intrinsic and the corresponding recombination rate is $R_{Auger} = C_p\left(np^2 - n_0 p_0^2\right) + C_n\left(pn^2 - p_0 n_0^2\right)$, where $C_p(C_n)$ is the Auger coefficient for the excitation of a hole (electron), by the recombination of an electron–hole pair. The Auger recombination is more important when the carrier concentrations are high, and thus may be substantial under intense illuminations, high forward voltages or high doping levels. The Auger coefficients of $Cu(In,Ga)Se_2$ are unknown. An interpolation from Auger coefficients of direct-bandgap III-V semiconductors as a function of bandgap [181] gives values of Auger recombination coefficient of Cu(In,Ga)Se2 (1.1–1.2 eV) around 10–30 $cm^6\ s^{-1}$. Thus, this mechanism is negligible in standard operation conditions. It should be noted that Auger recombination does not follow the formalism of equation (7.7).

The other recombination mechanisms are related to the presence of defects, and may be avoided by an optimum fabrication process. However, in $Cu(In,Ga)Se_2$ solar cells, even state-of-the-art devices are subjected to such recombination. We will review briefly the diode current related to the different recombination paths of 18. The case of tunneling-enhanced recombination is not treated here, and the reader is invited to refer to [24, 182].

7.2.3.2.5.1 Interface Based on the expression of recombination rate at the interface (equation (7.3)), supposing mid-gap defects, and highly asymmetric junction ($\varepsilon_D N_D \gg \varepsilon_A N_A$) and the absence of Fermi level pinning and tunneling, the saturation current due to interface recombination is [24]:

$$\begin{aligned} J_{0,IF} &= qS_p N_{A,a}\left(\frac{N_{c,w} N_{v,a}}{N_{D,w} N_{A,a}}\right) \exp\left(-\frac{E_g^{IF}}{kT}\right) \quad \text{for } \Delta E_c < 0 \\ J_{0,IF} &= qS_p N_{A,a}\left(\frac{N_{c,w} N_{v,a}}{N_{D,w} N_{A,a}}\right) \exp\left(-\frac{E_{g,a}}{kT}\right) \quad \text{for } \Delta E_c > 0 \end{aligned} \quad (7.8)$$

It should be noted that in the case of a negative conduction-band offset, E_g^{IF} is not the absorber bandgap $E_{g,a}$ but is smaller, $E_g^{IF} = E_{g,a} + \Delta E_c$ [24]. The ideality factor is 1, which is due to the fact that only the hole quasi-Fermi level can vary with applied voltage.

In the case where the Fermi Level at the interface is pinned by interface charges above midgap, and when electrons are majority carriers at the interface, the interface current density becomes [182, 183]:

$$J_{0,IF} = qS_p N_{v,a} \exp\left(-\frac{\phi_b^{p,0}}{kT}\right) \tag{7.9}$$

where $\phi_b^{p,0}$ is the hole barrier at zero applied voltage, defined by the level of Fermi-level pinning, S_p is the hole recombination velocity at the interface, in the case where all interface states are empty of holes. As $\phi_b^{p,0}$ is temperature independent, it can be assimilated to an activation energy. The diode ideality factor is 1.

7.2.3.2.5.2 Space-Charge Region Equation (7.2) is general and describes recombination processes through traps. In the space-charge region with midgap defects, we can obtain the expression of the saturation current as [24]:

$$J_{0,SCR} = \frac{\pi/2 \times kT}{F_m} \left(\frac{N_{c,a} N_{v,a}}{\tau_{n0,a} \tau_{p0,a}}\right)^{1/2} \exp\left(-\frac{E_{g,a}}{2kT}\right) \tag{7.10}$$

where F_m is the electric field at the position of maximum recombination, $\tau_{n0,a}$ and $\tau_{p0,a}$ are electron and hole minimum lifetimes (lifetime as minority carriers in a semiconductor of the same defect density). The bandgap of the absorber $E_{g,a}$ is the activation energy and the ideality factor is 2. The physical reason behind this ideality factor is that both electron and hole quasi-Fermi levels vary with applied voltage.

7.2.3.2.5.3 Quasineutral Region In the quasineutral region, the diode current depends on the collection of carriers that governs the minority-carrier concentration. To discriminate between back contact and quasineutral region recombination, we suppose that the absorber thickness is high compared to the electron diffusion length. In the case of a deep defect, the saturation current is [24]:

$$J_{0,QNR} = q \frac{n_i^2}{\tau_{n0,a} N_{A,a}} \int_{QNR} \eta_c(z)\, dz \tag{7.11}$$

where $\eta_c(z)$ is the collection function, i.e. the probability for generated charge carriers to be collected at the contact, which depends on the position z in the cell. The ideality factor is 1. In the case of large applied voltages, the hypothesis of minority carriers is no longer valid and the quasi-Fermi levels of both carriers move symmetrically to the defect level, and thus the ideality factor is expected to increase and reach a value of 2 under high injection.

7.2.3.2.5.4 Back Contact Back-contact recombination is a dominant path if the diffusion length of minority carriers is large compared to the quasi neutral region thickness. In this limiting case, the saturation current is [24]:

$$J_{0,\mathrm{BC}} = q\frac{D_{\mathrm{n,a}}n_{\mathrm{i}}^2}{N_{\mathrm{A,a}}}\frac{1}{d_{\mathrm{QNR}}} \tag{7.12}$$

where d_{QNR} is the width of the quasineutral region, and $D_{\mathrm{n,a}}$ the diffusion coefficient of electrons in the absorber. This mechanism is scaled with the inverse of the absorber thickness, thus it becomes predominant for thin and ultrathin samples.

We have seen the different diode currents and their dependence on voltage. We shall discuss briefly the dependence on voltage of photocurrents.

7.2.4 Outlook

Several pathways are followed to improve CIGS solar cells in terms of efficiency and in terms of material usage (especially In). Two of them are presented below.

7.2.4.1 Ultrathin CIGS

The concern about reducing the CIGS absorber thickness has been introduced by early experimental studies in 1997–1998 in USA [184] and Japan [185], by using the classical structure of CIGS solar cells, the CIGS layer being deposited by a coevaporation process on Mo-covered glass substrates. Losses were identified on decreasing the thickness. Then, this issue has been undertaken in a very systematic way by the group at the University of Uppsala in Sweden in 2003 [186] and then at the National Renewable Energy Laboratory in USA in 2006 [187]. These two groups proposed the same approach for reducing the absorber thickness, consisting in directly growing the CIGS layer while adapting the existing process. Most significant results are shown in Figure 7.19.

One can observe that decreasing the thickness below 0.4 μm reduces markedly the performances of the cell, especially the short-circuit current due to insufficient absorption in the CIGS layer. The open-circuit voltage is also markedly affected. In these best of practice levels from the international state-of-the-art, ultrathin CIGS was not able to maintain sufficiently high efficiencies for ultrathin layers, as noted recently [188], in contrast to the IRDEP group effort, where efficiencies independent of thickness were achieved [67]. Moreover, from theoretical simulations, highly efficient (>16%) ultrathin CIGS solar cells would be possible if excellent light trapping in CIGS could be achieved, with for instance lambertian reflectors at the back contact [66, 189] and if electron mirror and p^+-doped electrical back contact can be achieved [189].

The lack of efficient reflection at the CIGS/Mo layer interface led to attempts to replace Mo layers with more reflective metallic layers like zirconium nitride [190] without significant results up to now. In parallel, the same group [188] was able to prepare 1.3–1.4 μm films on Mo with almost no efficiency losses, but experienced 10% losses for 0.8 μm films

(14.4% efficiency compared to 15.5%). Controlling the gallium gradient at the back contact was a key point in order to reduce the recombination by a back surface field effect for minority carriers, confirming previous studies on thinner films [186] and theoretical evaluations [189].

The strategic approach of starting to avoid molybdenum by completely re-engineering the back contact was introduced by the IRDEP group by using lift off and extremely precise etching procedures [65, 191, 192] that offered for the first time the possibility to deposit and optimise back contacts independently from the constraints coming from the deposition step. This led to several breakthroughs, with devices with no photocurrent losses down to 0.3 micrometres, using reflecting gold as proof of concept and efficiencies maintained around 10% [65, 189, 191, 192].

The spontaneous tendency of CIGS films to peel off from Mo has been more considered as a problem than an opportunity. A Japanese group, in parallel with the IRDEP group started to consider the lift-off technique as a new process for CIGS solar cells [193, 194] but their performances were significantly degraded. The approach of the IRDEP group, in contrast, has been proven to be successful [65], and represents up to now the best practice at the international level.

Advanced optical confinement studies CIGS cells are still at their beginning, as compared to what is their status in the other technologies (thin-film Si) and mostly concern the reduction of the optical losses at the front surface [195]. The first studies on thin and ultrathin devices have been carried out by the IRDEP group [189], and their partners in particular with plasmonic structures [196].

In conclusion, going to layers thinner than 1 μm requires replacing Mo by a more reflecting layer, either by growing the layer directly on an alternative substrate, or by removing it afterwards. The lift-off technique, which is under intense focus for other solar cells technologies (cSi, aSi, III-V), is taking off for the CIGS technology. Finally, it should be stressed that for CIGS, advanced optical confinement is still at an early stage.

7.2.4.2 Concentration on CIGS

The challenges that the photovoltaic community is facing are the terawatt level development, both from the point of view of material availability and fabrication facilities, in a world of finite natural and financial resources, and the competition with cheap fossil energy sources. Coupling two photovoltaic domains, thin films and concentrating photovoltaics, could help meet these challenges. Thin films enable easy fabrication and high throughput, whereas concentrating photovoltaics is raw material thrifty and enables high efficiencies. This axis of research has not been explored much up to now [197–199, 200], due to resistive limitations in thin-film solar cells. Thus, the advantages of concentration in terms of material savings and efficiency gains were limited. A new thin-film solar cell architecture that has inherently fewer resistive sources was needed and developed: microscale thin-film solar cells.

7.2.4.2.1 Principle of Concentration on Solar Cells

The use of concentrated sunlight can be interesting for solar cells as it reduces the amount of absorbing material necessary to produce a certain electric power.

If the sunlight is concentrated by a factor C, i.e. C suns illumination also noted $\times C$, then the photocurrent should be $J_{ph} = J_{ph,1sun} \times C$. If we neglect series and shunt resistances, and considering $\exp(qV_{oc}/kT) \gg 1$, the open-circuit voltage becomes:

$$V_{oc} = \frac{2kT}{q}\ln\left(\frac{-J_{02} + \sqrt{J_{02}^2 + 4J_{01}C \times J_{ph,1sun}}}{2J_{01}}\right) \tag{7.13}$$

Thus, the open-circuit voltage increases with the logarithm of concentration. At low illumination, the diode current of ideality factor 2 dominates and the slope of the $V_{oc} - \ln(J_{ph,1sun})$ is $2kT/q$, then at higher illuminations the diode current of ideality factor 1 dominates and the slope decreases to kT/q. Due to the increased open-circuit voltage, if series resistance and temperature elevation are negligible, the efficiency under concentration increases.

However, in real devices the efficiency reaches a maximum due to resistive losses that cannot be neglected when the current density is sufficiently high. This optimum concentration ratio can be calculated analytically in the one-diode formalism, and it can be shown that the maximum in efficiency occurs when $J_{ph}R_s = C \times J_{ph,1}R_s \sim AkT/q$, or $C \times J_{sc,1}R_s \sim AkT/q$, if the collection factor is close to 1 at short circuit [201]. One can see that the maximum in efficiency occurs at a concentration ratio inversely proportional to the solar cell series resistance. Thus, its series resistance is too high, the maximum in efficiency occurs at low concentration ratio and the efficiency increase that can be expected is limited. The optimisation of concentrator cells towards the lowest series resistance possible is thus crucial.

7.2.4.2.2 Experiments of Concentration on Cu(In,Ga)Se_2

A few experiments of concentration on Cu(In,Ga)Se_2 solar cells have been conducted [197–199, 200], but maximum efficiencies were obtained at low concentration ($C<15$). The record Cu(In,Ga)Se_2 solar cell under concentration was fabricated by the NREL in 2002, its performance at $\times$ 1 and $\times$ 14 are given in Table 7.5 and Figure 7.20. This cell was 0.1 cm^2, with a collecting grid. They claim that the efficiency at 14 suns was not yet limited by series resistance as they found the same efficiency at $\times$ 50. The concentration of $\times$ 14 is thus probably not the optimum efficiency, which lies between $\times$ 14 and $\times$ 50. The relatively low optimum concentration ratio is the sign of a series resistance of the order of 0.5 ohm cm^2.

An interesting feature highlighted in the works of [197, 198, 200], is that devices that behave poorly under low light intensity benefit more from concentration that good devices because of the higher slope of V_{oc} with concentration ratio, and also lower impact of shunt resistance (Figure 7.21 left).

Table 7.5 Record Cu(In,Ga)Se_2 solar cell under concentration. (*) The value of J_{sc}under 1 sun is inferred from that at 14 suns.

47. Concentration	V_{oc} (mV)	J_{sc} (mA/cm^2)	**FF** (%)	Efficiency (%)
1	647	36.4 (*)	76.3%	17.9%
14	736	510.1	80.5%	21.5%

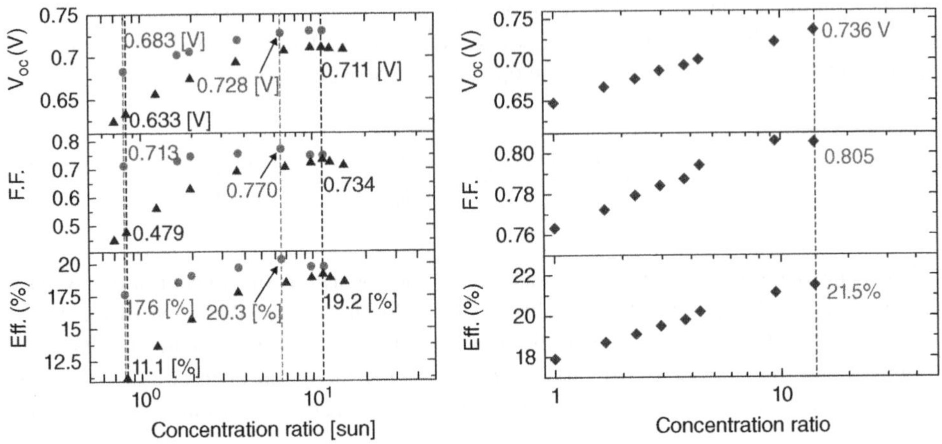

Figure 7.20 (left) Evolution of open-circuit voltage, efficiency and fill factor with concentration ratio of a high-performance Cu(In,Ga)Se_2 cell (red) and a poor device (blue). Reproduced from [200]. (right) Evolution of efficiency and open-circuit voltage with concentration ratio for the record Cu(In,Ga)Se2 solar cell (0.1 cm^2). (Data extracted from [197].)

7.2.4.2.3 Concentration on Cu(In,Ga)Se_2 Microcells

Because of series-resistance losses as well as thermal issues, it is beneficial to go down in scale, especially for thin-film solar cells [202,203]. Cu(In,Ga)Se_2 microcells can be fabricated with microelectronics processes [202, 203]. Their lateral dimensions can vary from a few micrometres to hundreds of micrometre. They performance under concentrated illumination were tested with laser and natural lights, both yielding similar results (Figure 7.21).The

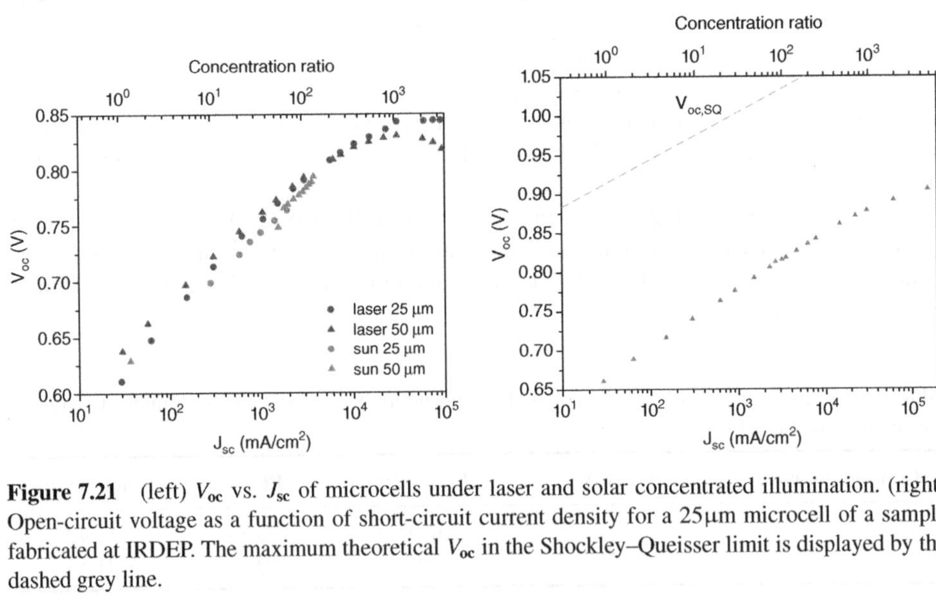

Figure 7.21 (left) V_{oc} vs. J_{sc} of microcells under laser and solar concentrated illumination. (right) Open-circuit voltage as a function of short-circuit current density for a 25μm microcell of a sample fabricated at IRDEP. The maximum theoretical V_{oc} in the Shockley–Queisser limit is displayed by the dashed grey line.

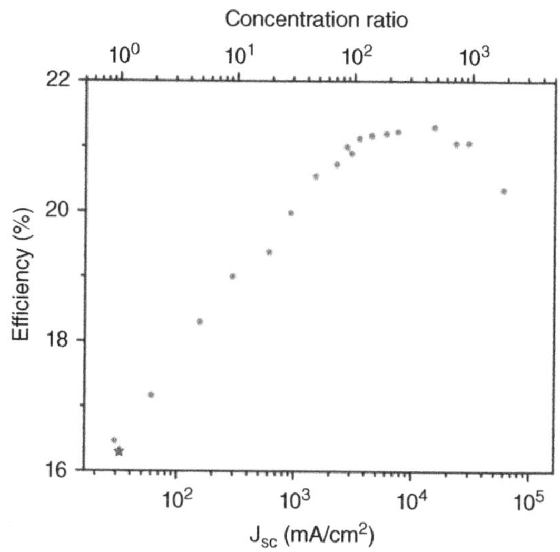

Figure 7.22 Efficiency as a function of concentration for the record cell of IRDEP sample of diameter 50 μm. The star corresponds to the measurement under AM1.5 spectrum with a solar simulator.

photocurrent versus incident power was found linear up to J_{sc} of a few hundred amperes per cm^2.

The highest V_{oc} measured on a microcell is 905 mV at $\times$ 4750. This corresponds to less than 130 mV difference with the V_{oc} of an ideal cell of the same optical bandgap (1.16 eV) at the same concentration, according to the Shockley–Queisser approach.

In Figure 7.22 is plotted the efficiency versus concentration for the champion cell fabricated at IRDEP. An efficiency of 21.3% was recorded under a $\times$475 equivalent solar concentration using a laser (efficiency at the same J_{sc}). This result was made possible by the low resistivity of the sample that enabled a maximum efficiency at $\times$475, which is a record for thin-film solar cells.

7.3 KESTERITES

The sustainability of PV production is a real question as seen in the introduction, and the development of a new PV technology based on abundant and preferably nontoxic elements would alleviate the pressure on all PV technologies in term of resources. The kesterite part is mainly based on ref. [204].

7.3.1 Advantages of CZTS

Recently, the CIGS scientific community started to work on $Cu_2ZnSn(S,Se)_4$ (CZTS), and the enthusiasm was kindled by the fast increase of record efficiency of CZTS-based solar cells, from the 2009 record of 6.7% [205] to the 2012 record of 11.1% [206], as shown

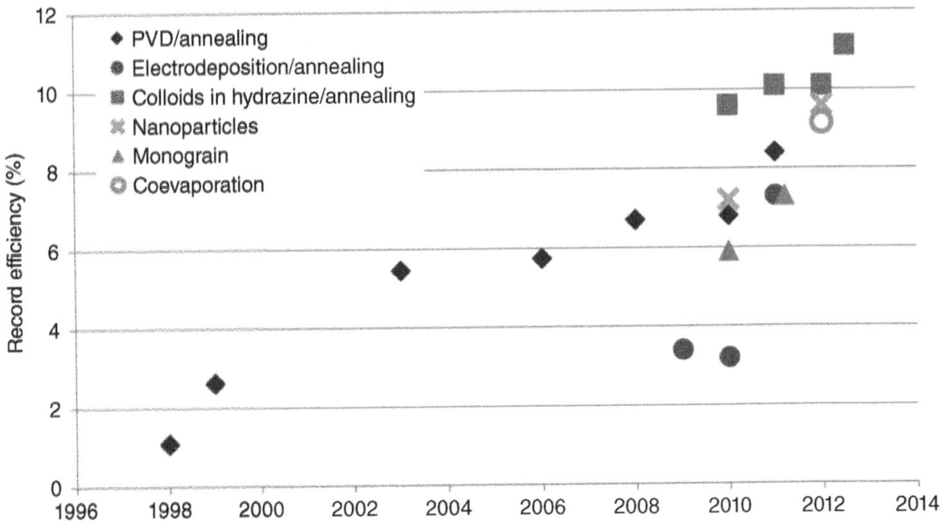

Figure 7.23 Record efficiencies through time of the CZTS solar cells.

in Figure 7.23. CZTS has numerous advantages that could lead to its massive use as an abundant, nontoxic, low-cost absorber for thin-film photovoltaic solar cells:

- It is a compound whose intrinsic point defects lead to p-type semiconductor behaviour.
- It has a direct bandgap and an absorption coefficient $> 10^4$ cm^{-1}, which is suitable for thin film photovoltaics applications [207–209].
- Its bandgap has been predicted [210] to be 1.0 eV for $Cu_2ZnSnSe_4$ and 1.5 eV for Cu_2ZnSnS_4, and evidences of the bandgap tunability have been found via the variations of V_{oc} [211], as well as direct measurements of the variation of the bandgap between 1.0 eV for $Cu_2ZnSnSe_4$ and 1.5 eV for Cu_2ZnSnS_4 [212–216]. This tunability is of particular interest for the manufacturing of absorbers with a bandgap between 1.1 and 1.5 eV, which allow theoretical efficiencies higher than 30% [217]. For Cu_2ZnSnS_4 nanoparticles, higher bandgaps (~1.76 eV) were measured [218].
- Its crystallographic structure can accept some shifts from the stoichiometric composition [219, 220].
- It includes Zn and Sn, which are, respectively, produced in quantities 20 000 and 500 times bigger than In [221].
- It is possible to make CZTS solar cells just by replacing CIGS by CZTS in CIGS solar cells, and such solar cells yielded efficiencies up to 11.1% [206]. The knowledge gathered on back contacts, buffer layers and window layer by CIGS scientists can therefore be used and adapted for CZTS solar cells. Nevertheless, for high-efficiency solar cells the processes used for CIGS solar cells will need to be tuned. The buffer layer, in particular, will have to be tailored in order to adjust the lattice matching, the valence- and conduction-band offsets with CZTS.
- The grain boundaries (GB) seem to have the same beneficial properties for CZTS as for CIGS, such as enhanced minority-carrier collection taking place at the GB [222–224].

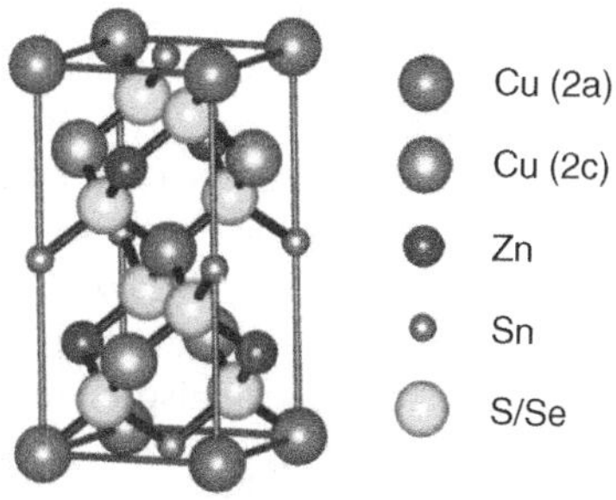

Figure 7.24 $Cu_2ZnSn(S,Se)_4$ compound in kesterite structure, according to [228]. (Reproduced with permission from Hall, 1978. Copyright © Mineralogical Association of Canada, 1978.)

7.3.2 Crystallographic and Optoelectronic Properties

This paper is focused on synthesis processes of CZTS, and thus will only briefly address the crystallographic and optoelectronic properties. For more information on crystal and band structure and defects in kesterites, the reader can find recent reviews dedicated to these topics in refs. [225, 226].

7.3.2.1 Crystalline Structure

In the literature, Cu_2ZnSnS_4 and $Cu_2ZnSnSe_4$ (as well as $Cu_2ZnGe(S,Se)_4$ [227]) are described by the structural models of two natural minerals: stannite (space group *I*-42*m*) [228–230] and kesterite (space group *I*-4, Figure 7.24) [228]. These crystal structures are very close, in both structures the cations are located on tetrahedral sites but their distributions on planes perpendicular to the *c*-axis are not the same. In addition, the position of the chalcogen atom is slightly different in these structures [228–230].

- kesterite: (Cu + Sn) / (Cu + Zn) / (Cu + Sn), chalcogen in position (*x*,*y*,*z*), Figure 7.24;
- stannite: Cu / (Zn + Sn) / Cu, chalcogen in position (*x*,*x*,*z*).

In the rest of the chapter, "kesterite" will be used for the crystallographic structure, and "kesterites" will be used for $Cu_2ZnSn(S,Se)_4$ compounds that are crystallised in kesterite-type structure. X will be used for the chalcogens S and Se when an effect or properties can apply to both of them (for example $Cu_2(S,Se)$ will be denoted Cu_2X.

Due to their structural similarities, kesterite and stannite are very difficult to distinguish by X-ray diffraction and Raman spectroscopy, and it is necessary to use neutron diffraction [231, 232] to tell them apart. According to *ab initio* calculations, the most stable crystalline structure is kesterite [233–237] and one study by *ab initio* calculations even suggests that all observations of stannite structure for CZTS compounds were due to partial disorder in the I-II (001) layer of the kesterite phase [227]. The Tallinn team observed 2 different PL bands in a CZTS crystal, and proposes that they were due to two different phases: kesterite and disordered kesterite [238]. One study suggested that this partial disorder leads to inhomogeneity at the nanometric scale (1–5 nm), creating peaks and troughs in the band structure of CZTS, diminishing carrier lifetime [239].

Recent neutron diffraction [232] and anomalous diffusion [240] characterisations show that CZTS crystallises in the kesterite structure, and synchrotron X-Ray experiments show that this structure is dominant for temperatures < 876 °C (transition to a sphalerite cubic structure) [232]. Raman characterisation suggests that the kesterite structure of the CZTS presents a highly disordered distribution of Cu and Zn atoms [241].

The structural difference between Cu_2ZnSnS_4 and $Cu_2ZnSnSe_4$ is not clear. It seems that the Se compound has larger lattice parameters as well as higher electric conductivity than the S compound [212]. The bandgap of CZTS can also be tuned by incorporation of other elements without changing the structure of the crystal:

- Ge ($Cu_2Zn(Sn,Ge)(S,Se)_4$), Ge-containing materials have bigger bandgap than their Ge-free counterparts [242];
- ($Cu_2ZnSn(O,S,Se)_4$), oxygen increases the bandgap. $Cu_2ZnSn(O,S)_4$ materials have bandgaps between 1.5 and 2.1 eV [243].

Concerning mechanical and thermophysical properties of CZTS, no experimental data are available, but *ab initio* simulations were performed [244, 245]. The conductivity properties have been studied [246], as have the elastic and optical properties [247].

Sulfur kesterites have XRD peaks very close to those of ZnS and Cu_2SnS_3 [231], and Raman spectroscopy is necessary to tell them apart [248]. Nevertheless, it is not possible to have a quantitative determination of the secondary phases by Raman. For such a purpose, XENAS has been used [249].

7.3.2.2 Intrinsic Point Defect Formation

The kesterite structure allows for several types of point defects, which make discrete energy levels appear, allowing p-type doping. As in the CIGS materials, the doping is obtained by stoichiometry variations and not by extrinsic doping. Nevertheless, according to *ab initio* calculations, the kesterite structure is less tolerant to these variations than the chalcopyrite structure [234, 250–253]. According to point-defect measurements, the kesterites are susceptible to higher carrier recombination than chalcopyrite [224]. Measurements of the deep defects of CZTS are reported in ref. [254].

Generally, the Sn–Se bond is much stronger than the Zn–Se and Cu–Se bond, leading to a high formation energy of V_{Sn} [253]. Contrary to CIGS, the accepting defect the most easily formed (and hence dominant accepting defect) is not V_{Cu} but Cu_{Zn} [234, 251, 255–257]. Nevertheless some authors suggest the contrary [224, 258, 259]. This accepting defect is not as shallow as V_{Cu}, possibly resulting in carrier freeze out, and could be the biggest issue of CZTS concerning high-efficiency devices [206]. Zn_{Cu} substitutions occur at the 2c site and Cu_{Zn} at the 2d site [260, 231] (see Figure 7.24). The accepting defects are easily formed, leading to difficult n-type doping [255]. The easy formation of Cu_{Zn} accepting defects also leads to low charge separation, a drawback that could be overcome thanks to the type-II band alignment of the CZTS / CdS interface [251]. The most active recombination centres are expected to be Cu_{Sn} (In_{Cu} for CIGS), and the lack of defect complex ($2V_{Cu} + In_{Cu}$ for CIGS) is thought to lead to the higher defect concentration observed [261].

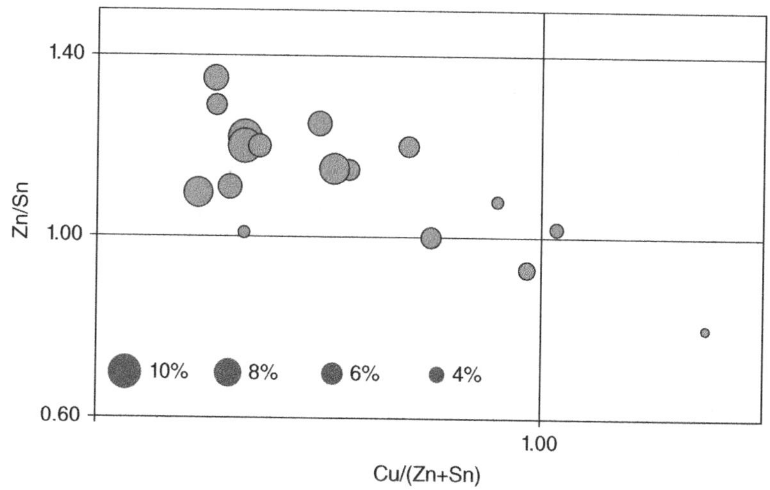

Figure 7.25 Efficiencies of kesterites as a function of composition.

7.3.2.3 Metal Composition

The single-phase composition region is much narrower for CZTS than for CIGS. According to the Tallinn team, single-phase CZTS monograin powders can be synthesised only from a precursor mixture comprising metal ratios of Cu/(Zn + Sn) = 0.92–0.95 and Zn/Sn = 1.0–1.03 [262]. Nevertheless, a Rietveld refinement of anomalous dispersion measurements revealed a Cu / (Zn + Sn) = 0.97, and Zn / Sn = 1.42 composition for kesterites crystals [240], suggesting that the single-phase composition region is not as narrow as the Tallinn team suggests.

Until now, only Zn-rich and Cu-poor materials gave high efficiency (see Figure 7.25), as was already noted by the Nagaoka team [263]. In Zn-rich growth conditions, the kesterite structure is tolerant to stoichiometry deviations by zinc excess and copper deficiency [263–266], in the form of Zn_{Cu} and V_{Cu} vacancies, [267, 253], but high Zn content also leads to structural disorder [268].

7.3.2.4 Bandgap Measurement

Absorbance measurements are not suitable for extracting the bandgap values, because the absorption coefficient derived from spectrophotometric data on thin films is determined by defect absorption and by measurement accuracy limitations [268]. In particular, ZnSe [269] and Zn-caused disorders [268] lead to incorrect bandgap determination when it is determined from absorbance measurements. PL measurements now confirm that the bandgap of $Cu_2ZnSnSe_4$ is indeed around 1.0 eV at room temperature [270], whereas before that publication the measured bandgap was around 1.5 eV. Suitable methods for measurement of CZTS bandgap are PL, TR-PL and EQE measurements.

7.3.3 Synthesis Strategies

7.3.3.1 One-Step or Two-Step Processes for Thin Films

For CIGS, the deposition techniques processes used to be classified as vacuum and nonvacuum techniques. For CZTS, the same approach was also used in different reviews [271–273]. Nevertheless, for CZTS, this classification is not accurate, because recent works showed that low-cost deposition does not mean low-cost solar modules [274], and because of the problem of binaries control during CZTS synthesis (see next section). We therefore decided to classify the processes into one-step or two-step processes:

- Two-step processes, where the needed elements are first incorporated during an ambient temperature step, followed by an annealing step. The chalcogen can be incorporated into the precursor or during the annealing step. These processes allow the use of fast and low-cost techniques for precursor deposition. Until now, and contrary to the CIGS, two-step processes yielded the best efficiencies (more than 10%) [275]. IBM worked on three such processes, the hybrid solution-particle technique [275], electrodeposition [276] and coevaporation [277]. Other teams investigated numerous other deposition techniques, such as codeposition by evaporation, nanoparticles deposition, sputtering, OA-CVD (open atmosphere chemical vapor deposition), nanoparticles, and sol-gel.
- One-step processes, where all the elements are incorporated simultaneously. This type of processes yields the better results for CIGS, but until now it was not the case for CZTS. The best efficiency until now is 9.15% by coevaporation [278]. Only a few groups published results for such processes, and very few had significant results in terms of solar cell efficiency: NREL [278], Tallinn University [279], HZB [280]. Other groups tried reactive sputtering and pulsed laser deposition (PLD). The method of Tallinn University also introduces a huge difference in cell architecture, where 50 μm CZTS monograins are synthesised in molten KI, wrapped in the buffer layer and attached to the substrate by epoxy glue [281].

These processes are different, but the two types of processes (one-step and two-step) reached high efficiencies (> 9%). Until now, it does not seem that one or the other is intrinsically better, because, as we will see in the next sections, the key point is temperature and atmosphere control during the one-step deposition or annealing step.

7.3.3.2 CZTS Formation

Similarly to CIGS, CZTS needs temperature between 500 and 600 °C to be synthesised. Nevertheless, the chemical reactions are not yet completely understood, even though a reaction path has been proposed [282]. There are successive reactions taking place in the bulk of the layer between the elements leading to binaries, ternaries and finally the quaternary.

At room temperature, only metal binaries can form, for example Cu_6Sn_5 [282, 283], Cu_5Zn_8 [284], Cu_3Sn and CuZn [285]. At higher temperature, between 200 and 450 °C, metal-chalcogenide binaries such as CuX, Cu_2X, and SnX form [286–288], and for temperature > 450 °C, Cu_xX binaries then react with Sn to form Cu_xSn_yS [82]. At higher temperature (between 550 and 580 °C [276, 285, 286, 289],), and for longer annealing time (~ 8 h) or

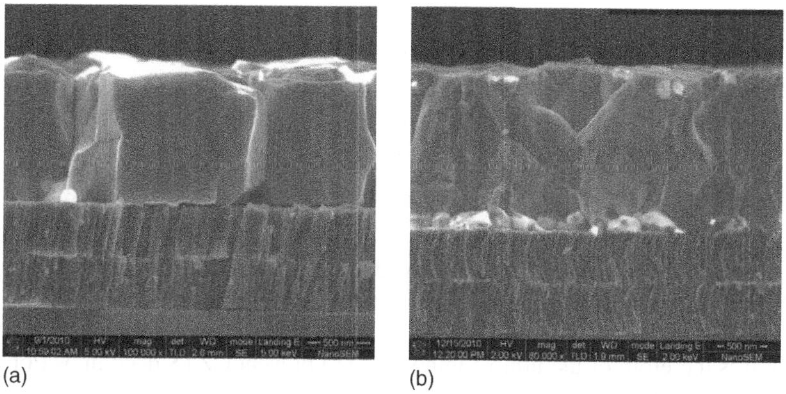

Figure 7.26 SEM cross section of CZTSe films grown at the same temperature and constant deposition fluxes, except Cu rates were adjusted to produce only (a) Cu-rich growth or (b) Cu-poor growth [261]. (Reproduced with permission from Repins, 2011. Copyright © Cambridge University Press, 2011.)

one-step deposition at 500 °C [278, 283], ZnX reacts with Cu_2SnX_3 to form Cu_2SnZnX_4 according to the equation proposed by [282]:

$$Cu_2SnX_3 + ZnX \rightarrow Cu_2ZnSnX_4 \tag{7.14}$$

In order to control which reactions take place during the annealing step of a two-step process, it is necessary to closely monitor the temperature of the layer, because, for example, too fast an annealing can lead to liquid Sn bubbles formation, preventing the formation of large grains [290]. For processes using elemental sulfur or selenium, fast annealing (a few minutes) is possible, whereas for processes using H_2S, which is less reactive than elemental sulfur, a long annealing time (2–3 h) is necessary for the binaries to react completely and to form large grains of CZTS [291, 292].

For one-step processes, Cu-rich growth conditions are needed at the beginning of the reaction in order to foster growth of large grains [100, 278, 293], as shown in Figure 7.26. Nevertheless, for both processes, the finished layer must be Zn-rich (see Section 7.7.3.2.3), and in particular Cu_2X-free [262] in order to get high photovoltaic efficiencies. If the layer is not Cu_2X-free at the end of the synthesis process, the subsequent cyanide etching leads to possible voids and defects [289]. If Zn-rich growth conditions are used, ZnX formation is promoted [250, 251].

Therefore, for one-step processes it is necessary to adapt the rate of deposition of each element with respect to the reaction path, as well as control deposition temperature, because higher temperatures can also increase grain size [261].

For two-step processes it is necessary to closely monitor growth conditions, and in particular prevent Cu_2X formation and increase ZnX reaction with Cu_2SnS_3.

7.3.3.3 Atmosphere Control

Unfortunately, contrary to CIGS, the elements involved in CZTS synthesis are prone to evaporation and sublimation. Zn sublimates at 430 °C [294], SnSe at 350 °C [288], SnS at

370 °C and Sn evaporates at 460 °C [295,296]). Moreover, high temperatures promote CZTS decomposition [297], according to the chemical equilibrium between CZTS and solid and gaseous binaries [298–300].

$$Cu_2ZnSnX_4(s) \leftrightarrow Cu_2X(s) + ZnX(s) + SnX(s) + {}^{1}\!/_{2}X_2(g) \quad (7.15)$$

$$SnX(s) \leftrightarrow SnX(g) \quad (7.16)$$

If the equilibrium of equation (7.15) is displaced towards the right member by SnX evaporation, ZnX and SnX formation and SnX and X_2 evaporation will occur, leading to a Cu_2X-rich layer, increasing layer resistance and preventing good photovoltaic efficiencies (see Section 7.**7.3.3.7**). It is therefore of the utmost importance to prevent CZTS decomposition as well as binaries losses. The key to that is atmosphere control.

The equilibrium of equation (7.15) can be displaced towards CZTS by saturating the atmosphere in one of the right-side members of the equation [293, 298, 300]. It is possible to counteract SnX evaporation by introducing gaseous SnX during the thermal treatment [299], but at the risk of SnX condensation at the end of the process [297]. It is also possible by increasing the partial pressure of the chalcogen [301,302], (but too high a partial pressure can worsen the crystallinity of the CZTS [303]), by performing the chalcogenation in a small closed volume [304], or by using a cap on the precursor layer during annealing [305]. Nevertheless, according to [300], the only way to stabilise CZTS during long annealing is to provide both high SnX and S_2 partial pressures. If the annealing takes place in a H_2 atmosphere, Zn loss can be prevented by using ZnS as a precursor [306]. Continuous evaporation of Zn towards the substrate prevents the decomposition of CZTS in a coevaporation process [278]. It also seems that the presence of MoX_2 at the back contact pushes the reaction towards the right member [307].

If atmosphere control is not good, annealing time is the next best lever on CZTS decomposition, because if CZTS is synthesised and cooled rapidly, only a small quantity of binaries will be lost. One paper compared slow and fast annealing [308]. It concluded that fast annealing is better than slow annealing because many different secondary phases were found with slow annealing, without further explanation. Nevertheless, in this case slow annealing could be linked to a badly controlled reaction atmosphere, leading to CZTS decomposition into binary and ternary phases.

Concerning the key point of atmosphere control, strategies of synthesis should be slightly different for one-step or two-step processes. For two-step processes, there seems to be two stages during the annealing step: one short (a few minutes) step of CZTS formation, and one longer (up to a few hours) step of grain growth. The control of the atmosphere should be very accurate in the second step, and the reaction atmosphere could also be completely different from the first step [293]. Fortunately, it is quite easy and inexpensive to design annealing equipments that provide good static atmosphere control (that is to say, equipments where nothing can either go in or go out). Nevertheless, vapor injection is trickier, as it requires carrier gas, pressure measurements, leading to an equipment resembling an evaporator.

For one-step synthesis, atmosphere control is not so easy, because reaction chambers are designed for controlling the deposition rate and not the partial pressure of each element. In order to prevent CZTS decomposition and binaries losses, Sn and chalcogen deposition rate should remain high during the cooling phase, in order to counteract Sn loss [278].

7.3.3.4 Metal Composition Gradient

The composition gradient either created by a variation of the rate of deposition of each element in a codeposition or by stacked elements, has an influence on CZTS film properties, because of Cu diffusion, Sn loss and chemical mechanisms of formation.

Concerning Cu, its diffusion can create voids at the back contact. Two-step processes, especially when the precursor layer is a codeposited metal layer or stacked metal layers with Cu as the first layer, are subject to these voids [289, 307]. Secondary phases are observed, preventing the total formation of CZTS [309].

As explained in the previous paragraph (see equation (7.2)), Sn evaporates if the atmosphere is not well controlled. Nevertheless, the loss of Sn could also be prevented by having the Cu layer on top [265, 293].

Concerning the reaction mechanism, for stacked metals, it is beneficial to have Cu and Sn in close contact for the formation of large CZTS grains [310]. If they are separated by Zn or Zn(S,Se), the reaction between copper chalcogenides and tin chalcogenides, which is necessary to form Cu_2SnX_3 [282], could be inhibited [311]. For stacked metals, Cu on top seems to foster formation of Cu_xS, which may result in larger grains and denser films [312]. For one-step processes, high Cu content at the beginning of the deposition promotes the formation of large grains [278].

Concerning the composition gradient,

- For two-step processes, the Cu must not be deposited first, and Cu and Sn should be in close contact. The possible combinations for stacks are therefore Mo/Zn(X)/Cu/Sn or Mo/Zn(X)/Sn/Cu. The codeposition seems also possible, but with a Cu-poor composition near the back contact.
- For one-step processes, the layer must be Cu rich at the beginning of the deposition in order to promote the growth of large grains.

7.3.3.5 Chalcogen Composition Gradient

A chalcogen composition gradient can be beneficial, as it creates a bandgap gradient. When inserting Se into CZTS, Se diffuses most easily in Cu-Zn-Sn-S precursors, and less easily in CZTS or Cu-ZnS-Sn precursors [313].

7.3.3.6 Sodium

Sodium seems to have the same effects on CZTS as on CIGS: it promotes the growth of larger grains, enhances conductivity and has a significant effect on film morphology, as was demonstrated by SLG/borosilicate substrate comparison and dipping in Na_2S [314, 315]. Nevertheless, contrary to CIGS, Na was not detected by XPS at the surface, meaning that the Na diffusion in CZTS is smaller than in CIGS [316].

7.3.3.7 Device Properties

CZTS devices are similar to CIGS devices. The architecture is usually SLG/Mo/CZTS/CdS/i-ZnO/ZnO:Al, as presented in Figure 7.27. Calculation showed that CdS has a suitable band offset with CZTS [317] and until now, the most efficient solar cells were obtained with

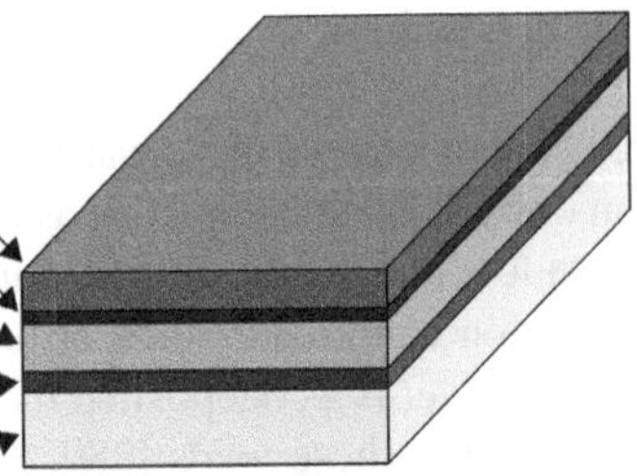

Figure 7.27 Classical architecture for CZTS solar cells.

a CdS buffer layer. According to SCAPS simulations, metals with higher work functions than Mo, such as Au, In, Ni, Co, Mo, Pd, Pt, Re and W could improve V_{oc}. [318].

Oxidation of the CZTS absorber by dipping in deionised water [205] or O_2 annealing [278] before deposition of CdS seems to improve efficiencies. As for CIGS, reducing the time between absorber synthesis and CdS deposition to the minimum is of the utmost importance [310].

A few variations of the architecture are known, especially for the buffer layer. Pure ZnS does not seem to be a suitable buffer layer because of too high a band offset [317], but good results were obtained with Zn(S,O,OH) [319] and In_2S_3 [320] by Solar Frontier. Nevertheless for Zn-based buffer layer the optimal composition of the absorber layer seems to be less Zn-rich (Zn/Sn $\sim$ 1) [319]. Other architectures were proposed: SLG / Mo / CZTS / a-Si / ITO [321], a superstrate architecture (glass / FTO / CZTS) [322, 323], another superstrate (glass / FTO / TiO_2 / In_2S_3 / CZTS / Mo) [324, 325] with 0.6% conversion efficiency, a flexible substrate (Al foil / Mo / CZTS / ZnS / i-ZnO / ITO / Al-Ni) [326], and replacement of Mo by ITO [323].

The literature is still quite poor concerning device properties, because very few high-efficiency devices were synthesised yet. The IBM team started to investigate this field [275, 277, 327, 328], concluding that their devices were limited by the recombination at the absorber / buffer interface, minority carrier lifetime and Schottky-type barrier in the back contact, which appears to be caused by secondary phases and/or the MoS_x interfacial layer between CZTS and Mo [329].

Concerning the absorber / buffer interface, it seems that the surface of the CZTS is Cu-poor, similarly to CIGS [316], or even Cu-free [330]. The KCN etching, which is also used in CIGS processes to remove binaries from the surface, seems to etch Cu preferentially and Sn to a lesser extent [293]. It also seems beneficial for removing Cu_2X that increase the resistivity of the layer [331, 332]. This causes a widening of the bandgap without type inversion, which could be an easy way to tailor bandgap alignments [333, 334]. It also seems that the electronic structure of a Sn-depleted surface is not favourable for the formation of well working p-n junction [279, 293]. In the case of $Cu_2ZnSnSe_4$, a thin layer of ZnSe at the surface of the absorber was found to be current blocking [335]. Even small amounts of tin binaries and ternaries at the interface were found to have a disastrous effect on the solar cell properties, because of a low conduction-band edge [332]. HCl etching has been found to remove ZnS completely and ZnSe to a lesser extent [336].

The conduction-band offset between CZTS and CdS was measured between 0.34 and 0.5 eV, which is somewhat above the optimal range of 0–0.4 eV and may contribute to lower J_{sc} and the fill factor in the CZTSSe devices [337, 338]. Nevertheless, the spike-like "type

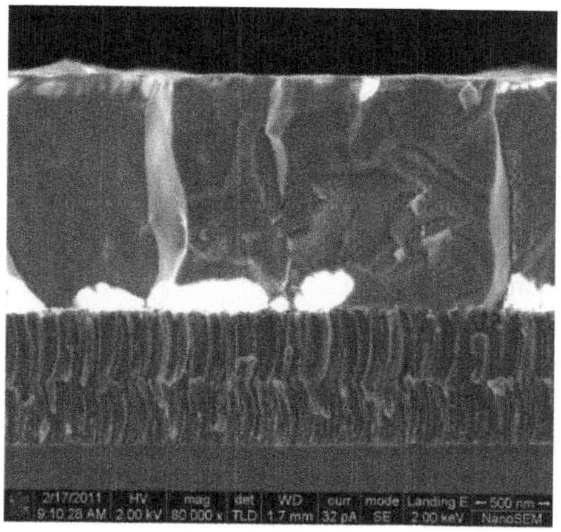

Figure 7.28 CZTSe with ZnSe segregation at the back contact [261]. (Reproduced with permission from Repins, 2011. Copyright © Cambridge University Press, 2011.)

1" band alignment at the interface decreases recombination. Better lattice matching between absorber and buffer could limit recombination at the interface, and using different buffers with smaller lattice such as In_2S_3 or Zn(S, O, OH) could improve the efficiency of the devices [339, 340]. ZnS does not seem to be a good buffer material because too high a CBO and ZnO creates a negative band offset, which decreases the V_{oc} [320]. In_2S_3 creates a slightly positive CBO with CZTS, which is promising combined with its good lattice matching to CZTS [320].

Concerning the back contact, ZnX was observed at the back contact [261, 310], as shown in Figure 7.28, and it seems that increasing the ZnX content strongly impairs efficiency [249]. It was nevertheless speculated that ZnS could be less detrimental than previously thought [277, 332] acting as a simple "dark" material. Binary losses that occur during annealing makes Cu_2X to diffuse towards the rear of the cell, could lead to a Cu_2X / MoX_2 mix, leading to high resistance at the back contact. For the selenide compound, the device performance was found to be reversely correlated to $MoSe_2$ thickness [206, 341]. The $MoSe_2$ thickness can be controlled be adjusting the Se pressure during precursor preparation [341] or during the selenisation step [302]. The formation of $MoSe_2$ can be averted by using a TiN layer, leading to much higher V_{oc} [341].

Concerning minority carrier lifetime, it might not be such a problem, because as in CIGS and CdTe, grain boundaries collect minority carriers and provide a current pathway for them to reach the n-type CdS and ZnO layers [222] and for coevaporated absorbers, high minority carrier diffusion length was measured (several hundreds of nm [277]). Nevertheless, it seems that grain boundaries are Cu-rich, which could increase recombination rates and thus diminishing V_{oc} [342]. Cu-rich absorbers were found to have lower V_{oc} than Cu-poor ones [262], probably in relationship with Cu_2X formation. Longer minority-carrier lifetimes were measured in low-bandgap materials, such as CZTSe, than in high-bandgap devices [342]. The difference between E_g in eV and V_{oc} in V, called the V_{oc} deficit, can therefore be decreased by diminishing the bandgap as well as tightly controlling the formation of Cu_2X or its removal by KCN etching.

For high efficiency, it is important to focus both on absorber quality and on interfaces between the absorber and the back contact and buffer. The Cu_xX binaries control seems to be a key point for both absorber and interface quality. Fortunately, the CIGS community has strong experience in such problems, and hopefully it will be possible to optimise these interfaces much faster than for the CIGS.

The field of CZTS solar cells is young, and until now it is difficult to point to certainties. Nevertheless, it seems that these assumptions start to be viewed as "common knowledge" in the community:

- CZTS has kesterite structure;
- the bandgap is tunable between 1.0 eV ($Cu_2ZnSnSe_4$) and 1.5 eV (Cu_2ZnSnS_4);
- the synthesis atmosphere must be controlled carefully;
- binary formation control is a key point for high-efficiency devices;
- Zn-rich materials yield better results (when using a CdS buffer layer).

Nevertheless, there are numerous questions still, such as

- What are the crystalline defects, how are they formed, what role do they play in the structure?
- What is the reaction path to CZTS, how to control binaries formation?
- What is the role of sodium during crystal growth?
- How should interfaces be tailored for best solar efficiency.

More and more scientists from the CIGS community now work on CZTS solar cells. The problems encountered in CZTS solar cells are quite similar to the ones of CIGS solar cells, and similar methods can be used to solve them. There is no clear superiority of two-step processes over one-step processes, even if until now the best results were obtained with one-step processes. For both types of processes, a focus on binary control formation by atmosphere control and adaptation of interfaces to kesterites materials are the key points for increasing photovoltaic efficiencies for both types of processes.

ACKNOWLEDGEMENTS

SD is thankful for financial support of the French Agence Nationale de la Recherche (ANR) under grant NovACEZ (ANR-10-HABISOL-008).

REFERENCES

[1] International Energy Agency, "*World Energy Outlook 2011.*" 2011.
[2] Greenpeace and the European Photovoltaic Industry Association, "*Solar Generation VI,*" 2011.
[3] European Climate Foundation, "*Roadmap 2050, a practical guide to a prosperous, low-carbon Europe,*" 2010.
[4] International Energy Agency, "*Technology Roadmap, Solar photovoltaic energy,*" IEA, 2009.
[5] A. Zaman and S. Lockman, "*Solar industry growth: you ain't seen nothing yet. The grid parity decade,*" Piper Jaffray Investment Research, 2011.
[6] C. Breyer, M. Görig, and J. Schmid, "Fuel-parity: impact of photovoltaics on global fossil fuel fired power plant business," presented at the *26th Symposium Photovoltaische Solarenergie*, Bad Staffelstein, 2011.
[7] "*La lettre du solaire.*" Cythelia, 2011.

[8] JRC Scientific and Technical reports, "*PV status report 2011.*" 2011.
[9] "Lux Research."
[10] EPIA, "*Solar photovoltaics competing in the energy sector, on the road to competitiveness.*" 2011.
[11] MiaSolé, "MiaSolé Achieves 15.7% Efficiency with Commercial-Scale CIGS Thin Film Solar Modules," *http://www.miasole.com/sites/default/files/MiaSole_release_Dec_02_2010.pdf*, 2010.
[12] P. Jackson, D. Hariskos, E. Lotter, S. Paetel, R. Wuerz, R. Menner, W. Wischmann, and M. Powalla, "New world record efficiency for Cu(In,Ga)Se2 thin-film solar cells beyond 20%," *Progress in Photovoltaics: Research and Applications*, **19**, pp. 894–897, 2011.
[13] Gloeckler et al., CdTe at the threshold to 20% efficiency, Proceedings of the 39th IEEE photovoltaic specialist conference, (2013)
[14] Avancis Press Release, "World record in efficiency: AVANCIS again achieves best value for thin-film modules," *http://www.avancis.de/en/press/press-releases/view/meldung/wirkungsgrad-weltrekord-avancis-erzielt-erneut-bestwert-fuer-duennschichtmodule/*, 2011.
[15] Solar Frontier, "Solar Frontier Sets New Efficiency World Record," *http://www.solar-frontier.com/news/179*, 2012.
[16] SoloPower Press release, "SoloPower sets record for flexible CIGS solar panel efficiency," *http://solopower.com/2012/03/solopower-sets-record-for-flexible-cigs-solar-panel-efficiency/*, 2012.
[17] V. Bermudez, "Développements industriels du CIGS électrodéposé," presented at the *Journées nationales du photovoltaïque*, Dourdan, France, 2011.
[18] J. L. Shay, S. Wagner, and H. M. Kasper, "Efficient CuInSe2/CdS solar cells," *Applied Physics Letters*, **27**, no. 2, p. 89, 1975.
[19] V. Fthenakis, "Sustainability of photovoltaics: The case for thin-film solar cells," *Renewable and Sustainable Energy Reviews*, **13**, no. 9, pp. 2746–2750, 2009.
[20] C. Candelise, J. F. Spiers, and R. J. K. Gross, "Materials availability for thin film (TF) PV technologies development: A real concern?," *Renewable and Sustainable Energy Reviews*, 2011.
[21] A. Zuser and H. Rechberger, "Considerations of resource availability in technology development strategies: The case study of photovoltaics," *Resources, Conservation and Recycling*, **56**, no. 1, pp. 56–65, Nov. 2011.
[22] "Indium." USGS Minerals information.
[23] U. Rau and H. W. Schock, "Cu(In,Ga)Se_2 solar cells," in *Clean Electricity From Photovoltaics*, 1, Published by Imperial College Press and Distributed by World Scientific Publishing Co., 2001, pp. 277–345.
[24] R. Scheer and H. W. Schock, *Chalcogenide Photovoltaics: Physics, Technologies, and Thin Film Devices*. Weinheim, Germany: Wiley-VCH Verlag & Co, 2011.
[25] W. Shafarman and L. Stolt, "Cu(In,Ga)Se2 solar cells," in *Handbook of Photovoltaic Science and Engineering*, John Wiley and Sons, 2003.
[26] R. Klenk and M. C. Lux-Steiner, "Chalcopyrite Based Solar Cells," in *Thin Film Solar Cells*, J. Poortmans and V. Arkhipov, Eds. John Wiley & Sons, Ltd, 2006, pp. 237–275.
[27] G. Gordillo, M. Grizález, and L. C. Hernandez, "Structural and electrical properties of DC sputtered molybdenum films," *Solar Energy Materials and Solar Cells*, **51**, no. 3–4, pp. 327–337, 1998.
[28] J. H. Yoon, K. H. Yoon, W. M. Kim, J. K. Park, Y. J. Baik, T. Y. Seong, and J. Jeong, "High-temperature stability of molybdenum (Mo) back contacts for CIGS solar cells: a route towards more robust back contacts," *Journal of Physics D: Applied Physics*, **44**, p. 425302, 2011.
[29] P. Salomé, J. Malaquias, P. Fernandes, and A. F. Cunha, "Mo bilayer for thin film photovoltaics revisited," *Journal of Physics D: Applied Physics*, **43**, p. 345501, 2010.

[30] F. Kessler and D. Rudmann, "Technological aspects of flexible CIGS solar cells and modules," *Solar Energy*, **77**, no. 6, pp. 685–695, Dec. 2004.

[31] D. Brémaud and A. N. Tiwari, "Flexible Cu(In,Ga)Se2 Solar Cells and Technologies," in *Proccedings of the 22nd European Photovoltaic Solar Energy Conference*, Milano, 2007.

[32] D. Boyd and D. Thompson, *Kirk-Othmer Encyclopaedia of Chemical Technology*, 3rd edn, **11**. Wiley & Sons, 1980.

[33] M. Martínez, C. Guillén, A. Morales, and J. Herrero, "Arrangement of flexible foil substrates for CuInSe2-based solar cells," *Surface and Coatings Technology*, **148**, no. 1, pp. 61–64, 2001.

[34] F. Kessler, D. Herrmann, and M. Powalla, "Approaches to flexible CIGS thin-film solar cells," *Thin Solid Films*, **480–481**, pp. 491–498, 2005.

[35] J. R. Tuttle, A. Szalaj, and J. Keane, "A 15.2% AMO/1433 W/kg thin-film Cu(In,Ga)Se2 solar cell for space applications," in *Photovoltaic Specialists Conference, 2000. Conference Record of the Twenty-Eighth IEEE*, 2000, pp. 1042–1045.

[36] F. Pianezzi, A. Chirilă, P. Blösch, S. Seyrling, S. Buecheler, L. Kranz, C. Fella, and A. N. Tiwari, "Electronic properties of Cu(In,Ga)Se_2 solar cells on stainless steel foils without diffusion barrier," *Progress in Photovoltaics: Research and Applications*, p. n/a–n/a, 2012.

[37] T. Nakada, T. Yagioka, K. Horiguchi, T. Kuraishi, and T. Mise, in *Proceedings of the 24th European Photovoltaic Solar Energy Conference*, Hamburg, Germany, 2009, p. 2425.

[38] M. Powalla, D. Hariskos, E. Lotter, M. Oertel, J. Springer, D. Stellbogen, B. Dimmler, and R. Schäffler, "Large-area CIGS modules: processes and properties," *Thin Solid Films*, **431–432**, pp. 523–533, 2003.

[39] A. Chirilă, S. Buecheler, F. Pianezzi, P. Bloesch, C. Gretener, A. R. Uhl, C. Fella, L. Kranz, J. Perrenoud, S. Seyrling, R. Verma, S. Nishiwaki, Y. E. Romanyuk, G. Bilger, and A. N. Tiwari, "Highly efficient Cu(In,Ga)Se2 solar cells grown on flexible polymer films," *Nature Material*, **10**, no. 11, pp. 857–861, 2011.

[40] A. Tiwari, EMPA Press release, http://www.empa.ch/plugin/template/empa/*/131441, (2013)

[41] T. Wada, N. Kohara, T. Negami, and M. Nishitani, "Chemical and Structural Characterization of Cu(In,Ga)Se2/Mo Interface in Cu(In,Ga)Se2 Solar Cells," *Japanese Journal of Applied Physics*, **35**, no. 10A, p. L1253, 1996.

[42] M. Powalla and B. Dimmler, "Scaling up issues of CIGS solar cells," *Thin Solid Films*, **361**, pp. 540–546, 2000.

[43] D. Rudmann, A. F. da Cunha, M. Kaelin, F. Kurdesau, H. Zogg, A. N. Tiwari, and G. Bilger, "Efficiency enhancement of Cu(In,Ga)Se_2 solar cells due to post-deposition Na incorporation," *Applied Physics Letters*, **84**, no. 7, pp. 1129–1131, 2004.

[44] M. Kaelin, D. Rudmann, and A. N. Tiwari, "Low cost processing of CIGS thin film solar cells," *Solar Energy*, **77**, no. 6, pp. 749–756, Dec. 2004.

[45] S. Taunier, J. Sicx-Kurdi, P. P. Grand, A. Chomont, O. Ramdani, L. Parissi, P. Panheleux, N. Naghavi, C. Hubert, M. Ben-Farah, J. P. Fauvarque, J. Connolly, O. Roussel, P. Mogensen, E. Mahé, J. F. Guillemoles, D. Lincot, and O. Kerrec, "Cu(In,Ga)(S,Se)2 solar cells and modules by electrodeposition," *Thin Solid Films*, **480–481**, no. 0, pp. 526–531, 2005.

[46] L. Assmann, J. C. Bernède, A. Drici, C. Amory, E. Halgand, and M. Morsli, "Study of the Mo thin films and Mo/CIGS interface properties," *Applied Surface Science*, **246**, no. 1–3, pp. 159–166, 2005.

[47] X. Zhu, Z. Zhou, Y. Wang, L. Zhang, A. Li, and F. Huang, "Determining factor of MoSe2 formation in Cu(In,Ga)Se2 solar Cells," *Solar Energy Materials and Solar Cells*, **101**, pp. 57–61, 2012.

[48] W. N. Shafarman and J. E. Phillips, "Direct current-voltage measurements of the Mo/$CuInSe_2$ contact on operating solar cells," in *Photovoltaic Specialists Conference, 1996., Conference Record of the Twenty Fifth IEEE*, 1996, pp. 917–919.

[49] P. Bommersbach, L. Arzel, M. Tomassini, E. Gautron, C. Leyder, M. Urien, D. Dupuy, and N. Barreau, "Influence of Mo back contact porosity on co-evaporated Cu(In,Ga)Se2 thin film

properties and related solar cell," *Progress in Photovoltaics: Research and Applications*, **21**, no. 3, pp. 332–343, 2013.

[50] J. F. Guillemoles, U. Rau, L. Kronik, H. W. Schock, and D. Cahen, "Cu(In,Ga)Se2 solar cells: stability based on chemical flexibility," *Advanced Materials*, **11**, no. 11, pp. 957–961, 1999.

[51] N. Kohara, S. Nishiwaki, Y. Hashimoto, T. Negami, and T. Wada, "Electrical properties of the Cu(In,Ga)Se2/ MoSe2/Mo structure," *Solar Energy Materials and Solar Cells*, **67**, no. 1–4, pp. 209–215, Mar. 2001.

[52] S. Niki, M. Contreras, I. Repins, M. Powalla, K. Kushiya, S. Ishizuka, and K. Matsubara, "CIGS absorbers and processes," *Progress in Photovoltaics: Research and Applications*, **18**, no. 6, pp. 453–466, 2010.

[53] M. Ohring, *The Materials Science of Thin Films*. Academic Press, 1992.

[54] J. H. Scofield, A. Duda, D. Albin, B. L. Ballard, and P. K. Predecki, "Sputtered molybdenum bilayer back contact for copper indium diselenide-based polycrystalline thin-film solar cells," *Thin Solid Films*, **260**, no. 1, pp. 26–31, 1995.

[55] T. Wada, N. Kohara, S. Nishiwaki, and T. Negami, "Characterization of the Cu(In,Ga)Se2/Mo interface in CIGS solar cells," *Thin Solid Films*, **387**, no. 1–2, pp. 118–122, May 2001.

[56] K. Orgassa, H. W. Schock, and J. H. Werner, "Alternative back contact materials for thin film Cu (In, Ga) Se2 solar cells," *Thin Solid Films*, **431**, pp. 387–391, 2003.

[57] T. Kubart, T. Polcar, L. Kopecký, R. Novák, and D. Nováková, "Temperature dependence of tribological properties of MoS_2 and $MoSe_2$ coatings," *Surface and Coatings Technology*, **193**, no. 1–3, pp. 230–233, 2005.

[58] D. Abou-Ras, "Structural and chemical analysis of buffer layers in Cu (In, Ga) Se 2 thin-film solar cells," PhD thesis, Departement of Physics, Federal Institute of Technology, Zürich, 2005.

[59] D. Abou-Ras, D. Mukherji, G. Kostorz, D. Brémaud, M. Kälin, D. Rudmann, M. Döbeli, and A. N. Tiwari, "Dependence of the MoSe2 Formation on the Mo Orientation and the Na Concentration for Cu(In,Ga)Se2 Thin-Film Solar Cells," *MRS Online Proceedings Library*, **865**, pp. 287–292, 2005.

[60] D. Abou-Ras, G. Kostorz, D. Bremaud, M. Kälin, F. V. Kurdesau, A. N. Tiwari, and M. Döbeli, "Formation and characterisation of MoSe2 for Cu(In,Ga)Se2 based solar cells," *Thin Solid Films*, **480–481**, no. 0, pp. 433–438, Jun. 2005.

[61] R. J. Matson, O. Jamjoum, A. D. Buonaquisti, P. E. Russell, L. L. Kazmerski, P. Sheldon, and R. K. Ahrenkiel, "Metal contacts to $CuInSe_2$," *Solar Cells*, **11**, no. 3, pp. 301–305, 1984.

[62] T. Nakada, Y. Hirabayashi, T. Tokado, D. Ohmori, and T. Mise, "Novel device structure for Cu(In,Ga)Se2 thin film solar cells using transparent conducting oxide back and front contacts," *Solar Energy*, **77**, no. 6, pp. 739–747, 2004.

[63] P. J. Rostan, J. Mattheis, G. Bilger, U. Rau, and J. H. Werner, "Formation of transparent and ohmic ZnO:Al/$MoSe_2$ contacts for bifacial Cu(In,Ga)Se_2 solar cells and tandem structures," *Thin Solid Films*, **480–481**, pp. 67–70, Jun. 2005.

[64] T. Nakada, "Microstructural and diffusion properties of CIGS thin film solar cells fabricated using transparent conducting oxide back contacts," *Thin Solid Films*, **480–481**, no. 0, pp. 419–425, 2005.

[65] F. Erfurth, Z. Jehl, M. Bouttemy, N. Dahan, P. Tran-Van, I. Gerard, A. Etcheberry, J. J. Greffet, M. Powalla, G. Voorwinden, and others, "Mo/Cu (In, Ga) Se< sub> 2</sub> back interface chemical and optical properties for ultrathin CIGSe solar cells," *Applied Surface Science*, **258**, no. 7, pp. 3058–3061, 2011.

[66] J. Malmstrom, "On generation and recombination in Cu(In,Ga)Se2 thin film solar cells," Uppsala University, Uppsala, Sweden, 2005.

[67] Z. Jehl Li Kao, N. Naghavi, F. Erfurth, J. F. Guillemoles, I. Gérard, A. Etcheberry, J. L. Pelouard, S. Collin, G. Voorwinden, and D. Lincot, "Towards ultrathin copper indium gallium diselenide

solar cells: proof of concept study by chemical etching and gold back contact engineering," *Progress in Photovoltaics: Research and Applications*, p. n/a–n/a, 2012.

[68] R. A. Mickelsen and W. S. Chen, "High photocurrent polycrystalline thin-film CdS/CuInSe2 solar cella," *Applied Physics Letters*, **36**, no. 5, p. 371, 1980.

[69] M. Contreras, B. Egaas, K. Ramanathan, J. Hiltner, A. Swartzlander, F. Hasoon, and R. Noufi, "Progress toward 20% efficiency in Cu(In,Ga)Se2 polycrystalline thin film solar cells," *Progress in Photovoltaics: Research and Applications*, **7**, pp. 311–316, 1999.

[70] V. K. Kapur, U. V. Choudary, and A. K. P. Chu, "Process of forming compound semiconductor material," 4581108.

[71] I. Repins, M. A. Contreras, B. Egaas, C. DeHart, J. Scharf, C. L. Perkins, B. To, and R. Noufi, "19.9% efficient ZnO/CdS/CuInGaSe2 solar cell with 81.2% fill factor," *Progress in Photovoltaics: Research and Applications*, **16**, no. 3, pp. 235–239, 2008.

[72] A. M. Gabor, J. R. Tuttle, D. S. Albin, M. A. Contreras, R. Noufi, and A. M. Hermann, "High-efficiency $CuInxGa_{1-x}Se_2$ solar cells made from (Inx,Ga1−x)2Se3 precursor films," *Applied Physics Letters*, **65**, no. 2, p. 198, 1994.

[73] G. Voorwinden, R. Kniese, and M. Powalla, "In-line Cu(In,Ga)Se2 co-evaporation processes with graded band gaps on large substrates," *Thin Solid Films*, **431–432**, no. 0, pp. 538–542, 2003.

[74] E. Wallin, T. Jarmar, U. Malm, M. Edoff, and L. Stolt, "Influence of the average Se-to-metal overpressure during co-evaporation of $Cu(InxGa_{1-x})Se_2$," *Thin Solid Films*, **519**, no. 21, pp. 7237–7240, 2011.

[75] A. M. Gabor, J. R. Tuttle, M. H. Bode, A. Franz, A. L. Tennant, M. A. Contreras, R. Noufi, D. G. Jensen, and A. M. Hermann, "Band-gap engineering in Cu (In, Ga) Se2 thin films grown from (In, Ga) 2Se3 precursors," *Solar Energy Materials and Solar Cells*, **41**, pp. 247–260, 1996.

[76] S. M. Schleussner, T. Törndahl, M. Linnarsson, U. Zimmermann, T. Wätjen, and M. Edoff, "Development of gallium gradients in three-stage $Cu(In,Ga)Se_2$ co-evaporation processes," *Progress in Photovoltaics: Research and Applications*, **20**, no. 3, pp. 284–293, 2012.

[77] A. Brummer, V. Honkimäki, P. Berwian, V. Probst, J. Palm, and R. Hock, "Formation of $CuInSe_2$ by the annealing of stacked elemental layers—analysis by in situ high-energy powder diffraction," *Thin Solid Films*, **437**, no. 1–2, pp. 297–307, Aug. 2003.

[78] F. Hergert, S. Jost, R. Hock, and M. Purwins, "A crystallographic description of experimentally identified formation reactions of $Cu(In,Ga)Se_2$," *Journal of Solid State Chemistry*, **179**, no. 8, pp. 2394–2415, Aug. 2006.

[79] EMPA, "Record efficiency of 18.7% for flexible CIGS solar cells on plastics," *http://www.empa.ch/plugin/template/empa/3/107447/—/l=2/changeLang=true/lartid=107447/orga=/type=/theme=/bestellbar=/new_abt=/uacc=*, 2011.

[80] E. Wallin, U. Malm, T. Jarmar, O. L. Edoff, and L. Stolt, "World-record Cu (In, Ga) Se_2-based thin-film sub-module with 17.4% efficiency," *Progress in Photovoltaics: Research and Applications*, **20**, no. 7, pp. 851–854, 2012.

[81] Manz Press Release, "Manz plans to acquire CIGS modules innovation line from Würth Solar." 2011.

[82] B. Basol, "*Commercialization of High Efficiency Low Cost CIGS Technology Based on Electroplating, Final Technical Progress Report*," National Renewable Energy Laboratory, 2009.

[83] Nanosolar Press release, "Nanosolar Secures $20 Million in New Funds," *http://www.nanosolar.com/company/blog*, 2012.

[84] Avancis GmbH & Co. KG, "PowerMax Strong Datasheet," *http://www.avancis.de/fileadmin/media/portal/produkt/Data_sheet_PowerMax_STRONG_2012_03_EN.pdf*, 2012.

[85] D. Lu, B. Sang, Y. Deng, B. J. Stanbery, and L. Eldada, "Copper Indium Gallium Selenide photovoltaic modules manufactured by reactive transfer," in *Proceedings of the 35th IEEE Photovoltaic Specialists Conference*, Honolulu, HI, 2010, pp. 000175–000178.

[86] A. Neisser, "Manufacturing of Large-Area Cu(In,Ga)Se_2 Solar Modules Fast Ramp Up to More than 12% Module Efficiency in Mass Production – Road Map to 14%," presented at the *European Photovoltaic Solar Energy Conference and Exhibition*, Hamburg, 2011.

[87] C. J. Hibberd, E. Chassaing, W. Liu, D. B. Mitzi, D. Lincot, and A. N. Tiwari, "Non-vacuum methods for formation of Cu(In, Ga)(Se, S)$_2$ thin film photovoltaic absorbers," *Progress in Photovoltaics: Research and Applications*, **18**, no. 6, pp. 434–452, 2010.

[88] T. K. Todorov, O. Gunawan, T. Gokmen, and D. B. Mitzi, "Solution-processed Cu(In,Ga)(S,Se)2 absorber yielding a 15.2% efficient solar cell," *Progress in Photovoltaics: Research and Applications*, p. n/a–n/a, 2012.

[89] D. Lincot, "Electrodeposition of semiconductors," *Thin Solid Films*, **487**, no. 1, pp. 40–48, 2005.

[90] D. B. Mitzi, M. Yuan, W. Liu, A. J. Kellock, S. J. Chey, V. Deline, and A. G. Schrott, "A High-Efficiency Solution-Deposited Thin-Film Photovoltaic Device," *Advanced Materials*, **20**, no. 19, pp. 3657–3662, 2008.

[91] D. Lincot, J. F. Guillemoles, S. Taunier, D. Guimard, J. Sicx-Kurdi, A. Chaumont, O. Roussel, O. Ramdani, C. Hubert, J. P. Fauvarque, N. Bodereau, L. Parissi, P. Panheleux, P. Fanouillere, N. Naghavi, P. P. Grand, M. Benfarah, P. Mogensen, and O. Kerrec, "Chalcopyrite thin film solar cells by electrodeposition," *Solar Energy*, **77**, no. 6, pp. 725–737, 2004.

[92] B. Basol, M. Pinarbasi, S. Aksu, J. Wang, Y. Matus, T. Johnson, Y. Han, M. Narasimham, and B. Metin, "Electroplating based technolgy for roll-to-roll manufacturing," in *Proceedings of the 23rd European Photovoltaic Solar Energy Conference*, Valencia, 2008.

[93] R. N. Bhattacharya, W. Batchelor, J. F. Hiltner, and J. R. Sites, "Thin-film CuIn(1-x)Ga(x)Se2 photovoltaic cells from solution-based precursor layers," *Applied Physics Letters*, **75**, no. 10, pp. 1431–1433, 1999.

[94] T. Todorov and D. B. Mitzi, "Direct Liquid Coating of Chalcopyrite Light-Absorbing Layers for Photovoltaic Devices," *European Journal of Inorganic Chemistry*, 2010, no. 1, pp. 17–28, 2010.

[95] V. Kapur, R. Kemmerle, A. Bansal, J. Haber, J. Schmitzberger, P. Le, D. Guevarra, V. Kapur, and T. Stempien, "Manufacturing of 'ink based' CIGS solar cells/modules," in *Photovoltaic Specialists Conference, 2008. PVSC '08. 33rd IEEE*, 2008, pp. 1–5.

[96] J. van Duren, D. Jackrel, C. Leidholm, A. Pudov, M. Robinson, and Y. Roussillon, "The next generation in thinfilm photovoltaic process technology," in *Conference Record of the Seventeenth International Photovoltaic Science and Engineering Conference*, Fukuoka, 2007.

[97] R. Ortega-Borges and D. Lincot, "Mechanism of Chemical Bath Deposition of Cadmium Sulfide Thin Films in the Ammonia-Thiourea System," *Journal of The Electrochemical Society*, **140**, no. 12, pp. 3464–3473, 1993.

[98] J. Kessler, K. O. Velthaus, M. Ruckh, R. Laichinger, H. W. Schock, D. Lincot, Ortega, and J. Vedel, "Chemical Bath Deposition of CdS on CIS, etching effects and growth kinetics," in *Proceedings of the 6th Int. Photovoltaic Science and Engineering Conference (PVSEC-6)*, 1992, pp. 1005–1010.

[99] T. Nakada and A. Kunioka, "Direct evidence of Cd diffusion into Cu(In, Ga)Se[sub 2] thin films during chemical-bath deposition process of CdS films," *Applied Physics Letters*, **74**, no. 17, p. 2444, 1999.

[100] K. Hiepko, J. Bastek, R. Schlesiger, G. Schmitz, R. Wuerz, and N. A. Stolwijk, "Diffusion and incorporation of Cd in solar-grade Cu(In,Ga)Se_2 layers," *Applied Physics Letters*, **99**, no. 23, p. 234101, 2011.

[101] N. Naghavi, D. Abou-Ras, N. Allsop, N. Barreau, S. Bücheler, A. Ennaoui, C.-H. Fischer, C. Guillen, D. Hariskos, J. Herrero, R. Klenk, K. Kushiya, D. Lincot, R. Menner, T. Nakada, C. Platzer-Björkman, S. Spiering, A. N. Tiwari, and T. Törndahl, "Buffer layers and transparent conducting oxides for chalcopyrite Cu(In,Ga)(S,Se)$_2$ based thin film photovoltaics: present status and current developments," *Progress in Photovoltaics: Research and Applications*, **18**, no. 6, pp. 411–433, 2010.

[102] J. Lindahl, J. T. Wätjen, A. Hultqvist, T. Ericson, M. Edoff, and T. Törndahl, "The effect of Zn1−xSnxOy buffer layer thickness in 18.0% efficient Cd-free Cu(In,Ga)Se_2 solar cells," *Progress in Photovoltaics: Research and Applications*, p. n/a–n/a, Jul. 2012.

[103] D. Hariskos, S. Spiering, and M. Powalla, "Buffer layers in Cu(In,Ga)Se_2 solar cells and modules," *Thin Solid Films*, **480–481**, pp. 99–109, 2005.

[104] S. Siebentritt, "Alternative buffers for chalcopyrite solar cells," *Solar Energy*, **77**, no. 6, pp. 767–775, Dec. 2004.

[105] U. Rau and M. Schmidt, "Electronic properties of ZnO/CdS/Cu(In,Ga)Se-2 solar cells aspects of heterojunction formation," *Thin Solid Films*, **387**, no. 1–2, pp. 141–146, 2001.

[106] P. O. Grabitz, U. Rau, and J. H. Werner, "Electronic inhomogeneities and the role of the intrinsic ZnO layer in Cu(In,Ga)Se2 thin film solar cells," in *Proceedings of the 20th European Photovoltaic Solar Energy Conference*, 2005, pp. 1771–1774.

[107] N. Naghavi, G. Renou, V. Bockelee, F. Donsanti, P. Genevee, M. Jubault, J. F. Guillemoles, and D. Lincot, "Chemical deposition methods for Cd-free buffer layers in CI(G)S solar cells: Role of window layers," *Thin Solid Films*, **519**, no. 21, pp. 7600–7605, 2011.

[108] O. Kluth, G. Schöpe, B. Rech, R. Menner, M. Oertel, K. Orgassa, and H. Werner Schock, "Comparative material study on RF and DC magnetron sputtered ZnO:Al films," *Thin Solid Films*, **502**, no. 1–2, pp. 311–316, 2006.

[109] D. Herrmann, M. Oertel, R. Menner, and M. Powalla, "Analysis of relevant plasma parameters for ZnO:Al film deposition based on data from reactive and non-reactive DC magnetron sputtering," *Surface and Coatings Technology*, **174–175**, no. 0, pp. 229–234, 2003.

[110] J. F. Guillemoles, L. Kronik, D. Cahen, U. Rau, A. Jasenek, and H. W. Schock, "Stability issues of Cu(In,Ga)Se-2-based solar cells," *Journal of Physical Chemistry B*, **104**, no. 20, pp. 4849–4862, 2000.

[111] K. D. Becker and S. Wagner, "Temperature-dependent nuclear magnetic resonance in CuInVI2 chalcopyrite-structure compounds," *Physical Review B*, **275**, pp. 240–249, 1983.

[112] K. Gartsman, L. Chernyak, V. Lyahovitskaya, D. Cahen, V. Didik, V. Kozlovsky, R. Malkovich, E. Skoryatina, and V. Usacheva, "Direct evidence for diffusion and electromigration of Cu in CuInSe2," *Journal of Applied Physics*, **82**, no. 9, pp. 4282–4285, 1997.

[113] I. Lubomirski, K. Gartsman, and D. Cahen, "Space Charge Effects on Dopant Diffusion Coefficient Measurements in Semiconductors," *Journal of Applied Physics*, **83**, pp. 4678–4683, 1997.

[114] B. Dimmler, E. Gross, D. Hariskos, F. Kessler, E. Lotter, M. Powalla, J. Springer, U. Stein, G. Voorwinden, M. Gaeng, and S. Schleicher, "CIGS thin films module technology: towards commercialisation," Vienna, 1998, I, pp. 419–423.

[115] H. S. Ullal, K. Zweibel, and B. v. Roedern, "Current status of polycrystalline thin film PV technologies," Anaheim, 1997, p. 301.

[116] J. Wennerberg, J. Kessler, M. Bodegard, and L. Stolt, "Damp heat testing of high performance CIGS solar cells," Vienna, 1998, I, p. 1161.

[117] D. Hariskos, G. Bilger, D. Braunger, M. Ruckh, and H. W. Schock, "Investigations on the mechanism of the oxidation of Cu(In,Ga)Se2," presented at the *ICTMC-11*, University of Salford, 1998, E, p. 707.

[118] F. A. Kroger, *The Chemistry of Imperfect Crystals*, 1st edn, North-Holland, 1964.

[119] W. Walukiewicz, "Mechanism of Fermi-level stabilization in semiconductors," *Physical Review B*, **37**, no. 9, pp. 4760–4763, 1988.

[120] S. B. Zhang, S. H. Wei, and A. Zunger, "A phenomenological model for systematization and prediction of doping limits in II-VI and I-III-VI2 compounds," *Journal of Applied Physics*, **83**, p. 3192, 1998.

[121] S. B. Zhang, S. H. Wei, A. Zunger, and H. Katayama-Yoshida, "Defect physics of the CuInSe2 chalcopyrite semiconductor," *Physical Review B*, **57**, no. 16, pp. 9642–9656, 1998.

[122] B. Canava, J. F. Guillemoles, K. Leo, H. W. Schock, R. D. Tomlinson, M. Yakushev, S. Zott, and S. Zweigart, "Extrinsic doping of CIGS: Optoelectronic studies of Cd-implanted films," Barcelona, 1997, p. 2150.

[123] H. W. Schock and K. W. Bogus, "Development of CIS solar cells for space applications," Vienna, 1998, III, p. 3586.

[124] T. Hisimatsu, T. Aburaya, and S. Matsuda, "Irradiation tests of recent terrestrial solar cells for MDS-I," Vienna, 1998, III, p. 3568.

[125] A. Polity, R. Krause-Rehberg, T. E. M. Staab, M. J. Puska, J. Klais, H. J. Möller, and B. K. Meyer, "Study of defects in electron irradiated CuInSe2 by positron lifetime spectroscopy," *Journal of Applied Physics*, **83**, p. 71, 1998.

[126] M. Yakushev, G. Lippold, A. E. Hill, R. D. Pilkington, and R. D. Tomlinson, "An RBS-Channeling and Raman study of damage in hydrogen implanted CuInSe2 single crystals," *Crystal Research and Technology*, **31**, pp. 357–360, 1996.

[127] L. Kronik, D. Cahen, U. Rau, R. Herberholtz, H. W. Schock, and J. F. Guillemoles, "Oxidation, sodium and annealing effects on Cu(In,Ga)Se2 films and solar cells," Vienna, 1998, I, p. 453.

[128] U. Rau, D. Braunger, R. Herberholz, H. W. Schock, J.-F. Guillemoles, L. Kronik, and D. Cahen, "Oxygenation and air-annealing effects on the electronic properties of Cu(In,Ga)Sesub 2] films and devices," *Journal of Applied Physics*, **86**, no. 1, pp. 497–505, 1999.

[129] I. Lyubomirski, M. K. Rabinal, V. Lyakhovitskaya, and D. Cahen, "Room Temperature Detection of Mobile Impurities in Compound Semiconductors by Transient Ion Drift Measurements," *Journal of Applied Physics*, **81**, pp. 6684 – 6692, 1997.

[130] R. Herberholz, H. W. Schock, U. Rau, J. H. Werner, T. Haalboom, T. Gödecke, F. Ernst, C. Beilharz, K. W. Benz, and D. Cahen, "New aspects of phase segregation and junction formation in CuInSe2," Anaheim, 1997, p. 323.

[131] U. Rau, M. Schmitt, J. Parisi, W. Riedl, and F. Karg, "Persistent photoconductivity in Cu(In,Ga)Se[sub 2] heterojunctions and thin films prepared by sequential deposition," *Applied Physics Letters*, **73**, no. 2, pp. 223–225, Juillet 1998.

[132] U. Rau and H. W. Schock, "Electronic properties of Cu(In,Ga)Se2 heterojunction solar cells–recent achievements, current understanding, and future challenges," *Applied Physics A: Materials Science & Processing*, **69**, no. 2, pp. 131–147, Aug. 1999.

[133] U. Rau, A. Jasenek, R. Herberholtz, H. W. Schock, J. F. Guillemoles, D. Lincot, and L. Kronik, "The inherent stability of Cu(In,Ga)Se2 solar cells," Vienna, 1998, I, p. 428.

[134] A. Klein and W. Jaegerman, "Fermi level dependent defect formation in Cu chalcopyrite semiconductors," *Applied Physics Letters*, **74**, p. 2283, 1999.

[135] L. Stolt, J. Hedstrom, J. Kessler, M. Ruckh, K. O. Velthaus, and H. W. Schock, "Zno/Cds/Cuinse2 Thin-Film Solar-Cells with Improved Performance," *Applied Physics Letters*, **62**, no. 6, pp. 597–599, 1993.

[136] D. Braunger, S. Zweigart, and H. Schock, "The influence of Na and Ga on the incorporation of the chalcogen in polycrystalline Cu (In, Ga)(S, Se) 2 thin-films for photovoltaic applications'," in *Proc. 2nd. World Conf. Photovoltaic Solar Energy Conversion*, 1998, pp. 1113–1116.

[137] S. Lany and A. Zunger, "Anion vacancies as a source of persistent photoconductivity in II-VI and chalcopyrite semiconductors," *Physical Review B*, **72**, no. 3, 2005.

[138] S. Lany and A. Zunger, "Light- and bias-induced metastabilities in Cu(In,Ga)Se2 based solar cells caused by the (VSe-VCu) vacancy complex," *Journal of Applied Physics*, **100**, no. 11, p. 113725, 2006.

[139] M. Igalson and R. Bacewicz, "Defect relaxation in CuInSe2," *Crystal Research and Technology*, **31**, p. 445, 1996.

[140] M. Igalson and H. W. Schock, "The metastable changes of the trap spectra of CuInSe2-based photovoltaic devices," *Journal of Applied Physics*, **80**, no. 10, pp. 5765–5769, 1996.

[141] D. W. Niles, M. Al-Jassim, and K. Ramanathan, "Direct observation of Na and O impurities at grain surfaces of CuInSe thin films," *Journal of Vacuum Science & Technology A: Vacuum, Surfaces, and Films*, **17**, p. 291, 1999.

[142] D. W. Niles, K. Ramanathan, F. Hasoon, R. Noufi, B. J. Tielsch, and J. E. Fulghum, "Na impurity chemistry in photovoltaic CIGS thin films: Investigation with x-ray photoelectron spectroscopy," *Journal of Vacuum Science & Technology A: Vacuum, Surfaces, and Films*, **15**, p. 3044, 1997.

[143] M. Contreras, B. Egaas, P. Dippo, J. Webb, J. Granata, K. Ramanathan, S. Asher, A. Swartzlander, and R. Noufi, "On The Role of Na and Modifications to Cu (In, Ga)Se2, Absorber Materials Using Thin-MF (M = Na, K, Cs) Precursor Layers," in *Conference record of the IEEE photovoltaic specialists conference*, **26**, pp. 359–362 1997.

[144] U. Rau, M. Schmitt, F. Engelhardt, O. Seifert, J. Parisi, W. Riedl, J. Rimmasch, and F. Karg, "Impact of Na and S incorporation on the electronic transport mechanisms of Cu(In,Ga)Se-2 solar cells," *Solid State Commun.*, **107**, no. 2, pp. 59–63, 1998.

[145] S. H. Wei, S. B. Zhang, and A. Zunger, "Effects of Ga addition to CuInSe2 on its electronic, structural, and defect properties," *Applied Physics Letters*, **72**, no. 24, pp. 3199–3201, 1998.

[146] G. Hanna, A. Jasenek, U. Rau, and H. W. Schock, "Influence of the Ga-content on the bulk defect densities of Cu(In,Ga)Se-2," *Thin Solid Films*, **387**, no. 1–2, pp. 71–73, 2001.

[147] S. Hegedus, "Thin film solar modules: The low cost, high throughput and versatile alternative to Si wafers," *Progress in Photovoltaics: Research and Applications*, **14**, pp. 393–411, 2006.

[148] S. Schuler, S. Nishiwaki, J. Beckman, N. Rega, S. Brehme, S. Siebentritt, and M. C. Lux-Steiner, "Estimation of diffusion lengths in CuInSe2 based cells using the photocurrent-capacitance method," in *Proceedings of the 29th IEEE Photovoltaic Specialists Conference*, Piscataway, 2002, p. 504.

[149] C. H. Champness, "Estimation of diffusion lengths in CuInSe2 based cells using the photocurrent-capacitance method," in *Proceedings of the 29th IEEE Photovoltaic Specialists Conference*, Piscataway, 2002, pp. 732–735.

[150] L. S. Yip and I. Shih, "Photovoltaic cells with efficiency exceeding 10% using monocrystalline CuInSe2 substrates," in *Proceedings of the 1st World Conference on Photovoltaic Energy Conversion*, Piscataway, **1**, pp. 210–213 1994.

[151] I. Visoly-Fisher, S. R. Cohen, K. Gartsman, A. Ruzin, and D. Cahen, "Understanding the beneficial role of grain boundaries in polycrystalline solar cells from single-grain-boundary scanning probe microscopy," *Advanced Functional Materials*, **16**, no. 5, pp. 649–660, 2006.

[152] U. Rau, K. Taretto, and S. Siebentritt, "Grain boundaries in Cu(In, Ga)(Se, S)2 thin-film solar cells," *Applied Physics A*, **96**, no. 1, pp. 221–234, 2009.

[153] T. Dullweber, U. Rau, M. A. Contreras, R. Noufi, and H.-W. Schock, "Photogeneration and carrier recombination in graded gap Cu(In, Ga)Se2 solar cells," *Electron Devices, IEEE Transactions on DOI - 10.1109/16.887004*, **47**, no. 12, pp. 2249–2254, 2000.

[154] M. A. Contreras, L. M. Mansfield, B. Egaas, J. Li, M. Romero, R. Noufi, E. Rudiger-Voigt, and W. Mannstadt, "Wide bandgap Cu(In,Ga)Se2 solar cells with improved energy conversion efficiency," *Progress in Photovoltaics: Research and Applications*, p. n/a, 2012.

[155] M. Igalson, M. Pawłowski, and D. Prządo, "Influence of electric field on photoluminescence of Cu (In, Ga) Se 2-based solar cells," *Opto-Electronics Review*, pp. 1–4, 2011.

[156] J. Pettersson, C. Platzer-Björkman, U. Zimmermann, and M. Edoff, "Baseline model of graded-absorber Cu(In,Ga)Se2 solar cells applied to cells with Zn(1-x)MgxO buffer layers," *Thin Solid Films*, **519**, no. 21, pp. 7476–7480, 2011.

[157] Z. Djebbour, A. Darga, A. Migan Dubois, D. Mencaraglia, N. Naghavi, J. F. Guillemoles, and D. Lincot, "Admittance spectroscopy of cadmium free CIGS solar cells heterointerfaces," *Thin solid films*, **511**, pp. 320–324, 2006.

[158] Z. Djebbour, A. M. Dubois, A. Darga, D. Mencaraglia, C. Bazin, J. P. Connolly, J. F. Guillemoles, D. Lincot, B. Canava, and A. Etcheberry, "Comparison of optical and electrical gap of electrodeposited CuIn(S,Se)2 determined by spectral photo response and I–V–T measurements," *Thin solid films*, **515**, no. 15, pp. 6233–6237, 2007.

[159] J. Serhan, Z. Djebbour, D. Mencaraglia, F. Couzinié-Devy, N. Barreau, and J. Kessler, "Influence of Ga content on defects in CuInxGa1−xSe2 based solar cell absorbers investigated by sub gap modulated photocurrent and admittance spectroscopy," *Thin Solid Films*, **519**, no. 21, pp. 7312–7316, Aug. 2011.

[160] J. F. Guillemoles, "Contribution à la maîtrise des propriétés semiconductrices du CuInSe2 electrodepose et à l'amélioration de ses propriétés photovoltaïques," UPMC, 1994.

[161] D. Abou-Ras, S. Schorr, and H. W. Schock, "Grain-size distributions and grain boundaries of chalcopyrite-type thin films," *Journal of Applied Crystallography*, **40**, no. 5, pp. 841–848, 2007.

[162] H. Du, I. Shih, and C. H. Champness, "Monocrystalline CuInSe2 photovoltaic cell of superior performance," *Journal of Vacuum Science & Technology A: Vacuum, Surfaces, and Films*, **22**, no. 3, pp. 1023–1026, 2004.

[163] D. Abou-Ras, C. T. Koch, V. Küstner, P. A. van Aken, U. Jahn, M. A. Contreras, R. Caballero, C. A. Kaufmann, R. Scheer, T. Unold, and H.-W. Schock, "Grain-boundary types in chalcopyrite-type thin films and their correlations with film texture and electrical properties," *Thin Solid Films*, **517**, no. 7, pp. 2545–2549, 2009.

[164] M. Gloeckler, J. R. Sites, and W. K. Metzger, "Grain-boundary recombination in Cu(In,Ga)Se-2 solar cells," *Journal of Applied Physics*, **98**, no. 11, p. 113704, 2005.

[165] U. Rau, K. Taretto, and S. Siebentritt, "Grain boundaries in Cu(In, Ga)(Se, S)2 thin-film solar cells," *Applied Physics A: Materials Science & Processing*, **96**, no. 1, pp. 221–234, Jul. 2009.

[166] M. Nichterwitz, D. Abou-Ras, K. Sakurai, J. Bundesmann, T. Unold, R. Scheer, and H. W. Schock, "Influence of grain boundaries on current collection in Cu(In,Ga)Se2 thin-film solar cells," *Thin Solid Films*, **517**, no. 7, pp. 2554–2557, 2009.

[167] A. Rockett, "Surface analysis of chalcopyrite materials for photovoltaics," *Progress in Photovoltaics: Research and Applications*, p. n/a–n/a, 2012.

[168] D. Cahen and R. Noufi, "Defect chemical explanation for the effect of air anneal on CdS/CuInSe2 solar cell performance," *Applied Physics Letters*, **54**, no. 6, p. 558, 1989.

[169] L. Kronik, D. Cahen, and H. W. Schock, "Effects of Sodium on Polycrystalline Cu(In,Ga)Se2 and Its Solar Cell Performance," *Advanced Materials*, **10**, no. 1, pp. 31–36, 1998.

[170] M. J. Hetzer, Y. M. Strzhemechny, M. Gao, M. A. Contreras, A. Zunger, and L. J. Brillson, "Direct observation of copper depletion and potential changes at copper indium gallium diselenide grain boundaries," *Applied Physics Letters*, **86**, no. 16, 2005.

[171] R. Herberholz, V. Nadenau, U. Rühle, C. Köble, H. W. Schock, and B. Dimmler, "Prospects of wide-gap chalcopyrites for thin film photovoltaic modules," *Solar energy materials and solar cells*, **49**, no. 1, pp. 227–237, 1997.

[172] W. K. Metzger, I. L. Repins, and M. A. Contreras, "Long lifetimes in high-efficiency Cu(In,Ga)Se2 solar cells," *Applied Physics Letters*, **93**, no. 2, p. 022110, 2008.

[173] M. Burgelman, P. Nollet, and S. Degrave, "Modelling polycrystalline semiconductor solar cells," *Thin Solid Films*, **361**, pp. 527–532, 2000.

[174] M. Troviano and K. Taretto, "Analysis of internal quantum efficiency in double-graded bandgap solar cells including sub-bandgap absorption," *Solar Energy Materials and Solar Cells*, **95**, no. 3, pp. 821–828, 2011.

[175] W. Shockley and W. T. Read Jr, "Statistics of the recombinations of holes and electrons," *Physical Review*, **87**, no. 5, p. 835, 1952

[176] H. Wilhelm, H.-W. Schock, and R. Scheer, "Interface recombination in heterojunction solar cells: Influence of buffer layer thickness," *Journal of Applied Physics*, **109**, no. 8, p. 084514, 2011.

[177] A. O. Pudov, A. Kanevce, H. A. Al-Thani, J. R. Sites, and F. S. Hasoon, "Secondary barriers in CdSCuIn(1−x)GaxSe$_2$ solar cells," *Journal of Applied Physics*, **97**, no. 6, p. 064901, 2005.

[178] G. Brown, A. Pudov, B. Cardozo, V. Faifer, E. Bykov, and M. Contreras, "Quantitative imaging of electronic nonuniformities in Cu(In,Ga)Se$_2$ solar cells," *J. Appl. Phys.*, **108**, no. 7, p. 074516, 2010.

[179] O. Lundberg, M. Edoff, and L. Stolt, "The effect of Ga-grading in CIGS thin film solar cells," *Thin Solid Films*, **480–481**, no. 0, pp. 520–525, 2005.

[180] J. H. Werner, J. Mattheis, and U. Rau, "Efficiency limitations of polycrystalline thin film solar cells: case of Cu(In,Ga)Se2," *Thin Solid Films*, **480–481**, no. 0, pp. 399–409, 2005.

[181] K. A. Bulashevich and S. Y. Karpov, "Is Auger recombination responsible for the efficiency rollover in III-nitride light-emitting diodes?," *Physica Status Solidi C*, **5**, no. 6, pp. 2066–2069, 2008.

[182] U. Rau, A. Jasenek, H. W. Schock, F. Engelhardt, and T. Meyer, "Electronic loss mechanisms in chalcopyrite based heterojunction solar cells," *Thin Solid Films*, **361**, pp. 298–302, 2000.

[183] R. Scheer, "Activation energy of heterojunction diode currents in the limit of interface recombination," *Journal of Applied Physics*, **105**, no. 10, p. 104505, 2009.

[184] W. Shafarman, in *Proceeding of the 26th IEEE Photovoltaic Specialists Conference*, Richland, 1997, pp. 331–334.

[185] T. Negami, S. Nishiwaki, Y. Hashimoto, and N. Kohara, "Effect of the absorber thickness on performance of Cu(In,Ga)Se2 solar cells," in *Proceedings of the 2nd World conference on Photovoltaic Solar Energy Conversion*, Vienna, Austria, 1998, pp. 1181–1184.

[186] O. Lundberg, M. Bodegard, J. Malmstrom, and L. Stolt, "Influence of the Cu(In,Ga)Se-2 thickness and Ga grading on solar cell performance," *Prog. Photovoltaics*, **11**, no. 2, pp. 77–88, 2003.

[187] K. Ramanathan, R. Noufi, B. To, D. L. Young, R. Bhattacharya, M. A. Contreras, R. G. Dhere, and G. Teeter, "Processing and properties of sub-micron CIGS solar cells," in *Conference Record of the 2006 IEEE 4th World Conference on Photovoltaic Energy Conversion, Vols 1 and 2*, 2006, pp. 380–383.

[188] M. Edoff, S. Schleussner, E. Wallin, and O. Lundberg, "Technological and economical aspects on the influence of reduced Cu(In,Ga)Se2 thickness and Ga grading for co-evaporated Cu(In,Ga)Se2 modules," *Thin Solid Films*, **519**, no. 21, pp. 7530–7533, 2011.

[189] Z. Jehl Li Kao, "Elaboration of ultrathin copper indium gallium diselenide based solar cells," Université Paris Sud Orsay, Orsay, France, 2012.

[190] S. Schleussner, T. Kubart, T. Törndahl, and M. Edoff, "Reactively sputtered ZrN for application as reflecting back contact in Cu(In,Ga)Se2 solar cells," *Thin Solid Films*, **517**, no. 18, pp. 5548–5552, 2009.

[191] M. Bouttemy, P. Tran-Van, I. Gerard, T. Hildebrandt, A. Causier, J. L. Pelouard, G. Dagher, Z. Jehl, N. Naghavi, G. Voorwinden, B. Dimmler, M. Powalla, J. F. Guillemoles, D. Lincot, and A. Etcheberry, "Thinning of Cu (In, Ga) S2 solar cells: Part I: Chemical processing in acidic bromine solutions," *Thin Solid Films*, **519**, no. 21, pp. 7207–7211, 2011.

[192] Z. Jehl, F. Erfurth, N. Naghavi, L. Lombez, I. Gerard, M. Bouttemy, P. Tran-Van, A. Etcheberry, G. Voorwinden, B. Dimmler, W. Wischmann, M. Powalla, J.-F. Guillemoles, and D. Lincot, "Thinning of CIGS solar cells: Part II: cell characterisations," *Thin Solid Films*, **519**, pp. 7212–7215, 2011.

[193] T. Anegawa, Y. Oda, T. Minemoto, and H. Takakura, "Comparison of lift-off processes and rear-surface characterization of Cu(In,Ga)Se2 thin films for solar cells," *Journal of Crystal Growth*, 311, no. 3, pp. 742–745, 2009.

[194] T. Minemoto, Y. Abe, T. Anegawa, S. Osada, and H. Takakura, "Lift-Off Process for Flexible Cu(In,Ga)Se2 Solar Cells," *Japanese Journal of Applied Physics*, **49**, no. 4, p. 04DP06, 2010.

[195] L. Aé, D. Kieven, J. Chen, R. Klenk, T. Rissom, Y. Tang, and M. C. Lux-Steiner, "ZnO nanorod arrays as an antireflective coating for Cu(In,Ga)Se2 thin film solar cells," *Progress in Photovoltaics: Research and Applications*, **18**, no. 3, pp. 209–213, 2010.

[196] I. Massiot, F. Pardo, N. Bardou, P. Lalanne, J.-L. Pelouard, S. Collin, C. Sauvan, and C. Colin, "High-efficient ultra-thin solar cells," in *Proceedings of the 25th European Photovoltaic Solar Energy Conference*, Valencia, Spain, 2010, pp. 265 – 268.

[197] J. S. Ward, K. Ramanathan, F. S. Hasoon, T. J. Coutts, J. Keane, M. A. Contreras, T. Moriarty, and R. Noufi, "A 21.5% efficient Cu(In,Ga)Se2 thin-film concentrator solar cell," *Progress in Photovoltaics: Research and Applications*, **10**, no. 1, pp. 41–46, 2002.

[198] J. R. Tuttle, "The Performance of Cu(In,Ga)Se2-Based Solar Cells in Conventional and Concentrator Applications," in *Material Research Society Proceedings*, San Francisco, CA, USA, 1996, 426, pp. 143–151.

[199] J. Wennerberg, J. Kessler, J. Hedstrom, L. Stolt, B. Karlsson, and M. Ronnelid, "Thin film PV modules for low-concentrating systems," *Solar Energy*, **69**, no. 1–6, pp. 243–255, 2000.

[200] Y. Hirai, H. Nagashima, Y. Kurokawa, and A. Yamada, "Experimental and Theoretical Evaluation of Cu(In,Ga)Se_2 Concentrator Solar Cells," *Japanese Journal of Applied Physics*, **51**, no. 1, p. 014101, 2012.

[201] E. Sanchez and G. L. Araujo, "Mathematical analysis of the efficiency-concentration characteristic of a solar cell," *Solar Cells*, **12**, no. 3, pp. 263–276, 1984.

[202] M. Paire, A. Shams, L. Lombez, N. Pere-Laperne, S. Collin, J.-L. Pelouard, J.-F. Guillemoles, and D. Lincot, "Resistive and thermal scale effects for Cu(In, Ga)Se2 polycrystalline thin film microcells under concentration," *Energy Environmental Science*, **4**, no. 12, pp. 4972–4977, 2011.

[203] M. Paire, L. Lombez, N. Pere-Laperne, S. Collin, J.-L. Pelouard, D. Lincot, and J.-F. Guillemoles, "Microscale solar cells for high concentration on polycrystalline Cu(In,Ga)Se[sub 2] thin films," *Applied Physics Letters*, **98**, no. 26, p. 264102, 2011.

[204] S. Delbos, "Kësterite thin films for photovoltaics: a review," *EPJ Photovoltaics*, **3**, p. 35004, Aug. 2012.

[205] H. Katagiri, K. Jimbo, W. S. Maw, K. Oishi, M. Yamazaki, H. Araki, and A. Takeuchi, "Development of CZTS-based thin film solar cells," *Thin Solid Films*, **517**, no. 7, pp. 2455–2460, Feb. 2009.

[206] T. K. Todorov, J. Tang, S. Bag, O. Gunawan, T. Gokmen, Y. Zhu, and D. B. Mitzi, "Beyond 11% Efficiency: Characteristics of State-of-the-Art Cu2ZnSn(S,Se)4 Solar Cells," *Advanced Energy Materials*, **3**, no. 1, pp. 34–38, 2013.

[207] K. Ito and T. Nakazawa, "Electrical and Optical Properties of Stannite-Type Quaternary Semiconductor Thin Films," *Japanese Journal of Applied Physics*, **27**, pp. 2094–2097, 1988.

[208] C. P. Chan, H. Lam, and C. Surya, "Preparation of Cu2ZnSnS4 films by electrodeposition using ionic liquids," *Solar Energy Materials and Solar Cells*, **94**, no. 2, pp. 207–211, 2010.

[209] W. Xinkun, L. Wei, C. Shuying, L. Yunfeng, and J. Hongjie, "Photoelectric properties of Cu2ZnSnS4 thin films deposited by thermal evaporation," *Journal of Semiconductors*, **33**, no. 2, p. 022002, 2012.

[210] S. Chen, X. G. Gong, A. Walsh, and S. H. Wei, "Crystal and electronic band structure of CuZnSnX (X = S and Se) photovoltaic absorbers: First-principles insights," *Applied Physics Letters*, **94**, p. 041903, 2009.

[211] K. Timmo, M. Altosaar, J. Raudoja, K. Muska, M. Pilvet, M. Kauk, T. Varema, M. Danilson, O. Volobujeva, and E. Mellikov, "Sulfur-containing Cu2ZnSnSe4 monograin powders for solar cells," *Solar Energy Materials and Solar Cells*, **94**, no. 11, pp. 1889–1892, Nov. 2010.

[212] J. He, L. Sun, S. Chen, Y. Chen, P. Yang, and J. Chu, "Composition dependence of structure and optical properties of Cu2ZnSn(S,Se)4 solid solutions: An experimental study," *Journal of Alloys and Compounds*, **511**, no. 1, pp. 129–132, 2012.

[213] M. Moynihan, G. Zoppi, R. Miles, and I. Forbes, "Investigating synthesis of Cu2ZnSn(Se1-x,Sx)4 for values of $0 \leq x \leq 1$ by S for Se substitution and direct sulphidisation of metallic precursors," presented at the *26th European Photovoltaic Solar Energy Conference and Exhibition*, Hamburg, 2011.

[214] J. He, L. Sun, N. Ding, H. Kong, S. Zuo, S. Chen, Y. Chen, P. Yang, and J. Chu, "Single-step preparation and characterization of Cu2ZnSn(SxSe1-x)4 thin films deposited by Pulsed Laser Deposition method," *Journal of Alloys and Compounds*, **529**, pp. 34–37, 2012.

[215] E. Mellikov, D. Meissner, M. Altosaar, M. Kauk, J. Krustok, A. Öpik, O. Volobujeva, J. Iljina, K. Timmo, I. Klavina, J. Raudoja, M. Grossberg, T. Varema, K. Muska, M. Ganchev, S. Bereznev, and M. Danilson, "CZTS Monograin Powders and Thin Films," *AMR*, **222**, pp. 8–13, Apr. 2011.

[216] S. Levcenco, D. Dumcenco, Y. P. Wang, Y. S. Huang, C. H. Ho, E. Arushanov, V. Tezlevan, and K. K. Tiong, "Influence of anionic substitution on the electrolyte electroreflectance study of band edge transitions in single crystal Cu2ZnSn(SxSe1−x)4 solid solutions," *Optical Materials*, **34**, no. 8, pp. 1362–1365, Jun. 2012.

[217] W. Shockley and H. J. Queisser, "Detailed Balance Limit of Efficiency of p-n Junction Solar Cells," *Journal of Applied Physics*, **32**, no. 3, pp. 510–519, Mar. 1961.

[218] R. Saravana Kumar, B. deul Ryu, S. Chandramohan, J. K. Seol, S.-K. Lee, and C.-H. Hong, "Rapid synthesis of sphere-like Cu2ZnSnS4 microparticles by microwave irradiation," *Materials Letters*.

[219] L. Choubrac, A. Lafond, C. Guillot-Deudon, Y. Moëlo, and S. Jobic, "Structure flexibility of the Cu2ZnSnS4 absorber in low cost photovoltaic cells: from the stoechiometric to the Cu-poor compounds.," *Inorganic Chemistry*, **51**, no. 6, pp. 3346–3348, 2012.

[220] A. Lafond, L. Choubrac, C. Guillot-Deudon, P. Deniard, and S. Jobic, "Crystal Structures of Photovoltaic Chalcogenides, an Intricate Puzzle to Solve: the Cases of CIGSe and CZTS Materials," *Zeitschrift für anorganische und allgemeine Chemie*, p. n/a–n/a, 2012.

[221] USGS, "Commodity Statistics and Information," *USGS Minerals Information*, 2010. [Online]. Available: http://minerals.usgs.gov/minerals/pubs/commodity/.

[222] J. B. Li, V. Chawla, and B. M. Clemens, "Investigating the Role of Grain Boundaries in CZTS and CZTSSe Thin Film Solar Cells with Scanning Probe Microscopy," *Advanced Materials*, **24**, Issue pages, no. 6, pp. 720–723, 2012.

[223] M. J. Romero, I. Repins, G. Teeter, M. A. Contreras, M. Al-Jassim, and R. Noufi, "A Comparative Study of the Defect Point Physics and Luminescence of the Kesterites Cu2ZnSnS4 and Cu2ZnSnSe4 and Chalcopyrite Cu (In, Ga) Se2," 2012.

[224] M. J. Romero, H. Du, G. Teeter, Y. Yan, and M. M. Al-Jassim, "Comparative study of the luminescence and intrinsic point defects in the kesterite Cu2ZnSnS4 and chalcopyrite Cu(In,Ga)Se2 thin films used in photovoltaic applications," *Physical Review B*, 84, no. 16, p. 165324, Oct. 2011.

[225] S. Siebentritt and S. Schorr, "Kesterites—a challenging material for solar cells," *Progress in Photovoltaics: Research and Applications*, **20**, no. 5, pp. 512–519, 2012.

[226] T. Wada, S. Nakamura, and T. Maeda, "Ternary and multinary Cu-chalcogenide photovoltaic materials from CuInSe2 to Cu2ZnSnS4 and other compounds," *Progress in Photovoltaics: Research and Applications*.

[227] S. Chen, X. G. Gong, A. Walsh, and S. H. Wei, "Electronic structure and stability of quaternary chalcogenide semiconductors derived from cation cross-substitution of II-VI and I-III-VI 2 compounds," *Physical Review B*, **79**, p. 165211, 2009.

[228] S. R. Hall, J. T. Szymanski, and J. M. Stewart, "Kesterite, Cu2(Zn,Fe)SnS4, and stannite, Cu2(Fe, Zn)SnS4) structurally similar but distinct minerals," *Canadian Mineralogist*, **16**, no. 2, p. 131, 1978.

[229] L. O. Brockway, "The crystal structure of stannite, Cu2FeSnS4," *Z. Kristallogr.*, **89**, pp. 434–441, 1934.

[230] P. Bonazzi, L. Bindi, G. P. Bernardini, and S. Menchetti, "A model for the mechanism of incorporation of Cu, Fe and Zn in the stannite-kesterite series, Cu2FeSnS4-Cu2ZnSnS4," *Canadian Mineralogist*, **41**, no. 3, p. 639, 2003.

[231] S. Schorr, H. J. Hoebler, and M. Tovar, "A neutron diffraction study of the stannite-kesterite solid solution series," *European Journal of Mineralogy*, **19**, no. 1, p. 65, 2007.

[232] S. Schorr, "The crystal structure of kesterite type compounds: A neutron and X-ray diffraction study," *Solar Energy Materials and Solar Cells*, **95**, no. 6, pp. 1482–1488, Jun. 2011.

[233] J. Paier, R. Asahi, A. Nagoya, and G. Kresse, "Cu2ZnSnS4 as a potential photovoltaic material: A hybrid Hartree-Fock density functional theory study," *Physical Review B*, **79**, no. 11, p. 115126, 2009.

[234] A. Walsh, S. Chen, X. G. Gong, and S. H. Wei, "Crystal structure and defect reactions in the kesterite solar cell absorber Cu2ZnSnS4 (CZTS): Theoretical insights." [Online]. Available: http://web.mac.com/aronwalsh/publications/10/icps_czts_10.pdf.

[235] A. Walsh, S. H. Wei, S. Chen, and X. G. Gong, "Design of quaternary chalcogenide photovoltaic absorbers through cation mutation," in *34th IEEE Photovoltaic Specialists Conference*, 2009.

[236] A. Walsh, S. Chen, S.-H. Wei, and X.-G. Gong, "Kesterite Thin-Film Solar Cells: Advances in Materials Modelling of Cu2ZnSnS4," *Advanced Energy Materials*, **2**, no. 4, pp. 400–409, 2012.

[237] Y. Zhang, X. Sun, P. Zhang, X. Yuan, F. Huang, and W. Zhang, "Structural properties and quasiparticle band structures of Cu-based quaternary semiconductors for photovoltaic applications," *Journal of Applied Physics*, **111**, no. 6, pp. 063709–063709–6, Mar. 2012.

[238] M. Grossberg, J. Krustok, J. Raudoja, and T. Raadik, "The role of structural properties on deep defect states in Cu2ZnSnS4 studied by photoluminescence spectroscopy," *Applied Physics Letters*, **101**, no. 10, pp. 102102–102102, 2012.

[239] B. G. Mendis, M. D. Shannon, M. C. J. Goodman, J. D. Major, R. Claridge, D. P. Halliday, and K. Durose, "Direct observation of Cu, Zn cation disorder in Cu2ZnSnS4 solar cell absorber material using aberration corrected scanning transmission electron microscopy," *Progress in Photovoltaics: Research and Applications*, pp. 1–11, 2012.

[240] H. Nozaki, T. Fukano, S. Ohta, Y. Seno, H. Katagiri, and K. Jimbo, "Crystal structure determination of solar cell materials: Cu2ZnSnS4 thin films using X-ray anomalous dispersion," *Journal of Alloys and Compounds*, **524**, no. 0, pp. 22–25, May 2012.

[241] X. Fontané, V. Izquierdo-Roca, E. Saucedo, S. Schorr, V. O. Yukhymchuk, M. Y. Valakh, A. Pérez-Rodríguez, and J. R. Morante, "Vibrational properties of stannite and kesterite type compounds: Raman scattering analysis of Cu2(Fe,Zn)SnS4," *Journal of Alloys and Compounds*, **539**, pp. 190–194, 2012.

[242] G. M. Ford, Q. Guo, R. Agrawal, and H. W. Hillhouse, "Earth Abundant Element Cu2Zn(Sn(1−x)Gex)S4 Nanocrystals for Tunable Band Gap Solar Cells: 6.8% Efficient Device Fabrication," *Chemistry of Materials*, **23**, pp. 2626–2629, Apr. 2011.

[243] K. Yang and M. Ichimura, "Fabrication of Cu–Zn–Sn–S–O Thin Films by the Electrochemical Deposition Method and Application to Heterojunction Cells," *International Journal of Photoenergy*, 2012, 2012.

[244] X. He and H. Shen, "First-principles study of elastic and thermo-physical properties of kesterite-type Cu2ZnSnS4," *Physica B: Condensed Matter*, **406**, pp. 4604–4607, Dec. 2011.

[245] X. He and H. Shen, "First-principles calculation of some mechanical and thermo-physical properties of kesterite-type Cu2ZnSnSe4," *Physica Scripta*, **85**, no. 3, p. 035302, Mar. 2012.

[246] V. Kosyak, M. A. Karmarkar, and M. A. Scarpulla, "Temperature dependent conductivity of polycrystalline Cu2ZnSnS4 thin films," *Applied Physics Letters*, **100**, p. 263903, 2012.

[247] I. Camps, J. Coutinho, M. Mir, A. F. da Cunha, M. J. Rayson, and P. R. Briddon, "Elastic and optical properties of Cu2ZnSn(SexS(1−x))4 alloys: density functional calculations," *Semiconductor Science and Technology*, **27**, no. 11, p. 115001, Nov. 2012.

[248] P. A. Fernandes, P. M. P. Salomé, and A. F. da Cunha, "Growth and Raman scattering characterization of Cu2ZnSnS4 thin films," *Thin Solid Films*, **517**, no. 7, pp. 2519–2523, Feb. 2009.

[249] J. Just, D. Lützenkirchen-Hecht, R. Frahm, S. Schorr, and T. Unold, "Determination of secondary phases in kesterite Cu2ZnSnS4 thin films by x-ray absorption near edge structure analysis," *Applied Physics Letters*, **99**, no. 26, pp. 262105–262105–3, Dec. 2011.

[250] A. Nagoya, R. Asahi, R. Wahl, and G. Kresse, "Defect formation and phase stability of Cu2ZnSnS4 photovoltaic material," *Physical Review B*, **81**, no. 11, p. 113202, 2010.

[251] S. Chen, X. G. Gong, A. Walsh, and S. H. Wei, "Defect physics of the kesterite thin-film solar cell absorber Cu2ZnSnS4," *Applied Physics Letters*, **96**, p. 021902, 2010.

[252] P. J. Dale, K. Hoenes, J. Scragg, and S. Siebentritt, "A review of the challenges facing kesterite based thin film solar cells," in *Photovoltaic Specialists Conference (PVSC), 2009 34th IEEE*, 2010, pp. 002080–002085.

[253] T. Maeda, S. Nakamura, and T. Wada, "First principles calculations of vacancy formation in In-free photovoltaic semiconductor Cu2ZnSnSe4," *Thin Solid Films*, **519**, no. 21, pp. 7513–7516, 2011.

[254] E. Kask, T. Raadik, M. Grossberg, R. Josepson, and J. Krustok, "Deep defects in Cu2ZnSnS4 monograin solar cells," *Energy Procedia*, **10**, pp. 261–265, 2011.

[255] S. Chen, J. H. Yang, X. G. Gong, A. Walsh, and S. H. Wei, "Intrinsic point defects and complexes in the quaternary kesterite semiconductor Cu2ZnSnS4," *Physical Review B*, **81**, no. 24, p. 245204, 2010.

[256] T. Maeda, S. Nakamura, and T. Wada, "First Principles Calculations of Defect Formation in In-Free Photovoltaic Semiconductors Cu2ZnSnS4 and Cu2ZnSnSe4," *Japanese Journal of Applied Physics*, **50**, p. 04DP07, 2011.

[257] O. Gunawan, T. Gokmen, C. W. Warren, J. D. Cohen, T. K. Todorov, D. A. R. Barkhouse, S. Bag, J. Tang, B. Shin, and D. B. Mitzi, "Electronic properties of the Cu2ZnSn(Se,S)4 absorber layer in solar cells as revealed by admittance spectroscopy and related methods," *Applied Physics Letters*, **100**, no. 25, pp. 253905–253905–4, Jun. 2012.

[258] A. Jeong, W. Jo, S. Jung, J. Gwak, and J. Yun, "Enhanced exciton separation through negative energy band bending at grain boundaries of Cu2ZnSnSe4 thin-films," *Applied Physics Letters*, **99**, p. 082103, 2011.

[259] A. Nagaoka, K. Yoshino, H. Taniguchi, T. Taniyama, and H. Miyake, "Preparation of Cu2ZnSnS4 single crystals from Sn solutions," *Journal of Crystal Growth*, **341**, no. 1, pp. 38–41, Feb. 2012.

[260] T. Washio, H. Nozaki, T. Fukano, T. Motohiro, K. Jimbo, and H. Katagiri, "Analysis of lattice site occupancy in kesterite structure of Cu2ZnSnS4 films using synchrotron radiation x-ray diffraction," *Journal of Applied Physics*, **110**, p. 074511, 2011.

[261] I. Repins, N. Vora, C. Beall, S. H. Wei, Y. Yan, M. Romero, G. Teeter, H. Du, B. To, M. Young, and others, "Kesterites and Chalcopyrites: A Comparison of Close Cousins," in *MRS Proceedings*, 1324 2011.

[262] K. Muska, M. Kauk, M. Altosaar, M. Pilvet, M. Grossberg, and O. Volobujeva, "Synthesis of Cu2ZnSnS4 monograin powders with different compositions," *Energy Procedia*, **10**, pp. 203–207, 2011.

[263] H. Katagiri, K. Jimbo, M. Tahara, H. Araki, and K. Oishi, "The influence of the composition ratio on CZTS-based thin film solar cells," in *Materials Research Society Symposium Proceedings*, p. 1165 2009.

[264] A. Ennaoui, M. Lux-Steiner, A. Weber, D. Abou-Ras, I. Kötschau, H.-W. Schock, R. Schurr, A. Hölzing, S. Jost, R. Hock, T. Voß, J. Schulze, and A. Kirbs, "Cu2ZnSnS4 thin film solar cells from electroplated precursors: Novel low-cost perspective," *Thin Solid Films*, **517**, no. 7, pp. 2511–2514, Feb. 2009.

[265] P. A. Fernandes, P. M. P. Salomé, and A. F. da Cunha, "Precursors order effect on the properties of sulfurized Cu2ZnSnS4 thin films," *Semiconductor Science and Technology*, **24**, p. 105013, 2009.

[266] T. K. Todorov, K. B. Reuter, and D. B. Mitzi, "High-Efficiency Solar Cell with Earth-Abundant Liquid-Processed Absorber," *Advanced Materials*, **22**, 2010.

[267] K. Jimbo, R. Kimura, T. Kamimura, S. Yamada, W. S. Maw, H. Araki, K. Oishi, and H. Katagiri, "Cu2ZnSnS4-type thin film solar cells using abundant materials," *Thin Solid Films*, **515**, no. 15, pp. 5997–5999, 2007.

[268] M. Valentini, C. Malerba, F. Biccari, R. Chierchia, P. Mangiapane, E. Salza, and A. Mittiga, "Growth and characterization of Cu2ZnSnS4 thin films prepared by sulfurization of evaporated precursors," in *26th EUPVSEC*, Hamburg, 2011.

[269] S. J. Ahn, S. Jung, J. Gwak, A. Cho, K. Shin, K. Yoon, D. Park, H. Cheong, and J. H. Yun, "Determination of band gap energy (Eg) of Cu2ZnSnSe4 thin films: On the discrepancies of reported band gap values," *Applied Physics Letters*, **97**, p. 021905, 2010.

[270] D. Park, D. Nam, S. Jung, S. J. An, J. Gwak, K. Yoon, J. H. Yun, and H. Cheong, "Optical Characterization of Cu2ZnSnSe4 grown by thermal co-evaporation," *Thin Solid Films*, **519**, no. 21, pp. 7386–7389, 2011.

[271] H. Wang, "Progress in Thin Film Solar Cells Based on Cu2ZnSnS4," *International Journal of Photoenergy*, 2011, p. 801292, 2011.

[272] S. Ji and C. Ye, "Cu2ZnSnS4 as a New Solar Cell Material: The History and the Future," *Reviews in Advanced Sciences and Engineering*, **1**, no. 1, pp. 42–58, 2012.

[273] K. Ramasamy, M. A. Malik, and P. O'Brien, "Routes to copper zinc tin sulfide Cu(2)ZnSnS(4) a potential material for solar cells," *Chemical Communications (Cambridge)*, **48**, no. 46, pp. 5703–5714, Jun. 2012.

[274] M. Edoff, S. Schleussner, E. Wallin, and O. Lundberg, "Technological and economical aspects on the influence of reduced Cu(In,Ga)Se2 thickness and Ga grading for co-evaporated Cu(In,Ga)Se2 modules," *Thin Solid Films*, **519**, no. 21, pp. 7530–7533, Aug. 2011.

[275] D. A. R. Barkhouse, O. Gunawan, T. Gokmen, T. K. Todorov, and D. B. Mitzi, "Device characteristics of a 10.1% hydrazine-processed Cu2ZnSn(Se, S)4 solar cell," *Progress in Photovoltaics: Research and Applications*, **20**, no. 1, pp. 6–11, 2012.

[276] S. Ahmed, K. B. Reuter, O. Gunawan, L. Guo, L. T. Romankiw, and H. Deligianni, "A High Efficiency Electrodeposited Cu2ZnSnS4 Solar Cell," *Advanced Energy Materials*, **2**, pp. 253–259, 2012.

[277] B. Shin, O. Gunawan, Y. Zhu, N. A. Bojarczuk, S. J. Chey, and S. Guha, "Thin film solar cell with 8.4% power conversion efficiency using an earth-abundant Cu2ZnSnS4 absorber," *Progress in Photovoltaics: Research and Applications*, **21**, no. 1, pp. 72–76, 2013.

[278] I. Repins, C. Beall, N. Vora, C. DeHart, D. Kuciauskas, P. Dippo, B. To, J. Mann, W.-C. Hsu, and A. Goodrich, "Co-evaporated Cu2ZnSnSe4 films and devices," *Solar Energy Materials and Solar Cells*, **101**, pp. 154–159, 2012.

[279] M. Kauk, K. Muska, M. Altosaar, J. Raudoja, M. Pilvet, T. Varema, K. Timmo, and O. Volobujeva, "Effects of sulphur and tin disulphide vapour treatments of Cu2ZnSnS (Se) 4 absorber materials for monograin solar cells," *Energy Procedia*, **10**, pp. 197–202, 2011.

[280] B.-A. Schubert, B. Marsen, S. Cinque, T. Unold, R. Klenk, S. Schorr, and H.-W. Schock, "Cu2ZnSnS4 thin film solar cells by fast coevaporation," *Progress in Photovoltaics: Research and Applications*, **19**, no. 1, pp. 93–96, 2011.

[281] E. Mellikov, D. Meissner, T. Varema, M. Altosaar, M. Kauk, O. Volobujeva, J. Raudoja, K. Timmo, and M. Danilson, "Monograin materials for solar cells," *Solar Energy Materials and Solar Cells*, **93**, no. 1, pp. 65–68, Jan. 2009.

[282] F. Hergert and R. Hock, "Predicted formation reactions for the solid-state syntheses of the semiconductor materials Cu2SnX3 and Cu2ZnSnX4 (X= S, Se) starting from binary chalcogenides," *Thin solid films*, **515**, no. 15, pp. 5953–5956, 2007.

[283] A. J. Cheng, M. Manno, A. Khare, C. Leighton, S. Campbell, and E. Aydil, "Imaging and phase identification of Cu2ZnSnS4 thin films using confocal Raman spectroscopy," *Journal of Vacuum Science & Technology A: Vacuum, Surfaces, and Films*, **29**, p. 051203, 2011.

[284] O. Volobujeva, J. Raudoja, E. Mellikov, M. Grossberg, S. Bereznev, and R. Traksmaa, "Cu2ZnSnSe4 films by selenization of Sn-Zn-Cu sequential films," *Journal of Physics and Chemistry of Solids*, **70**, no. 3–4, pp. 567–570, 2009.

[285] R. Schurr, A. Hölzing, S. Jost, R. Hock, and Others, "The crystallisation of Cu2ZnSnS4 thin film solar cell absorbers from co-electroplated Cu-Zn-Sn precursors," *Thin Solid Films*, **517**, no. 7, pp. 2465–2468, 2009.

[286] M. Ganchev, J. Iljina, L. Kaupmees, T. Raadik, O. Volobujeva, A. Mere, M. Altosaar, J. Raudoja, and E. Mellikov, "Phase Composition of selenized Cu2ZnSnSe4 thin films determined by

X-ray diffraction and Raman spectroscopy," *Thin Solid Films*, **519**, no. 21, pp. 7394–7398, 2011.

[287] K. Maeda, K. Tanaka, Y. Nakano, and H. Uchiki, "Annealing Temperature Dependence of Properties of Cu2ZnSnS4 Thin Films Prepared by Sol–Gel Sulfurization Method," *Japanese Journal of Applied Physics*, **50**, no. 5, p. 05FB08, 2011.

[288] A. Redinger and S. Siebentritt, "Coevaporation of Cu2ZnSnSe4 thin films," *Applied Physics Letters*, **97**, no. 9, pp. 092111–092111–3, Sep. 2010.

[289] H. Yoo and J. Kim, "Growth of Cu2ZnSnS4 thin films using sulfurization of stacked metallic films," *Thin Solid Films*, **518**, no. 22, pp. 6567–6572, 2010.

[290] J. Ge, Y. Wu, C. Zhang, S. Zuo, J. Jiang, J. Ma, P. Yang, and J. Chu, "Comparative study of the influence of two distinct sulfurization ramping rates on the properties of Cu2ZnSnS4 thin films," *Applied Surface Science*, **258**, no. 19, pp. 7250–7254, Jul. 2012.

[291] K. Maeda, K. Tanaka, Y. Fukui, and H. Uchiki, "Influence of H2S concentration on the properties of Cu2ZnSnS4 thin films and solar cells prepared by sol-gel sulfurization," *Solar Energy Materials and Solar Cells*, **95**, no. 10, pp. 2855–2860, 2011.

[292] K. Maeda, K. Tanaka, Y. Nakano, Y. Fukui, and H. Uchiki, "H2S Concentration Dependence of Properties of Cu2ZnSnS4 Thin Film Prepared under Nonvacuum Condition," *Japanese Journal of Applied Physics*, **50**, no. 5, p. 05FB09, 2011.

[293] J. J. Scragg, "Studies of Cu2ZnSnS4 films prepared by sulfurisation of electrodeposited precursors," PhD thesis, University of Bath, Department of Chemistry, Bath, 2010.

[294] G. Teeter, H. Du, J. E. Leisch, M. Young, F. Yan, S. Johnston, P. Dippo, D. Kuciauskas, M. Romero, P. Newhouse, S. Asher, and D. Ginley, "Combinatorial study of thin-film Cu2ZnSnS4 synthesis via metal precursor sulfurization," in *35th IEEE-PVSEC*, 2010.

[295] A. Weber, R. Mainz, and H. W. Schock, "On the Sn loss from thin films of the material system Cu–Zn–Sn–S in high vacuum," *Journal of Applied Physics*, **107**, no. 1, p. 013516, 2010.

[296] H. Yoo, J. Kim, and L. Zhang, "Sulfurization temperature effects on the growth of Cu2ZnSnS4 thin film," *Current Applied Physics*, **12**, no. 4, pp. 1052–1057, 2012.

[297] X. Yin and H. Gong, "Heat-field-stimulated decomposition reaction in Cu2ZnSnS4," *Acta Materialia*.

[298] A. Redinger, D. M. Berg, P. J. Dale, R. Djemour, L. Gutay, T. Eisenbarth, N. Valle, and S. Siebentritt, "Route Toward High-Efficiency Single-Phase Cu2ZnSn(S,Se) Thin-Film Solar Cells: Model Experiments and Literature Review," *IEEE Journal of Photovoltaics*, **1**, no. 2, pp. 200–206, Oct. 2011.

[299] A. Redinger, D. M. Berg, P. J. Dale, and S. Siebentritt, "The Consequences of Kesterite Equilibria for Efficient Solar Cells," *Journal of the American Chemical Society*, p. 156, 2011.

[300] J. J. Scragg, T. Ericson, T. Kubart, M. Edoff, and C. Platzer-Björkman, "Chemical Insights into the Instability of Cu2ZnSnS4 Films during Annealing," *Chemistry of Materials*, **23**, no. 20, pp. 4625–4633, Oct. 2011.

[301] C. M. Fella, G. Ilari, A. R. Uhl, A. Chirilă, Y. E. Romanyuk, and A. N. Tiwari, "Non-vacuum deposition of Cu2ZnSnSe4 absorber layers for thin film solar cells," in *26 EUPVSEC*, Hamburg, 2011.

[302] C. M. Fella, A. R. Uhl, Y. E. Romanyuk, and A. N. Tiwari, "Cu2ZnSnSe4 absorbers processed from solution deposited metal salt precursors under different selenization conditions," *Physica Status Solidi A*, **209**, no. 6, 2012.

[303] J. Ge, S. Zuo, J. Jiang, J. Ma, L. Yang, P. Yang, and J. Chu, "Investigation of Se supply for the growth of Cu2ZnSn(SxSe(1-x))4(x$\approx$ 0.02–0.05) thin films for photovoltaics," *Applied Surface Science*, **258**, no. 20, pp. 7844–7848, 2012.

[304] F. Biccari, R. Chierchia, M. Valentini, P. Mangiapane, E. Salza, C. Malerba, C. L. A. Ricardo, L. Mannarino, P. Scardi, and A. Mittiga, "Fabrication of Cu2ZnSnS4 solar cells by sulfurization of evaporated precursors," *Energy Procedia*, **10**, pp. 187–191, 2011.

[305] S. Guha, D. B. Mitzi, T. K. Todorov, and K. Wang, "Annealing Thin Films," US 2012/0070936.

[306] P. Salomé, J. Malaquias, P. Fernandes, M. Ferreira, J. Leitão, A. da Cunha, J. González, F. Matinaga, G. Ribeiro, and E. Viana, "The influence of hydrogen in the incorporation of Zn during the growth of Cu2ZnSnS4 thin films," *Solar Energy Materials and Solar Cells*, **95**, no. 12, pp. 3482–3489, 2011.

[307] K. Wang, B. Shin, K. B. Reuter, T. Todorov, D. B. Mitzi, and S. Guha, "Structural and elemental characterization of high efficiency Cu2ZnSnS4 solar cells," *Applied Physics Letters*, **98**, p. 051912, 2011.

[308] R. Juškėnas, S. Kanapeckaitė, V. Karpavičienė, Z. Mockus, V. Pakštas, A. Selskienė, R. Giraitis, and G. Niaura, "A two-step approach for electrochemical deposition of Cu–Zn–Sn and Se precursors for CZTSe solar cells," *Solar Energy Materials and Solar Cells*, **101**, pp. 277–282, Jun. 2012.

[309] S. W. Shin, S. M. Pawar, C. Y. Park, J. H. Yun, J.-H. Moon, J. H. Kim, and J. Y. Lee, "Studies on Cu2ZnSnS4 (CZTS) absorber layer using different stacking orders in precursor thin films," *Solar Energy Materials and Solar Cells*, **95**, no. 12, pp. 3202–3206, Dec. 2011.

[310] L. Grenet, S. Bernardi, D. Kohen, C. Lepoittevin, S. Noël, N. Karst, A. Brioude, S. Perraud, and H. Mariette, "Cu2ZnSn(S1−xSex)4 based solar cell produced by selenization of vacuum deposited precursors," *Solar Energy Materials and Solar Cells*, **101**, pp. 11–14, 2012.

[311] H. Yoo and J. Kim, "Comparative study of Cu2ZnSnS4 film growth," *Solar Energy Materials and Solar Cells*, **95**, no. 1, pp. 239–244, 2011.

[312] R. Chalapathy, G. S. Jung, and B. T. Ahn, "Fabrication of Cu2ZnSnS4 films by sulfurization of Cu/ZnSn/Cu precursor layers in sulfur atmosphere for solar cells," *Solar Energy Materials and Solar Cells*, **95**, no. 12, pp. 3216–3221, 2011.

[313] P. M. P. Salomé, J. Malaquias, P. A. Fernandes, M. S. Ferreira, A. F. da Cunha, J. P. Leitão, J. C. González, and F. M. Matinaga, "Growth and characterization of Cu2ZnSn(S,Se)4 thin films for solar cells," *Solar Energy Materials and Solar Cells*, **101**, pp. 147–153, Jun. 2012.

[314] W. Hlaing Oo, J. Johnson, A. Bhatia, E. Lund, M. Nowell, and M. Scarpulla, "Grain Size and Texture of Cu2ZnSnS4 Thin Films Synthesized by Cosputtering Binary Sulfides and Annealing: Effects of Processing Conditions and Sodium," *Journal of Electronic Materials*, **40**, no. 11, pp. 2214–2221.

[315] T. Prabhakar and N. Jampana, "Effect of sodium diffusion on the structural and electrical properties of Cu2ZnSnS4 thin films," *Solar Energy Materials and Solar Cells*, **95**, no. 3, pp. 1001–1004, 2011.

[316] M. Bär, B. A. Schubert, B. Marsen, S. Krause, S. Pookpanratana, T. Unold, L. Weinhardt, C. Heske, and H. W. Schock, "Native oxidation and Cu-poor surface structure of thin film Cu2ZnSnS4 solar cell absorbers," *Applied Physics Letters*, **99**, p. 112103, 2011.

[317] A. Nagoya, R. Asahi, and G. Kresse, "First-principles study of Cu2ZnSnS4 and the related band offsets for photovoltaic applications," *Journal of Physics: Condensed Matter*, **23**, p. 404203, Oct. 2011.

[318] M. Patel and A. Ray, "Enhancement of output performance of Cu2ZnSnS4 thin film solar cells—A numerical simulation approach and comparison to experiments," *Physica B: Condensed Matter*, **407**, no. 21, pp. 4391–4397, Nov. 2012.

[319] N. Sakai, H. Hiroi, and H. Sugimoto, "Development of cd-free buffer layer for Cu2ZnSnS4 thin-film solar cells," in *37th IEEE PVSC Conference*, 2011.

[320] D. A. R. Barkhouse, R. Haight, N. Sakai, H. Hiroi, H. Sugimoto, and D. B. Mitzi, "Cd-free buffer layer materials on Cu2ZnSn(SxSe1-x)4: Band alignments with ZnO, ZnS, and In2S3," *Applied Physics Letters*, **100**, no. 19, pp. 193904 –193904–5, May 2012.

[321] F. Jiang, H. Shen, W. Wang, and L. Zhang, "Preparation and Properties of Cu2ZnSnS4 Absorber and Cu2ZnSnS4 / Amorphous Silicon Thin-Film Solar Cell," *Applied Physics Express*, **4**, no. 7, p. 074101, 2011.

[322] P. K. Sarswat and M. L. Free, "Demonstration of a sol-gel synthesized bifacial CZTS photoelectrochemical cell," *Physica Status Solidi A*, **208**, no. 12, pp. 2861–2864, 2011.

[323] P. K. Sarswat, M. Snure, M. L. Free, and A. Tiwari, "CZTS thin films on transparent conducting electrodes by electrochemical technique," *Thin Solid Films*, **520**, no. 6, pp. 1694–1697, 2012.

[324] Q.-M. Chen, Z.-Q. Li, Y. Ni, S.-Y. Cheng, and X.-M. Dou, "Doctor-bladed Cu2ZnSnS4 light absorption layer for low-cost solar cell application," *Chinese Physics B*, **21**, no. 3, p. 038401, Mar. 2012.

[325] Q. Chen, S. Cheng, S. Zhuang, and X. Dou, "Cu2ZnSnS4 solar cell prepared entirely by non-vacuum processes," *Thin Solid Films*.

[326] Q. Tian, X. Xu, L. Han, M. Tang, R. Zou, Z. Chen, M. Yu, J. Yang, and J. Hu, "Hydrophilic Cu2ZnSnS4 nanocrystals for printing flexible, low-cost and environmentally friendly solar cells," *CrystEngComm*, **14**, no. 11, pp. 3847–3850, May 2012.

[327] O. Gunawan, T. K. Todorov, and D. B. Mitzi, "Loss mechanisms in hydrazine-processed Cu2ZnSn(Se,S)4 solar cells," *Applied Physics Letters*, **97**, no. 23, p. 233506, 2010.

[328] K. Wang, O. Gunawan, T. Todorov, B. Shin, S. J. Chiy, N. A. Bojarczuk, D. Mitzi, and S. Guha, "Thermally evaporated Cu2ZnSnS4 solar cells," *Applied Physics Letters*, **97**, no. 14, pp. 143508–143508–3, Oct. 2010.

[329] B. Shin, K. Wang, O. Gunawan, K. B. Reuter, S. J. Chey, N. A. Bojarczuk, T. Todorov, B. Mitzi, and S. Guha, "High Efficiency Cu2ZnSn(S(x)Se(1-X))4 Thin Film Solar Cells by Thermal Co-Evaporation," in *37th IEEE PVSC Conference*, Seattle, 2011, 1.

[330] M. Bär, B. A. Schubert, B. Marsen, R. G. Wilks, M. Blum, S. Krause, S. Pookpanratana, Y. Zhang, T. Unold, and W. Yang, "Cu2ZnSnS4 thin-film solar cell absorbers illuminated by soft x-rays," *Journal of Materials Research*, **27**, no. 8, pp. 1097–1104, 2012.

[331] T. Tanaka, T. Sueishi, K. Saito, Q. Guo, M. Nishio, K. M. Yu, and W. Walukiewicz, "Existence and removal of Cu2Se second phase in coevaporated Cu2ZnSnSe4 thin films," *Journal of Applied Physics*, **111**, no. 5, pp. 053522–053522–4, Mar. 2012.

[332] N. Vora, J. Blackburn, I. Repins, C. Beall, B. To, J. Pankow, G. Teeter, M. Young, and R. Noufi, "Phase identification and control of thin films deposited by co-evaporation of elemental Cu, Zn, Sn, and Se," *Journal of Vacuum Science & Technology A: Vacuum, Surfaces, and Films*, **30**, no. 5, p. 051201, 2012.

[333] M. Bär, B.-A. Schubert, B. Marsen, S. Krause, S. Pookpanratana, T. Unold, L. Weinhardt, C. Heske, and H.-W. Schock, "Impact of KCN etching on the chemical and electronic surface structure of Cu2ZnSnS4 thin-film solar cell absorbers," *Applied Physics Letters*, **99**, p. 152111, 2011.

[334] M. Bär, B.-A. Schubert, B. Marsen, R. G. Wilks, S. Pookpanratana, M. Blum, S. Krause, T. Unold, W. Yang, L. Weinhardt, C. Heske, and H.-W. Schock, "Cliff-like conduction band offset and KCN-induced recombination barrier enhancement at the CdS/Cu2ZnSnS4 thin-film solar cell heterojunction," *Applied Physics Letters*, **99**, no. 22, p. 222105, 2011.

[335] J. Timo Wätjen, J. Engman, M. Edoff, and C. Platzer-Björkman, "Direct evidence of current blocking by ZnSe in Cu2ZnSnSe4 solar cells," *Applied Physics Letters*, **100**, no. 17, pp. 173510–173510–3, Apr. 2012.

[336] A. Fairbrother, E. García-Hemme, V. Izquierdo-Roca, X. Fontané, F. A. Pulgarín-Agudelo, O. Vigil-Galán, A. Pérez-Rodríguez, and E. Saucedo, "Development of a Selective Chemical Etch To Improve the Conversion Efficiency of Zn-Rich Cu2ZnSnS4 Solar Cells," *Journal of the American Chemical Society*, 2012.

[337] R. Haight, A. Barkhouse, O. Gunawan, B. Shin, M. Copel, M. Hopstaken, and D. B. Mitzi, "Band alignment at the Cu2ZnSn (SxSe1- x) 4/CdS interface," *Applied Physics Letters*, **98**, p. 253502, 2011.

[338] J. Li, M. Wei, Q. Du, W. Liu, G. Jiang, and C. Zhu, "The band alignment at CdS/Cu2ZnSnSe4 heterojunction interface," *Surface and Interface Analysis*, p. n/a–n/a, 2012.

[339] J. M. Raulot, C. Domain, and J. F. Guillemoles, "Ab initio investigation of potential indium and gallium free chalcopyrite compounds for photovoltaic application," *Journal of Physics and Chemistry of Solids*, **66**, no. 11, pp. 2019–2023, 2005.

[340] A. Opanasyuk, D. Kurbatov, M. Ivashchenko, I. Y. Protsenko, and H. Cheong, "Properties of the Window Layers for the CZTSe and CZTS Based Solar Cells," *Journal of Nano- and Electronic Physics*, **4**, no. 1, p. 01024, 2012.

[341] B. Shin, Y. Zhu, N. A. Bojarczuk, S. Jay Chey, and S. Guha, "Control of an interfacial MoSe2 layer in Cu2ZnSnSe4 thin film solar cells: 8.9% power conversion efficiency with a TiN diffusion barrier," *Applied Physics Letters*, **101**, no. 5, pp. 053903–053903–4, Jul. 2012.

[342] S. Bag, O. Gunawan, T. Gokmen, Y. Zhu, T. K. Todorov, and D. B. Mitzi, "Low band gap liquid-processed CZTSe solar cell with 10.1% efficiency," *Energy Environ. Sci.*, **5**, no. 5, pp. 7060–7065, Apr. 2012.

8 Printed Organic Solar Cells

***Claudia Hoth*[1], *Andrea Seemann*[1], *Roland Steim*[1], *Tayebeh Ameri*[2], *Hamed Azimi*[2] *and Christoph J. Brabec*[1,2]**
[1] *Konarka Technologies Germany GmbH, Germany*
[2] *Konarka Technologies Austria GmbH, Austria*

8.1 INTRODUCTION

Solution-processed bulk-heterojunction solar cells have gained serious attention during the last few years and are becoming established as one of the future photovoltaic technologies for lowest-cost power production. The first products became realistic after the OPV (organic photovoltaics) efficiency has rather rapidly increased from a ~1% in 1999 to almost 7% in 2009 and to beyond 12% in 2013. This chapter reviews a few of the essential technologies around the OPV science that are relevant for the realisation of first products [1–4].

Most of the review articles in the field of OPV are very focused on the fundamental science, the materials and their optical and electrical properties as well as device physics. These disciplines are certainly essential to continue driving this technology towards higher efficiencies, and also fascinating from a scientific point of view. However, realisation of a photovoltaic product requires more than excellent materials for outstanding good cells. It requires a whole product technology to be developed around the cell technology. This chapter reviews the state-of-the-art in some of these technology disciplines, which have a major contribution to the progress of OPV as a commercial solar technology.

Section 8.2 will discuss various aspects of morphology formation in organic bulk-heterojunction solar cells. The relevance of processing conditions, of ink formulation and the use of additives will be reviewed. Section 8.3 reviews the progress in selected interface materials, which are decisive not only for the solar cell performance but also for the long-time stability of the solar cell. Section 8.4 presents and reviews the concept of multilayer solar cells (tandem solar cells) for OPV. Section 8.5 explains the relevant design criteria for electrodes suitable for OPV. Vacuum deposition vs. solution processing of electrodes makes a major impact on the costs of a solar cell as well as on the productivity of the production process. Nonindium-based or non-TCO-based electrodes is another aspect that gets addressed in this section. Finally, Section 8.6 concludes by reviewing the overall production process and here especially the printing processes for the semiconductor layer.

Despite great progress in fundamental aspects of organic photovoltaics, including transport, charge generation, numerical simulation of *J–V* curves, models to correlate material properties with solar cell performance as well as financial forecasts and life-cycle analysis studies, this review will focus more on performance and lifetime of bulk-heterojunction composites from polymer fullerene blends. It is the purpose of this contribution to give a state-of-art-report on the plastic solar cells based on conjugated polymers from the post-P3HT

Solar Cell Materials: Developing Technologies, First Edition. Edited by Gavin Conibeer and Arthur Willoughby.

era between 2005–2010, where the fundamentals for OPV were developed. This period was especially important for the progress of OPV – not only was the basis of the materials classes laid out, but also the relevant device architectures and device concepts. Since this review was majorly designed under educational aspects, it appears very worthwhile to review this period in more detail. Results from other organic solar cells like dye-sensitised liquid electrolyte [5] or solid-state polymer electrolyte cells [6], pure dye cells [7, 8] or small-molecule cells [9], mostly based on heterojunctions between phatolcyanines and fullerenes or perylenes [10] will not be discussed.

8.2 MATERIALS AND MORPHOLOGY

One of the most challenging parts to further increase the efficiency potential of organic solar cells undoubtedly is the design of organic semiconductors that offer better photon harvesting across the entire solar spectrum, better charge-transport properties, better environmental stability, and, the right morphology between the donor and acceptor [11–15].

The beginning part of the current section is devoted to a brief overview on promising donor and acceptor (D/A) materials with potential to push the efficiency of organic solar cells beyond 7%. Despite the great progress of several different organic/hybrid approaches like solution-processed or evaporated small molecules, polymer–polymer blends or organic–inorganic blends, this review will focus exclusively on bulk-heterojunction composites from polymer–fullerene blends.

The primary excitations in organic semiconductors are Coloumb bound electron–hole pairs, so-called excitons. In an excitonic solar cell, the external quantum efficiency is governed by the product of the photon absorption (η_{A}), the exciton diffusion efficiency (η_{ED}, the fraction of photogenerated excitons that reaches a donor/acceptor interface before they want to recombine) and ultimately the charge-carrier collection efficiency (η_{CC}, the likelihood for free charge carriers that are dissociated at the donor/acceptor interface to reach the respective anode and cathode electrodes).

$$\eta_{EQE} = \eta_{A} \times \eta_{ED} \times \eta_{CC} \tag{8.1}$$

The quite large exciton binding energy in organic semiconductors can be overcome by using semiconductor couples with different electron affinity. Here, one semiconductor serves as the donor-type molecule, while the second semiconductor is of an acceptor type. The quantum efficiency of charge generation for such a donor–acceptor couple can be as high as one for a composite with properly chosen electronic levels. The charge dissociation happens at the interface of the two semiconductors. The simplest solar cell configuration for such a material composite is a bilayer. A bilayer configuration allows generation of excitons over all the semiconductor phases, but charge separation occurs exclusively at the planar interface of the two materials. As such, the exciton diffusion length L_{exc} should be as large as possible ($L_{exc} = \sqrt{D_{exc}\tau_{exc}}$ with D_{exc} the exciton diffusion coefficient and τ_{exc} the natural exciton lifetime) Unfortunately, most organic semiconductor materials are amorphous in nature and have exciton diffusion lengths in the single-digit nm regime [16–19]. Such short exciton diffusion lengths are a severe limitation and typically restrict the power efficiency η_{EQE} for bilayer cells to the low single-digit efficiency regime.

A revolutionary concept to tackle this issue was the intermixing of two D/A materials in one layer, in a structure known as a bulk heterojunction (BHJ). In this configuration, the arrangement of the two semiconductor components can take place in a way that provides a large interface between the donor and acceptor phases and obviously reduces the requirement of large exciton diffusion lengths. Bulk-heterojunction composites can have a charge-generation efficiency of unity once the average separation of the two components is in the dimension of twice the exciton diffusion length. The desired case for an efficient oBHJ photovoltaic (PV) is an interpenetrating network formed from nanoscale phase separation of D/A domains to guarantee both: on the one hand, an efficient free charge-carrier generation, and on the other hand, a continuous pathway for an efficient transport of the positive and negative carriers toward their relative electrodes [20–24].

Following the section on the alternative materials for organic photovoltaics (OPVs), an overview is given regarding the influential parameters on morphology and some recent experimental studies are discussed in more detail. The current state-of-the-art in morphology study through novel characterisation tools is reviewed and subsequently the importance of providing a 3-dimensional image of film organisation is stressed. The final sections discuss some theoretical models for predicting and understanding the morphology and, finally, techniques to control the morphology are addressed.

8.2.1 Organic Semiconductors

The first years of OPV were dominated by poly[2-methoxy,5-(2′-ethyl-hexyloxy)-*p*-phenylene vinylene) (MEH-PPV)/C_{60} composites, which were later substituted by the better processable combination of poly[2-methoxy-5-(3′,7′-dimethyloctyloxy)-1,4-phenylene vinylene] (MDMO-PPV)/1-(3-methoxycarbonyl) propyl-1-phenyl[6,6]C_{61} (PCBM) [25–32]. Because of the rather large gap and low mobility of the PPV-type polymers, efficiencies remained at 3% at best [33–36], and the general interest in this class of material has faded. During the last 5 years research efforts became focused on poly-alkyl-thiophenes, and here especially on P3HT. In 2002 the first encouraging results for P3HT:PCBM solar cells with a ratio 1:3 in weight were published [37]. The short-circuit current density was the largest ever observed in an OSC (8.7 mA cm^{-2}) at that time, and resulted from an EQE showing a maximum of 76% at 550 nm. This paper appeared to be a starting point for a rapid development for the P3HT:PCBM blend, followed quickly by the first explicit reports on efficiency enhancement in P3HT:PCBM cells due to thermal annealing [38]. The main development over recent years has consisted in understanding and optimising the processing of the active layer and especially the device annealing conditions that appeared to be mandatory to achieve high efficiencies until recently. The optimum P3HT:PCBM ratio for that is about 1:1 in weight, and the two best-suited solvents for this blend are chlorobenzene (CB) and ortho-dichlorobenzene (DCB).

The efficiency limitations of organic solar cells have been described earlier [39, 40] discussing the importance of the bandgap, the HOMO and LUMO levels of the donor and the acceptor molecules. The maximum short-circuit current is determined by the smaller optical bandgap of the two materials and the V_{oc} is proportional to the difference between the HOMO level of the donor material and the LUMO level of the acceptor compound. For an efficient charge generation in the donor–acceptor blend a certain offset of the HOMO and LUMO-levels (ΔE_{HOMO}, ΔE_{LUMO}) is required [41], which is believed to be a few hundred meV. The chemistry of conjugated polymers offers powerful methods to tune the HOMO and LUMO levels and to modify the bandgap of the material. In the so-called donor–acceptor approach [11, 42] alternating electron-rich (donor D) and electron-poor (acceptor A) units are coupled together to form the polymer backbone. For such a $(\text{-D-A-})_n$ polymer a second resonance structure $(\text{-D}^+\text{-A}^-)_n$ gains importance with respect to the neutral structure and increases the double-bond character of the single bonds in the polymer. This consequent reduction of the bond-length alternation effectively modifies the HOMO and LUMO level and the bandgap of the polymer. Several promising candidates have been synthesised and a noncomprehensive selection of materials will be discussed in the next section. At this point we would like to stress that a favourable arrangement of the HOMO and LUMO levels of the donor and the acceptor materials is a prerequisite for highly efficient solar cell. In addition, an optimised nanomorphology of the donor–acceptor composite as well as sufficient charge transport (charge carrier mobilities in range of 0.001 $cm^2/V\ s$) is necessary for high power-conversion efficiencies.

8.2.1.1 Promising Donor Materials

Figure 8.1 summarises a selection of promising structures that were discussed in the initial phase of the post-P3HT period [43–48]. Most of the structures are from the material classes

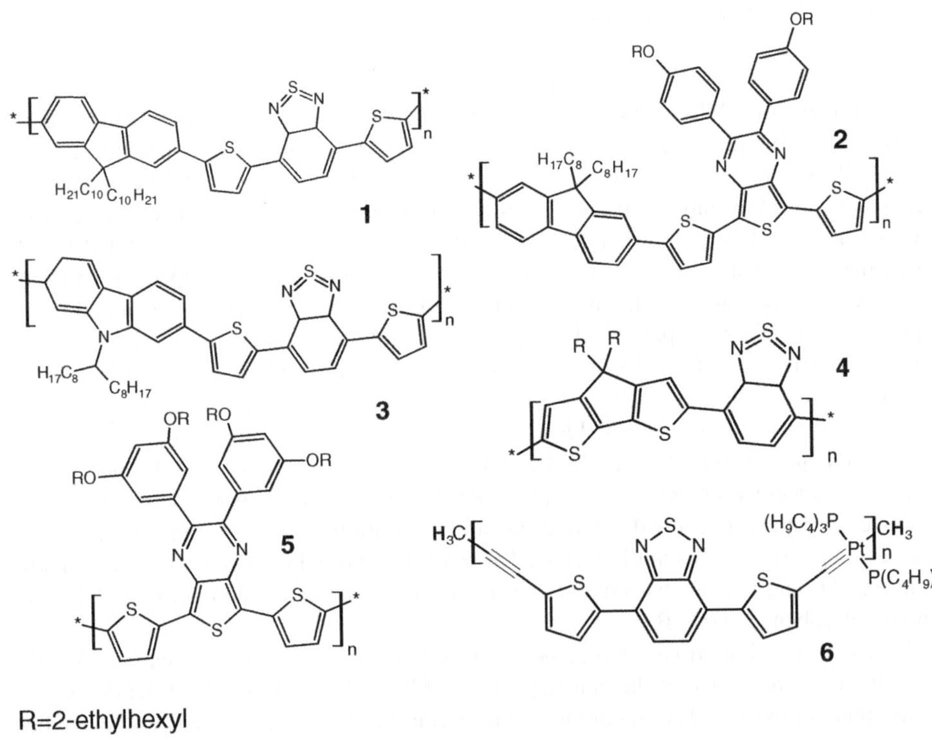

Figure 8.1 Promising polymers for OPV devices (1) poly[9,9-didecanefluorene-alt-(bis-thienylene) benzothiadiazole], [43] (2) APFO-Green 5, [44] (3) poly[*N*-9′-heptadecanyl-2,7-carbazole-*alt*-5,5-(4′,7′-di-2-thienyl-2′,1′,3′-benzothiadiazole)], [45] (4) poly[2,6-(4,4-bis-(2-ethylhexyl)-4*H*-cyclopenta[2,1-b;3,4-b′]-dithiophene)-*alt*-4,7-(2,1,3-benzothiadiazole)], [46] (5) poly{5,7-di-2-thienyl-2,3-bis(3,5-di(2-ethylhexyloxy)phenyl)-thieno[3,4-b]pyrazine}, [47] (6) platinum(II) polyyne polymer [48].

of the thiophene-, fluorine-, carbazole- and cyclopentadithiophene-based copolymers. In addition, one typical low-bandgap polymer and a metallated conjugated polymer are discussed. All compounds have been tested in bulk-heterojunction solar cells in combination with PCBM. These materials have an efficiency potential between 7–10%, and, for some of these material classes, efficiencies beyond 7% were reported.

8.2.1.1.1 Fluorene-Based Copolymers

In past years several different polyfluorene copolymers were prepared and tested in solar cells. Polyfluorenes quickly became the leading donor–acceptor structures, and, most of the acceptor units relevant for OPV applications were designed and tested for polyfluorenes [43, 44, 49–51]. Andersson et al. prepared more than 10 different compounds called APFO polymers. This class of polymers offers a sufficient large variability in the position of the HOMO/LUMO levels, and polymers with a low bandgap showing a photosensitivity down to 1 μm (polymer 2, Figure 8.1) were demonstrated. The APFO-family is a successful demonstration of the donor–acceptor approach and illustrates the high potential of this material class for organic solar cells. Bulk-heterojunction solar cells based on a blend of polymer 1 (Figure 8.1) and PCBM were reported with a hero efficiency of 4.2% (AM1.5 corrected for the spectral mismatch). The external and internal quantum efficiency [43] of these devices was found to have maximum values close to 60% and 75%, respectively, though the good performance of this polymer is mainly attributed to the high V_{oc} of ~1 V, which can compensate the rather low short-circuit current (7.7 mA/cm^2) and fill factor (54%). A high open-circuit voltage is a typical feature of fluorene-based polymer devices as the polymers often have a low-lying HOMO level. An interesting variation of polymer 1 in Figure 8.1 is obtained by replacing the fluorene unit by dibenzosilole [52, 53]. Replacing the bridging C atom of the fluorene by a Si atom is motivated by the expectation of a positive impact on the charge-transport properties. This idea is supported by the work of Wang et al. [53] reporting an uncertified power-conversion efficiency of 5.4% for a alternating copolymer of 2,7-silafluorene and 4,7-di(2′-thienyl)-2,1,3-benzothiadiazole PCBM mixture.

8.2.1.1.2 Carbazole-Based Copolymers

A few recent reports [45, 54] described the use of a carbazole-copolymers in solar cells. This material class appears to have identical electrical and optical properties as the polyfluorene class. Blouin et al. tested polymer 3 from Figure 8.1 in bulk-heterojunction solar cells with PCBM. Best device performance was in the range of 3.6% (measured at 90 mW/cm^2, AM1.5, not certified or verified by external quantum efficiency (EQE) measurement) with a high V_{oc} of 890 mV and a high FF (63%). Overall, this specific polymer performed very similar to the polyfluorene or poly-silafluorene pendants (structure 1 in Figure 8.1). Further work from the Heeger group in cooperation with the Leclerc group demonstrated that polycarbazole-based polymers can be as efficient as polyfluorene-based polymers, and a record efficiency of 6% was presented by Heeger et al. and later certified and repeated by several groups [55].

8.2.1.1.3 Cyclopentadithiophene-Based Copolymers

Cyclopentadithiophene-based polymers have attracted a lot of attention in the last two years [46, 56–58], with poly[2,6-(4,4-bis-(2-ethylhexyl)-4*H*-cyclopenta[2,1-b;3,4-b′]-dithiophene)-*alt*-4,7-(2,1,3-benzothiadiazole)] [PCPDTBT, structure 4 Figure 8.1] as the most prominent candidate of this novel class of copolymers. This polymer is a true

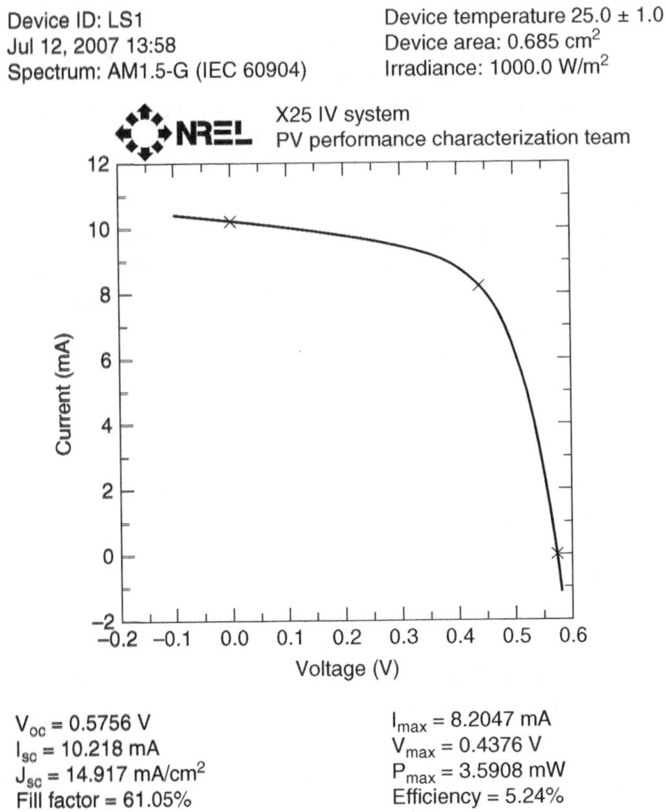

Figure 8.2 NREL certificate of the device Si-PCPDTBT-type-based device LS1 submitted to NREL by Konarka. (Reproduced from NREL.)

low-bandgap material (E_g~1.45 eV) as well as an excellent charge transporter [59] with high hole mobility, thereby fulfilling all the requirements for highly efficient solar cells. When PCBM is blended into PCPDTBT, an unfavourably intimate nanomorphology is formed and the composites typically suffer from short carrier lifetimes and considerable recombination [46]. It takes the use of additives like alkanedithioles to form a more course nanomorphology. Heeger and coworkers [1] investigated the use and function of these additives in great detail, and reported solar cells with uncertified efficiencies beyond 5% for PCPDTBT:PCBM composites.

As an alternative to the processing additives, minor modification in the backbone of PCPDTBT also allowed to optimise the device performance. Figure 8.2 shows an NREL certificate for a PCPDTBT-type polymer, where the C atom bridging the two thiophenes was substituted by a Si atom [60]. Solar cells using a blend of $PC_{70}BM$ and the Si-bridged PCPDTBT-type polymer deliver a short-circuit current of ~15 mA/cm^2 and a V_{oc} of 575 mV resulting, together with a FF of 61%, in an efficiency of ~5.2%. The EQE of the certified device reaches ~63% at ~780 nm with an estimated internal quantum efficiency (IQE) of 85% at the same wavelength. Alternative acceptor units allowed the efficiency of bridged bithiophene polymers to be raised into the 7–8% regime.

The only drawback of PCPDTBT is the rather high-lying HOMO level (~ –5.2 eV) that does not allow open-circuit voltages higher than 600–700 mV when mixed with PCBM. The current research is therefore focused on two strategies to overcome this limitation. On the one hand, synthetic efforts are strengthened to design novel bridged bithiophene copolymers with lower-lying HOMO levels, on the other hand, novel acceptors with higher-lying LUMO levels are investigated [61].

8.2.1.1.4 Metallated Conjugated Polymers

Metallated conjugated polymers have attracted a lot of attention as emitter materials in light-emitting diodes (PLED) [62–65]. The metal atom integrated into the polymer backbone can increase the mixing of the first excited singlet and triplet states leading to higher electroluminescence quantum efficiencies of PLEDs. In contrast, metallated conjugated polymers have rarely been tested as donor materials in bulk-heterojunction solar cells [66, 67]. In early reports, power-conversion efficiencies significantly below 1% were published. Recently, Wong et al. [48] demonstrated highly efficient bulk-heterojunction solar cells using polymer 6 in Figure 8.1 as donor and PCBM as acceptor material. The authors report ~5% power-conversion efficiency with EQEs as high as 87% at 570 nm. Several groups raised serious doubts that the reported efficiencies were significantly overestimated, [48, 68] and a verification by an independent institution is still missing today. Nevertheless, the concept to design polymers involving triplet states and long-living triplet excitons in charge generation could become interesting for a next-generation organic PV material.

8.2.1.2 Promising Acceptor Materials

PCBM [30] was first reported in solar cell applications in 1995 [25] and since then, no significantly better acceptor has been found. The ideal acceptor material for a bulk-heterojunction solar cell should have a strong absorption complementary to the absorption profile of the donor. Furthermore, the LUMO level offset of the donor to the acceptor needs to be optimised to ensure efficient charge transfer and a high open-circuit voltage at the same time. Finally, the acceptor needs to exhibit sufficient electron mobility in composites with the donor. Several acceptor molecules have been tested in bulk-heterojunction solar cells, among them conjugated polymers, fullerenes, carbon nanotubes, perylenes and inorganic semiconducting nanoparticles [69]. So far, only derivatives of C_{60} and C_{70} were reported to give highly efficient BHJ devices, despite the fact that the position of the HOMO and LUMO levels and the optical absorption are not ideal for most of the donor polymers [40].

A significant number of other C_{60} and C_{70} derivatives have been synthesised to improve the processability, to vary the HOMO–LUMO levels or to influence the morphology in blends with conjugated polymers. [70–72] Despite all these valuable efforts, it is the shift of the acceptor LUMO level that can give the biggest boost in efficiency. In the case of P3HT:PCBM blends the acceptor level offset is ~1 eV. More than 50% of the available energy after photoexcitation is lost. A reduction of the LUMO offset would directly be translated in an increased open-circuit voltage. A novel acceptor with a ~ 600 mV higher-lying LUMO level compared to PCBM could theoretically double the efficiency of poly-3-hexylthiophene (P3HT)-based bulk-heterojunction solar cells. Up to now only small shifts (<100 meV) of the LUMO level of derivatised C_{60} have been achieved by attaching electron-donating group to the carbon cage [70] but, at the time this review is being written, several groups (Hummelen et al., Plextronics, Nano-C, etc.) reported independently from each other and successfully demonstrated an exciting pathway to utilise fullerene multiadducts, which have 100–200 mV higher-lying LUMO values compared to pristine C_{60} [61]. Record efficiencies beyond 6% were reported for P3HT-based cells with bis-indene C_{60} and C_{70} adducts.

8.2.2 Control of Morphology in oBHJ Solar Cells

The interplay between a number of intrinsic and extrinsic parameters determines the final morphological arrangement in the active layer. The intrinsic properties include the inherent properties of the donor and acceptor (the chemical structure, molecular weight and others) and the interaction between the two components such as their crystallinity and miscibility, while the extrinsic variables are connected with the device fabrication and depend on several factors such as donor–acceptor molar ratio, selection of solvent, coating techniques and conditions, thermal treatment, etc. [73].

8.2.2.1 Molecular Weight and Regioregularity

As an example of a material intrinsic influence on morphology, we refer to the strong impact of regioregularity (RR) degree (defined as the percentage of monomers adopting a head-to-tail configuration rather than a head-to-head) and molecular weight on morphology of P3HT:PCBM. It has been reported by several groups that those factors alter the absorption feature and charge-carrier mobility of the device due to the different organisation of polymer chains and domains. Regioregularity in P3HT leads to the formation of lamellar-type structures (crystallites) in the solid state. The higher RR degree enables closer packing of these lamellae such that the optical and charged excitations adopt some interchain character. But even though the influence of RR on chain packing and ordering is well described in pristine P3HT film, its role in P3HT:PCBM blend film was ambiguous for quite some time. For the blend films, Kim et al. [74] ascribed the beneficial effect of higher RR and annealing on solar cell performance to the enhanced intraplane stacking and under this circumstance, the large increase in the J_{sc} was explained by partly increasing absorption in parallel with fewer deep traps in the high RR blend films. Recently, Frechet et al. [75] have reported that high RR is not always a prerequisite for obtaining a high solar cell performance and even a lower RR could be beneficial to the thermal stability of the device as a result of less crystallisation-induced phase segregation of PCBM. These findings are still somewhat in contrast to those of other groups [76]. Optical microscope images of P3HT:PCBM films with different degrees of RR confirm the formation of many large needlelike PCBM crystals that get longer as RR increases (as shown in Figure 8.3). The two-dimensional grazing-incidence X-ray scattering (GIXS) patterns (Figure 8.4), show that the PCBM is highly crystalline in the higher-RR P3HT blend film and moreover suggests that at higher RR the P3HT stacks are less oriented in the axis normal to the substrate due to the presence of large PCBM crystallites.

Molecular weight (MW) is found to have a similar influence on the solid-state morphology of P3HT as regioregularity. In the blend of P3HT-PCBM, a higher MW causes enhanced intermolecular ordering (π-stacking) of the P3HT phase. As molecular weight increases, more extended crystalline regions can form and offer better charge-carrier mobility and subsequently better device performance [77]. In another study, the influence of MW has been shown to be in connection with the annealing treatment and the ideal morphology has been reported by using P3HT with an optimum ratio between high-MW and low-MW components

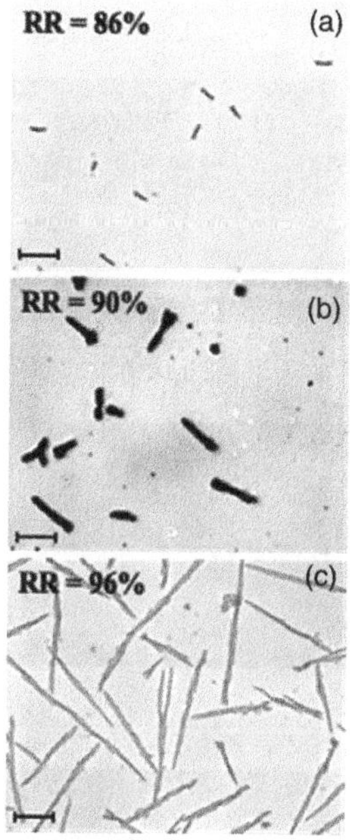

Figure 8.3 Optical microscopy images of (a) 86, (b) 90, and (c) 96% RR P3HT-PCBM blends at a 55:45 weight ratio after 3 h of annealing at 150 °C. Dark areas are PCBM-rich regions. Scale bar = 50 μm [75]. (Reproduced with permission from Woo, 2008. Copyright & 2008, American Chemical Society.)

(high/low ratio = 1:4). The corresponding ideal morphology is comprised of highly ordered crystalline regions formed by low-MW P3HT embedded and interconnected by a high-MW P3HT matrix [78]. In this case, the corresponding system avoids an unfavourable fullerene distribution and could be superior in morphology. In fact here, morphological fate will be determined by an intricate interplay of several factors such as temperature, molecular weight and solubility.

8.2.2.2 Influence of the Solvent

Depending on the type of solvent, the solubility of each D/A materials as well as their interaction and miscibility could differ significantly. In addition, the boiling temperature of the solvent plays a crucial role in the film formation and morphological organisation of oBHJ PVs. The strong influence of solvent on MDMO-PPV (poly(2-methoxy-5-(3′,7′-dimethyloctyloxy))-1,4-phenylene vinylene) and $PC_{60}BM$ systems has been proved by several investigations [32,79]. An enormous improvement of PV performance has been reported by using a chlorobenzene (CB) rather than toluene (about threefold increase from 0.9% to 2.5% in PCE with a concurrent improvement in the FF and J_{sc}) [32]. Further studies via TEM and AFM, verified the large influence of solvent on the size of the phase separated PCBM-rich domains in spin-coated MDMO-PPV:PCBM films. Figure 8.5 shows that replacing chlorobenzene with toluene, the average size of PCBM-rich domains reduced from ± 100 nm to ± 15 nm, providing a shorter distance for excitons to diffuse and separate at the D/A

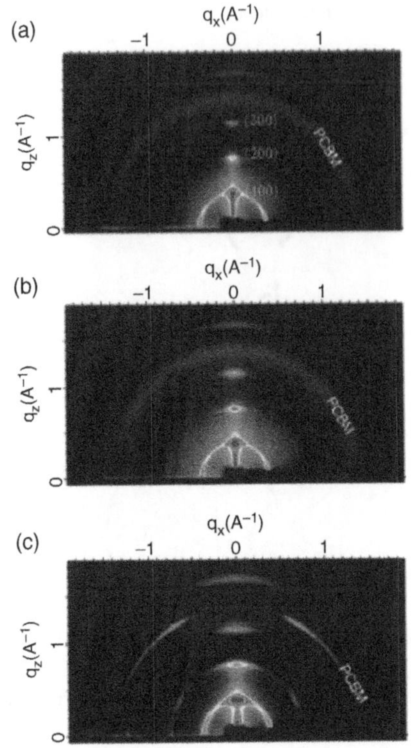

Figure 8.4 GIXS patterns of films of (a) 86, (b) 90, and (c) 96% RR P3HT-PCBM blends at a 55:45 weight ratio after 1 h of annealing at 150 °C. The vertical section at $q_x \approx 0$ is not the true specular direction (i.e. $q_x = 0$) but is tilted from this [75]. (Reproduced with permission from Woo, 2008. Copyright © 2008, American Chemical Society.)

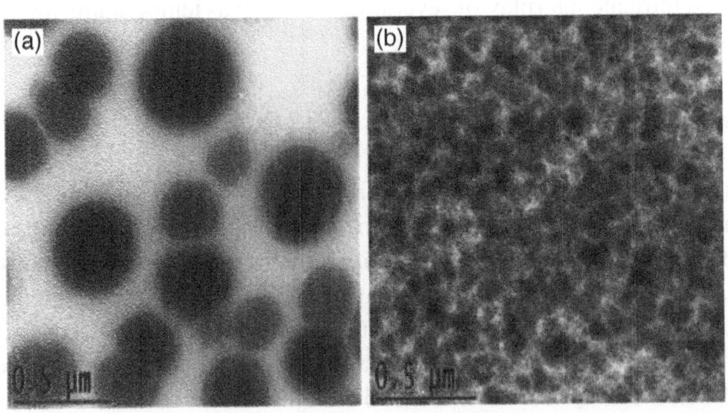

Figure 8.5 The 1.8 μm × 1.8 μm TEM images of spin-coated 1:4 MDMO-PPV:PCBM thin films from a toluene (a) and chlorobenzene (b) solution on a PET substrate [80]. (Reproduced with permission from Martens, 2003. Copyright © Elsevier, 2003.)

interfaces. This discrepancy in morphology is mainly attributed to the higher solubility of fullerene in chlorobenzene [80, 81].

A similar observation has been reported by Yang and coworkers [82] for poly(2-methoxy-5-(2′-ethylhexyloxy)-1,4-phenylenevinylene) (MEH-PPV) with buckminsterfullerene (C_{60}). According to the AFM images, bicontinuous interpenetrating structures are formed for the films processed from aromatic solvents like xylene or dichlorobenzene (DCB) and in this case both MEH-PPV and C_{60} show features less than 20 nm in width, while films coated from nonaromatic solvents like tetrahydrofuran (THF) exhibit domains about 40 nm in diameter.

Recently, some notable results have been achieved by using a mixture of solvents [83–86]. A study on the blend of polyfluorene copolymer, poly(2,7-(9,9-dioctyl-fluorene)-alt-5,5-(4′,7′-di-2-thienyl-2′,1′,3′-benzothiadiazole)) (APFO-3), with [6,6]-phenyl-C61-butyric acid methyl ester, revealed an enhancement in the photoresponsivity of devices made from a mixture of chloroform and chlorobenzene (CFCB) while devices processed from chloroform mixed with xylene or toluene (CFXY and CFTO, respectively) exhibited a reduction in photocurrent density, compared to devices made form neat chloroform. Time-resolved spectroscopy suggests an excitation-intensity-dependent nongeminate recombination of free charge carriers together with the fact that charge mobility, which is involved in formation of free charge carriers, is influenced by mixing solvents. An AFM study shows smaller domains and a smoother surface for the films based on CFCB as compared to the CF, CFXY and CFTO films [83].

8.2.2.3 Thermal Annealing

A large number of studies have demonstrated improved PCE of some polymer–fullerene composite solar cells upon thermal treatment. A substantial improvement in the PCE of P3HT:PCBM-based solar cells has been obtained by thermal annealing and its favourable influence on the morphology and solar cell performance is well described by a large number of studies [38, 87–90]. Chirvase et al. [87] ascribed the enhancement of the P3HT:PCBM optical absorption to the molecular diffusion of fullerene out of the polymer matrix upon annealing. The growth of the PCBM clusters leads to formation of percolation paths and, therefore, improves the photocurrent. Further studies by Erb et al. suggest that the significant increase in optical absorption is due to the aggregation state of P3HT, from amorphous (nonannealed) to crystalline (annealed). The grazing-incidence X-ray diffraction (XRD) of the annealed P3HT:PCBM film revealed one single peak at $2\theta = 5.4°$ pertaining to the polymer crystallinity in the *a*-axis orientation (main chain parallel and side chains perpendicular to the substrates, as shown in Figure 8.6). The main effect of annealing is described as redistribution of fullerene; isolated molecules of PCBM begin to diffuse into larger aggregates and in the PCBM-free regions, the P3HT aggregates can be converted into P3HT crystallites (Figure 8.7). Complementary transport studies show that annealed and nonannealed devices can have comparable values of the mobility–lifetime product ($\mu\tau$) for the majority carrier and attributes the efficiency gain due to annealing to the reduced recombination losses by an optimised phase-structured morphology alongside an increased spectral photocurrent contribution in the near-IR [88, 91]. Note that these studies mainly referred to the majority carriers and not to the balance of the ambipolar carrier properties.

The conventional bright-field TEM images of P3HT:PCBM blend films confirmed the presence of fibrillar P3HT crystals even prior to thermal treatment. TEM studies showed that controlled annealing results in an increase of the length of these fibers, while the width

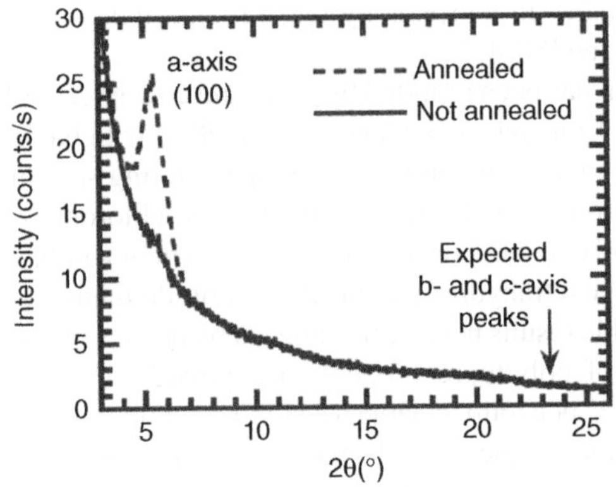

Figure 8.6 Diffractogram (grazing incidence) of P3HT:PCBM composite film deposited on glass/ITO/PEDOT-PSS. [88]. (Reproduced with permission from Erb, 2005. Copyright © John Wiley & Sons, Ltd, 2005.)

of them remains almost constant as compared to the nonannealed films. Comparison of TEM images and SAED patterns of non- and annealed films of P3HT:PCBM blend films (as shown in Figures 8.8 and 8.9) indicates that the longer fibrillar-like P3HT crystals enhances the formation of a P3HT network within the composite film and hinders the excessive diffusion of PCBM molecules and their large-scale crystallisation and therefore only small PCBM crystals are formed during the annealing treatment (dark domains between the P3HT fibrillar-like crystals denote to the PCBM crystals). In fact, after annealing, a large interfacial area, necessary for efficient charge generation, is not hindered by the presence of long fibers since P3HT favourably tends to crystallise in one direction [92].

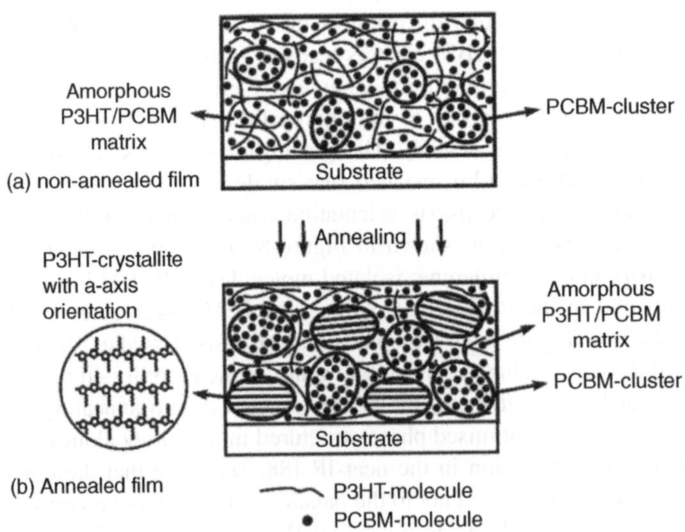

Figure 8.7 Structural changes of P3HT:PCBM films upon annealing (schematic) [88]. (Reproduced with permission from Erb, 2005. Copyright © John Wiley & Sons, Ltd, 2005.)

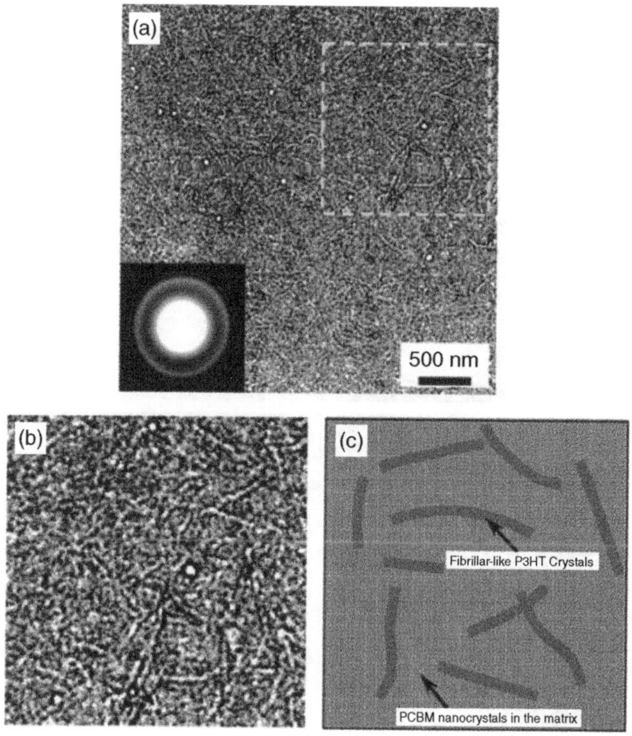

Figure 8.8 BF TEM images show the overview (a) and the zoom in (b), and the corresponding schematic representation (c) of the pristine photoactive layer of a P3HT:PCBM plastic solar cell. The inset in Figure 8.8a is the corresponding SAED pattern [92]. (Reproduced with permission from Yang, 2005. Copyright © American Chemical Society, 2005.)

Savenije et al. reported that annealing of P3HT/PCBM composites not only changes the morphology and optical absorption but even results in changes in the redox and photoconductive properties of the blend layer. Enhanced photoconductivity is mainly ascribed to the increase in the hole mobility. This study suggested that in the annealed P3HT:PCBM samples, after photoinduced charge separation, holes are localised on the crystalline fibrils and since these fibrils are enclosed by nonordered region, this would act as energetic barriers for the positive charge carriers and therefore the release time of the holes will be the rate-limiting step for recombination or trapping [93]. The significant increase in the current density of annealed P3HT:PCBM devices is also explained by more balanced electron and hole transports as a consequence of better morphology. The influence of space-charge buildup that strongly limits the photocurrent in nonannealed devices is greatly weakened by an increase in the electron mobility and a dramatic increase in the hole mobility upon an optimised annealing treatment [90].

Besides P3HT:PCBM, some other systems such as PCPDTBT:PCBM [94] and some polymer–polymer-based solar cells [95–97] have shown higher PCE after annealing. Although the improvement is not as outstanding as reported for P3HT-based devices, these are all strong motivations for investing more efforts into the study of morphology when the composite is subject to thermal treatment. In addition, the long-term thermal stability that can

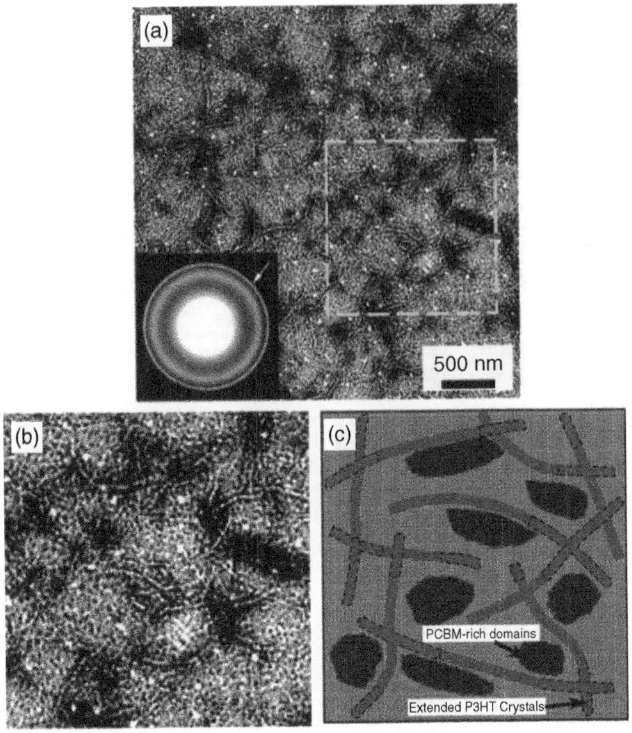

Figure 8.9 BF TEM images show the overview (a) and the zoom in (b), and the corresponding schematic representation (c) of the thermal annealed photoactive layer. The inset in Figure 8.9a is the corresponding SAED pattern. The arrow is to indicate the increased intensity of (020) Debye–Scherrer ring from P3HT crystals compared to the SAED pattern shown in the inset of Figure 8.8a [92]. (Reproduced with permission from Yang, 2005. Copyright © American Chemical Society, 2005.)

be guaranteed under harsh condition like thermal annealing, points to the further importance of this study.

8.2.2.4 Compositional Dependence of Morphology

The blending ratio of the donor with the acceptor as well as the overall concentration of the applied solution can have an importance influence on charge transport, photophysical properties and morphological organisation of the system. As pointed out, for the case of P3HT:PCBM, the best device performance is typically reported for a blend ratio between 1:0.8 and 1:1 with a concentration of 1 wt% P3HT (10 mg mL^{-1}). Time-of-flight (TOF) measurements on P3HT-PCBM blend films with different wt/wt ratios verified that it is only when these ratios are balanced that nondispersive electron and hole transport is found [98]. AFM studies revealed that films cast from solutions with higher concentrations or smaller P3HT:PCBM ratios exhibited "overgrown" PCBM crystals after high-temperature annealing, leading to degradation of the morphology and poor device performance [89]. In contrast to the semicrystalline P3HT:PCBM, many systems such as MEH-PPV:PCBM, MDMO-PPV:PCBM require high fullerene content to provide a proper nanoscale phase separation

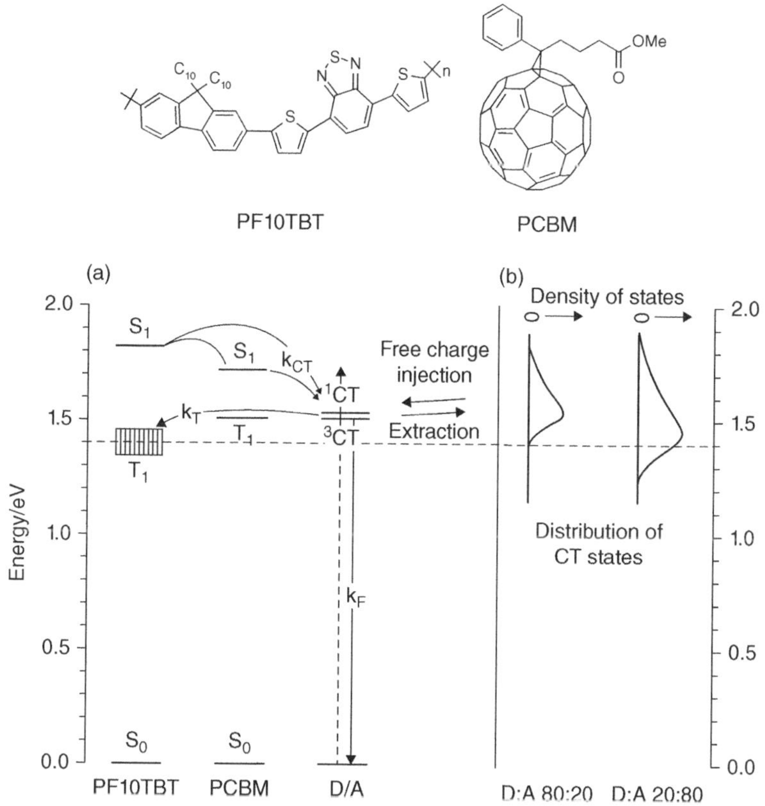

Figure 8.10 Chemical structures of PF10TBT and PCBM (top). (a) Jablonski diagram showing the singlet and triplet excited-state energies of PF10TBT and PCBM, and the CT state, relative to their ground states. The dashed arrow represents absorption from a ground-state D/A complex. The CT state can also be populated from S1 states of PF10TBT or PCBM by charge transfer (b) Schematic energetic distribution of the CT states for two different PF10TBT:PCBM ratios [103]. (Reproduced with permission from Tang, 1986. Copyright © 1986, American Institute of Physics.)

and an efficient charge transport [99] (in these cases a blending ratio of 1:4 is reported as an optimal ratio that is far beyond the percolation threshold). Recently, the existence of the intermediate state between the exciton and the free charges in some blend systems based on polyfluorene has been reported [100–103]. In this matter, identifying and understanding the nature of such an intermediate state that is often revealed to be largely compositional dependent could provide us with invaluable information especially on the effect of morphology on charge generation. As an example, Veldman et al. confirmed the existence of a charge transfer (CT) state in the blend of a fluorene copolymer (PF10TBT) "poly[2,7-(9,9-dialkylfluorene)-alt-5,5-(4′,7′-di-2-thienyl-2′,1′,3′-benzothiadiazole)] (PF*n*TBT, with *n* being the number of carbons of the alkyl substituents, Figure 8.10)" with PCBM. In this study, the CT emission is used as a probe to assess the dissociation of CT excited states at the D/A interface in PV devices, as a function of PCBM concentration. The formation and decay of emissive CT excited states via electron transfer from photoexcited PF10TBT or PCBM is confirmed by steady-state and time-resolved photoluminescence measurements. It is shown

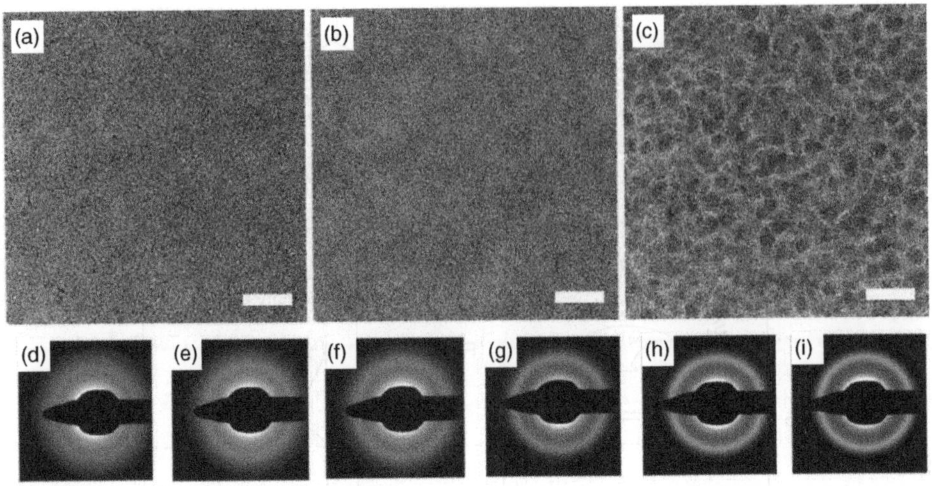

Figure 8.11 TEM images of PF10TBT:PCBM films containing 20 (a), 50 (b), and 80 (c) wt% PCBM. The size of each image is width × height = 1.25 × 1.25 μm^2, the horizontal scale bar in images in a–c is 200 nm. Panels d–i show the SAED patterns of PF10TBT films containing 0, 10, 20, 50, 80, and 100 wt% PCBM, respectively, [103]. (Reproduced with permission from Tang, 1986. Copyright © 1986, American Institute of Physics.)

that the maximum of the CT emission shifts to lower energy and decreases in intensity with higher PCBM content. At higher PCBM content, such a reduction in energy of CT states is explained with the higher relative permittivity of PCBM ($\varepsilon_r = 4.0$) compared to that of the polymer ($\varepsilon_r = 3.4$), stabilising the energy of CT states and of the free charge carriers. Figure 8.10 schematically shows how at higher PCBM concentrations (80 wt% PCBM), the low-energy tail of the energetic distribution of CT states drops below the PF10TBT triplet energy ($E(T1) = 1.4 \pm 0.05$ eV), hindering charge recombination to that neutral PF10TBT triplet excited state. Very similar to that observed for MEH-PPV:PCBM and MDMO-PPV:PCBM blends, TEM measurements demonstrate that nanoscale (>10 nm) phase separation occurs at high fullerene concentration. SAED patterns indicate that nanocrystalline PCBM clusters are even present at low PCBM concentration and they just get larger as PCBM concentration increases (see Figure 8.11). These clusters with high local electron mobility enable the efficient dissociation of short-lived CT excitons that may otherwise recombine radiatively or nonradiatively. Therefore the improved device efficiency of the PF10TBT:PCBM devices at higher PCBM content is attributed to the larger phase separation and a concurrent improvement in separation of bound electron–hole pairs at the D/A interface [103].

From analysis of the carrier-recombination dynamics, the existence of such an intermediate charge-transferred state is also suggested in PCPDTBT:PCBM blends. The yield of long-lived mobile carriers originating from this intermediate state was found to be highly dependent on the ratio of the components. Studies indicate that PCPDTBT:PCBM composites blended in a molar ratio of 1 to 3.3 exhibit the highest yield of long-lived mobile carriers and have the smallest contribution from the intermediate charge transferred state, consistent with the observation that the best solar cell efficiency is obtained using the 1:3.3 PCPDTBT:PCBM bulk-heterojunction material [94].

8.2.3 Monitoring Morphology

8.2.3.1 Scanning Techniques

Phase separation and phase control are the key to higher performance in oBHJ solar cells, and research needs to provide precise information on the size and chemical composition of each phase to understand and predict the mechanisms of phase separation. The desired information needs to be attained with nm resolution, and many characterisation techniques have been established and developed so far to address this issue. These techniques can be classified into methods presenting information on the surface and on the bulk of the composite. The surface morphology of thin films has been studied primarily by atomic force microscopy (AFM) in contact mode and tapping mode TM and these studies have confirmed the significant impact of phase separation on the PV performance [104, 105].

Transmission electron microscopy (TEM) [80, 103] and scanning electron microscopy (SEM) [106] have been frequently used to inspect the structure of phase-separated domains and record the cross-sectional image of polymer–fullerene blends. The bulk structure of blend films can be visualised by means of TEM with high spatial resolution and based on differences in material density, the polymer- and fullerene-rich domains can be identified. Despite the great powers that conventional TEM possesses, this technique provides only limited information on the local blend composition. McNeill et al. [107] have used scanning transmission X-ray microscopy (STXM) to quantitatively determine the local blend composition in a polymer-methanofullerene blend. They realised that in a prototypical blend of MDMO-PPV:PCBM, the relative morphology formed from 100%-pure PCBM-rich domains surrounded by an intermixed domain contains roughly equal proportions by weight of polymer and methanofullerene. Raman spectroscopy has also been used to provide information on chemical composition of the domains [108–110]. Moreover, various techniques were applied to study how individual nanostructures contribute to the device performance. For instance, near-field scanning optical microscopy (NSOM) has been widely used to measure local fluorescence and photocurrents with ~200 nm resolution [110–113]. Scanning Kelvin probe microscopy (SKPM) has been employed as a tool to measure the surface potential and the photoinduced surface photovoltage with a lateral resolution better than 100 nm [114–116]. Coffey et al. have shown by means of time-resolved electrostatic force microscopy (EFM) with a resolution of 100 nm and 100 μs that the measured EFM charge accumulation rates correlate with the external quantum efficiencies (EQEs) measured for photodiodes prepared from poly-(9,9′-dioctylfluorene-co-benzothiadiazole)(F8BT)/poly-(9,9′-dioctylfluorene-co-bis-N,N′-(4-butylphenyl)-bis-N,N′-phenyl-1,4-phenylenediamine (PFB) blends.

In Figure 8.12, EFM and AFM images of F8BT:PFB blend film refer to the slower charging at the interfaces between polymer domains and associate the bulk of the photocurrent with regions away from the visible domain interfaces. In this case, the observation interestingly suggests that most of the photocurrent results from the domain centres rather than domain interfaces [117]. This correlation between local photocurrents and device performance has further been seen for MDMO-PPV:PCBM blends through photoconductive atomic force microscopy (pcAFM) providing higher resolution of about 20 nm while all the previously applied methods had been suffering from the insufficient resolution [118].

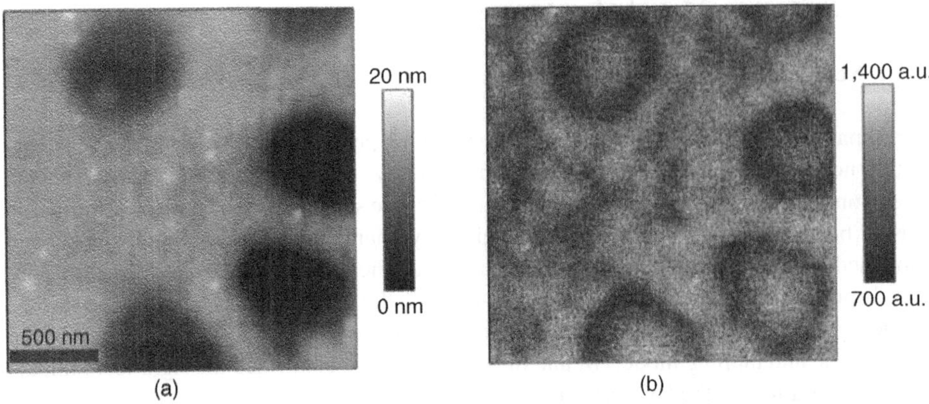

Figure 8.12 Charging rate image of a polymer blend film (a) AFM height image of 50:50 film of F8BT:PFB spin coated from xylenes. The recessed domains are enriched in PFB (film thickness: 90 nm). (b) EFM photoinduced charging rate map generated by plotting the inverse exponential time constant for photoinduced charging at each point in (a). The dark rings indicate regions of slower charging [118]. (Reproduced with permission from Coffey, 2007. Copyright © 2007, American Chemical Society.)

8.2.3.2 3-Dimensional Mapping of the Bulk Morphology

Normal and parallel to the substrate, morphology organisation can be completely different and the components could distribute across the film thickness in an inhomogeneous way. Cross-sectional images acquired by AFM, SEM and TEM can hint on the vertical organisation of the single component across the film. As an example, Moon et al. reported the observation of "columnlike" structures in the defocused cross-sectional TEM images of RR-P3HT:PCBM blend films providing an efficient charge transport through the film thickness [119]. Another technique that can provide detailed information on the three-dimensional (3D) organisation is secondary ion mass spectroscopy (SIMS) where the film is gradually sputtered by a scanning ion beam and the fragments simultaneously analysed by mass spectrometry [120–123]. Depth profiling of the chemical composition is attainable here but this technique cannot give a microscopic image and suffers from the low spatial resolution. Electron tomography (ET) is found to be a promising technique to provide a full three-dimensional picture of film morphology and uses a series of 2D pictures to reconstruct a three-dimensional image; data collection is performed by tilting the specimen in the electron beam of the TEM. This method has been recently utilised in OPVs with a resolution in the range of 5–20 nm, matching to the dimension of phase separation [124, 125]. For the first time Young and Loos implemented this technique in OPVs and provided a 3D morphological picture for the MDMO-PPV:PCBM system, giving information on PCBM-rich domain sizes as well as connectivity of the PCBM domain networks within the active layer [124]. Most recently, electron tomography is exploited to visualise the 3D organisation of RR P3HT:PCBM layers, comparing the film morphology prior to and after the annealing of the samples. This study substantiated the formation of genuine 3D network formed with high crystallinity in thermally annealed P3HT:PCBM layers and the favourable vertical segregation of P3HT and fullerene through the thickness of the active layer inferred from the measurement. Additionally, the overall crystallinity of P3HT in the blend film has been quantified via this technique; investigation showed that P3HT takes up about 58% of the total volume of the annealed blend layer, with about 60% of that being crystalline P3HT nanorods and the other 40% being more amorphous [126]. Deployment of techniques such as electron tomography is crucial at this formative stage of OPV technology where the traditional techniques cannot provide a full understanding of the correlation between the morphology and the device performance.

8.2.4 Numerical Simulations of Morphology

Besides the experimental efforts, the first computational studies have been performed to investigate the intricacy of morphology and its impact on overall OPV performance. Many of these investigations were based on a continuum approximation [127–129] that is unable to describe carrier interactions accurately and limits morphological effects to bulk parameters [130]. Monte Carlo (MC) modeling is a commonly used method to simulate the dissociation of columbic charge pairs and the hopping movement and interactions of the photoinduced charges carriers in an energetically disordered system [131]. Watkins et al. [131] have used a dynamical Monte Carlo model to comparatively study the effect of an ordered and disordered morphology on the internal quantum efficiency (IQE) of PV cells. This study indicated that a checkered morphology exhibits a peak IQE 1.5 times higher than a disordered blend and suggested a tradeoff between η_{ED} and η_{CC} with increasing interface area. Young and Forrest have developed the method by taking into account the influence of optical interference and variations in charge-carrier mobilities [132]. Although the above-mentioned studies provide valuable information, none of them incorporate the effects of chemical structure. Hopping transport in such systems has commonly been analysed through a Gaussian disorder model (GDM) [133] that has a physical meaning only for a cubic lattice of hopping sites. Nelson and coworkers have presented a new Monte Carlo model that includes the influence of chain connectedness and complexity on morphology and charge transport. For a polymer blend system, they varied the interaction energies between the chains and produced two different contrasting morphologies consisting of well-dispersed, loosely coiled polymer chains (heterophilic) and aggregated, tightly coiled chains (homophilic), representative of morphologies of polymer films following different types of process treatment (solvent, annealing). Studies showed that homophilic blends have a lower threshold for charge percolation and more dispersive transport than heterophilic blends. They have further pointed to the fact that morphology influences both transport and charge recombination such that the optimum composition and morphology is a sensitive function of charge recombination and separation rates [134]. Beside the aforementioned efforts, it is worthwhile mentioning the study of Peumans et al. [24] in which they have modeled the morphological evolution of a mixed small-molecule solar cell system upon thermal annealing by using the cellular automata approach [135] and the appropriate choice of interfacial energies between the D/A phases. They have provided a 3-dimensional image of morphology and found a clear consistency between the simulated images and the recorded cross-sectional SEM pictures.

8.2.5 Alternative Approaches to Control the Morphology

In recent years, a variety of techniques have been explored to bring the morphology under control; here some of the efforts are briefly discussed. As already mentioned, P3HT:PCBM

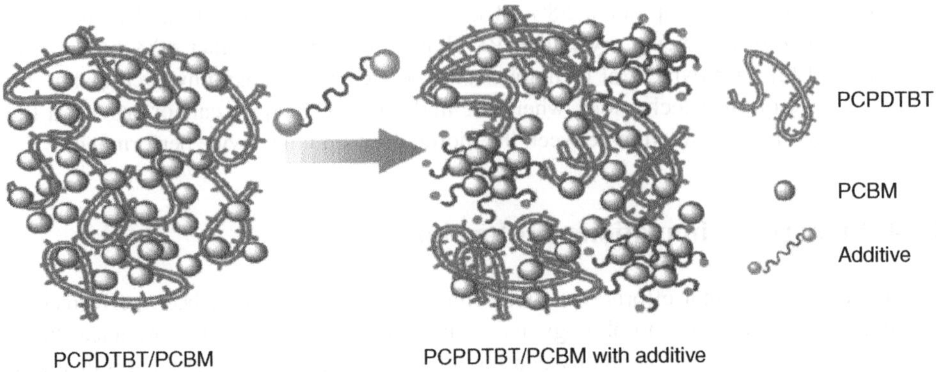

Figure 8.13 Schematic depiction of the role of the processing additive in the self-assembly of PCPDTBT:PCBM blend [139]. (Reproduced with permission from Lee, 2008. Copyright © 2008, American Chemical Society.)

blends require thermal annealing in order to self-organise into a conformation ensuring optimum charge carrier creation and extraction. But also other ways of controlling the morphology have been proposed and have proven to be highly effective. Slow drying was reported as one of the methods to improve the order in P3HT in blends with PCBM [98]. The improved order [136] was reflected in higher hole mobility [137], higher fill factors and a reduced serial resistance. Additives were reported as an alternative method to create better order in blends of PCPDTBT:PCBM and P3HT:PCBM. Alkylthiols of different length [138], like n-hexylthiol, n-octylthiol, or n-dodecylthiol but also oleic acids [1] were added into P3HT:PCBM solutions and allowed the formation of thin films with slightly enhanced hole mobility and significantly enhanced charge-carrier lifetime due to enlarged P3HT domains with higher crystallinity. Nevertheless, some thermal annealing was still necessary to give the highest possible performance. In the case of PCPDTBT:PCBM, applying alkanedithiol resulted in an increase of efficiency from 2.8% to 5.5% [1]. Morphology evolution was explained by the fact that alkanedithiol selectively dissolves the fullerene. As a result, three separate phases are formed during the process of liquid–liquid phase separation and drying: a fullerene–alkanedithiol phase, a polymer aggregate phase, and a polymer–fullerene phase, as schematically depicted in Figure 8.13 [139].

The additive approach is actually very close to the technique called miniemulsions, described earlier and in more detail by others [140–142]. In that approach a mixture of P3HT in water, surfactants and a solvent was rigorously sonicated before letting the solvent evaporate. Such dispersions typically have a particle distribution between 70–200 nm and give homogeneous films [143] upon spin coating. FET mobilities for such nanoparticluar films were found to be of the order of 10^{-4}–10^{-3} cm^2/V s. Solar cell fabrication was hardened by the absence of a well-performing, water-soluble fullerene, and only allowed the investigations of bilayer-type devices that had moderate performance [144]. A third, quite similar approach to control the nanomorphology of P3HT:PCBM blends requires the addition of "nonsolvents" into the P3HT:PCBM solution [145, 146]. This phenomenon is attributed to the aggregation of the polymer into nanoparticulates, similar to the miniemulsion mentioned above. The most recent approach [147, 148] grew fibers by slow cooling of P3HT solutions, with the crystalline fibers being isolated from the amorphous material by centrifugation and filtration. The fibers were reformulated in dispersions with PCBM and used for solar cell processing. The best results (efficiency up to 3.6% under 100 mW cm^{-2}) were obtained for a mixture of 75% P3HT fibers and 25% disorganised P3HT, the latter being suspected to be necessary to fill the gaps present in the nanostructure layer and to ensure intimate contact between the donor fibers and the PCBM domains. And last but not least is the surface patterning technique that can modify film morphology on length scales smaller than the lateral dimensions of monolayer surface patterns, this could eventually lead to approach length scales relevant to studies of charge separation [149].

8.3 INTERFACES IN ORGANIC PHOTOVOLTAICS

In this section we discuss the impact of interface materials on the performance of organic solar cells. The general course of power generation in organic photovoltaics (OPV) has been introduced already in Section 8.2. In short, we distinguish between the following processes: Light is absorbed (either by the polymer or the fullerene), excitons are created, split into free carriers, transported and extracted to electrodes. The interface between the active layer (i.e. blend of donor and acceptor) and the electrode materials play a crucial role. Depending on the material system used the main functions of interface materials are and may impact:

1. the open-circuit voltage (V_{oc});
2. the polarity of the device;
3. light propagation and distribution in the case of optical spacer materials;
4. the physical, chemical or environmental interaction of the polymer and electrode;
5. the contact selectivity.

These 5 points will be elaborated in more detail in the following sections.

8.3.1 Origin of V_{oc}

The V_{oc} originates from the difference of the highest occupied molecular orbital (HOMO) of the donor material and the lowest unoccupied molecular orbital (LUMO) of the acceptor material when these two materials are in thermal equilibrium. Interface materials may reduce the open-circuit voltage, but cannot increase it beyond the HOMO–LUMO difference. We have investigated this effect on solar cells from poly(2-methoxy-5-(3′,7′-dimethyloctyloxy)-1,4-phenylenevinylene) (MDMO-PPV) blended with highly soluble fullerene derivatives as the active layer. The investigated structure was ITO/poly(3,4-ethylenedioxythiophene) poly(styrene-sulfonate) (Pedot:PSS)/MDMO-PPV:fullerene/metal electrode. The results are visualised in Figure 8.14. A reduction in the acceptor strength by 160 mV (reduction potential changed from –0.53 V to –0.69 V vs. NHE) resulted in an increase of V_{oc} from 0.56 V to 0.76 V. Further trials showed that V_{oc} scales nearly linear with the acceptor strength with a scaling factor of $\sim$1. On the other hand, varying the work function of the metal electrode from 5.1 eV for Au to 2.87 eV for Ca/Ag results in an increase of V_{oc} from 0.650 V to 0.814 V, and the corresponding scaling factor was estimated to be 0.1 or lower. This contact

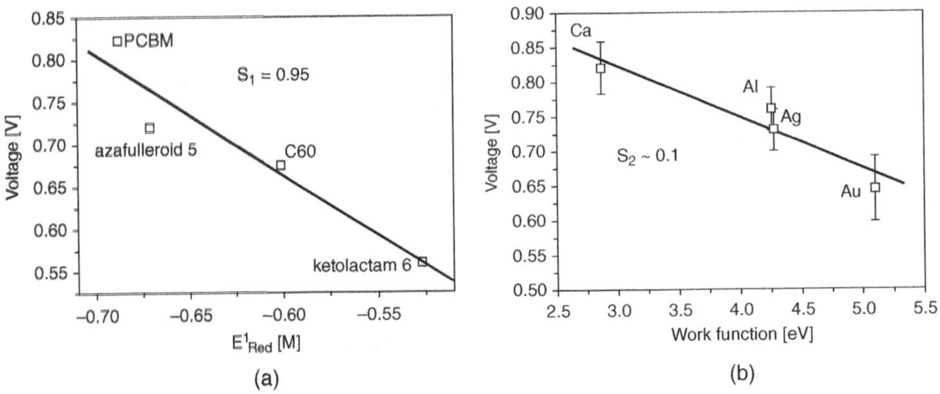

Figure 8.14 (a) V_{oc} versus acceptor strength and (b) V_{oc} versus negative electrode work function. The slopes S1 and S2 of the linear fits to the data are given [71]. (Reproduced with permission from Brabec, 2001. Copyright © John Wiley & Sons, Ltd, 2001.)

behaviour could be explained by the Fermi-level pinning mechanism, a phenomena well known in semiconductor physics [71, 150, 151].

Various metals were investigated to form ohmic contacts with fullerenes. Among them, Ca, Al and Ag were reported to follow the Fermi-level pinning principle [71, 152]. In contrast, for nonohmic contacts it was found that the V_{oc} is influenced by the work function of the interface materials [152]. Palladium is such a metal. It forms a nonohmic contact to fullerenes, and can reduce the V_{oc} by up to 0.5 V as compared to LiF/Al.

A similar trend holds for the contact formation between the donor and the anode. Again, the V_{oc} was found to correlate linearly to the HOMO level of the donor. [40] This linear correlation can be influenced by using Pedot:PSS as the interface material between ITO and the semiconductor blend. For organic light-emitting diodes (OLEDs) it was shown that a variation in the work function of Pedot:PSS from 4.7 eV to 5.3 eV correlates linearly with the built-in potential of the OLED [153]. This linear correlation between the Pedot:PSS work function and the V_{oc} also holds for OPV cells [154, 158].

In summary, the V_{oc} of a well-optimised solar cell should be governed by the HOMO of the donor material and the LUMO of the acceptor material. Significant deviations from that behaviour can be found when the electrode materials do not pin to the semiconductor bands. In such a case, the V_{oc} will be smaller than the theoretical maximum given by the semiconductors.

8.3.2 Determination of Polarity-Inverted and Noninverted Structure

A bulk-heterojunction composite can be regarded as one intrinsic semiconductor with a high quantum efficiency for charge generation. If one can neglect vertical phase segregation between the fullerene and the polymer, such a composite does not have a preferential direction to extract electrons or holes. As such, it is clear that the choice of the electrode and interface materials will decide the polarity of the device. The polarity of the solar cell is thus determined by the interface materials or more precisely by the interface formed between the donor

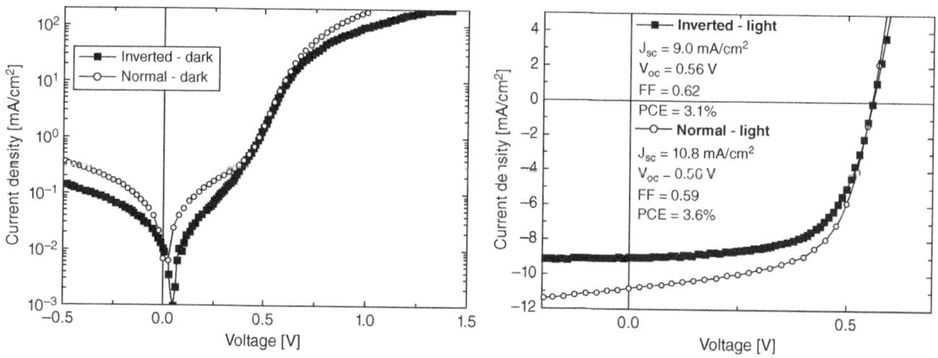

Figure 8.15 *J–V* curves in a semilogarithmic representation for the normal (open circles) and inverted (solid squares) OPV in the dark (upper plot) and under illumination (bottom plot) [157]. (Reproduced with permission from Waldauf, 2006. Copyright © 2006, American Institute of Physics.)

material at the anode and the acceptor material at the cathode [155–159]. To demonstrate that the polarity of a device is determined by the electrode materials, the *J–V* curves of a cell with "normal" structure ITO/Pedot:PSS/P3HT:PCBM/LiF/Al is compared to an "inverted" cell structure ITO/TiO_x/P3HT:PCBM/Pedot:PSS/Al in Figure 8.15. Note that the inverted device was contacted with reverse polarity to the normal device. As expected, Pedot:PSS forms for both device structures the anode, i.e. the contact where holes are extracted or injected. The cathode, on the other hand, is either given by the LiF/Al contact or by the ITO/TiO_x layer. [157] Both architectures give quite comparable performance. Small differences in the photocurrent density are usually explained by a vertical phase segregation within a bulk-heterojunction composite [160, 161]. In summary, it is essential to note that the polarity of a device and herewith also the charge extraction of the device is controlled by the choice of the interface materials.

8.3.3 Optical Spacer

The maximum thickness and thus absorption of the active layer is given by the transport properties of the donor–acceptor blend. Due to these transport limitations, the active-layer thickness of organic solar cells is typically smaller than half the wavelength of the visible light. It is this thickness regime where interference effects play an important role. For such optically thin structures light is reflected at the nontransparent electrode and can pass the active layer twice. The electrical field strength is 0 at the reflecting metal electrode and has its maximum somewhere in the bulk, depending on the refractive index and the thickness of the layers. The optical spacer is a concept that can help to increase the absorption for thin-film structures. To do so, a transparent layer, ideally with a high refractive index, is inserted between the active layer and the reflecting metal electrode. Such a thin and transparent layer of some 10 nm is enough to significantly change the position of the local absorption maximum. Once optimised, such an optical spacer layer can help to shift the maximum of absorption into the active layer. An exemplary result for an OPV cell structure with an optical spacer is glass/ITO/PEDOT/P3HT:PCBM/TiO_x/Al and is presented in Figure 8.16. Here, TiO_x was used as the optical spacer [161].

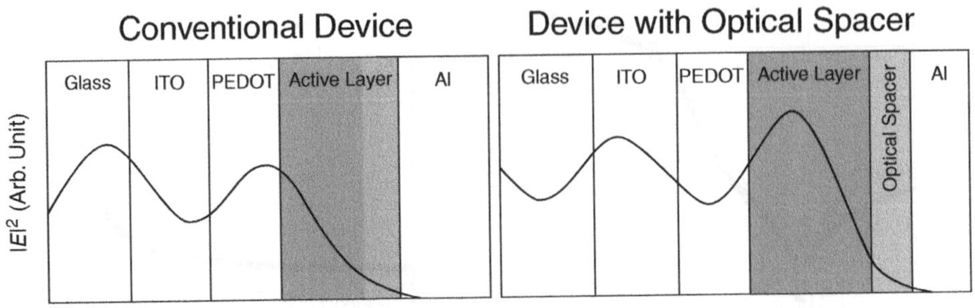

Figure 8.16 Schematic representation of the spatial distribution of the squared optical electric field strength $|E|^2$ inside the devices with a structure of ITO/PEDOT/active layer/Al (left) and ITO/PEDOT/active layer/optical spacer/Al (right). ITO: indium tin oxide; PEDOT: poly(3,4-ethylenedioxylenethiophene) [161]. (Reproduced with permission from Kim, 2006. Copyright © John Wiley & Sons, Ltd, 2006.)

8.3.4 Protection Layer between the Electrode and the Polymer

An interfacial layer also serves as physical and chemical barrier for the organic layers. Metal electrodes are often evaporated on top of the device. Without interfacial layer the hot atoms can damage the organic layer or diffuse into the active layer and short the cell.

8.3.5 Selective Contact

More importantly, interfacial layers are essential to form selective contacts. In an ideal case all single carriers created within the active layer are transported to their respective electrode (i.e. electrons by the acceptor to the cathode, holes by the donator to the anode. In a bulk-heterojunction system the generation of free carriers is distributed equally over the active layer (i.e. the splitting of excitons at the acceptor/donor interfaces). Thus, interfacial layers are essential to block carriers of one sort and enable the facile extraction of carriers of the other sort. A suitable interfacial layer also prevents recombination at the interface. In general, there are 4 different types of materials for interfacial layers. In the following we will discuss some of the more prominent interface materials.

8.3.6 Interface Material Review for OPV Cells

Most of the interface materials are wide-bandgap semiconductors, and the position of their valence and conduction bands relative to that of the organic semiconductor may be enough to distinguish between a p-type and an n-type contact. Figure 8.17 summarises a few of the popular candidates. A few of them are discussed in more detail in the next sections.

8.3.6.1 p-Doped Polymers

For polymer OPV cells PEDOT:PSS is widely used. [162–165]. PEDOT:PSS forms a selective contact for holes. It is solution deposited and has high transparency. Since PEDOT:PSS is

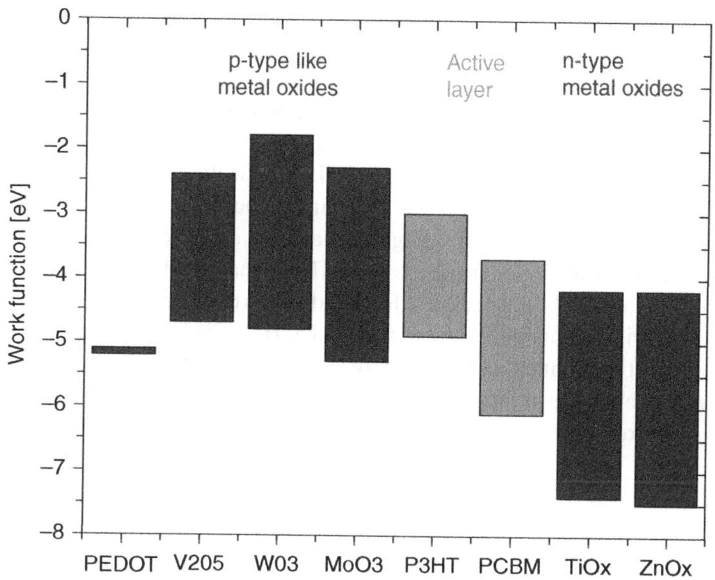

Figure 8.17 HOMO and LUMO level of varying semiconductor materials [167–169, 171]. Data Extracted from Kyaw, 2008; Shrotriya, 2006; Tao, 2009; Han, 2009.

an acid (PH ~ 1), byproducts of PEDOT or side reactions between the acid and the other solar cell materials can cause environmental degradation. One of the alternatives to PEDOT:PSS is the sulfonated poly(diphenylamine) [166].

8.3.6.2 p-Type-Like Metal Oxides

p-Type-like metal oxides can form highly selective interface material at the anode. Many of these metal oxides, like MoO_3 [167, 168], V_2O_5 [168] or WO_3 [169–171] need to be deposited by thermal evaporation, and only recently, solution-based, alternative processing methods were successfully demonstrated for these oxides.

8.3.6.3 n-Doping Layers

The most prominent interface material for the cathode is a combination of LiF and Al [33]. Incorporation of a thermally evaporated LiF layer between the active layer and the electrode does even convert an Au electrode into a cathode. In the literature there are different explanations for the operation of LiF. LiF is suggested to either modify the work function of the cathode, or to dope the active layer, or to form a dipole towards the active layer [172–176].

Other interfacial layers for the cathode like TiO_x [157, 161, 177, 178], ZnO_x [159, 179, 181], Poly (ethylene oxide) [180] and polyoxyethylene tridecyl ether [177] can be solution deposited.

8.3.6.4 SAM

Self-assembled monolayers (SAM) with a permanent dipole moment are reported to modify the work function of metal electrodes. As a consequence the performance of OPV cells is modified as well. Yip et al. [181] investigated the effect of SAMs with negative and positive net dipole in OPV cells structure ITO/PEDOT:PSS/P3HT:PCBM/ZnO/SAM/metal electrode. The SAMs investigated are parasubstituted benzoic acids with various dipole moments. The work function of the investigated materials and a schematic illustration of the mode of operation are summarised in Figure 8.18. The formation of an ohmic contact between the metal electrode and ZnO is necessary for good performance. Thus, modification of the cathode by SAMs with negative dipole moments (-OCH3, -CH3, -H) supports the formation of an electron-selective contact and does increase device performance. As a consequence, SAMs with positive dipole moment (-SH, -CF3, -CN) decrease device performance by forming a less-electron-selective contact [181].

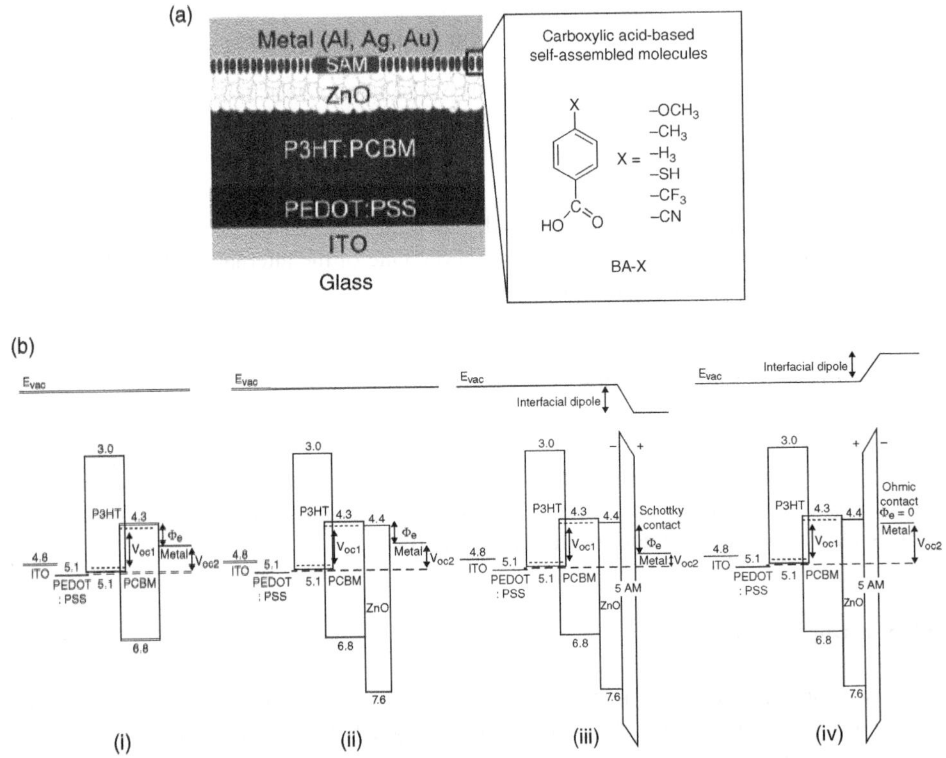

Figure 8.18 (a) Device architecture of the polymer solar cell with self-assembled-monolayer-modified ZnO/metal bilayer cathodes, and the chemical structures of the self-assembling molecules used in this study. (b) Schematic illustration of the energy-level diagram of the devices with (i) a metal-only cathode; (ii) a ZnO/metal bilayer cathode; (iii) a ZnO/SAM/metal cathode with an interfacial dipole directed towards the metal surface; and (iv) a ZnO/SAM/metal cathode with an interfacial dipole directed away from the metal surface [181]. (Reproduced with permission from Yip, 2008. Copyright © John Wiley & Sons, Ltd, 2008.)

	Relative work function	Surface energy (mJ/m^2)	V_{oc}(V)	J_{sc}(V) (mA/Cm^2)	FF	η(%)
Untreated	4.7	69.2				
As prepared			0.39	5.38	0.35	0.73
Annealed			0.36	5.98	0.35	0.75
NH_2 SAM	4.35	46.5				
As prepared			0.38	3.63	0.38	0.53
Annealed			0.55	5.71	0.30	0.95
CH_3 SAM	3.87	31.4				
As prepared			0.41	1.46	0.35	0.21
Annealed			0.57	6.82	0.32	1.24
CF_3 SAM	5.16	28.7				
As prepared			0.57	6.20	0.37	1.29
Annealed			0.60	13.87	0.38	3.15

Figure 8.19 Surface energies of the various SAM-treated ITO substrates and the electrical properties of the corresponding OPV devices [183].

Similar results are reported for modifications of the interface TiO_x/ bulk heterojunction (cathode) [182] and ITO/ bulk heterojunction (anode) [183, 184]. Kim et al. investigated devices with SAM-modified ITO electrodes of the structure ITO/SAM/P3HT:PCBM/Al. Figure 8.1 summarises the device results. An increase in work function of the SAM-modified ITO result in increased device efficiency. A decrease in work function results in lower device efficiency. The surface energy of the SAM-modified ITO changes with varying work function [184]· The device performance is highest for the work function of the SAM modified ITO that is close to PEDOT:PSS. The same is reported from another group using 4-chlorobenzoylchloride (CBC), 4-chlorobenzenesulfonyl chloride(CBS), and 4-chlorophenyldichlorophosphate (CBP) (–COCl, –SO_2Cl, and –PO_2Cl_2, respectively) modified SAMs [184].

8.4 TANDEM TECHNOLOGY

In this section, we aim to review the most important and recent developments that have been reported on organic tandem solar cells. In the first part, we introduce some brief theoretical considerations addressing the potential of single and tandem solar cells. Then, we present and discuss the main experimental achievements reported in the literature so far. Finally, we propose some design rules for efficient material combinations in bulk-heterojunction (BHJ) organic tandem solar cells.

8.4.1 Theoretical Considerations

8.4.1.1 The Single-Junction Solar Cell

In 1961, Shockley and Queisser derived a simple calculation in order to evaluate the thermodynamical limitation of the energy conversion efficiency of a solar cell [185]. The two

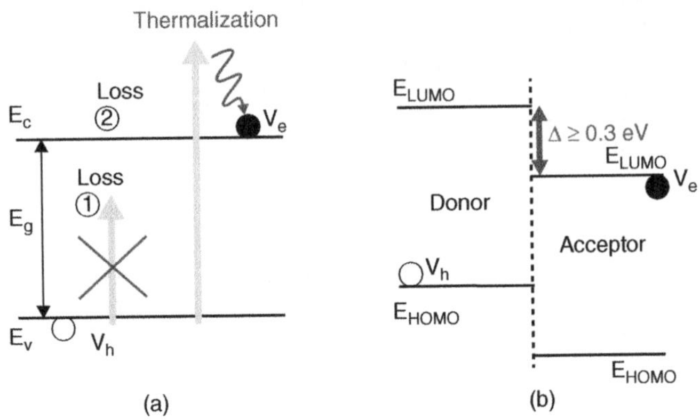

Figure 8.20 (a) Description of the absorption and thermalisation losses occurring in a solar cell. (b) Band-diagram representation of a donor–acceptor organic solar cell.

basic phenomena taken into account are schematically represented in Figure 8.20a and are described below:

- Only photons having an energy larger than the bandgap can be absorbed by the photovoltaic active material and contribute to the photovoltaic conversion.
- Hot charge carriers created upon photon absorption relax down to the conduction band of the photoactive materials, and give rise to the so-called "thermalisation".

The direct consequence of these two assumptions is that the maximum charge-extraction potential at open circuit (V_{oc}) is given by the bandgap of the photoactive material.

The ultimate efficiency under AM1.5G versus the bandgap is displayed in Figure 8.21 (case 1). One can clearly see that $\eta(E_G)$ shows a maximum of 48% for E_G close to 1.1 eV. For bandgap energies larger than this value, the efficiency is mostly limited by a reduction in the number of absorbed photons, namely by a reduction in the short-circuit current. For bandgap energies lower than this value, the limitation in efficiency is mostly related to a reduction in the V_{oc}. Indeed, Shockley and Queisser showed that taking into account additional phenomena like radiative or nonradiative recombination of hole–electron pairs reduces the maximum efficiency down to 30.1% for a single device [185].

8.4.1.2 The Specific Case of Organic Solar Cells

As already described extensively in the literature [186], typical organic solar cells are based on a charge transfer occurring at the interface between two distinct materials named donor and acceptor, respectively (See Figure 8.20b). To allow the separation of the excitons diffusing to this interface, the energy difference between the lower unoccupied molecular orbital (LUMO) levels of the two materials has to be larger than the exciton binding energy. It is shown experimentally that 0.3 eV is a typical value below which the charge transfer may not occur. Assuming a 0.3 eV loss in the V_{oc} of the device, the maximum efficiency achievable by an organic solar cell is reduced to 36% (case 2 in Figure 8.21). For this calculation and all the following, we have assumed that only the donor material absorbs the light.

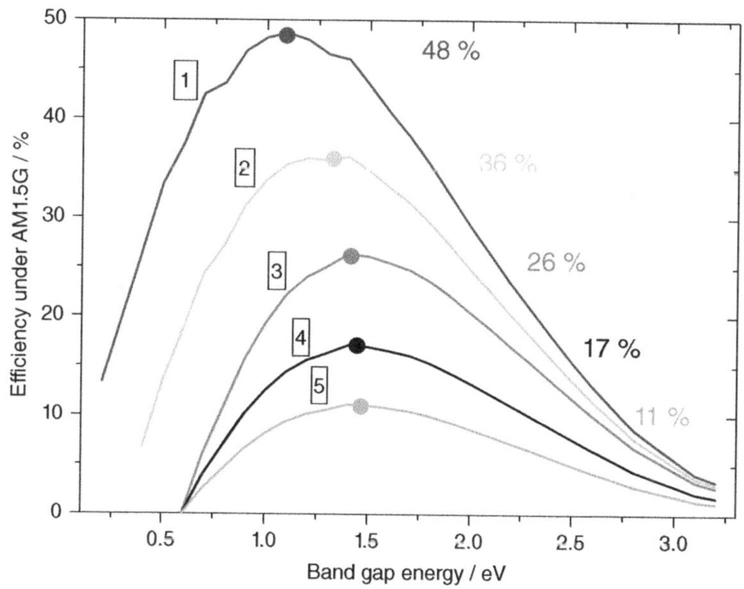

Figure 8.21 Efficiency versus bandgap energy of the photoactive material, for various loss mechanisms. The origins of the loss mechanisms are explained in the section below.

A detailed screening of the literature reveals that most of the time the V_{oc} observed in a device is not exactly given by the energy difference between the highest occupied molecular orbital (HOMO) of the donor and the LUMO of the acceptor. Indeed, an additional loss of 0.3 eV is often reported, in such a way that the V_{oc} obeys the following empirical equation [40]:

$$V_{oc} = \frac{1}{e}\left(\left|E_{\text{HOMO}}^{\text{Donor}}\right| - \left|E_{\text{LUMO}}^{\text{Acceptor}}\right|\right) - 0.3 \qquad (8.2)$$

This loss, whose origin is still under debate [187], further reduces the maximum energy-conversion efficiency of the device down to 26% (case 3 in Figure 8.21). Finally, by assuming some charge-carrier transport losses (FF = 65%, case 4 in Figure 8.21), and some EQE restrictions (EQE = 65%, case 5 in Figure 8.21), one arrives at a practical efficiency of nearly 11%, which should be achievable with today's materials. This efficiency is expected for a donor having an ideal bandgap energy of 1.5 eV and an ideal LUMO difference of 0.3 eV.

8.4.1.3 Tandem Solar Cells

As explained at the beginning of this section, the two major losses occurring in solar cells are the subbandgap transmission and the thermalisation of hot charge carriers [188]. One way to circumvent both effects simultaneously is the realisation of a tandem solar cell.

The fundamental (detailed balance) limit of the performance of tandem structures has been studied extensively by De Vos [189]. This author demonstrated that stacking several subcells in series allows theoretical efficiencies way beyond the Shockley–Queisser limitation, mostly due to an enhancement of the electrochemical potential of charge-carrier extraction. While the maximum efficiency of a single cell under nonconcentrated sun light is calculated to

Table 8.1 The optimal set of bandgaps (E_{gi}) for tandem structures with n stacked cells in unconcentrated sunlight [189]. (Reproduced with permission from Kim, 2007. Copyright © 2007, American Institute of Physics.)

n	η (%)	E_{g1} (eV)	E_{g2} (eV)	E_{g3} (eV)	E_{g4} (eV)
1	30	1.3	-	-	-
2	42	1.9	1.0	-	-
3	49	2.3	1.4	0.8	-
4	53	2.6	1.8	1.2	0.8

be around 30%, this value is raised to 42% for a tandem comprising two subcells having a bandgap of 1.9 and 1.0 eV, respectively, and to 49% for a tandem comprising 3 subcells having a bandgap of 2.3, 1.4 and 0.8 eV, respectively (Table 8.1). Under the highest possible light concentration, these efficiencies are 40% (one cell), 55% (two cells) and 63% (three cells). This model also predicted the ideal efficiency of a stack with an infinite number of solar cells: Such a tandem system could convert 68% of the nonconcentrated sun light, and 86% of the concentrated sun light. Experimentally, efficiencies as high as 33.8% have been recently measured on a device based on GaInP/GaInAs/GaInAs under AM1.5G [190].

In the specific case of organic solar cells, the tandem approach allows two additional issues intrinsic to π conjugated organic molecules to be tackled. The first one is the quite poor charge-carrier mobility and lifetime that limits the distance over which the carriers can be transported [191]. This hinders the realisation of a thick active layer that would absorb the maximum amount of light. The second relates to the very nature of light absorption in those molecular materials. Indeed, the absorption spectrum of such materials is not made of a continuum like it is in inorganic semiconductors. It rather shows narrow and discrete peaks, as illustrated in Figure 8.22 [192]. Hence, a combination of various different materials can help to cover more efficiently the emission spectrum of the sun.

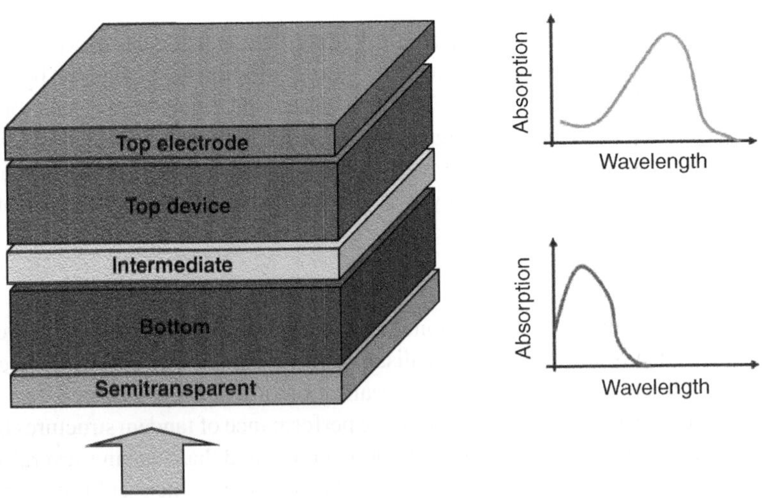

Figure 8.22 Schematic representation of an organic tandem device comprised of two subcells having different, complementary absorption spectra.

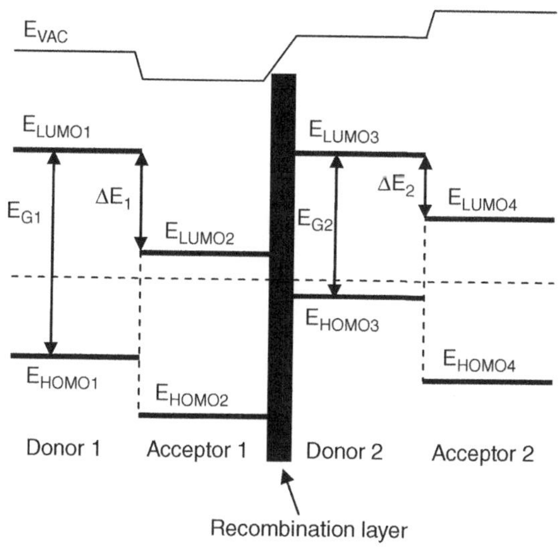

Figure 8.23 Simplified band diagram of a tandem cell made of two subcells connected in series via a recombination layer.

Figure 8.22 depicts a typical organic tandem cell comprised of two distinct devices stacked on top of each other, each of them being based on a donor–acceptor composite. The light that is not absorbed in the bottom device can further impinge on the top cell. Moreover, the thermalisation losses are lowered due to the usage of materials having different bandgaps. The two cells involved in the device can be connected either in series (two-terminal) or in parallel (three-terminal) depending on the nature of the intermediate layer and on the way the intermediate layer and the two electrodes are connected.

In the vast majority of reports though, the series connection is used. In this case, the energetic diagram of the device can be represented as in Figure 8.23. The bottom subcell comprises a donor and an acceptor that are either stacked on each other or mixed. The intermediate layer should ensure the alignment of the quasi-Fermi level of the acceptor of the bottom cell with the quasi-Fermi level of the donor of the top cell (or vice versa in a reversed architecture). In other words, the intermediate layer should allow the recombination of holes coming from one subcell with electrons coming from the other. Following the same principle, an infinite number of devices can be theoretically piled up this way.

According to Kirchhoff's law, this type of connection implies that the voltage across the whole device is equal to the sum of the voltages across each subdevice. In other words, at open-circuit voltage, we have in the case of a loss-free connection:

$$V_{\mathrm{oc1}} + V_{\mathrm{oc2}} + V_{\mathrm{oc3}} \ldots = V_{\mathrm{oc\ tandem}} \tag{8.3}$$

However, contrarily to what is often believed, the short-circuit current of the tandem cells is not equal to the smallest short-circuit current of the subcells. It entirely depends on the current-matching conditions and therefore a lot on the FF of the respective devices.

We have introduced a model assuming a series connection of two donor–acceptor solar cells stacked via an ideal intermediate layer [193] to predict the practical efficiency of

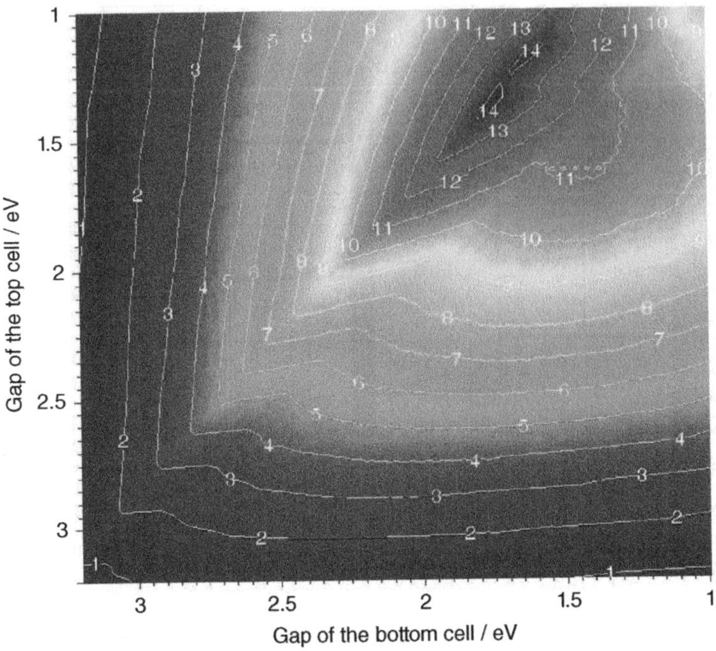

Figure 8.24 Efficiency of a donor–acceptor tandem organic solar cell made of two subcells stacked in series versus the bandgap energy of the top and bottom donors; both top and bottom donor–acceptor couples are assumed to have a LUMO offset of 0.3 eV. (Reproduced with permission from Tvingstedt, 2007. Copyright © 2007, American Institute of Physics.)

organic multijunction solar cells. We have fixed the LUMO offset between both donor–acceptor couples to be equal to 0.3 eV, which is the optimum value for bulk-heterojunction composites. We have then varied the bandgaps of both donors, assuming the FF of both subcells to be 0.65, a rectangular-shaped external quantum efficiency of both subcells to be 65%, and the internal quantum efficiency of the bottom cell to be 85%. Figure 8.24 shows the tandem efficiency isolines. This calculation reveals that, under the given assumptions, the most efficient tandem cells require materials with bandgap energy different by about 0.3 eV. A maximum value of almost 15% is achievable. This performance can be achieved by using a bottom donor having a bandgap of 1.6 eV with a top donor having a bandgap of 1.3 eV.

8.4.2 Review of Experimental Results

During the last 18 years, an increasing number of scientific reports have been published dealing with stacked or tandem organic solar cells. Several approaches for organic and hybrid organic solar cell architectures have been employed over the last 20 years. Depending on the materials used for the active layer and the separation or recombination layer(s), one can distinguish between three main categories:

(i) Tandem organic solar cells where both the bottom (in front of the light illumination) and the top (back) cells are based on low molecular weight evaporated molecules.

(ii) Hybrid tandem organic solar cells in which one of the cells, either bottom or top cell, is processed from solution while the other cell is based on vacuum-deposited low molecular weight materials.
(iii) Fully solution-processed tandem organic solar cells where both the bottom and top cells are processed from solution.

Besides the nature of the active layer, an important characteristic of the tandem device resides in the type of intermediate layer employed. As explained above, these intermediate layers are used as a recombination centre ensuring the alignment of the quasi-Fermi levels of the subcells stacked in series. Moreover, they can act as a protective layer to support the bottom cell during the top active layer deposition. In the earliest reports, this intermediate layer was often based on a thin, vacuum-processed metallic layer. But recently, an increasing number of organic tandem solar cells with fully solution-processed interlayers are reported.

In this section, the different types of organic tandem and multijunction photovoltaic cells are described and recent results obtained by various groups are reviewed.

8.4.2.1 The Evaporated Small-Molecule-Based Tandem Organic Solar Cells

The first organic solar cell was reported by Hiramoto et al. in 1990 [194]. This device comprised two stacked, series-connected subcells based on evaporated small molecules composed of metal-free phthalocyanine (H_2Pc) and perylene tetracarboxylic derivative (Me-PTC). In order to make an ohmic contact between the two subcells, an ultrathin (2 nm) Au interstitial layer was evaporated (see Figure 8.25): The tandem cell showed a V_{oc} of 0.78 V, which is about twice the V_{oc} of a single cell (0.44 V).

Some 10 years later, Forrest and his group followed the same approach, yet another kind of small-molecule material was utilised as the active layer. In 2002, Yakimov and Forrest [195] demonstrated high V_{oc} organic photovoltaic cells consisting of Cu-phthalocyanine (CuPc) as a donor and 3,4,9,10 perylenetetracarboxylic bis-benzimidazole (PTCBI) as an acceptor. In this case, an ultrathin (≈5 Å) discontinuous layer of Ag clusters served as the charge-recombination sites. The power-conversion efficiencies (η) under one sun (AM1.5

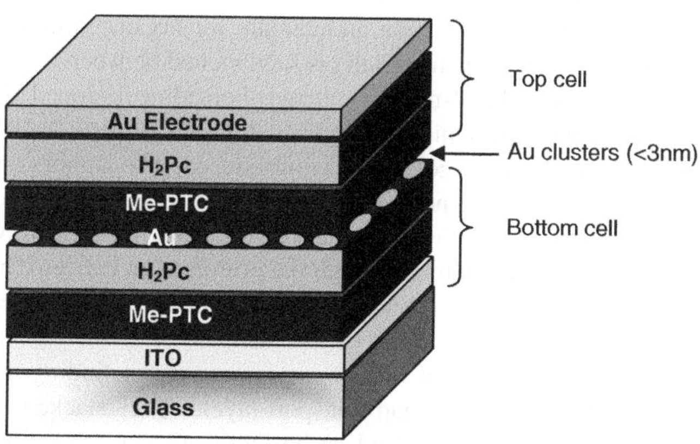

Figure 8.25 Schematic structure of the first tandem organic solar cell realised by Hiramoto et al.

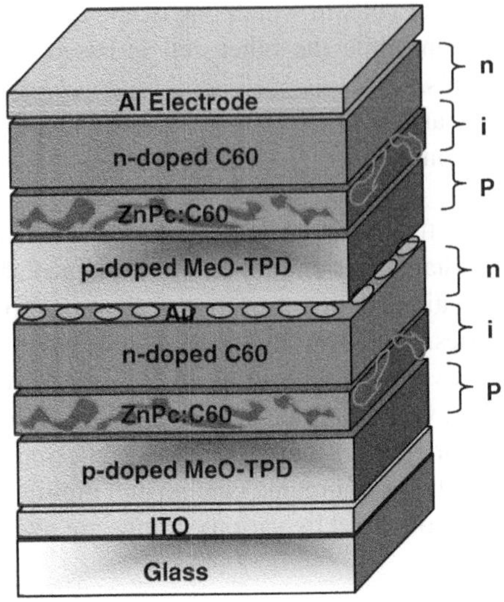

Figure 8.26 Schematic structure of the tandem organic solar cell realised by Maennig et al. based on multiple stacked p-i-n structures, each of them comprising a photovoltaic layer sandwiched between p- and n-type wide-gap transport layers.

illumination) was $\eta = 2.5\%$, more than twice that of a comparable single-junction cell based on the same materials, where $\eta = 1.1\%$.

Later, Xue et al. achieved a significant improvement in efficiency [196] by stacking two bulk heterojunctions [25] made of coevaporated CuPc and C_{60}. Optimising the optical absorption in each active layer allowed them to report an efficiency of up to 5.7%. This high performance, which was never certified or reproduced, was about 24% more efficient than the single CuPc:C_{60} devices. In this tandem stack, the recombination centres were Ag nanoclusters with 5 Å average thickness buried in a 50-Å thick 4,4′,4″-tris(3-methyl-phenyl-phenyl-amino)triphenylamine (m-MTDATA) p-doped with 5 mol% tetrafluoro-tetracyano-quinodimethane.

The evaporated small molecule route was also investigated by Maennig et al. [197]. They introduced a p-i-n-type heterojunction architecture for organic solar cells where the active layer of ZnPc:C_{60} (as the i-type materials) is sandwiched between two wide-bandgap layers; the p-type *N,N,N′,N′*-Tetrakis(4-methoxyphenyl)-benzidine (p-doped MeO-TPD) and the n-type C_{60} layers. Figure 8.26 depicts the device structure for two stacked p-i-n cells. Between individual cells an ultrathin gold layer is introduced. The authors believe that the gold interlayer may play a dual role here. First, it is expected to hinder interdiffusion of dopants that otherwise lead to a depletion region in cells without metal interlayer. Secondly, it provides gap states that assist tunneling through the p-n junction barrier. The tandem cell showed significantly higher power efficiency of 2.4% compared to the 1.95% efficient single p-i-n cell.

Later, Drechsel et al. [198] achieved further improvements by finding the appropriate dopants for both the p and n-type wide-gap transport layers. They stacked two p-i-n cells both based on a phthalocyanine-fullerene blend, as the photoactive layer, reaching efficiencies up to 3.8% as compared to 2.1% for the respective single p-i-n cell. The biggest boost in efficiency of small-molecule tandem cells was driven by the developments of Heliatek, who published certified efficiencies of >10% for double junction cells and >12% for triple-junction cells.

8.4.2.2 The Hybrid and Fully Solution-Processed Tandem Organic Solar Cells

In order to produce organic tandem solar cells employing solution processing, serious engineering challenges have to be met. As multiple layers have to be cast on top of each other, it has to be avoided that casting a further layer from solution dissolves the underneath layer. Researchers have developed several approaches to circumvent this intrinsic problem. These comprise employing sophisticated material sequences soluble in incompatible solvents or applying one or more layers by vacuum-deposition methods.

In the case of the solution-processed organic solar cells, the first tandem structures were reported by Kawano et al. in 2006 [199] based on conjugated Poly [2-methoxy-5-(3′,7′-dimethyloctyloxy)-1,4-Phenylene vinylene] (MDMO-PPV) blended with the fullerene derivative PCBM as the active layer for both top and bottom cells. The 20-nm thick ITO interlayer was deposited by DC magnetron sputtering at a relatively high pressure of 1 Pa of Ar without damaging the underneath active layer and substrate heating. Stacking two devices based on the same material combination led to 35% performance improvement compared to the single one ($\eta = 3.1\%$).

In the same year, Dennler et al. [200] and Lemmer and coworkers [201a] reported an organic tandem cell comprising two subcells with different active materials and different absorption spectra, i.e. a Pc/C60 layer in combination with a bottom P3HT/PCBM interdiffused bilayer. The authors of Ref. [194] using a 1-nm thick Au intermediate recombination layer, achieved a V_{oc} of 1.02 V, equal to the sum of the individual V_{oc}s of the subcells, and an efficiency of 2.3%. Spectral photocurrent measurements with two light sources showed that the photon harvesting in this tandem cell occurs over the entire visible range from 400 to beyond 800 nm. Alternative interlayers such as thermally evaporated oxides like WO_3 were investigated by Janssen et al. [201b].

The first tandem cell composed of two solution-processed subcells based on two different conjugated polymers was reported by Hadipour et al. in 2006 [202]. They were able to couple a wide-bandgap polymer with a low-bandgap polymer. The combination of a 110-nm poly(2,7-(9,9-dioctyl)-fluorene)-alt-5,5-(4′,7′-di-2-thienyl-2′,1′,3′-benzothiadiazole) (PFDTBT):PCBM (1:4) layer for the bottom cell (large bandgap) and a 90-nm poly(5,7-di-2-thienyl-2,3-bis(3,5-di(2-ethylexyloxy) phenyl)-thieno(3,4-b) pyrazine) (PTBEHT):PCBM layer for the top cell generated a broad spectral coverage. Since a gold middle electrode does not give an ohmic contact for electrons, a very thin layer of 0.5 nm LiF and 0.5 nm Al is added on top of the bottom active layer. LiF/Al provides an ohmic contact to PCBM while a 60-nm spin-coated PEDOT:PSS layer on top of the Au layer provides the ohmic contact to the polymer of the top cell. The performance of the tandem cell was 0.57%, which is 1.6 times higher than the performance of the bottom cell (0.35%) and 2.5 times higher than the performance of the top cell (0.23%), resulting in 0at least a 62% improvement compared to the more efficient single cell.

A significant innovation was brought by the work of Gilot et al., where an intermediate layer was entirely processed from solution like the whole tandem device [203]. The authors used a layer of zinc oxide nanoparticles [204] and a layer of PEDOT as the recombination centre. Importantly, they established a solvent combination that allows subsequent layers to be processed without affecting previously deposited layers. The blends of MDMO-PPV:PCBM and P3HT:PCBM were spin coated from chlorobenzene as the bottom and top active layers, respectively. The intermediate layers in this study: PEDOT was heavily p-doped and ZnO layer could be easily doped by exposure to UV irradiation (photodoping) for a few seconds [205]. Values of V_{oc} of 1.53 V and 2.19 V were achieved for a tandem and a triple-junction solar cell, respectively; corresponding closely to the sum of single cells' V_{oc}s. This successful deposition of fully solution-processed multijunction organic solar cells opened up the possibility of roll-to-roll production of tandem cells.

The breakthrough in solution-processed tandem cells was reported by Kim et al., demonstrating a high efficiency tandem device showing a 38% performance improvement versus the best single device [206]. These tandem devices were entirely solution processed except for the top evaporated electrode. For the bottom BHJ cell a 130-nm thick layer of poly[2,6-(4,4-bis-(2-ethylhexyl)-4H-cyclopenta[2,1-b;3,4-b′]dithiophene)-alt-4,7-(2,1,3-benzothiadiazole)] (PCPDTBT):PCBM was used and the top cell was made of a blend of P3HT:PC70BM with a thickness of 170 nm. The two polymer–fullerene layers were separated by a transparent TiO_x layer and a highly conductive PEDOT:PSS layer. The maximum, though unverified efficiency reported so far of any organic tandem solar cell up to 6.5% was achieved in this study at an illumination of 100 mW/cm^2 based on a 3% efficient bottom and a 4.7% efficient top cell.

A brief summary dealing with tandem organic solar cells from the early phases of tandem developments is represented in Table 8.2. After 2009 the development of fully solution-processed tandem cells was dominantly driven by Yang's group and Janssen's group, and fully solution-processed tandem cells with certified efficiencies beyond 10% were reported.

8.4.2.3 Specific Geometries for Organic Tandem Solar Cells

All the reviewed tandem cells so far were based on two-terminal devices, comprising two or more cells connected in series. However, several groups have suggested other approaches to realise multijunction devices. This part of the study will discuss some of these novel concepts.

Shrotriya and coworkers have reported a stacking method to overcome all processing difficulties. They superimposed two independent devices, and connected them either in series or in parallel [207]. To do so, the bottom cell required a semitransparent cathode that was realised with a 1-nm LiF/2-nm Al/12-nm Au layer sequence showing a maximum transparency of almost 75%. Both active layers were made of a blend of poly(2-methoxy-5-(2′-ethyl)-hexyloxy)-1,4-phenylene-vinylene (MEH-PPV) and PCBM.

In 2007 Hadipour et al. introduced an additional solution-processable, transparent, and insulating layer between the bottom and the top subcell that serves as an optical spacer and also allows the fabrication of a monolithic four-terminal device [208]. The cathode of the bottom cell (P3HT:PCBM) was evaporated with 3 nm of samarium (Sm) and 12 nm of Au. Then, 250 nm of polytrifluoroethylene (PTrFE), dissolved in methyl ethyl ketone (MEK), was spin coated onto the bottom cell to separate the two subcells, as suggested earlier by Perrson et al. [209]. The top cell was made of PTBEHT:PCBM on top of a 20-nm Au/50-nm PEDOT:PSS anode. They concluded that the most efficient connection for the active materials employed was the parallel one; with a V_{oc} of 0.59 V, a J_{sc} of 9.2 mA/cm^2 and a FF of 0.54.

Table 8.2 Nonexhaustive survey of reports dealing with solution-processed tandem organic solar cells. (Reproduced with permission from De Vos, 1980. Copyright © IOP Publishing, 1980.)

		Bottom cell					Top cell					Tandem cell				
Year	Intermediate layer	Active Materials	V_{oc} (V)	FF	J_{sc} (mA.cm^{-2}) (mW cm^{-2})	Eff (%)	Active Materials	V_{oc} (V)	FF	J_{sc} (mA.cm^{-2}) (mW.cm^{-2})	Eff (%)	V_{oc} (V)	FF	J_{sc} (mA.cm^{-2}) (mW.cm^{-2})	Eff (%)	Ref.
1990	2 nm Au	H2Pc/Me-PTC	0.44	–	2.7 (78)	–	as bottom	as bottom	as bottom	as bottom	as bottom	0.78	–	0.9 (78)	–	[194]
2002	0.5 nm Ag	CuPc/PTCBI	0.45	–	–	1.0	as bottom	as bottom	as bottom	as bottom	as bottom	0.9	0.43	6.5 (100)	2.6	[195]
2004	0.5 nm Ag	CuPc: C60	–	0.64	–	4.6	as bottom	as bottom	as bottom	as bottom	as bottom	1.03	0.59	9.7 (100)	5.7	[196]
2005	0.8 nm Au	ZnPc :C60	0.5	0.37	15.2 (130)	2.1	as bottom	as bottom	as bottom	as bottom	as bottom	0.99	0.47	10.8 (130)	3.8	[198]
2006	20 nm ITO + PEDOT:PSS	MDMO-PPV :PCBM	0.84	0.58	4.6 (100)	2.3	as bottom	as bottom	as bottom	as bottom	as bottom	1.34	0.56	4.1 (130)	3.1	[199]
2006	1 nm Au	P3HT: PCBM	0.55	0.55	8.5 (100)	2.6	ZnPc: C60	0.47	0.5	9.3 (100)	2.2	1.02	0.45	4.8 (100)	2.3	[200]
2006	0.5 nm LiF +0.5 nm Al +15 nm Au + 60 nm PEDOT:PSS	PFDTBT: PCBM	0.9	0.5	1.0 (100)	0.4	PTBEHT: PCBM	0.5	0.64	0.9 (100)	0.23	1.4	0.55	0.9 (100)	0.6	[202]
2007	30 nm ZnO + PEDOT	MDMO-PPV :PCBM	0.82	0.55	4.1 (100)	1.9	P3HT: PCBM	0.75	0.48	3.5 (100)	1.3	1.53	0.42	3.0 (100)	1.9	[203]
2007	8 nm TiOx + 25 nm PEDOT:PSS	PCPDTBT: PCBM	0.66	0.5	9.2 (100)	3.0	P3HT: PCBM	0.63	0.69	10.8 (100)	4.7	1.24	0.67	7.8 (100)	6.5	[206]

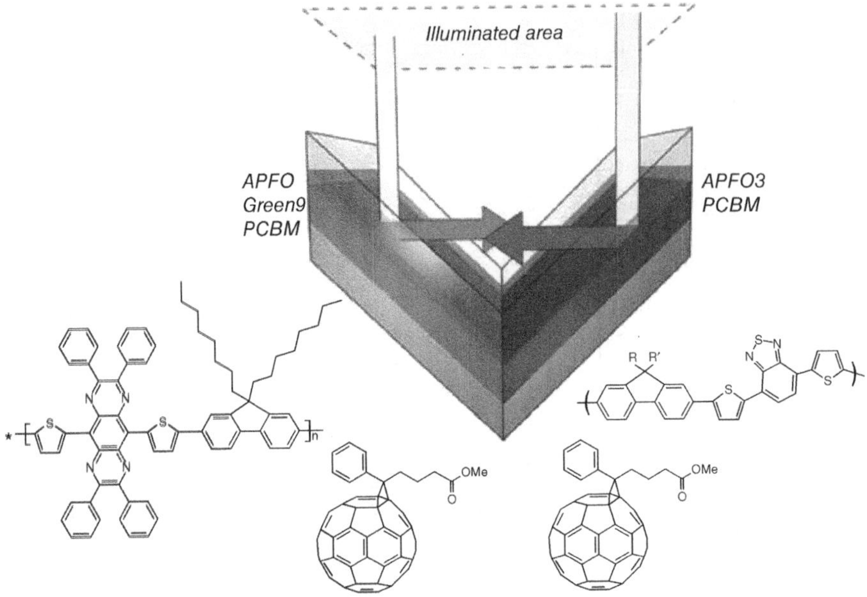

Figure 8.27 Sketch of the folded tandem cell realised by Tvingstedt et al. [210]. The chemical structures of the exploited alternating polyfluorenes, namely APFO3, APFO Green-9, and the acceptor molecule PCBM, are also shown.

Tvingstedt and colleagues proposed a novel geometrical modification of multijunction cells called a "folded reflective tandem device" [210]. The device scheme is depicted in Figure 8.27. This geometry enables the construction of tandem or multiple-bandgap solar cells with arbitrary electrical parallel or series connection. In addition, this geometry allowed the authors to benefit from three major advantages: First, the reflected light of one cell is directed toward the second one, which ideally has a complementary absorption spectrum. Secondly, the folded structures cause light trapping at high angles and absorb more photons from incoming solar light. Finally, the tilting of the cells enhances the light path within the active layer [211]. The used polymers were based on alternating fluorine copolymers (APFs) combined with PCBM, APFO3:PCBM for one cell and APFO-Green9:PCBM for the other cell. The conversion efficiency for a series connection increased from 2% up to 3.7% upon folding the V-shaped device from 0° to 70°. The advantage of this approach is that all problems related to multijunction stacking, extra transparent electrodes, and solvent incompatibility are simply avoided.

Zhang et al. demonstrated a simple alternative to a parallel interconnection in 2008 [212]. In this device structure, a PCBM layer is employed to simultaneously form a bilayer heterojunction PV subcell with the underlying CuPc layer and a BHJ photovoltaic subcell blended with P3HT. In comparison with the conventional tandem structure, the omission of the semitransparent intercellular connection layer reduces the complexity of the device processing and the light losses. According to the working principle of the bilayer PV cells, only the excitons created in a 5–10-nm thickness of CuPc can diffuse to the interface of the CuPc/PCBM where they are separated into free carriers. The enhanced $J_{sc} = 8.63$ mA/cm^2 and $\eta = 2.79\%$ of the tandem structure were nearly the sum of those of the standalone cells of CuPc/PCBM and P3HT:PCBM.

8.4.3 Design Rules for Donors in Bulk-Heterojunction Tandem Solar Cells

We have introduced a simple model that allows us to predict the efficiency increase that one can expect by going from single cells to tandem cells. This mode was already described in great detail elsewhere [193], and only the main findings will be briefly summarised here.

An ideal intermediate layer placed between the subcells is assumed to ensure a perfect alignment of the quasi-Fermi level of the PCBM of the bottom cell with the quasi-Fermi level of the donor of the top cell. The V_{oc} of the tandem cell is therefore calculated as being the exact sum of the V_{oc}s of the subcells. We assumed that the active layer of each cell absorbs the light between 350 nm and a wavelength (λ_1, λ_2) defined by the bandgap of the donor materials (E_{G1}, E_{G2}). In this wavelength range, the external quantum efficiency (EQE_1, EQE_2) of the subcells is considered as being 65% constant, the internal quantum efficiency (IQE) equal to 0.85%, and the fill factor (FF) equal to 0.65.

For each simulated donor, we have varied the bandgap between 1 and 3.15 eV, and the LUMO level between –4 and –3 eV. Thus, we deal with six variables, namely EQE_1, EQE_2, E_{LUMO1}, E_{LUMO2}, E_{G1}, and E_{G2}. Besides evaluating the potential efficiency of the tandem cells, we calculated R, the increase of efficiency of the tandem versus the best of the single cell:

$$R = \frac{E\,\text{max}_{\text{tan}} - \text{Max}[E\,\text{max}_{\text{bottom}}; E\,\text{max}_{\text{top}}]}{\text{Max}[E\,\text{max}_{\text{bottom}}; E\,\text{max}_{\text{top}}]} \tag{8.4}$$

And if $R \leq 0$, we set it to 0 since in this case there is no benefit to go from single to tandem devices.

One of the various cases we have considered is a tandem cell having on the bottom a device based on the most widely studied wide-bandgap polymer, namely P3HT. Blended with PCBM, this material has a potential to deliver efficiencies in the 5% regime at best. Figure 8.28 shows R versus the E_{G2} and E_{LUMO2}, as well as lines indicating the efficiency of the tandem device with a P3HT/PCBM bottom cell. One can note that for the very high tandem efficiencies, the top cell alone could yield even better performances ($R = 0$). It appears though that using a donor having a bandgap of about 1.7 eV and a LUMO level of –3.4 eV could significantly improve the potential of the respective subdevice. Such a cell would have an ultimate efficiency of 4.5% like the P3HT:PCBM one. Hence, combining these two 4.5% devices could allow fabrication of a 6.5% solar cell.

The second case we have considered is a tandem made of PCPDTBT:PCBM as the top cell, combined with a variable bottom cell. The optical bandgap of this material is assumed to be 1.55 eV. Combined with a E_{HOMO2} of –5.3 eV as measured by Mühlbacher et al. [46] this suggests $E_{LUMO2} = -3.75$ eV and an ideal $V_{oc2} = 0.7$ V. Thus, in that calculation, we have fixed E_{LUMO2} and E_{G2} and varied E_{LUMO1} and E_{G1}. The result is displayed in Figure 8.29. Interestingly, Figure 8.29 suggests that the ultimate efficiency of 10% identified by Scharber et al. [40] for a single cell based on a donor having [$E_{LUMO1} = -4.0$ eV; $E_{G1} = 1.8$ eV] can be pushed towards 13% by stacking a PCPDTBT:PCBM device on the top of it.

By plotting PCPDTBT on Figure 8.28 or P3HT on Figure 8.29, it appears that stacking those two materials would not allow increasing their individual potential. However, such tandem devices still yield higher efficiencies than the best single cells realised so far, as proven by Kim et al. [206]. This achievement relies on the fact that the PCPDTBT:PCBM

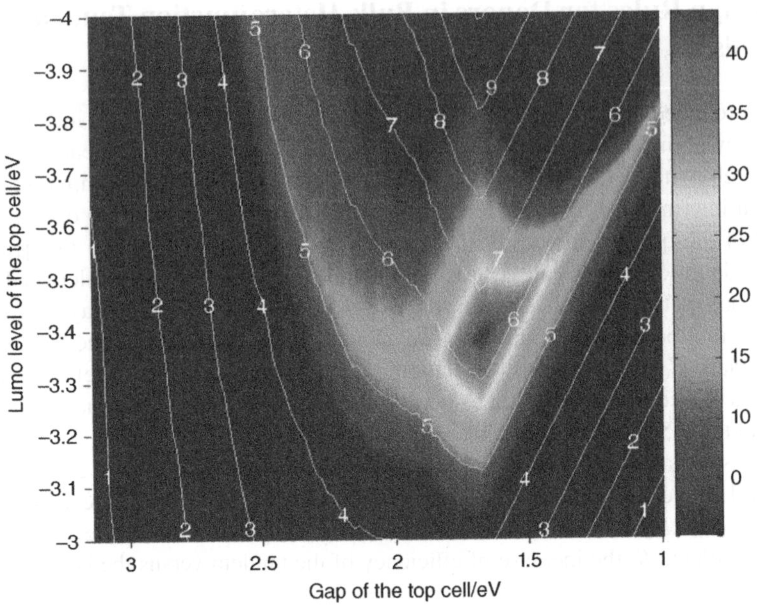

Figure 8.28 Percentage of increase of efficiency of the tandem cell over the best single cell for a device comprised of a P3HT:PCBM bottom subcell, and a top subcell based on a variable donor. The lines indicate the efficiency of the tandem devices.

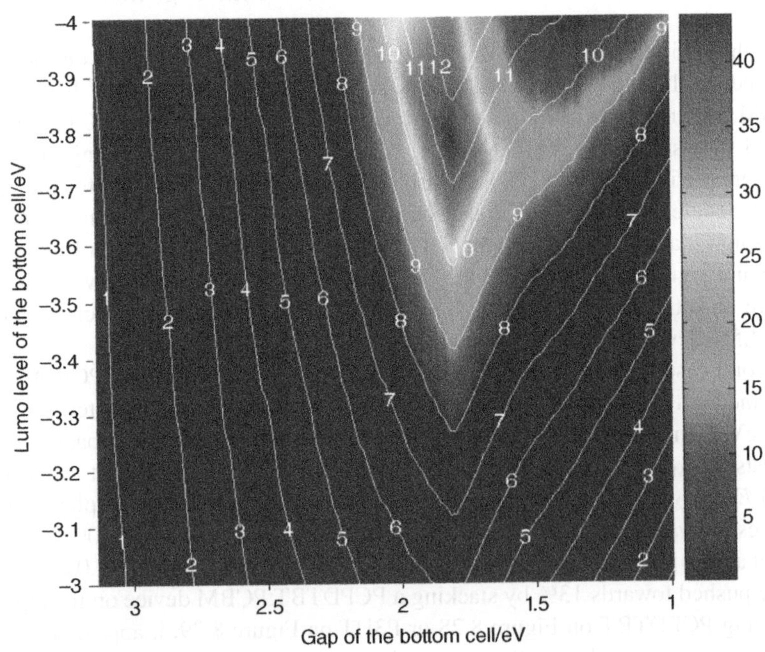

Figure 8.29 Percentage of increase in efficiency of the tandem cell over the best single cell for a device comprised of a PCPDTBT:PCBM top subcell, and a bottom subcell based on a variable donor. The lines indicate the efficiency of the tandem devices.

single cell did show an average EQE of only 30% over the range 350–800 nm instead of 65% assumed in the calculations. This observation suggests that realising tandem devices can be highly beneficial to enhance the performance of devices based on materials unable to deliver their full potential.

8.5 ELECTRODE REQUIREMENTS FOR ORGANIC SOLAR CELLS

Like any other solar cell, organic solar cells also require at least one light-transmissive electrode. The second electrode can be opaque or transparent, again depending on the specific application. The distinction between anode and cathode in bulk-heterojunction organic photovoltaics is not that trivial due to the nature of a bulk heterojunction and depends strongly on the architecture of the cell, as we already discussed in Section 8.3. It is therefore more common to define one electrode as the bottom electrode and the other one as the top electrode, which indicates the processing sequence for solar cell fabrication. Bottom electrodes are normally processed by vacuum-deposition processes like thermal evaporation or reactive and ceramic sputtering, as well as chemical deposition methods like CVD (chemical vapour deposition). Unfortunately, these methods frequently require in addition to the vacuum step also a high-temperature step to postanneal or postcrystallise the inorganic conductor. Both vacuum processing as well as high temperatures are rather unfavourable because of their considerable processing cost when compared to printing technologies. The top electrode is either a thermally evaporated metal layer or a thick, printed and opaque metal layer.

General requirements for OPV electrodes are defined by the material properties and the interaction with the semiconductor materials, but also need to take into account the compatibility to roll-to-roll processing, sufficient flexibility, good environmental stability – all in combination with low material costs. Additional benefits are gained if the electrode also acts as a barrier against water or oxygen and a good stability and can help to block the UV part of the visible light. Functionality sets the requirements for electrode material properties like conductivity, transmission and reflectance, and chemical stability is required to withstand the production process. We have already discussed the importance of ohmic contacts for optimised solar cell performance, and for organic semiconductors, where HOMO and LUMO levels are typically in the range of 5.5 to 3.5 eV, the electrodes need to cover that band regime.

In summary, the development of electrode materials requires the parallel optimisation of the electrode performance with the interaction to the semiconductor material and compatibility with the production process in combination with the cost factor. In the following sections, we will focus on a few technical key parameters for solar cells. Transparency is such a key parameter – a transparency of more than 80% (better 90%) over the whole solar spectra is required to keep optical losses to a reasonable minimum. The second key parameter is the surface conductivity, which should be as high as 10 Ω/sq to prevent electrical power losses. Due to the fact that the thin-film solar stack is only a few hundred nm thick, the bottom electrode must have very low surface roughness of a few nm rms to prevent local shunting. More qualitative key parameters are the chemical stability against organic and halogenated solvents like toluene, xylene, chlorobenzene, etc. as well as excellent adhesion of the bottom electrode to the substrate.

8.5.1 Materials for Transparent Electrodes

The most widely used materials for transparent electrodes are transparent conductive oxides (TCO) like indium tin oxide (ITO), doped zinc oxide (ZnO) or doped tin oxide (SnO). Most of these TCOs are deposited on the substrate as a bottom electrode by vacuum processing techniques like sputtering [213, 214]. Alternative deposition methods are pulsed laser deposition, plasma enhanced sputtering [214] or sol-gel deposition [215] techniques. The second big class of electrode materials is grid electrode composites. Grid electrodes are a combination of a highly conductive transparent layer, either from conducting polymers, nanoparticles or nanorods vertically stacked with a nontransparent metal grid. The metal grid serves the carrier collection and transport and is necessary since most transparent conductive coatings have insufficient conductivity at the required high transparency. Once these transparent coatings meet the 10 Ohm/sq regime, metallic grids will not be necessary. It is important to note that both electrode types can be used in the bottom and top electrode configuration. However, for processing reasons, most solar cell configurations use TCO electrodes as transparent bottom electrodes, while grid electrodes are frequently used as top electrodes.

8.5.1.1 Indium Tin Oxide (ITO)

Indium tin oxide (ITO) is a composite from two binary oxides, where 90% of indium oxide (In_2O_3) is doped with 10% tin oxide (SnO_2). Glass substrates coated with ITO are commercially available by different suppliers like Osram, Präzisions Glas & Optics, Kintec Company, Diamond Coatings, etc. Benchmark values for ITO on PET shows a high transparency of up to 80% over the visible spectra and a sheet resistance as low as 7 Ω/sq at a layer thickness of around 700 nm [216]. The intrinsic work function of ITO is between 4.4–4.9 eV. The chemical stability of ITO is reasonable and can be further improved by doping with other materials [224]. Unfortunately, highly conductive ITO is also quite crystalline, and as such, shows rather high brittleness. Another drawback of the highest-conductive ITO films are the rather large crystallites, which cause further surface roughness. But the biggest disadvantage of ITO is the cost. One kilogram of ITO costs around 1000 US$, which is due to the rareness of indium. Indium only exists in an average amount of 0.1 ppm in the earth's crust [217]. Due to the high usage of ITO in the display and touch panel industry, indium costs are expected to continue rising. Dopants other than indium may ease the cost situation: molybdenum doping was reported to give good-quality TCO with increased carrier mobility, while titanium doping led to good conductivity paired with high mobility as well [218, 219]. In summary, the high costs and the brittleness of ITO on PET are the main drivers to find alternative transparent electrode solutions for OPV.

8.5.1.2 Other TCO Systems

The variety of TCOs is huge. Many TCOs are composed of binary compounds such as SnO_2 and In_2O_3 or ternary compounds such as Cd_2SnO_4, $CdSnO_3$ and $CdIn_2O_4$ as well as combinations of the binary and ternary compounds [220, 224]. By combining these oxides the optical, electrical, chemical and physical properties can be controlled more easily than by altering the chemical compositions. An excellent and detailed overview on the most promising TCOs is given by Hilgendorff in his PhD thesis [221] and in the special issue of the MRS Bulletin "Transparent conducting oxides" [222]. Almost all TCOs are n-type, and only the group of delafossites are promising p-type oxides. Delafossites belong to the family of ternary oxides with the general formula ABO_2, where A is a cation linearly coordinated to two oxygen ions and occupied by a noble-metal cation (Pt, Pd, Cu or Ag). The B cation is located in distorted edge-shared BO_6 octahedral with a central metal cation (Ga, In, Al, Fe, Co, Y, La, Nd and Eu). Delafossites show the capability of a high conductivity, transparency and have the potential of being cheap [223].

The most frequently used binary compounds that are doped with impurities like antimony, fluorine, boron, aluminium or gallium are tin oxide (SnO_2), zinc oxide (ZnO) or cadmium oxide (CdO). They all show conductivity and transparency in the same range like ITO, but they are much cheaper [224]. Other problems like stability and surface adherence and surface roughness vary strongly for the different oxides and have to be optimised [225, 226, 236].

The work function of the most promising TCOs is in the same range as for ITO, some even show higher work functions, which is important if we think of the highest possible difference between the work function of the used electrode materials.

It is also possible to use the TCOs in a multilayer stack system, which was done by Yoo in a dye-sensitised solar such by using a ITO/ATO/TiO_2 triple layered substrate. The thickness of the ITO layer is responsible for the sheet resistance, so at a 500-nm ITO layer thickness the sheet resistance is around 2.4 Ω/sq, by decreasing the ITO layer thickness, the sheet resistance increases to 9.4 Ω/sq at 100 nm ITO thickness. But the performance of the cell is more influenced by the absence or presence of the TiO_x layer. With a TiO_x layer the conversion efficiency is around 6.29% compared to 4.57% when no TiO_x layer is deposited [227].

This kind of multilayer electrodes but with a slightly changed stacking, is already used in organic light-emitting diodes and is also a promising partner for organic solar cells. Here, layer stacks with the following structure dielectric/metal/dielectric (DMD) are used. In most cases silver or silver alloys are used as the metal layer. The dielectric layer consists very often of TCOs. The key parameter in this stack is the metal layer, which is responsible for the sheet resistance. These multilayer systems show a high flexibility due to a lower dielectric layer thickness, combined with a high transparency and conductivity [228, 229]. Unfortunately, a big problem is the environmental and thermal stability of these systems. There is a significant change in the surface roughness, sheet resistance and transmission when annealed in different atmospheres at elevated temperatures [230]. Most of these multilayer systems are prepared using the sputtering technology. Transmission and conductivity for different metal layer thicknesses and different insulators, like ITO, ZnO or Nb_2O_5 have been well investigated by different groups [231–234]. For the usage in solar cells, one of the biggest problems is the surface roughness and the wetting behaviour of such multilayer systems.

8.5.1.3 Grid-Based Electrode Systems

Organic solar cell operation requires electrodes to have a sheet resistance of 10 Ω/sq and a transmission in the visible range of around 90%. Optical losses will not only exist due to a low transmission of the selected material, but also due to the shadowing by the metal grid

Table 8.3 Average sheet resistance and average transmission in the visible range between 400–800 nm for selected TCOs or multilayer stack systems.

Material	Average sheet resistance (Ω/sq)	Average transmission (%)	Comments	References
ITO	10	90		
ZnO (doped with Al, Ga, Zr)	10–50	88–91	Depends on doping concentration	[224–226, 235, 236]
SnO (doped with Sb or F)	2–60	~85–90	Depends on doping concentration	[237]
ITO/ATO/TiO_2	2–10	~90		[227]
ZnO/Ag/ZnO	3–10	~85–90	Strongly depends on layer thickness	[234]
ITO/Ag/ITO	6	~90		[229, 232, 233]

itself. The optical loss O_l of a metal finger of a grid can be easily calculated by using the width of the finger, w, and the distance between two fingers, d

$$O_l = \frac{w}{d} \tag{8.5}$$

In Figure 8.30 the shadowing of a grid is plotted against the distance between two grid fingers for a finger width of 100 μm. From the plot it is obvious that the optical loss is less than 10% when the grid distance is less than 1 mm.

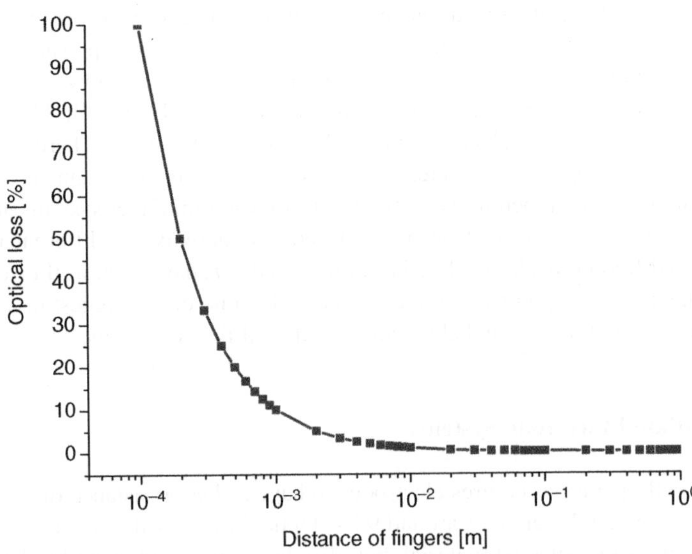

Figure 8.30 Optical loss against distance of fingers, calculated for a finger width of 100 μm.

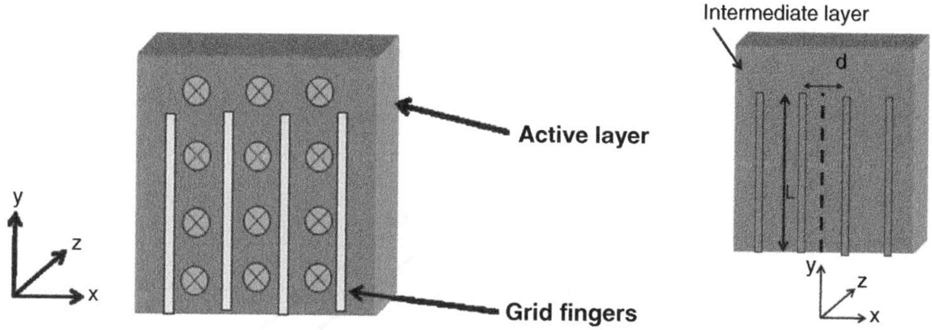

Figure 8.31 Current flow in the active layer and intermediate layer.

The calculation for the electrical losses is unfortunately a little more complicated, but is similar to the calculation for a grid used in a silicon solar cell. We have to make two different assumptions. The first assumption is that the generation of charge carriers is homogeneous over the whole area inside the active layer, and that due to a low $\mu\tau$ product a charge transport J is just possible in one direction (the z direction) (see Figure 8.31).

$$\vec{J} = j_o \cdot \hat{z} \tag{8.6}$$

As the second assumption, we suppose that the transport in the xy direction is so poor that an intermediate layer, a high conductive material, is necessary to provide the transport of the charge carriers to the grid. The transport is zero directly between two grid fingers and then increases linearly with the half distance of the grid distance d.

$$j_x(x) = j_0 \cdot x \tag{8.7}$$

With these two assumptions we are now able to calculate the power loss and the relative power loss by integration over the x direction, where j_0 is the short-circuit current, l is the length of the grid finger and R_{sq} is the sheet resistance of the intermediate layer.

$$P = J^2 \cdot R \tag{8.8}$$

$$\mathrm{d}P = J^2 \cdot \mathrm{d}R \tag{8.9}$$

$$P = j^2 \cdot l \cdot R_{sq} \int_0^{d/2} x^2 \mathrm{d}x \tag{8.10}$$

$$P = \frac{j_0^2 \cdot l \cdot R_{sq} \cdot d^3}{24} \tag{8.11}$$

and a relative power loss of

$$\frac{\Delta P}{P} = \frac{j_0 \cdot R_{sq} \cdot d^2}{V \cdot 24} \tag{8.12}$$

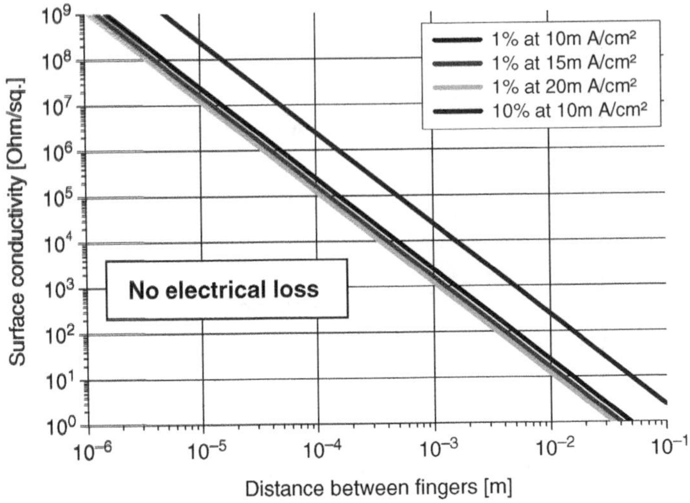

Figure 8.32 Surface conductivity plotted against the distance between fingers. A constant value of the relative power loss of 1% and 10% has been calculated for different short-circuit current densities.

This means that the power loss is increasing in the third exponent with the distance *d* between the grid fingers. In Figure 8.32 the relative power loss is set to a constant value of 1% or 10% for three different short-circuit current densities. We plot the surface resistance against the distance between two fingers for these constant relative power loss values. As can be seen when we have a sheet resistance of less than 100 Ω/sq we can easily work with a grid distance between 1 mm, when the sheet resistance is even lower, something around 10 °/sq we can even go to 1 cm grid distance, without having any electrical losses [238, 239].

The fabrication of this metal grid can be done by thermal evaporation through a mask, by inkjet printing or even hot embossing [240], as well as by screen printing [241] or using nanoimprinting lithography [242, 243].

8.5.1.4 PEDOT:PSS and Other Highly Conductive Polymers

The best known and mostly used highly conductive polymer is poly (3,4-ethylenedioxythiophene):poly(styrenesulfonate) (PEDOT:PSS). PEDOT:PSS is a mixture of two ionomers. The part of the sulfonyl groups are deprotonated and carry a negative charge, whereas the PEDOT is a conjugated polymer and carry a positive charge. Together, the charged macromolecules form a macromolecular salt. PEDOT:PSS shows a slight blue colour. PEDOT:PSS is mainly commercially available at H.C. Starck under the trade name Clevios and Agfa with the trade name Orgacon. It is used in almost all bulk-heterojunction solar cells as a charge-carrier sensitive layer, which leads to enhanced injection conditions. The problem of PEDOT:PSS is its acidity, which leads to etching of the TCO under environmental conditions. But this material could also be used as an intermediate layer in a grid electrode system due to a high conductivity combined with a medium to high transmission in the visible range. By adding surfactants or changing the morphology the conductivity of this material can be changed over a wide range [244]. PEDOT:PSS can be deposited by using spray-coating, spin-coating or doctor-blading technologies. The conductivity and transmission of PEDOT:PSS vary strongly from the kind of deposition method and the used solvent. But transmissions around 85% with a sheet resistance of 50–100 Ω/sq are possible. The performance of these cells depends strongly on the quality of the active layer. But in most cases the performance is similar to the reference cell, which is mostly an ITO-based electrode system [155, 245–247].

The oldest known highly conductive polymer is polyaniline (Pani), which can also be used as an interlayer. The high conductivity is reached by a protonic doping of Pani. The conductivity is around 10–200 S/cm when processed from solution [248]. The publication list of the synthesis and properties of Pani was increasing in the 1980s, but the usage of Pani in organic solar cells has not been published yet. Pani offers good ambient stability properties, it is nonacid, and can be also used by all deposition methods like for PEDOT:PSS. Pani shows a green colour, and is nowadays mostly used as a corrosion protection layer [249].

8.5.1.5 Nanoparticles and Nanotubes

The usage of nonpolymeric materials as the intermediate layer for a grid electrode offers also a big innovative potential. Most investigations so far were done on carbon nanotubes. Carbon nanotubes also provide a wide range of conductivity and transmission and an intrinsic work function in the range of 4.5–5.1 eV, which can be tuned by n- or p-type doping. The big advantage of carbon nanotubes is that the processing can be done at room temperature and ambient conditions, which is compatible with the polymeric substrates. Another advantage is that the conductivity remains unchanged when the layers are bent. The disadvantages of carbon nanotubes are the price level of SWNT, the stability of the SWNT films to air exposure over several 100 h as well as the restriction to process highly diluted formulations. Depending on the preparation of the carbon nanotube films the sheet resistance ranges from 20 Ω/sq to 1000 Ω/sq with a transmission between 80–85% [250, 251]. As for PEDOT:PSS most cells with carbon nanotubes work in the same range or slightly better like the reference cell with an ITO-electrode in the various publications [250–254]. Carbon nanotubes also seem to bring an increase in performance when applied as an interlayer between ITO and PEDOT:PSS, or between PEDOT:PSS and P3HT:PCBM, or even mixed inside the P3HT:PCBM blend [255].

ITO nanoparticles provide the possibility to process the films out of solution under ambient conditions and at room temperatures. They further can be used as bottom as well as top electrodes. Unfortunately, the highest conductivity is reached after a high-temperature annealing step for a long time, which is incompatible with the film substrate [256, 257].

Further, silver needles or nanoparticles offer the possibility to process electrodes from solution. No further annealing step seems to be necessary for an optimised solution or ink. These films of silver needles show a transparency of around 80–90% in the visible range and a sheet resistance in the range of 10–200 Ω/sq. Cells with silver needles show a higher short-circuit current compared with ITO reference devices, due to the higher transparency, and an improved optical path length due to scattering of incident light [258].

8.5.2 Materials for Nontransparent Electrodes

Nontransparent electrodes need to show surface conductivities of the order of 1 Ohm/sq or lower and therefore require high conductive materials. Furthermore, solar cells can benefit if the opaque electrode also has high reflectivity or other light-management features to confine light within the active layer. Ideally, these electrodes also have specific barrier functionality against oxygen and water diffusion. Finally, packaging requires a good adhesion of the printed layers to the bottom ones. Good adhesion is specifically important for module bus bars as well as module interconnection regimes

8.5.2.1 Evaporated Materials (LiF/Ag, Ca/Ag, Ag, Au, etc.)

The assortment of materials that can be evaporated and used as an electrode is large. The advantage of evaporation is that a shadow mask can be used and so a certain cell layout can be applied. The most used combination of materials is lithium fluoride in combination with silver or aluminium, where lithium fluoride acts to reduce the work function of the electrode. Another combination is calcium with silver or aluminium. Unfortunately, these electrode combinations are not environmentally stable and need to be handled under oxygen and water exclusion. The usage of silver or gold directly on top of the active layer typically leads to sufficient power losses. Both Au and Ag make nonrectifying contacts and typically lead to FF and V_{oc} losses due to surface recombination. The work functions of Au (5.1 eV) and Ag (4.6 eV) seem to be responsible for this lower performance when compared to Al (4.3 eV) in a polymer–fullerene solar cell. However, that loss can be reduced by using so-called intermediate layers (see Section 8.3), which to some extent can fine tune the work function. This intermediate layer could be, for example, PEDOT:PSS. One other tactic is to place a self-assembled monolayer of ZnO between the metal and the active layer material or by introducing a MoO_3 buffer layer [259,260]. The diffusion of silver or gold into the active layer material is suppressed as well in the case of intermediate layers [261].

The disadvantages of evaporated electrodes are the rather complex transfer of that technology to a roll-to-roll process. High-speed and low-cost roll-to-roll evaporation is principally possible. However, the quality of these layers as well as the challenge to pattern the electrode layers in register to the rest of the module stack remain to be solved.

8.5.2.2 Printable Materials (Ag Pastes, Organometallic Materials, etc.)

The advantage of printable electrode materials like Ag pastes or organometallic materials is in their compatibility to roll-to-roll processing. Ag pastes are already used for the conventional silicon solar cell industry. These pastes need two adjustments to become compatible for OPV: (i) first, the ink and the solvents need to be chemically compatible to the cell stack, and (ii), the drying temperature regime needs to be less than a few minutes @ 150 °C.

Ag pastes can be processed via screen printing, gravure and flexoprinting, pad printing or by using a simple paint brush [241].

Organometallic materials are normally formulated in high boiling solvent like tetradecane or ethylene glycol/ethanol and polymer mixtures of transition metals like Au, Cu, Ag, Pt. These organometallic materials have mostly been used in inkjet systems so far, but screen printing is also possible once the viscosity is adapted. These electrodes typically show the highest conductivity after longer periods of high-temperature annealing. In former times the temperature has been above 200 °C, which is not compatible with the polymeric substrate or the active layer materials. But recent investigations made it possible to decrease the transformation temperature to below 200 °C. The conductivities of organometallic materials are lower than for sputtered materials, for example a vacuum-deposited Ag layer has resistivity of around 2 μΩ cm and for organometallic Ag layers the resistivity ranges from 40–500 μΩ cm, dependent on the transformation temperature [240, 262–264]. The layer thicknesses range from 100 nm to 1 μm depending on the deposition method.

8.6 PRODUCTION OF ORGANIC SOLAR CELLS

Organic semiconductors can be processed at low temperatures from solution and this opens up the application of low-cost and scalable manufacturing processes such as printing or coating from roll-to-roll with higher productivity. Printing as a fabrication method has evolved to one of the most prominent production technologies for organic electronics. This elegant process may ultimately allow deposition of materials from solution directly in defined spots onto flexible plastic substrates.

In this section we discuss and analyse a series of production technologies for organic photovoltaics such as blade coating, spray coating, inkjet printing, gravure and flexographic printing, curtain and slot-die coating and outline the capability for the production from roll-to-roll. Depending on the printing and coating technologies, the ink requirements differ in their characteristics such as viscosity, vapour pressure, surface tension and boiling point, which are primarily defined by the solvent properties. The selection of appropriate organic solvents is determined by the solubility, drying characteristics and rheological behaviour. Different printing technologies require adequate ink characteristics. In this section, we extensively discuss the printing and coating methods and highlight the solar cell device performance of differently printed and coated photovoltaic devices.

The first reported organic solar cell devices were fabricated by spin coating [21, 25]. In this well-established laboratory method the fluid is deposited onto the centre of the substrate that is subsequently spun at high velocity (up to 10 000 rpm). Centripetal acceleration forces spreading of the ink on the surface of the substrate resulting in high-quality thin films. The final film thickness depends not only on the nature of the fluid such as viscosity, drying rate, solid content and surface tension, but strongly on the spinning parameters like spinning speed, acceleration and time interval. Despite low viscosities and boiling points and thus, high vapour pressures, spin coating requires little ink-formulation effort. A huge limitation of the spin-coating method is identified by the production of solely large-area films with over 90% material waste. Furthermore, spin coating has low potential to be integrated in roll-to-roll production. However, most of the research activities have been done by spin coating, mainly due to the homogeneity of the film thickness, reduced particles, reduced number of pinholes and with record power-conversion efficiencies (PCE) on glass substrates up to 5% [32, 38, 89, 98].

Another technology to deposit thin films from solution is proven by blade coating [265, 266]. Figure 8.33 demonstrates the principle of blade coating.

A coating knife (doctor blade) applies a defined amount of solution evenly over the substrate by moving the knife along the substrate with an adjustable velocity in the range of 5 to 70 mm/s depending on the desired film thickness. This leads to a large-area wet film that dries during the vaporisation of the solvent. The drying and resulting quality and thickness of the solid-state film depends not only on the solvent vehicle, but strongly on the blading parameters, such as temperature of the table, slit height, defined by the gap between the knife and the substrate, accompanied by the solution volume. Different drying kinetics may lead to distinct solid-state topographies and, eventually, morphologies over spin-coated

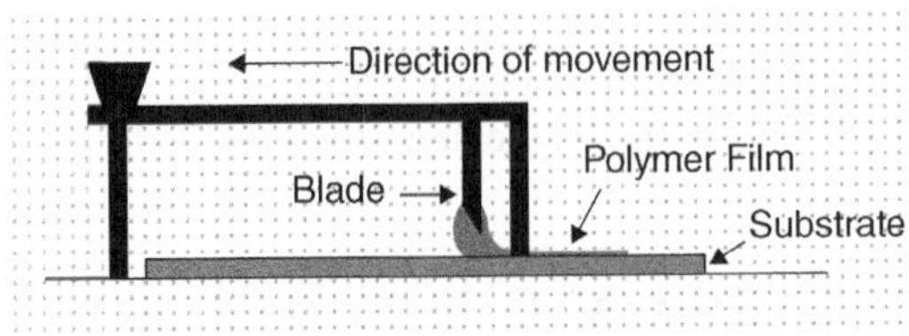

Figure 8.33 Principle of doctor blading.

films. Large-area films can be coated with lateral resolutions in the submillimetre regime and low material waste compared to spin coating. Blade coating allows a huge choice of organic solvents and therefore, little ink formulation effort is required. This fast method to obtain thin films is compatible with roll-to-roll processes and was previously reported [267] as an elegant technology for the fabrication of photovoltaic devices. The authors evaluated the quality of doctor blading over spin coating and found that blade-coated devices show as high performances as spin-coated ones. A slight increase in power-conversion efficiency (PCE) reflects a 10% higher J_{sc} owing to a gradient in the film thickness of blade-coated layers, which is frequently observed for small-area coating trials. However, similar quantum efficiencies for charge generation and charge separation for both techniques are found. The current–voltage (*J–V*) behaviour of solar cell devices in [267] was analysed [268] to examine the origin of the slight increase in device performance of doctor-bladed over spin-coated OPVs. The investigated spin-coated and doctor-bladed devices based on pristine o-xylene solvent show similar ideality factors n, but different saturation currents J_0 resulting in different morphologies for the two types of processing [267]. The distinct drying kinetics of the wet film processed by spin coating or doctor blading involves different solid-state morphologies, while not losing performance. Optimised solvent formulations (oDCB/mesitylene), and thus improved drying with homogeneous film formation, lead to high-efficiency doctor-bladed devices as demonstrated in Figure 8.34. More details of the ink formulation development can be found elsewhere [269].

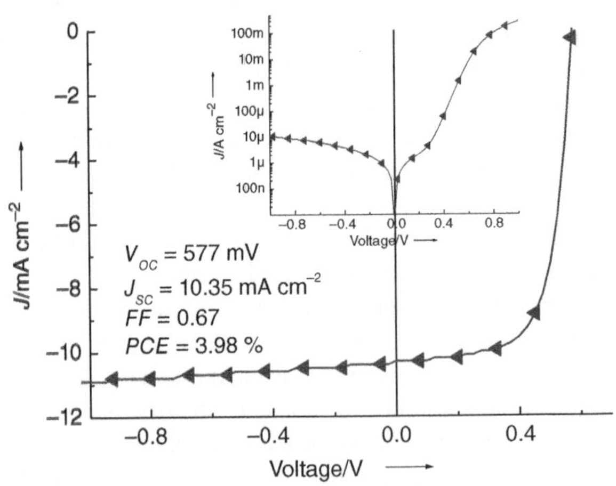

Figure 8.34 Device performance of doctor-bladed solar cell on glass with optimised solvent formulation (oDCB/mesitylene).

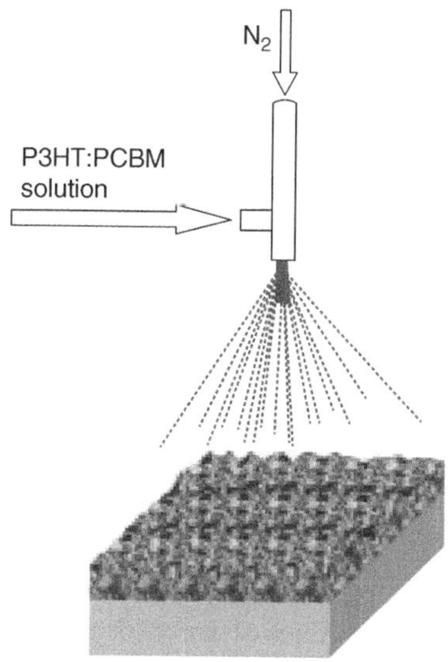

Figure 8.35 Principle of spray coating. (Reproduced with permission from Vak, 2007. Copyright © 2007, American Institute of Physics.)

These investigations show a clear benefit of a roll-to-roll compatible coating technology with comparable or even better device performances to spin coating [267].

Thin film deposition by spray coating [270, 271] was recently published as a suitable method for the fabrication of organic photovoltaics. Vak et al. previously showed that spray coating the active layer from chlorobenzene solutions gives highly efficient devices [272]. Furthermore, Green et al. presented spray-deposited P3HT:PC_{61}BM films from a variety of common organic solvents with different boiling points [273]. Even efficiencies in the range of 3.2% based on chlorobenzene have been demonstrated using an ultrasonic spray deposition with droplet volume in the picolitre regime, but these devices were processed in an inert atmosphere [274], which limits the compatibility with roll to roll processes. A schematic representation of the spray-coating method is shown in Figure 8.35 [272].

In the spray-coating method, the film formation occurs stepwise via single droplets. The transfer gas pressure forces single droplets with high velocity onto the substrate. The droplets dry immediately as they hit the surface of the substrate. This morphology evolution is entirely different compared to other printing techniques such as inkjet printing [269], where the film formation is based on the spreading of droplets and combining with adjacent droplets to form a liquid bulk that dries during evaporation of the solvent formulation. Typically, the surface roughness is in the tens of nanometres owing to the transfer of rather large droplets in the micrometre regime that individually dry on the substrate without forming a liquid bulk.

The spray-coating technique requires inks having high vapour pressures for rapid drying and low viscosities at room temperature, since the inks are processed at ambient conditions and are not heated during spraying. Highly viscous inks require higher temperatures to decrease the viscosity and adapt the fluid properties on the airbrush setup. The nozzle-to-substrate distance depends on the vapour pressure of the ink and has to be adjusted. For inks with too low drying rates or too close nozzle-to-substrate distance, liquid droplets are immediately pushed sideward by the pressure gas of the airbrush, resulting in nonuniform

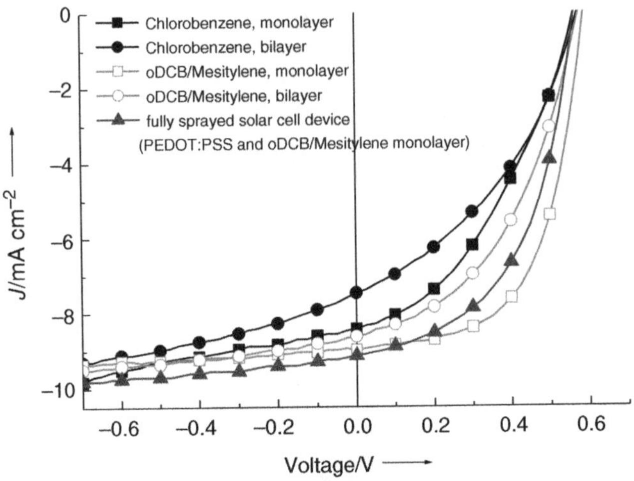

Figure 8.36 Device performance of spray-coated devices on glass based on either chlorobenzene or oDCB/mesitylene. (Data extracted from [275])

wetting. On the other hand, for organic solvents with too high evaporation rates and too large a nozzle-to-substrate distance, the spray-deposited droplets may be dry prior to reaching the substrate surface and a film deposition is inhibited. In all printing and coating technologies, the surface tension plays a major role in successful coatings and is responsible not only for the wetting and spreading on the substrate, but also for the interaction with the printhead, nozzle, blade, etc. "Low wets high", in fact the low surface tension for organic semiconductor solutions guarantees decent wetting with low contact angles on PEDOT:PSS substrates owing to higher surface tensions and the hygroscopic behaviour of PEDOT:PSS [275]. The authors of this study previously presented photovoltaic devices with spray-coated P3HT:PCBM films based on oDCB/mesitylene [269] compared to pristine chlorobenzene. The *J–V* characteristics are depicted in Figure 8.36.

Spray-coated monolayers were compared with spray-coated double layers, whereas the bilayers reveal strong limitations in fill factor (FF) and serial resistances (R_s) owing to the high film thickness of 400 nm. Moreover, the bilayers indicate limitations in the mobility–lifetime ($\mu\tau$) product, leading to losses in photocurrent and thus, lower J_{sc}. In contrast, 250-nm thick monolayers perform significantly better, yielding a V_{oc} of 588 mV, J_{sc} of 9.0 mA cm^{-2}, and a high FF of 0.59, corresponding to a PCE of 3.1%. Monolayer devices show a much weaker field dependence than thicker bilayer ones. In accordance with earlier studies [98], the slowly dried oDCB/mesitylene films have a larger $\mu\tau$ product, sufficient to extract all carriers under J_{sc} conditions. These results indicate that morphological limitations can be overcome by appropriate ink formulations. Surprisingly, the high surface roughness in the range of tens of nanometres for spray-coated P3HT:PCBM films over spin-coated or doctor-bladed ones (few nanometres) does not affect the device performance adversely. To further prove the potential of spray coating as a fabrication method, a fully sprayed organic photovoltaic device is reported in [275] with a spray-coated hole-extraction layer PEDOT:PSS as well as photoactive P3HT:PCBM film. The fully spray-coated solar cell reveals an outstanding high device performance with V_{oc} of 560 mV, J_{sc} of 9.1 mA cm^{-2}, FF of 0.52 and PCE of 2.7% [275].

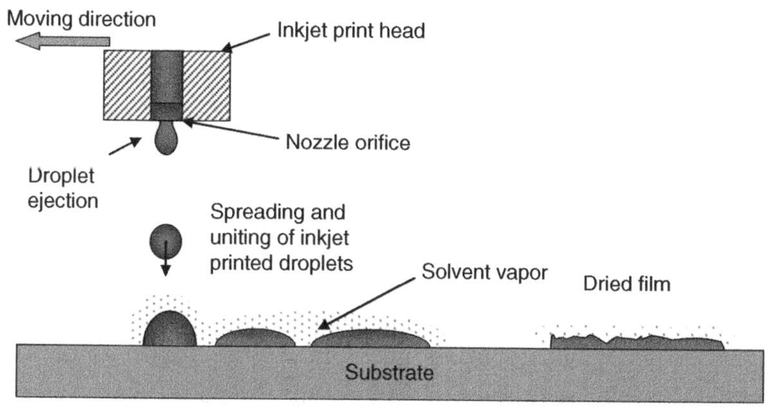

Figure 8.37 Principle of inkjet printing. (Copyright Wiley-VCH Verlag GmbH & Co. KGaA. Reproduced with permission.)

Inkjet printing has emerged as one of the most important patterning methods not only in the paper industry, but has also gained huge prestige in organic electronics. Nowadays three types of inkjet printers are commercially available that differ in the formation of droplets [276]: continuous inkjet, drop on demand (DOD) inkjet and bubble jet. Inkjet printheads position a fluid accurately in defined spots on a substrate, and thus the DOD technology has become more important since a droplet is created and ejected on demand leading to a predefined printing pattern without postpatterning of the coated films, as is necessary for large-area coating methods, e.g. blade or spray coating. Functional films can be inkjet printed with high resolution of a few micrometres owing to the formation of droplets in the picolitre regime. Inkjet printing has been widely implemented in the field of organic light-emitting diodes and field effect transistors [277], but only a few reports are found on inkjet-printed organic solar cells [84, 269, 278–280].

Figure 8.37 [269] shows the principle of inkjet printing. The inkjet fluid is transferred from a writing head onto a substrate without direct contact of the surface [281], which makes inkjet printing so promising due to the compatibility with various substrates. As demonstrated in the scheme in Figure 8.37, a droplet is ejected at the nozzle orifice. As it hits the surface of the substrate, the droplet wets and spreads on the substrate due to the interaction of surface energies. Combining of adjacent droplets to a liquid bulk leads to film formation. Individual drying kinetics of the wet film depends strongly on the solvent vehicle and the corresponding solvent properties of the single components. Inkjet printing inks are generally low-viscous ones in the range of 1 to 10 mPa s, but certainly higher viscosity inks up to 40 mPa s can be printed with high frequencies by heatable ink reservoirs or temperature-controlled printheads to decrease the viscosity at the nozzle orifice during droplet formation. Too high viscosities and dispersions with large particles lead to jetting problems and nozzle clogging. The uniformity of inkjet-printed layers is determined by the temperature of the substrate and the choice of organic solvents. In the design of inkjet inks, the vapour pressure is of great interest. The nozzle orifice is open to the atmosphere and consequently, high vapour pressure solvents may evaporate even in the nozzle. This leads to an increase of viscosity at the nozzle affecting the droplet characteristics such as velocity, volume and angularity (angle deviation) and even materials precipitate in the jet resulting in serious nozzle clogging [269]. The surface tension controls the spreading and wetting of the ink on the substrate as

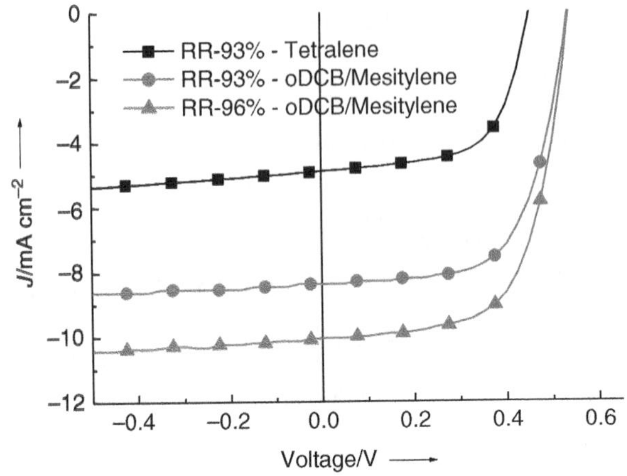

Figure 8.38 Device performance of inkjet-printed devices on glass based on oDCB/mesitylene. Increased regioregularity of P3HT (i.e. reduced structural defects) yields improved photovoltaic performance. (Data extracted from [84, 269])

well as the interaction with the printhead. A well-optimised ink formulation gives excellent droplet formation owing to suitable rheology of the ink and well-adjusted driving conditions of the printhead. Thus, inkjet printing requires huge effort to properly design inks for reliable jetting and excellent film quality. Film thicknesses can be varied by changing the droplet volume and droplet spacing. Recently, Hoth et al. [269] studied the morphological properties of inkjet-printed P3HT:PCBM blend layer and highlighted the device performance on glass substrates achieved with an adequate inkjet formulation oDCB/mesitylene over pristine solvent tetralene. Outstanding high device performance was obtained with V_{oc} of 540 mV, J_{sc} of 8.4 mA cm^{-2} and a FF of 0.64, corresponding to PCE of 2.9% with inkjet-printed P3HT:PCBM film based on oDCB/mesitylene.

With higher regioregular (96%) P3HT and the optimised inkjet solvent formulation oDCB/mesitylene to control the drying and film formation, Hoth et al. [84] were able to gain suitable gelation time, improved morphological properties of the active layer yielding excellent device performance of 3.5% with V_{oc} of 537 mV, J_{sc} of 10.05 mA cm^{-2} and a FF of 0.64 for inkjet-printed P3HT:PCBM bulk-heterojunction solar cells. Figure 8.38 represents the J–V characteristics of inkjet-printed OPVs.

Gravure printing demonstrates one of the simplest industrial printing technologies. A schematic representation is shown in Figure 8.39 [282].

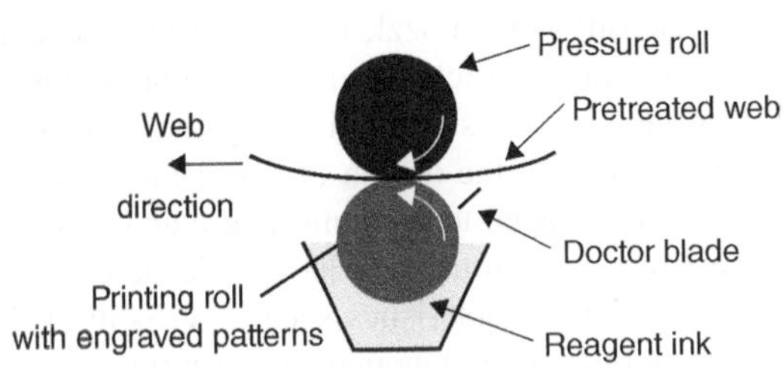

Figure 8.39 Principle of gravure printing.

The printing pattern is engraved into a printing roll and filled with ink by rotating in an ink bath. The excessive ink is peeled off by a flexible coating knife prior to touching the surface of the substrate. The remaining ink in the engraved pattern is directly transferred onto the flexible plastic substrate. Imprinting occurs and the substrate absorbs the ink from the engraved cells [282]. This high-throughput and cost-effective roll-to-roll printing technology is capable of printing films with low film thicknesses ranging from 1 to 50 micrometres and small patterns with line width less than 100 micrometres at high speeds up to 15 m/s [283]. These properties in combination with low-temperature processing and appropriate patterning resolution make this technology as one of the most promising methods for the mass production of organic solar cells. Tuomikoski and Suhonen [282] previously reported on the first gravure-printed organic solar cells with film qualities of gravure-printed 100-nm P3HT:PCBM comparable to that of spin-coated ones. Despite the decent diode properties, the reduced J_{sc} of the gravure-printed device might reflect a poor morphology owing to unfavourable drying kinetics. A fully gravure printed organic solar cell with gravure-printed hole-transporting layers as well as photoactive P3HT:PCBM film was lately reported in [284] with efficiencies as high as 1.68% under AM 1.5 illumination. These studies show that efficient devices can be fabricated by gravure printing and evaluate gravure printing as a manufacturing method for the mass production of OPVs.

Curtain coating and slot-die coating have emerged as promising technologies for the fabrication of organic photovoltaics from roll-to-roll. The curtain-coating technique uses a gravity-fed slot orifice to apply a fluid onto the substrate by a falling ink curtain. This technology produces high-quality films with extremely high line speeds up to 1000 m/min. The thickness of the coating is affected by the rheological properties of the ink, the coating speed and the die gap, and thus the ink volume. This method requires huge ink formulation effort since the curtain during coating lacks high stability without defects.

Contrary to curtain coating, in slot-die coating the ink is squeezed via a slot onto the substrate. However, both technologies appear rather comparable to the doctor-blading technique. The inset in Figure 8.40 shows the principle of slot-die coating. A pump-fed slot

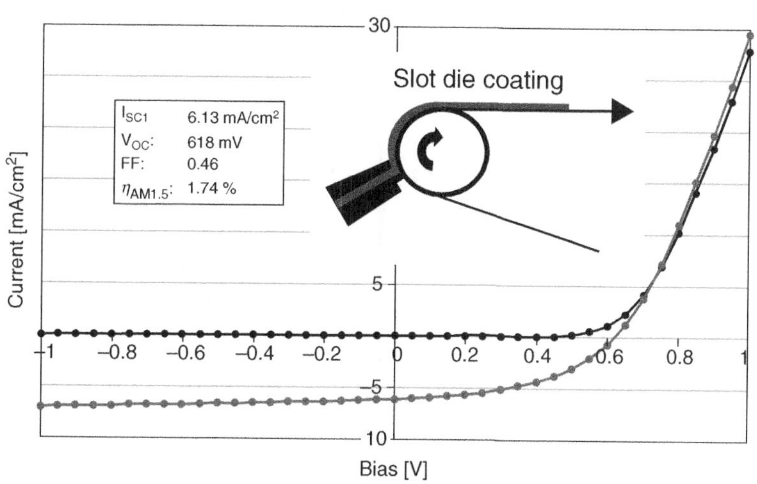

Figure 8.40 Dark (black) and light (red) *J–V* characteristics of a slot-die-coated device. The inset shows the principle of slot-die coating. (Data extracted from [287])

orifice directly applies a fluid to the flexible substrate, which is supported by a back up roll. Apart from the ink properties and temperature control of the substrate, the thicknesses of the coating can be varied by the distance and angle between die and substrate, the pump speed, the die gap and web speed. Slot-die coating can handle inks with a wide range of viscosities. The uniformity, and hence the quality of the coated films require fluid properties with high shelf life and consistency in concentration, viscosity and temperature. Moreover, the substrate should be of high quality that ensures reliable surface properties over the whole area. As aforementioned, the drying kinetics is of great interest in the slot-die-coating technique. Basically, thin films exhibit high drying rates, favouring the usage of fast web speeds leading to high throughputs. However, apart from the numerous advantages of slot-die coating, the adjustment of the pump pressure is decisive, since the material loss increases significantly at higher web speeds and die systems possess high costs. Nevertheless, slot-die coating is ascertained as a high-throughput and economical fabrication technology with good prospects for the future OPV mass production. Organic solar cells with slot-die-coated PEDOT:PSS films and P3HT:PCBM active layers on ITO-coated polyester films were previously demonstrated in [285–287]. The *J–V* characteristics of a slot-die-coated solar cell device are presented in Figure 8.40 [287].

The authors achieved outstandingly high power-conversion efficiencies up to 1.74% for a fully printed organic solar cell and these promising findings show the feasibility of slot-die coating as a suitable roll-to-roll manufacture for organic photovoltaics.

Transferring a laboratory coating method to a roll-to-roll printing or coating technology may entail challenges not only in fluid stability and compatibility of the ink, but also in quality and morphology of printed or coated films. Therefore, flexible organic solar cells were fabricated with identical device architectures by a roll-to-roll technique and compared to lab-scale doctor-bladed devices on an ITO-coated polyester carrier [288].

The *J–V* characteristics in Figure 8.41 indicate similar photovoltaic behaviour under AM 1.5 illumination for both fabrication types. The roll-to-roll processed solar cell reveals a V_{oc} of

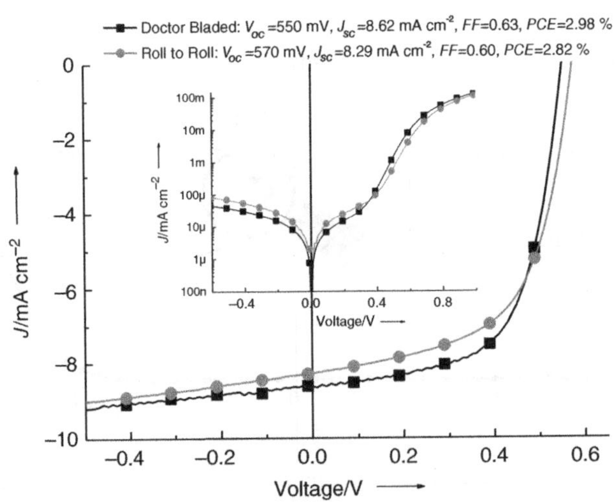

Figure 8.41 *J–V* characteristics of a roll-to-roll processed device (gray circles) compared to a lab-scale-doctor bladed solar cell (black aquares) under AM1.5 illumination. The inset shows the *J–V* behaviour in the dark.

570 mV, J_{sc} of 8.29 mA cm^{-2}, FF of 0.60 and PCE of 2.82%, whereas the doctor-bladed device has a V_{oc} of 550 mV, J_{sc} of 8.62 mA cm^{-2}, FF of 0.63 leading to a PCE of 2.98%. Although processed from different technologies, both devices show identical ideality values n of 1.6 and saturation currents J_0 (3–8 $\times$ 10^{-9} mA cm^{-2}), suggesting similar phase separation and aggregation, and thus bulk morphologies of the P3HT:PCBM blend composites. The reduced J_{sc} of the roll-to-roll processed device is a result of a slightly decreased $\mu\tau$ product, whereas the lower V_{oc} of the doctor-bladed solar cell is partly attributed to a reduced built-in potential. However, the roll-to-roll fabricated OPV indicates a similar device performance to a lab-scale blade-coated solar cell. These investigations demonstrate the high potential of a roll-to-roll manufacturing to produce high-quality thin films, which are comparable to lab-scale methods, for the fabrication of OPVs.

Also, currently slot-die coating is the most reliable method to produce organic solar cells, and all record efficiencies of large-area processed modules were demonstrated with slot-die coated modules.

8.7 SUMMARY AND OUTLOOK

The fundamentals of OPV were developed in the period between 2005 and 2010. During this period, the 3rd generation of OPV materials was developed, tandem concepts were successfully implemented, the essential classes of electrode and interface materials were reported, and, the key production method for large-area volume processing was found. Although efficiencies at those times just hit the 7% level for single-junction cells, the know-how and knowledge gained in that period set the basis for the efficiency records of over 12%, as demonstrated recently. In 2009 we predicted that OPV will hit the 10% milestone between 2010 and 2012. By today, the expectations are that the 15% milestone will be hit before 2015. In parallel, there is visible progress on lifetime. The whole technology platform around the organic solar cells was developed outstandingly rapidly. Electrode materials, packaging materials, production processes – all were developed and transferred to flexible substrates within only a few years.

The first reports on indoor, accelerated and outdoor degradation can be summarised such that the lifetime of flexible organic solar cells are already good enough for the first consumer electronics application. Indeed, by 2009, the first commercial OPV product, a solar bag with integrated battery, made it onto the public market [289].

REFERENCES

[1] J. Peet, J. Y. Kim, N. E. Coates, W. L. Ma, D. Moses, A. J. Heeger, and G. C. Bazan, *Nature Mater.* 2007, **6**, 497–500.

[2] M. A. Green, K. Emery, Y. Hisikawa, and W. Warta, *Prog. Photovolt.: Res. Appl.* 2008, **16**, 81.

[3] R. Gaudiana, and C. J. Brabec, *Nature Photon.* 2008, **2**, 287.

[4] J. Hou, H. Chen, S. Zhang, G. Li, and Y. Yang, *J. Am. Chem. Soc.* 2008, **130**, 16144.

[5] B. O'Regan, and M. Grätzel, *Nature* 1991, **353**, 737.

[6] U. Bach, D. Lupo, P. Comte, J. E. Moser, F. Weissörtel, J. Salbeck, H. Spreitzer, and M. Grätzel, *Nature* 1998, **395**, 583.

[7] V. Y. Merrit, *IBM J. Res. Develop.* 1978, **22**, 353.

[8] D. L. Morel, A. K. Gosh, T. Feng, E. L. Stogryn, P. E. Purwin, R. F. Shaw, and C. Fishman, *Appl. Phys. Lett.* 1978, **32**, 495.
[9] A. P. Piechowski, G. R. Bird, D. L. Morel, and E. L. Stogryn, *J. Phys. Chem.* 1984, **88**, 923.
[10] C. W. Tang, *Appl. Phys. Lett.* 1986, **48**, 183.
[11] J. Roncali, *Chem. Rev.* 1997, **97** (1), 173–206.
[12] Y. Liang, S. Xiao, D. Feng, and L. Yu, *J. Phys. Chem. C* 2008, **112**, 21.
[13] R. Kroon, M. Lenes, J. C. Hummelen, P. W. M. Blom, and B. De Boer, *Polym. Rev.* 2008, **48**, 531–582.
[14] J. Hou, M.-H. Park, S. Zhang, Y. Yao, L.-M. Chen, J.-H. Li, and Y. Yang, *Macromolecules* 2008, **41**, 16.
[15] D. P. Hagberg, T. Marinado, K. M. Karlsson, K. Nonomura, P. Qin, G. Boschloo, T. Brinck, A. Hagfeldt, and L. Sun, *J. Org. Chem.* 2007, **72**, 9550–9556.
[16] H. R. Kerp, H. Donker, R. B. M. Koehorst, T. J. Schaafsma, and E. E. van Faassen, *Chem. Phys. Lett.* 1998, **298**, 302–308.
[17] D. E. Markov, E. Amsterdam, P. W. M. Blom, A. B. Sieval, and J. C. Hummelen, *J. Phys. Chem. A* 2005, **109**, 5266–5274.
[18] A. Huijser, T. J. Savenije, J. E. Kroeze, and L. D. A. Siebbeles, *J. Phys. Chem. B* 2005, **109**, 20166–20173.
[19] M. Stössel, G. Wittmann, J. Staudigel, F. Steuber, J. Blässing, W. Roth, H. Klausmann, W. Rogler, J. Simmerer, A. Winnacker, M. Inbasekaran, and E. P. Woo, *J. Appl. Phys.* 2000, **87**, 9.
[20] C. W. Tang, *Appl. Phys. Lett.* 1986, **48**, 183.
[21] G. Yu, A. J. Heeger, *J. Appl. Phys.* 1995, **78**, 4510.
[22] J. J. M. Halls, C. A. Walsh, N. C. Greenham, E. A. Marseglia, R. H. Friend, S. C. Moratti, and A. B. Holmes, *Nature* 1995, **376**, 498.
[23] C. J. Brabec, N. S. Sariciftci, and J. C. Hummelen, *Adv. Funct. Mater.* 2001, **11**, 15.
[24] P. Peumans, S. Uchida, and S. R. Forrest, *Nature* 2003, **425**, 158.
[25] G. Yu, J. Gao, J. C. Hummelen, F. Wudl, and A. J. Heeger, *Science* 1995, **270**, 1789.
[26] Y. Wang, *Nature* 1992, **356**, 585.
[27] S. Morita, A. A. Zakhidov, and K. Yoshino, *Solid State Commun.* 1992, **82**, 249.
[28] N. S. Sariciftci, L. Smilowitz, A. J. Heeger, and F. Wudl, *Science* 1992, **258**, 1474.
[29] N. S. Sariciftci, D. Baun, C. Zhang, V. I. Srdanov, A. J. Heeger, G. Stucky, and F. Wudl, *Appl. Phys. Lett.* 1993, **62**, 585.
[30] J. C. Hummelen, B. W. Knight, F. LePeq, F. Wudl, J. Yao, and C. L. Wilkins, *J. Org. Chem.* 1995, **60**, 532.
[31] V. D. Mihailetchi, J. K. J. van Duren, P. W. M. Blom, J. C. Hummelen, R. J. Janssen, J. M. Kroon, M. T. Rispens, W. J. H. Verhees, and M. M. Wienk, *Adv. Funct. Mater.* 2003, **13**, 43.
[32] S. E. Shaheen, C. J. Brabec, N. S. Sariciftci, F. Padinger, T. Fromherz, and J. C. Hummelen, *Appl. Phys. Lett.* 2001, **78**, 841.
[33] C. J. Brabec, S. E. Shaheen, C. Winder, N. S. Sariciftci, and P. Denk, *Appl. Phys. Lett.* 2002, **80**, 1288.
[34] P. W. M. Blom, M. J. M. De Jong, and S. Breedijk, *App. Phys. Lett.* 1997, **71**, 930.
[35] V. D. Mihailetchi, L. J. A. Koster, P. W. M. Blom, C. Merzer, B. De Boer, J. K. J. van Duren, and R. A. J. Janssen, *Adv. Funct. Mater.* 2005, **15**, 795.
[36] C. Melzer, E. J. Koop, V. D. Mihailetchi, and P. W. M. Blom, *Adv. Funct. Mater.* 2004, **14**, 865.
[37] P. Schilinsky, C. Waldauf, and C. J. Brabec, *Appl. Phys. Lett.* 2002, **81**, 3885.
[38] F. Padinger, R. S. Rittberger, and N. S. Sariciftci, *Adv. Funct. Mater.* 2003, **13**, 85.
[39] L. J. A. Koster, V. D. Mihailetchi, P. W. And M. Blom, *Appl. Phys. Lett.* 2006, **88**, 093511.
[40] M. C. Scharber, D. Mühlbacher, M. Koppe, P. Denk, C. Waldauf, A. J. Heeger, and C. J. Brabec, *Adv. Mater.* 2006, **18**, 789.
[41] J.-L. Bredas, D. Beljonne, V. Coropceanu, and J. Cornil, *Chem. Rev.* 2004, **104**, 4971.

[42] H. A. M. van Mullekom, J. A. J. M. Vekemans, E. E. Havinga, and E. W. Meijer, *Mater. Sci. Eng.* 2001, **32**, 1.
[43] L. H. Slooff, S. C. Veenstra, J. M. Kroon, D. J. D. Moet, J. Sweelssen, and M. M. Koetse, *Appl. Phys. Lett.* 2007, **90**, 143506.
[44] F. Zhang, W. Mammo, L. M. Andersson, S. Admassie, M. R. Andersson, and O. Inganäs, *Adv. Mater.* 2006, **18**, 2169–2173.
[45] N. Blouin, A. Michaud, and M. Leclerc, *Adv. Mater.* 2007, **19**, 2295.
[46] D. Mühlbacher, M. Scharber, M. Morana, Z. Zhu, D. Waller, R. Gaudiana, and C. Brabec, *Adv. Mater.* 2006, **18**, 2884.
[47] M. M. Wienk, M. G. R. Turbiez, M. P. Struijk, M. Fonrodona, and R. A. J. Janssen, *Appl. Phys. Lett.* 2006, **88**, 153511.
[48] W.-Y. Wong, X.-Z. Wang, Z. He, A. B. Djurisic, C.-T. Yip, K-Y. Cheung, H. Wang, C. S. K. Mak, and W.-K. Chan, *Nature Mater.* 2007, **6**, 521.
[49] Q. Zhou, Q. Hou, L. Zheng, X. Deng, G. Yu, and Y. Cao, *J. Appl. Phys.* 2004, **84**, 1653.
[50] X. Wang, E. Perzon, F. Oswald, F. Langa, S. Admassie, M. R. Andersson, and O. Inganäs, *Adv. Funct. Mater.* 2005, **15**, 1665.
[51] S. Admassie, O. Inganäs, W. Mammo, E. Perzon, and M. R. Andersson, *Synth. Met.* 2006, **156**, 614.
[52] P.-L. T. Boudreault, A. Michaud, and M. Leclerc, *Macromolecules Rapid. Commun.* 2007, **28**, 2176.
[53] E. Wang, L. Wang, L. Lan, C. Luo, W. Zhuang, J. Peng, and Y. Cao, *Appl. Phys. Lett.* 2008, **92**, 333007.
[54] N. Blouin, A. Michaud, D. Gendron, S. Wakim, E. Blair, R. Neagu-Plesu, M. Belletete, G. Duricher, Y. Tao, and M. Leclerc, *J. Am. Chem. Soc.* 2008, **130**, 732.
[55] See press release Konarka 2008 (www.konarka.com)
[56] Z. Zhu, D. Waller, R. Gaudiana, M. Morana, D. Mühlbacher, M. Scharber, and C. J. Brabec, *Macromolecules* 2007, **40**, 1981.
[57] P. Coppo, and M. L. Turner, *J. Mater. Chem.* 2005, **15**, 1123.
[58] A. Tsami, T. W. Bunnagel, T. Farrell, M. C. Scharber, S. A. Choulis, C. J. Brabec, and U. Scherf, *J. Mater. Chem.* 2007, **17**, 1353.
[59] M. Morana, M. Wegscheider, A. Bonanni, N. Kopidakis, S. Shaheen, M. C. Scharber, Z. Zhu, D. Waller, R. Gaudiana, and C. J. Brabec, *Adv. Funct. Mater.* 2008, **18**, 1757.
[60] M. C. Scharber, M. Koppe, J. Gao, F. Cordella, M. A. Loi, P. Denk, M. Morana, H.-J. Egelhaaf, K. Forberich, G. Dennler, R. Gaudiana, D. Waller, Z. Zhu, X. Shi, and C. J. Brabec, *Adv. Mater.* 2010, **22**, 367.
[61] M. Lenes, G.-J. A. H. Wetzelaer, F. B. Kooistra, S. C. Veenstra, J. C. Hummelen, and P. W. M. Blom, *Adv. Mater.* 2008, **20**, **11**, 2116–2119.
[62] H. F. Wittmann, R. H. Friend, M. S. Khan, and J. J. Lewis, *J. Chem. Phys.* 1994, **101**, 2693.
[63] V. Marin, E. Holder, R. Hoogenbom, and U. S. Schubert, *Chem. Soc. Rev.* 2007, **36**, 618.
[64] A. J. Sandee, C. K. Williams, N. R. Evans, J. E. Davies, C. E. Boothby, A. Köhler, R. H. Friend, and A. B. Holmes, *J. Am. Chem. Soc.* 2004, **125**, 7041.
[65] B. Liang, L. Wang, Y. Xu, H. Shi, and Y. Cao, *Adv. Funct. Mater.* 2007, **17**, 3580.
[66] A. Kohler, H. F. Wittman, R. H. Friend, M. S. Khan, and J. Lewis, *Synth. Met.* 1996, **77**, 147.
[67] F. Guo, Y.-G. Kim, J. R. Reynolds, and K. S. Schanze, *Chem. Commun.* 2006, **17**, 1887–1889.
[68] J. Gilot, M. M. Wienk, and R. A. J. Janssen, *Nature Mater.* 2007, **6**, 704.
[69] H. Hoppe, and S. Sariciftci, *Adv. Polym. Sci.* 2007, **12**, 121.
[70] F. B. Kooistra, J. Knol, F. Kastenberg, L. M. Popescu, W. J. H. Verhees, J. M. Kroon, and J. C. Hummelen, *Org. Lett.* 2007, **9**, 551.
[71] C. J. Brabec, A. Cravino, D. Meissner, N. S. Sariciftci, T. Fromherz, M. T. Rispens, L. Sanchez, and J. C. Hummelen, *Adv. Funct. Mater.* 2001, **11**, 374.
[72] L. Zheng, Q. Zhou, X. Deng, M. Yuan, G. Yu, and Y. Cao, *J. Phys. Chem. B* 2004, **108**, 11921.

[73] B. C. Thompson, and J. M. J. Frechet, *Angew. Chem. Int. Ed.* 2008, **47**, 58–77.

[74] Y. Kim, S. Cook, S. M. Tuladhar, S. A. Choulis, J. Nelson, J. R. Durrant, D. D. C. Bradley, M. Giles, I. McCulloch, C.-S. Ha, and M. Ree, *Nature Mater.* 2006, **5**, 197–203.

[75] C. H. Woo, B. C. Thompson, B. J. Kim, M. F. Toney, and J. M. J. Frechet, *J. Am. Chem. Soc.* 2008, **130**, 16324–16329.

[76] M. Koppe, C. J. Brabec, S. Heiml, A. Schausberger, W. Duffy, M. Heeney, and I. McCulloch, *Macromolecules* 2009, **42**, **13**, 4661–4666.

[77] P. Schilinsky, U. Asawapirom, U. Scherf, M. Biele, and C. J. Brabec, *Chem. Mater.* 2005, **17**, 2175–2180.

[78] W. L. Ma, J. Y. Kim, K. Lee, and A. J. Heeger, *Macromolecules Rapid Commun.* 2007, **28**, 1776–1780.

[79] H. Hoppe, M. Niggemann, C. Winder, J. Kraut, R. Hiesgen, A. Hinsch, D. Meissner, and N. S. Sariciftci, *Adv. Funct. Mater.* 2004, **14**, 1005.

[80] T. Martens, J. D'Haen, T. Munters, Z. Beelen, L. Goris, J. Manca, M. D'Olieslaeger, D. Vanderzande, L. De Schepper, and R. Andriessen, *Synth. Met.* 2003, **138**, 243–247.

[81] X. Yang, J. K. J. van Duren, R. A. J. Janssen, M. A. J. Michels, and J. Loos, *Macromolecules* 2004, **37**, 2151–2158.

[82] J. Liu, Y. Shi, and Y. Yang, *Adv. Funct. Mater.* 2001, **11**, 6.

[83] F. Zhang, K. G. Jespersen, C. Björström, M. Svensson, M. R. Andersson, V. Sundström, K. Magnusson, E. Moons, A. Yartsev, and O. Inganäs, *Adv. Funct. Mater.* 2006, **16**, 667–674.

[84] C. N. Hoth, P. Schilinsky, S. A. Choulis, and C. J. Brabec, *Nano Lett.* 2008, **8**, 2806–2813.

[85] C. Yin, M. Schubert, S. Bange, B. Stiller, M. Castellani, D. Neher, M. Kumke, and H.-H. Horhold, *J. Phys. Chem. C* 2008, **112**, 14607–14617.

[86] K. Kawano, J. Sakai, M. Yahiro, and C. Adachi, *Sol. Energy Mater. Sol. Cells* 2009, **93**, 4, 514–518.

[87] D. Chirvase, J. Parisi, J. C. Hummelen, and V. Dyakonov, *Nanotechnology* 2004, **15**, 1317.

[88] T. Erb, U. Zhokhavets, G. Gobsch, S. Raleva, B. Stühn, P. Schilinsky, C. Waldauf, and C. J. Brabec, *Adv. Funct. Mater.* 2005, **15**, 1193–1196.

[89] W. L. Ma, C. Yang, X. Gong, K. Lee, and A. J. Heeger, *Adv. Funct. Mater.* 2005, **15**, 1617–1622.

[90] V. D. Mihailetchi, H. Xie, B. de Boer, L. J. A. Koster, and P. W. M. Blom, *Adv. Funct. Mater.* 2006, **16**, 699–708.

[91] P. Schilinsky, C. Waldauf, J. A. Hauch, and C. J. Brabec, *J. Appl. Phys.* 2004, **95**, 2816.

[92] X. Yang, J. Loos, S. C. Veenstra, W. J. H. Verhees, M. M. Wienk, J. M. Kroon, M. A. J. Michels, and R. A. J. Janssen, *Nano Lett.* 2005, **5**, 4, 579–583.

[93] T. J. Savenije, J. E. Kroeze, X. Yang, and J. Loos, Thin *Solid Films* 2006, **511–512**, 2–6.

[94] I.-W. Hwang, C. Soci, D. Moses, Z. Zhu, D. Waller, R. Gaudiana, C. J. Brabec, and A. J. Heeger, *Adv. Mater.* 2007, **19**, 2307–2312.

[95] S. C. Veenstra, W. J. H. Verhees, J. M. Kroon, M. M. Koetse, J. Sweelssen, J. J. A. M. Bastiaansen, H. F. M. Schoo, X. Yang, A. Alexeev, J. Loos, U. S. Schubert, and M. M. Wienk, *Chem. Mater.* 2004, **16**, 2503–2508.

[96] M. M. Mandoc, W. Veurman, J. Sweelssen, M. M. Koetse, and P. W. M. Blom, *Appl. Phys. Lett.* 2007, **91**, 073518.

[97] T. Kietzke, R. Yee C. Shin, D. Ayuk Mbi Egbe, Z.-K. Chen, and A. Sellinger, *Macromolecules* 2007, **40**, 4424–4428.

[98] G. Li, V. Shrotriya, J. Huang, Y. Yao, T. Moriarty, K. Emery, and Y. Yang, *Nature Mater.* 2005, **4**, 864.

[99] H. Hoppe, and N. S. Sariciftci, *J. Mater. Chem.* 2006, **16**, 45–61.

[100] H. J. Snaith, A. C. Arias, A. C. Morteani, C. Silva, and R. H. Friend, *Nano Lett.* 2002, **2**, 12, 1353–1357.

[101] M. A. Loi, S. Toffanin, M. Muccini, M. Forster, U. Scherf, and M. C. Scharber, *Adv. Funct. Mater.* 2007, **17**, 2111–2116.

[102] J. J. Benson-Smith, L. Goris, K. Vandewal, K. Haenen, J. V. Manca, D. Vanderzande, D. D. C. Bradley, and J. Nelson, *Adv. Funct. Mater.* 2007, **17**, 451–457.

[103] D. Veldman, Ö. Ipek, S. C. J. Meskers, J. Sweelssen, M. M. Koetse, S. C. Veenstra, J. M. Kroon, S. S. van Bavel, J. Loos, and R. A. J. Janssen, *J. Am. Chem. Soc.* 2008, **130**, 24, 7721–7735.

[104] E. Moons, *J. Phys. Condens. Matter* 2002, **14**, 12235–12260.

[105] H. Hoppe, T. Glatzel, M. Niggemann, W. Schwinger, F. Schaeffler, A. Hinsch, M. C. Lux-Steiner, and N. S. Sariciftci, *Thin Solid Films* 2006, **511–512**, 587–592.

[106] C.-P. Chen, S.-H. Chan, T.-C. Chao, C. Ting, and B.-T. Ko, *J. Am. Chem. Soc.* 2008, **130**, 38, 12828–12833.

[107] C. R. McNeill, B. Watts, L. Thomsen, W. J. Belcher, A. L. David Kilcoyne, N. C. Greenham, and P. C. Dastoor, *Small* 2006, **2**, 12, 1432–1435.

[108] D. L. Gerrard, and H. J. Bowley, *Anal. Chem.* 1988, **60**, 12, 368–377.

[109] A. F. Nogueira, B. S. Lomba, M. A. Soto-Oviedo, C. Roque, D. Correia, P. Corio, C. A. Furtado, and I. A. Hmmelgen, *J. Phys. Chem. C* 2007, **111**, 49.

[110] E. Klimov, W. Li, X. Yang, G. G. Hoffmann, and J. Loos, *Macromolecules* 2006, **39**, 4493–4496.

[111] J. Chappel, D. G. Lidzey, P. C. Jukes, A. M. Higgins, R. L. Thompson, S. O'Connor, I. Grizzi, R. Fletcher, J. O'Brien, M. Geoghegan, and R. A. L. Jones, *Nature Mater.* 2003, **2**, 616.

[112] A. Cadby, R. Dean, A. M. Fox, R. A. L. Jones, and D. G. Lidzey, *Nano Lett.* 2005, **5**, 11, 2232–2237.

[113] C. R. McNeill, H. Frohne, J. L. Holdsworth, and P. C. Dastoor, *Nano Lett.* 2004, **4**, 12, 2503–2507.

[114] M. Chiesa, L. Bürgi, J.-S. Kim, R. Shikler, R. H. Friend, and H. Sirringhaus, *Nano Lett.* 2005, **5**, 4, 559–563.

[115] H. Hoppe, T. Glatzel, M. Niggemann, A. Hinsch, M. Ch. Lux-Steiner, and N. S. Sariciftci, *Nano Lett.* 2005, **5**, 2, 269–274.

[116] V. Palermo, M. B. J. Otten, A. Liscio, E. Schwartz, P. A. J. de Witte, M. A. Castriciano, M. M. Wienk, F. Nolde, G. De Luca, J. J. L. M. Cornelissen, R. A. J. Janssen, K. Müllen, A. E. Rowan, R. J. M. Nolte, and P. Samori, *J. Am. Chem. Soc.* 2008, **130**, 44, 14605–14614.

[117] D. C. Coffey, and D. S. Ginger, *Nature Mater.* 2006, **5**, 735.

[118] D. C. Coffey, O. G. Reid, D. B. Rodovsky, G. P. Bartholomew, and D. S. Ginger, *Nano Lett.* 2007, **7**, **3**, 738–744.

[119] J. S. Moon, J. K. Lee, S. Cho, J. Byun, and A. J. Heeger, *Nano Lett.* 2009, **9**, 1, 230–234.

[120] A. Bernasik, J. Rysz, A. Budkowski, K. Kowalski, J. Camra, and J. Jedlinaski, *Macromolecules Rapid Commun.* 2001, **22**, 11.

[121] C. W. T. Bulle-Lieuwma, W. J. H. van Gennip, J. K. J. van Duren, P. Jonkheijm, R. A. J. Janssen, and J. W. Niemantsverdriet, *Appl. Surface Sci.* 2003, **203–204**, 547–550.

[122] C. M Björström, A. Bernasik, J. Rysz, A. Budkowski, S. Nilsson, M. Svensson, M. R. Andersson, K. O Magnusson, and E. Moons, *J. Phys.: Condens. Matter* 2005, **17**, L529–L534.

[123] J. K. J. van Duren, X. Yang, J. Loos, C. W. T. Bulle-Lieuwma, A. B. Sieval, J. C. Hummelen, and R. A. J. Janssen, *Adv. Funct. Mater.* 2004, **14**, 5, 425–434.

[124] X. Yang, and J. Loos, *Macromolecules* 2007, **40**, 5, 1353–1362.

[125] B. V. Andersson, A. Herland, S. Masich, and O. Inganäs, *Nano Lett.* 2009, **9**, 2, 853–855.

[126] S. S. van Bavel, E. Sourty, G. de With, and J. Loos, *Nano Lett.* 2009, **9**, 2, 507–513.

[127] P. Peumans, A. Yakimov, and S. R. Forrest, *J. Appl. Phys.* 2003, **93**, 7, 3693.

[128] H. H. P. Gommans, M. Kemerink, J. M. Kramer, and R. A. J. Janssen, *Appl. Phys. Lett.* 2005, **87**, 122104.

[129] C. M. Martin, V.M. Burlakov, and H.E. Assender, *Solar Energy Mater. Solar Cells* 2006, **90**, 900–915.

[130] R. A. Marsh, C. Groves, and N. C. Greenham, *J. Appl. Phys.* 2007, **101**, 083509.

[131] P. K. Watkins, A. B. Walker, and G. L. B. Verschoor, *Nano Lett.* 2005, **5**, 9, 1814–1818.

[132] F. Yang, and S. R. Forrest, *ACS Nano* 2008, **2**, 5, 1022–1032.

[133] H. Bässler, *Phys. Status Solidi B* 1993, **175**, 15.
[134] J. M. Frost, F. Cheynis, S. M. Tuladhar, and J. Nelson, *Nano Lett.* 2006, **6**, 8, 1674–1681.
[135] D. Raabe, *Annu. Rev. Mater. Res.* 2002, **32**, 36–76.
[136] G. Li, Y. Yao, H. Yang. V. Shrotriya, G. Yang, and Y. Yang, *Adv. Funct. Mater.* 2007, **17**, 1636.
[137] V. D. Mihailetchi, H. Xie, B. de Boer, L. M. Popescu, J. C. Hummelen, P. W. M. Blom, and J. J. A. Koster, *Appl. Phys. Lett.* 2006, **89**, 012107.
[138] J. Peet, C. Soci, R. C. Coffin, T. Q. Nguyen, A. Mihailovsky, D. Moses, and G. C. Bazan, *Appl. Phys. Lett.* 2006, **89**, 252105.
[139] J. K. Lee, W. L. Ma, C. J. Brabec, J. Yuen, J. S. Moon, J. Y. Kim, K. Lee, G. C. Bazan, and A. J. Heeger, *J. Am. Chem. Soc.* 2008, **130**, 11, 3619–3623.
[140] K. Landfester, *Macromolecules Rapid. Commun.* 2001, **22**, 896.
[141] K. Landfester, *Adv. Mater.* 2001, **13**, 765.
[142] K. Landfester, R. Montenegro, U. Scherf, R. Günter, U. Asawapirom, S. Patil, D. Neher, and T. Kietzke, *Adv. Mater.* 2002, **14**, 651.
[143] T. Piok, S. Gamerith, C. Gadermaier, H. Plank, F. P. Wenzl, S. Patil, R. Montenegro, T. Kietke, D. Neher, U. Scherf, L. Landfester, and E. J. W. List, *Adv. Mater.* 2003, **15**, 800.
[144] C. J. Brabec, *Public annual report of the BMBF project EKOS*, 03N2023A-E, Jülich, 2004.
[145] M. M. Bouman, E. E. Havinga, R. A. J. Janssen, and E. W. Meijer, *Mol. Cryst. Liq. Cryst. Sci. Technol. A* 1994, **256**, 439.
[146] A. Moulé, and K. Meehrholz, *Adv. Mater.* 2008, **20**, 2, 240–245.
[147] E. Mena-Osteritz, A. Meyer, B. M. W. Langeveld-Voss, R. A. J. Janssen, E. W. Meijer, and P. Bäuerle, *Angew. Chem.* 2000, **112**, 2792.
[148] S. Malik, and A. K. Nandi, *J. Polym. Sci. Part B: Polym. Phys.* 2002, **40**, 2073.
[149] L. Y. Park, A. M. Munro, and D. S. Ginger, *J. Am. Chem. Soc.* 2008, **130**, 47, 15916–15926.
[150] C. J. Brabec, A. Cravino, D. Meisner, N. S. Sariciftci, M. T. Rispens, L. Sanchez, J. C. Hummelen, and T. Fromherz, *Thin Solid Films* 2002, **403–404**, 368.
[151] L. J. Brillson, *Surf. Sci. Rep.* 1982, **2**, 145
[152] V. D. Mihailetchi, P. W. M. Blom, J. C. Hummelen, and M. T. Rispens, *J. Appl. Phys.* 2003, **94**, 6849.
[153] A. Petr, F. Zhang, H. Petseng, M. Knupfer, and L. Dunsch, *Chem. Phys. Lett.* 2004, **385**, 140.
[154] H. Frohne, S. E. Shaheen, C. J. Brabec, D. C. Muller, N. S. Sariciftci, and K. Meerholz, *Chem. Phys. Chem.* 2002, **3**, 795.
[155] M. Glatthaar, M. Niggemann, B. Zimmermann, P. Lewer, M. Riede, A. Hinsch, and J. Luther, *Thin Solid Films* 2005, **491**, 298.
[156] Y. Sahin, S. Alem, R. de Bettignies, and J.-M. Nunzi, *Thin Solid Films* 2005, **476**, 340.
[157] C. Waldauf, M. Morana, P. Denk, P. Schilinsky, K. Coakley, S. A. Choulis, and C. J. Brabec, *Appl. Phys. Lett.* 2006, **89**, 233517.
[158] G. Li, C.-W. Chu, V. Shrotriya, J. Huang, and Y. Yang, *Appl. Phys. Lett.* 2006, **88**, 253503.
[159] M. S. White, D. C. Olson, S. E. Shaheen, N. Kopidakis, and D. S. Ginley, *Appl. Phys. Lett.* 2006, **89**, 143517.
[160] Z. Xu, L.-M. Chen, G. Yang, C.-H. Huang, J. Hou, Y. Wu, G. Li, C.-S. Hsu, and Y. Yang, *Adv. Funct. Mater.* 2009, **19**, 8, 1227–1234.
[161] J. Y. Kim, S. H. Kim, H. H. Lee, K. Lee, W. Ma, X. Gong, and A. J. Heeger, *Adv. Mater.* 2006, **18**, 572.
[162] S. E. Shaheen, C. J. Brabec, N. S. Sariciftci, F. Padinger, T. Fromberz, and J. C. Hummelen, *Appl. Phys. Lett.* 2001, **78**, 841.
[163] Y. Cao, G. Yu, C. Zhang, R. Menon, and A. J. Heeger, *Synth. Met.* 1997, **87**, 171.
[164] S. A. Carter, M. Angelopoulos, S. Karg, P. J. Brock, and J. C. Scott, *Appl. Phys. Lett.* 1997, **70**, 2067.
[165] J. C. Carter, I. Grizzi, S. K. Heeks, D. J. Lacey, S. G. Latham, P. G. May, O.R.d.l. Panos, K. Pichler, C.R. Towns, and H. F. Wittman, *Appl. Phys. Lett.* 1997, **71**, 34.

[166] C. Y. Li, T. C. Wen, and T-F. Guo, *J. Mater. Chem.* 2008, **18**, 4478.
[167] A. K. K. Kyaw, X. W. Sun, C. Y. Jiang, G. Q. Lo, D. W. Zhao, and D. L. Kwong, *Appl. Phys. Lett.* 2008, **93**, 221107.
[168] V. Shrotriya, G. Li, Y. Y. Yao, C. W. Chu, and Y. Yang, *Appl. Phys. Lett.* 2006, **88**, 073508.
[169] C. Tao, S. Ruan, G. Xie, X. Kong, L. Shen, F. Meng, C. Liu, X. Zhang, W. Dong, and W. Chen, *Appl. Phys. Lett.* 2009, **94**, 043311.
[170] M. Y. Chan, C. S. Lee, S. L. Lai, M. K. Fung, F. L. Wong, H. Y. Sun, K. M. Lau, and S. T. Lee, *Appl. Phys. Lett.* 2006, **100**, 094506.
[171] S. Han, W. S. Shin, M. Seo, D. Gupta, S.-J. Moon, and S. Yoo, *Org. Electron.* 2009, **10**, 5, 791–797.
[172] S. K. M. Jonsson, E. Carlegrim, F. Zhang, W. R. Salaneck, and M. Fahlman, *Jpn. J. Appl. Phys. Part 1* 2005, **44**, 3695.
[173] T. M. Brown, R. H. Friend, I. S. Millard, D. J. Lacey, J. H. Burroughes, and F. Cacialli, *Appl. Phys. Lett.* 2000, **77**, 3096.
[174] L. S. Hung, C. W. Tang, and M. G. Mason, *Appl. Phys. Lett.* 1997, **70**, 152.
[175] S. E. Shaheen, G. E. Jabbour, M. M. Morrell, Y. Kawabe, B. Kippelen, N. Peyghambarian, M. F. Nabor, R. Schlaf, E. A. Mash, and N. R. Armstron, *J. Appl. Phys.* 1998, **84**, 2324.
[176] M. Matsumura, and K. Furukawa, Y. Jinde, *Thin Solid Films* 1998, **331**, 96.
[177] R. Steim, S. A. Choulis, P. Schilinsky, and C. J. Brabec, *Appl. Phys. Lett.* 2008, **92**, 093303.
[178] K. Lee, J. Y. Kim, S. H. Park, S. H. Kim, S. Cho, and A. J. Heeger, *Adv. Mater.* 2007, **19**, 2445.
[179] J. Gilot, I. Barbu, M. M. Wienk, and R. A. J. Janssen, *Appl. Phys. Lett.* 2007, **91**, 113520.
[180] F. Zhang, M. Ceder, and O. Inganäs, *Adv. Mater.* 2007, **19**, 1835.
[181] H. L. Yip, S. K. Hau, N. S. Baek, H. Ma, and A. K. Y. Jen, *Adv. Mater.* 2008, **20**, 2376.
[182] S. K. Hau, H. L. Yip, O. Acton, N. S. Baek, H. Ma, and A. K. Y. Jen, *J. Mater. Chem.* 2008, **18**, 5113.
[183] J. S. Kim, J. H. Park, J. H. Lee, J. Jo, D. Kim, and K. Cho, *Appl. Phys. Lett.* 2007, **91**, 112111–3.
[184] S. Khodabakhsh, B. M. Sanderson, J. Nelson, and T. S. Jones, *Adv. Funct. Mater.* 2006, **16**, 95–100.
[185] S. Shockley, and H. J. Queisser, *J. Appl. Phys.* 1961, **32**, 510.
[186] G. Dennler, N. S. Sariciftci, C. J. Brabec, *Conjugated Polymer-Based Organic Solar Cells*, Hadziioannou and Malliaras (eds.) *Semiconducting Polymers*, 2nd edn, Wiley-VCH Verlag GmbH, Weinheim, 2006.
[187] (a) A. Cravino, *Appl. Phys. Lett.* 2007, **91**, 243502; (b) K. Vandewal, A. Gadisa, W. D. Oosterbaan, S. Bertho, F. Banishoeib, I. V. Severen, L. Lutsen, T. J. Cleij, D. Vanderzande, and J. V. Manca, *Adv. Funct. Mater.* 2008, **18**, 2064.
[188] P. Würfel, *Physics of Solar Cells*, Weinheim, Germany, Wiley VCH, 2004.
[189] A. De Vos, *J. Phys. D: Appl. Phys.* 1980, **13**, 839.
[190] (a) M. A. Green, K. Emery, Y. Hisikawa, and W. Warta, *Prog. Photovolt.: Res. Appl.* 2007, **15**, 425; (b) M. Wanlass, P. Ahrenkiel, D. Albin, J. Carapella, A. Duda, K. Emery, D. Firedman, J. Geisz, K. Jones, A. Kibbler, J. Kiehl, S. Kurtz, W. McMahon, T. Moriarty, J. Olson, A. Ptak, M. Romero, and S. Ward, *4th World Conference on Photovoltaic Energy Conversion (WCEP-4)*, Hawaii, May 2006, 729.
[191] P. W. Blom, V. D. Mihailetchi, L. J. A. Koster, and D. E. Markov, *Adv. Mater.* 2007, **19**, 1551.
[192] J. Grieschner, J. Cornil, and H.-J. Egelhaaf, *Adv. Mater.* 2007, **19**, 173–191.
[193] G. Dennler, M. C. Scharber, T. Ameri, P. Denk, K. Forberich, C. Waldauf, and C. J. Brabec, *Adv. Mater.* 2008, **20**, 579.
[194] M. Hiramoto, and M. Suezaki, M. Yokoyama, *Chem. Lett.* 1990, 327.
[195] A. Yakimov and S. R. Forrest, *Appl. Phys. Lett.* 2002, **80**, 1667.
[196] J. Xue, S. Ushida, B. P. Rand, and S. T. Forrest, *Appl. Phys. Lett.* 2004, **85**, 5757.

[197] B. Maennig, J. Drechsel, D. Gebeyehu, P. Simon, F. Kozlowski, A. Werner, F. Li, S. Grundmann, S. Sonntag, M. Koch, K. Leo, M. Pfeiffer, H. Hoppe, D. Meissner, N. S. Sariciftci, I. Riedel, V. Dyakonov, and J. Parisi, *Appl. Phys. A* 2004, **79**, 1.
[198] J. Drechsel, B. Maennig, K. Kozlowski, M. Pfeiffer, and K. Leo, *Appl. Phys. Lett.* 2005, **86**, 244102.
[199] K. Kawano, N. Ito, T. Nishimori, and J. Sakai, *Appl. Phys. Lett.* 2006, **88**, 73514.
[200] G. Dennler, H.-J. Prall, R. Koeppe, M. Egginger, R. Autengruber, and N. S. Sariciftci, *Appl. Phys. Lett.* 2006, **89**, 73502.
[201] (a) A. Colsmann, J. Junge, C. Kayser, and U. Lemmer, *Appl. Phys. Lett.* 2006, **89**, 203506; (b) A. G. F. Janssen, T. Riedl, S. Hamwi, H.-H. Johannes, and W. Kowalsky, *Appl. Phys. Lett.* 2007, **91**, 073519.
[202] A. Hadipour, B. de Boer, J. Wildeman, F. B. Kooistra, J. C. Hummelen, M. G. R. Turbiez, M. M. Wienk, R. A. J. Janssen, and P. W. M. Blom, *Adv. Funct. Mater.* 2006, **16**, 1897.
[203] J. Gilot, M. M. Wienk, and R. A. J. Janssen, *Appl. Phys. Lett.* 2007, **90**, 143512.
[204] (a) C. Pacholski, A. Kornowski, and H. Weller, *Angew. Chem. Int. Ed.* 2002, **41**, 1188; (b) W. J. E. Beek, M. M. Wienk, M. Kemerink, X. Yang, and R. A. J. Janssen, *J. Phys. Chem. B* 2005, **109**, 9505.
[205] F. Verbakel, S. C. J. Meskers, and R. A. J. Janssen, *Appl. Phys. Lett.* 2006, **89**, 102103.
[206] J. Y. Kim, K. Lee, N. E. Coates, D. Moses, T.-Q. Nguyen, M. Dante, and A. J. Heeger, *Science* 2007, **317**, 222.
[207] V. Shrotriya, E. Hsing-En, G. Li, Y. Yao, and Y. Yang, *Appl. Phys. Lett.* 2006, **88**, 64104.
[208] A. Hadipour, B. de Boer, and P. W. M. Blom, *J. Appl. Phys.* 2007, **102**, 074506.
[209] N.-K. Persson, and O. Inganäs, *Sol. Energy. Mater. Sol. Cells* 2006, **90**, 3491.
[210] K. Tvingstedt, V. Andersson, F. Zhang, and O. Inganas, *Appl. Phys. Lett.* 2007, **91**, 123514.
[211] G. Dennler, K. Forberich, M. C. Scharber, C. J. Brabec, I. Tomis, K. Hingerl, and T. Fromherz, *J. Appl. Phys.* 2007, **102**, 054516.
[212] C. Zhang, S. W. Tong, C. Jiang, E. T. Kang, D. S. H. Chan, and C. Zhu, *Appl. Phys. Lett.* 2008, **92**, 083310.
[213] A. L. Dawar, and J. C. Joshi, *J. Mater. Sci.* 1984, **19**, 1.
[214] R. F. Bunshah (ed.) *Handbook of Deposition Technologies for Films and Coatings: Science, Technology and Applications*, 2nd edn, Materials Science and Process Technology Series. Park Ridge, N.J.: Noyes Publications, 1994. ISBN 0815513372.
[215] C. J. Brinker, and G. W. Scherer, *Sol-Gel Science: The Physics and Chemistry of Sol-Gel Processing*, Academic Press, 1st edn, 12 May 1990, ISBN-10: 0121349705.
[216] C. Guillen, and J. Herrero, *Thin Solid Films*, 2005, **480–481**, 19.
[217] B. Szyszka, *Vakuum in Forschung und Praxis*, 2001, **Nr. 1**, 38–45.
[218] M. F. A. M. van Hest, M.S. Dabney, J.D. Perkins, and D.S. Ginley, *Thin Solid Films* 2006, **496**, 70–74.
[219] M. F. A. M. van Hest, M. S. Dabney, J. D. Perkins, D. S. Ginley, and M. P. Taylor, *Appl. Phys. Lett.* 2005, **87**, 032111.
[220] T. Minami, *Semicond. Sci. Technol.* 2005, **20**, S35–S44.
[221] M. Hilgendorff, Dissertation 1998, Julius-Maximilians-Universität Würzburg.
[222] D. S. Ginley, and C. Bright, *MRS Bull.* 2000, **25**, 8, 15–18.
[223] M. A. Marquardt, N.A. Ashmore, and D. P. Cann, *Thin Solid Films* 2006, **496**, 146–156.
[224] T. Minami, *MRS Bull.* 2000, **25**, 8, 38–44.
[225] J. Owen, M. S. Son, and K.-H. Yoo, B. D. Ahn and S. Y. Lee, *Appl. Phys. Lett.* 2007, **90**, 033512.
[226] X. Jiang, F. L. Wong, M. K. Fung, and S. T. Lee, *Appl. Phys. Lett.* 2003, **83**, 9, 1875–1877.
[227] B. Yoo, K. Kim, S. H. Lee, W. M. Kim, and N.-G. Park, *Sol. Energy Mater. Sol. Cells* 2008, **92**, 873–877.
[228] K.-H. Choi, H.-J. Nam, J.-A. Jeong, S.-W. Cho, H.-K. Kim, J.-W. Kang, D.-G. Kim, and W.-J. Cho, *Appl. Phys. Lett.* 2008, **92**, 223302.

[229] S. Y. Ryu, J. H. Noh, B. H. Hwang, C. S. Kim, S. J. Jo, J. T. Kim, H. S. Hwang, H. K. Baik, H. S. Jeong, C. H. Lee, S. Y. Song, S. H. Choi, and S. Y. Park, *Appl. Phys. Lett.* 2008, **92**, 023306.
[230] D. R. Sahu, and J.-L. Huang, *Thin Solid Films* 2007, **516**, 208–211.
[231] M. Fahland, T. Vogt, W. Schoenberger, and N. Schiller, *Thin Solid Films* 2008, **516**, 5777–5780.
[232] C. Guillen, and J. Herrero, *Sol. Energy Mater. Sol. Cells* 2008, **92**, 938–941.
[233] M. Bender, W. Seelig, C. Daube, H. Frankenberger, B. Ocker, and J. Stollenwerk, *Thin Solid Films* 1998, **326**, 67–71.
[234] D. R. Sahu, and J.-L. Huang, *Thin Solid Films* 2006, **515**, 876–879.
[235] H. Kim, J. S. Horwitz, W. H. Kim, A. J. Mäkinen, Z. H. Kafafi, and D. B. Chrisey, *Thin Solid Films* 2002, **420–421**, 539–543.
[236] S. Kobayakawa, Y. Tanaka, and A. Ide-Ektessabi, *Nucl. Instrum. Methods Phys. Res., Sect. B* 2006, **249**, 536–639.
[237] E. Elangoran, and K. Ramamurthi, *Cryst. Res. Technol.* 2003, **38**, 9, 779–784.
[238] A. Cheknane, B. Benyoucef, J.-P. Charles, R. Zerdoum, and M. Trari, *Sol. Energy Mater. Sol. Cells* 2005, **87**, 557–565.
[239] A. Götzberger, and B. Voß, J. Knobloch, *Sonnenenergie: Photovoltaik, Teubner Studienbücher*, 1997, 2nd edn.
[240] C. E. Hendriks, P. J. Smith, J. Perelaer, A. M. J. van den Berg, and U. S. Schubert, *Adv. Funct. Mater.* 2008, **18**, 1031–1038.
[241] F. C. Krebs, *Sol. Energy Mater. Sol. Cells* 2008, **92**, 715–726.
[242] M.-G. Kang, M.-S. Kim, J. Kim, and L. J. Guo, *Adv. Mater.* 2008, **20**, 4408–4413.
[243] K. Tvingstedt, and O. Inganäs, *Adv. Mater.* 2007, **19**, 2893–2897.
[244] A. M. Nardes, R. A. J. Janssen, and M. Kemerink, *Adv. Funct. Mater.* 2008, **18**, 865–871.
[245] S. Admassie, F. Zhang, A. G. Manoj, M. Svensson, M. R. Andersson, and O. Inganäs, *Sol. Energy Mater. Sol. Cells* 2006, **90**, 133–141.
[246] Y.-F. Lim, S. Lee, D. J. Herman, M. T. Lloyd, J. E. Anthony, and G. G. Malliaras, *Appl. Phys. Lett.* 2008, **93**, 193301.
[247] T. Aernouts, P. Vanlaeke, W. Geens, J. Poortmans, P. Heremans, S. Borghs, R. Mertens, R. Andriessen, and L. Leenders, *Thin Solid Films* 2004, **451–452**, 22–25.
[248] K. Fehse, G. Schwartz, K. Walzer, and K. Leo, *J. Appl. Phys.* 2007, **101**, 124509.
[249] http://www2.zipperling.de/Products/Pani/Welcome.de.html.
[250] E. Kymakis, E. Stratakis, and E. Koudoumas, *Thin Solid Films* 2007, **515**, 8598–8600.
[251] A. Du Pasquier, H. E. Unalan, A. Kanwal, S. Miller, and M. Chhowalla, *Appl. Phys. Lett.* 2005, **87**, 203511.
[252] X. Wang, L. Zhi, N. Tsao, Z. Tomovic, and J. Li, K. Müllen, *Angew. Chem.* 2008, **120**, 1–4.
[253] H. Peng, *J. Am. Chem. Soc.* 2008, **140**, 42–43.
[254] R. Jackson, B. Domercq, R. Jain, B. Kippelen, and S. Graham, *Adv. Funct. Mater.* 2008, **18**, 1–7.
[255] S. Chaudhary, H. Lu, A. M. Müller, C. J. Bardeen, and M. Ozkan, *Nano Lett.* 2007, **7**, 1973–1979.
[256] J. Ederth, P. Heszler, A. Hultaker, G. A. Niklasson, and C. G. Granqvist, *Thin Solid Films* 2003, **445**, 199–206.
[257] M. Gross, A. Winnacker, and P. J. Wellmann, *Thin Solid Films* 2007, **515**, 8567–8572.
[258] J.-Y. Lee, S. T. Connor, Y. Cui, and P. Peumans, *Nano Lett.* 2008, **8**, 2, 689–692.
[259] H.-L. Yip, S. K. Hau, N. S. Baek, and A. K.-Y. Jen, *Appl. Phys. Lett.* 2008, **92**, 193313.
[260] C. Tao, S. Ruan, X. Zhang, G. Xie, L. Shen, X. Kong, W. Dong, C. Liu, and W. Chen, *Appl. Phys. Lett.* 2008, **93**, 193307.
[261] M. O. Reese, A. J. Morfa, M. S. White, N. Kopidakis, S. E. Shaheen, G. Rumbles, and D. S. Ginley, *Sol. Energy Mater. Sol. Cells* 2008, **92**, 746–752.
[262] D. Kim, S. Jeong, H. Shin, Y. Xia, and J. Moon, *Adv. Mater.* 2008, **20**, 16, 3084–3089.
[263] S. Jeong, K. Woo, D. Kim, S. Lim, J. S. Kim, H. Shin, Y. Xia, and J. Moon, *Adv. Funct. Mater.* 2008, **18**, 679–686.

[264] T. H. J. van Osch, J. Perelaer, A. W. M. de Laat, and U. S. Schubert, *Adv. Mater.* 2008, **20**, 343–345.
[265] C. J. Brabec, F. Padinger, J. C. Hummelen, R. A. Janssen, and N. S. Sariciftci, *Synth. Met.* 1999, **102**, 861.
[266] F. Padinger, C. J. Brabec, T. Fromherz, J. C. Hummelen, and N. S. Sariciftci, *Optoelectron. Rev.* 2000, **8**, 4, 280.
[267] P. Schilinsky, C. Waldauf, and C. J. Brabec, *Adv. Funct. Mater.* 2006, **16**, 1669.
[268] C. Waldauf, M. C. Scharber, P. Schilinsky, J. A. Hauch, and C. J. Brabec, *J. Appl. Phys.* 2006, **99**, 104503.
[269] C. N. Hoth, S. A. Choulis, P. Schilinsky, and C. J. Brabec, *Adv. Mater.* 2007, **19**, 3973.
[270] T. Ishikawa, M. Nakamura, K. Fujita, and T. Tsutsui, *Appl. Phys. Lett.* 2004, **84**, 2424.
[271] X. L. Mo, T. Mizokuro, H. Mochizuki, N. Tanigaki, and T. Hiraga, *Jpn. J. Appl. Phys. Part 1* 2005, **44**, 656.
[272] D. Vak, S. Kim, J. Jo, S. Oh, S. Na, J. Kim, and D. Kim, *Appl. Phys. Lett.* 2007, **91**, 081102.
[273] R. Green, A. Morfa, A. J. Ferguson, N. Kopidakis, G. Rumbles, and S. E. Shaheen, *Appl. Phys. Lett.* 2008, **92**, 033301.
[274] K. X. Steirer, M. O. Reese, B. L. Rupert, N. Kopidakis, D. C. Olson, R. T. Collins, and D. S. Ginley, *Sol. Energy Mater. Sol. Cells* 2009, **93**, 447–453.
[275] C. N. Hoth, R. Steim, P. Schilinsky, S. A. Choulis, S. F. Tedde, O. Hayden, and C. J. Brabec, *Org. Electron.* 2009, **10**, 587–593.
[276] D. R. Gamota, P. Brazis, K. Kalyanasundaram, and J. Zhang, *Printed Organic and Molecular Electronics*, Kluwer Academic Publishers 2004, ed. T. Claypole, 320–322.
[277] P. Calvert, *Chem. Mater.* 2001, **13**, 3299–3305.
[278] V. G. Shah, and D. B. Wallace, *Proc. IMAPS 37th Annual International Symp. on Microelectronics*, Long Beach, CA, 14–18th Nov. 2004, p1.
[279] V. Marin, E. Holder, M. M. Wienk, E. Tekin, D. Kozodaev, and U. S. Schubert, *Macromolecules Rapid Commun.* 2005, **26**, 319–324.
[280] T. Aernouts, T. Aleksandrov, C. Girotto, J. Genoe, and J. Poortmans, *Appl. Phys. Lett.* 2008, **92**, 033306.
[281] Z. Nie, and E. Kumacheva, *Nature Mater.* 2008, **7**, 277.
[282] M. Tuomikoski, and R. Suhonen, *Proc. of TPE06*, 2nd Int. Symp. Techn. for Polym. Electronics, Rudolstadt 2006, **83**.
[283] X. Yin, and S. Kumar, *Chem. Eng. Sci.* 2006, **61**, 1142–1152.
[284] J. M. Ding, A. De la Fuente Vornbrock, C. Ting, and V. Subramanian, *Sol. Energy Mater. Sol. Cells* 2009, **93**, 459–464.
[285] K. Schultheis, L. Blankenburg, S. Sensfuss, and M. Schrödner, *Proc. of the Cintelliq Conference*, Frankfurt, 2007.
[286] M. Schrödner, K. Schultheis, L. Blankenburg, H. Schache, and S. Sensfuss, *Int. Symp. TPE08*, Rudolstadt, 2008.
[287] L. Blankenburg, K. Schultheis, H. Schache, S. Sensfuss, and M. Schrödner, *Sol. Energy Mater. Sol. Cells* 2009, **93**, 476–483.
[288] C. N. Hoth, S. Balasubramanian, P. Schilinsky, E. Zeira, and C. J. Brabec, unpublished results, 2008.
[289] http://www.neubers.de/vmchk/Solar-Taschen/View-all-products.html.

9 Third-Generation Solar Cells

Gavin Conibeer
School of Photovoltaic and Renewable Energy Engineering, University of New South Wales, Australia

9.1 INTRODUCTION

The two most important power-loss mechanisms in single-bandgap solar cells are the inability to absorb photons with energy less than the bandgap (5 in Figure 9.1) and thermalisation of photon energies exceeding the bandgap (1 in Figure 9.1). These two mechanisms alone amount to the loss of about half of the incident solar energy in solar cell conversion to electricity. As discussed in Chapter 2, Section 2.3.2, the maximum thermodynamic efficiency for the conversion of unconcentrated solar irradiance into electrical free energy in the radiative limit assuming detailed balance and a single threshold absorber was calculated by Shockley and Queisser in 1961 [Shockley and Queisser, 1961] to be about 31%.

Carrier thermalisation and the nonabsorption of subbandgap photons account for the loss of about half of the incident solar energy in solar cell conversion to electricity. Multiple energy threshold approaches are required to tackle the lost energy and thus to achieve the higher efficiency potential of third-generation PV goals. Such approaches do not in fact disprove the validity of the Shockley–Queisser limit, rather they avoid it by the exploitation of more than one energy level for which the limit does not apply. The limit that does apply is the thermodynamic one, shown in Figure 9.2, of 67% for unconcentrated irradiance. There are three types of approaches for applying multiple energy levels [Nelson, 2003]: (a) increasing the number of energy levels; (b) multiple carrier-pair generation per high-energy photon or single carrier-pair generation with multiple low-energy photons; and (c) capturing carriers before thermalisation. Of these, tandem cells, an implementation of strategy (a), are the only ones that have, as yet, been realised with efficiencies exceeding the Shockley–Queisser limit.

Third-generation approaches aim to achieve high efficiency for photovoltaic (PV) devices by circumventing the Shockley–Queisser limit for single-bandgap devices. The concept is to do this with only a small increase in areal costs and hence reduce the cost per Watt peak [Green, 2003]. Also, in common with the silicon- and CZTS-based second generation thin-film technologies, these will use abundant and nontoxic materials. Thus, these "third-generation" technologies will be compatible with large-scale implementation of photovoltaics. The approach differs from "first-generation" fabrication of high-quality and hence low defect single-crystal photovoltaic devices, which have high efficiencies that are approaching the limiting efficiencies for single-bandgap devices but that use energy- and time-intensive techniques. Thin-film or second-generation approaches seek to use foreign substrates to reduce the amount of absorber material used and low-cost vapour phase deposition to reduce the thermal budget. Hence, the overall cost of thin-film devices is lower than single-crystal

Solar Cell Materials: Developing Technologies, First Edition. Edited by Gavin Conibeer and Arthur Willoughby.

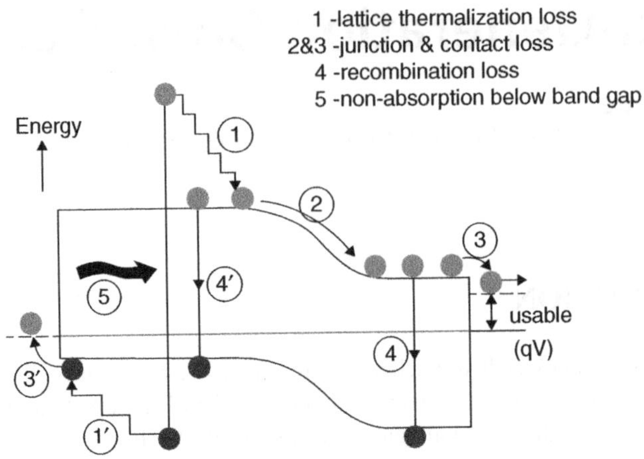

Figure 9.1 Loss processes in a standard solar cell: (1) – lattice thermalisation loss of carriers generated by above bandgap photons; (2) and (3) – junction and contact voltage losses; (4) – recombination loss (radiative recombination is unavoidable); and (5) – nonabsorption of photons below the bandgap.

ones, but the efficiencies are also lower because the nonmelt processing of materials leads to a much larger defect density. Third-generation approaches aim to enhance the efficiency of thin film devices through the use of multiple energy levels. Hence, third generation aims to decrease costs to well below the US$0.5–1/W level of second generation towards US$0.1–$0.2/W or better, by significantly increasing efficiencies, but maintaining the economic and environmental cost advantages of thin-film deposition techniques (see Figure 9.2 of the three PV generations) [Green, 2003].

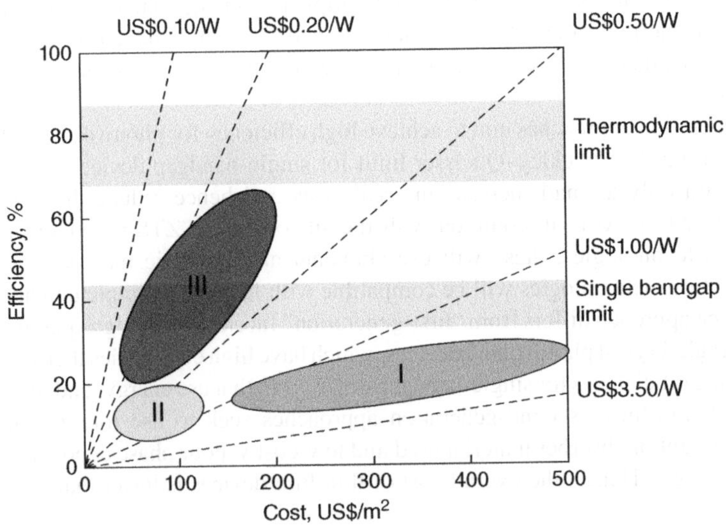

Figure 9.2 Efficiency and cost projections for first- (I), second- (II) and third-generation (III) PV technologies (wafer-based films, thin films, and advanced thin films, respectively).

In the last few decades, progress in synthesis of uniformly sized nanoparticles and nanocrystals has provided great potential to implement the above approaches. When semiconductor nanoparticles or nanocrystals are made very small, they behave as quantum dots (QDs) due to three-dimensional confinement of carriers [Boer, 1990]. Quantum confinement causes a material's effective bandgap to increase so as to allow engineering of the material bandgap to realise approach (a). Also, the relaxation dynamics of photoinduced carriers in semiconductor QDs arrays may be greatly affected by the size confinement. These include enhanced impact ionisation or exciton multiplication and greatly slowed relaxation and cooling of photogenerated hot electrons and holes [Nozik, 2002; Luque and Marti, 2007; Luque et al., 2010, Conibeer et al., 2006A]. The former can be applied to approach (b) to generate more photocurrent, the latter can be applied to approach (c) to increase photovoltage. QD materials may also be applied to intermediate band solar cells (which are related to upconversion) to increase photocurrent.

9.2 MULTIPLE-ENERGY-LEVEL APPROACHES

The concept of using multiple energy levels to absorb different sections of the solar spectrum can be applied in many different device structures. The ideal limiting efficiencies for these are often very similar and sometimes identical for a given number of energy levels. Hence their differences are manifest in the degree to which each overcomes nonidealities. This includes any inability of a particular cell design to select photon absorption at its optimum energy level in the cell, the presence of parasitic processes (usually associated with defects), and the ease of manufacture and the abundance of appropriate materials.

9.2.1 Tandem Cells

The tandem or multicolor cell is conceptually the easiest configuration to understand. It belongs to strategy (a) of increasing the number of energy levels. Solar cells consisting of p-n junctions in different semiconductor materials of increasing bandgap are placed on top of each other, such that the highest bandgap intercepts the sunlight first (see Figure 9.3). This approach was first suggested by Jackson in 1955 using both spectrum splitting and photon

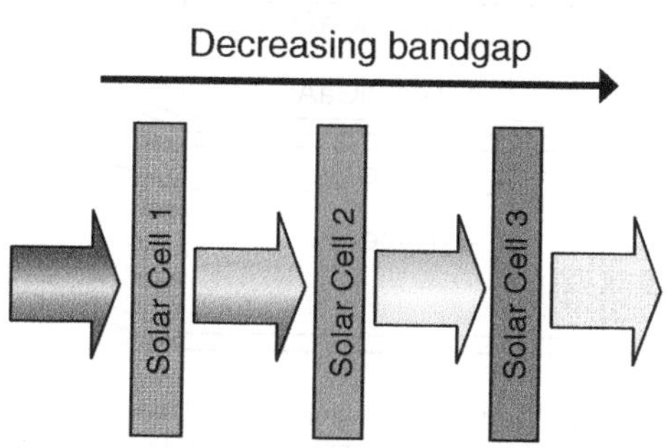

Figure 9.3 A tandem cell with the bandgap of each subcell decreasing from the front to the back, giving both spectrum splitting and photon selectivity.

selectivity [Jackson, 1955]. The particle balance limiting efficiency depends on the number of subcells in the device. For 1, 2, 3, 4, and ∞ subcells, the efficiency η is 31.0%, 42.5%, 48.6%, 52.5%, and 68.2% for unconcentrated light [Brown and Green, 2002; Green, 2003] and as also discussed in Section 2.3.4.

9.2.1.1 III-V Tandems

The highest-quality and hence highest-efficiency tandem devices are those made by epitaxial single-crystal III-V growth. A simplified generic III-V tandem device is shown in Figure 9.4. Such structures are grown monolithically by epitaxial growth process such as metal organic vapor phase epitaxy (MOVPE) or molecular beam epitaxy (MBE). These epitaxial techniques are very expensive but give very high quality crystalline material. The epitaxial growth requires control of the lattice parameter to a constant value; and bandgap control is required for a tandem cell. It is the flexibility of the III-V group of compounds that lends it to the growth of such cells, usually lattice matched on a germanium substrate. This requires most devices to be based on the AlAs/GaAs system which has a lattice parameter close enough to that of Ge, at 5.66 Å, to avoid dislocations. Nonetheless, the twin requirements of lattice parameter and bandgap control mean that ternary (or even quarternary) compounds are required for a three cell stack (e.g. GaInP/GaAs/GaInAs cells) [Takamoto et al., 2005; Wanlass et al., 2006]. Other researchers are investigating the InN/GaN/AlN system, these three compounds have a similar lattice parameter at about 3.2 Å and a bandgap range easily covering that of interest [Jani et al., 2006; Trybus et al., 2006]. Another approach that is proving to be very successful,

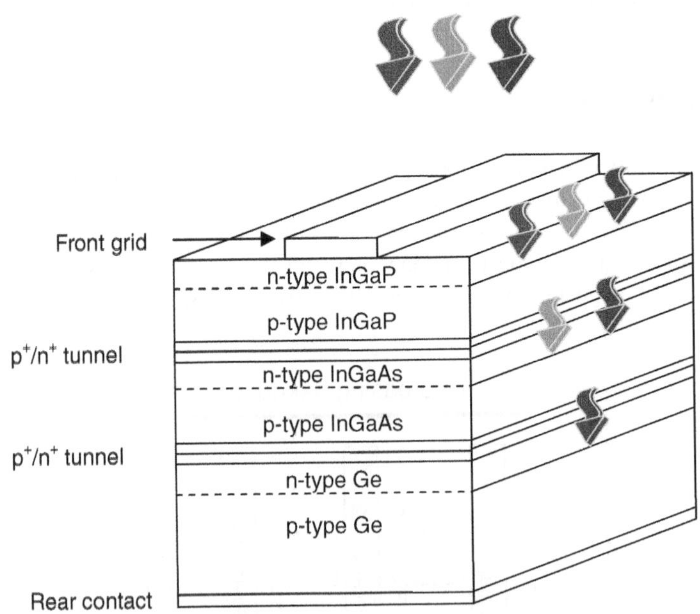

Figure 9.4 A simplified schematic of a three-bandgap III-V tandem solar cell on an active germanium substrate. The structure is monolithic with interconnection between each subshell via recombination tunnel junctions, such that the current must be the same in each cell. Increased efficiency is given by the summation of the voltages generated in each subcell.

is to avoid the requirement for strict lattice matching by growing a partially strained stack of cells approximately lattice matched to Ge. Such 'metamorphic' cells do have higher defect concentrations, but this is offset by the increased flexibility in bandgap design. The first device to exceed 40% was such a GaInP/GaInAs/Ge metamorphic device at 40.7%, under 240 suns [King et al., 2007]. The current world record efficiency for any photovoltaic device is held by Solar Junction with an improved metamorphic GaInP/GaAs/GaInNAs cell to give an efficiency of 44.0% at 942 suns [Green et al., 2013; www.sj-solar.com]. These efficiency values are significant fractions of the limiting efficiencies indicating very high radiative efficiencies. The next stage is to move onto four- or even five-bandgap cells. These not only have potentially higher efficiencies but also have higher voltage and lower current than 3-bandgap cells. This means that series-resistance losses are lower, an important consideration for concentrator cells [King et al., 2006].

9.2.1.2 Concentrator Systems

The expense of the growth techniques and of the compounds used means that such devices are usually designed for use in optical concentrator systems operating at a few hundred suns. This means that only a small area of the very efficient but also very expensive cell material is required at the optical focus of a relatively cheap concentrator. Potentially this can bring the cost per Watt of electricity generated down to low levels [King et al., 2006]. Concentration also gives the higher limiting efficiencies mentioned above because the sun effectively fills a larger fraction of the sky as far as the cell is concerned compared to no concentration.

This increases efficiency because the cell must clearly be able to absorb photons from the sun and hence must have an acceptance angle at least as large as the solid angle of the sun 3×10^{-5} steradians and it must therefore also be allowed to emit photons into this same solid angle, but it does not need to emit outside this solid angle. At a concentration ratio of 1, i.e. no concentration, the acceptance angle for emission is very much larger than this, in fact it fills the whole hemisphere, π st. The photons emitted into the hemisphere but away from the sun are unnecessarily emitted, in the sense that the cell does not have to accept at these angles in order to operate. As concentration angle is increased this acceptance angle for emission decreases and hence fewer photons emitted by the cell will be emitted away from the sun and the ratio of the number of photons emitted to those absorbed decreases and thus efficiency increases. The limit of this occurs when the angle over which the cell emits equals the angle of the sun. This is the maximum concentration ratio and equals $42\,600\times$. At this condition the sun fills the whole of the hemisphere as far as the cell is concerned and any light emitted is directed back towards the sun, giving the least possible loss through emitted photons. The limiting efficiency increases from the 31% for concentration ratio of 1 to 40.8% for maximum concentration.

However, another factor is that with increased concentration ratio the cell has to be pointed more and more accurately at the sun. And as the sun moves through the sky the cell must track the suns position with an increasing precision. This adds to the complexity of the system design and must be considered when assessing the advantages of a concentrating system.

Tandems are further well suited to concentrator systems because as the number of cells in the stack increases the voltage to current ratio increases, thus decreasing resistive losses in the high current densities of concentrator cells. However, concentrators require direct sunlight and hence do not work with an overcast sky, unlike flat-plate cell modules. They also make the overall system more complex and less modular. Hence, they tend to be more appropriate for large-scale systems located in inland areas in which cloud cover is low. Nonetheless, another important advantage for tandem devices, is that concentrator systems work best if optimised for operation in the middle of the day when sunlight is strongest. Over this period the spectrum varies the least and hence spectral sensitivity is less significant.

9.2.1.3 Thin-Film Tandems

An alternative approach to reducing the cost per Watt is to use material that is not of as high a quality as epitaxial III-V materials and hence has a higher defect density and lower efficiency, but that can be produced by much cheaper, low energy intensity deposition methods and uses elements and compounds that are not scarce or toxic. This thin-film approach thus tackles the twin requirements of third-generation devices, namely low cost per Watt and the use of nontoxic and abundant materials.

9.2.1.3.1 Amorphous Si and Micromorph Tandems

Amorphous silicon (a-Si) cells are used for single-junction cells, but tend to give efficiencies of only about 4–5% because of high defect concentrations associated with the lack of crystallinity [Meier et al., 2004]. These efficiencies can be boosted in tandem cells with amorphous silicon (a-Si:H) as a top cell[1] with one or two lower cells of an alloy with Ge (a-Si:Ge), which lowers the bandgap. These cells are in-series devices that are grown by thin-film processes such as chemical vapour deposition (CVD) or other vacuum deposition techniques. The lack of a need for crystallisation and the vapour-phase deposition mean that much less energy is required for the process, and the use of raw materials tends to be low for the thin layers deposited. Such two- or three-bandgap stacks can give efficiencies as high as 13% in the laboratory [Yang et al., 1998], but it is difficult to transfer these to production where efficiencies are around 8%.

Micromorph cells combine a microcrystalline silicon (μc-Si) Si cell in tandem with an a-Si cell connected by a defect recombination junction [Meier et al., 2002]. μc-Si has grains of order 30–40 nm in diameter and has a bulk Si bandgap of 1.1 eV, although the significant amount of grain-boundary material leads to significant band tails within the bandgap. Thus, the combination of 1.8 eV for a-Si and 1.1 eV is a good tandem combination. Devices with well over 10% efficiency have been fabricated and the cell is in production. Approaches to optimising the absorption of appropriate fractions of the solar spectrum are being investigated [Domine, 2008]. Micromorph cells are covered in greater detail in Chapter 5 in this volume in the section on Multijunction Cells.

9.2.1.3.2 Si Nanostructure Tandems

It is possible to retain both the advantages of crystalline material and of thin-film deposition but to avoid the high costs of epitaxial III-Vs by use of thin-film crystalline Si, which is crystallised by a postgrowth solid-phase crystallisation anneal [Aberle, 2006]. Such single-junction cells are now in production at efficiencies of 10% [Basore, 2006].

---

[1] a-Si:H is an alloy of amorphous silicon with at least 10 at% hydrogen. The optical density and the bandgap of this material tend to be higher – refractive index 4 and bandgap 1.8 eV – than those of bulk Si – refractive index 3.6 and bandgap 1.1 eV. Giving a-Si:H a significantly shorter wavelength absorption edge and a pseudodirect bandgap, meaning that it is suitable as a thin-layer top cell material in a tandem cell.

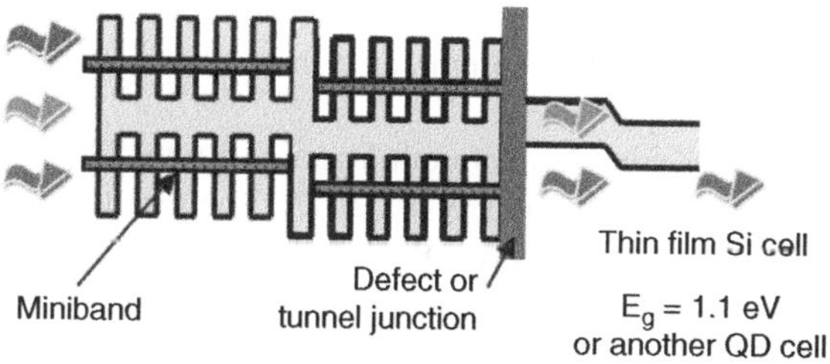

Figure 9.5 Band structure of a Si QDs/Si tandem cell: the nanostructure cell consists of Si QWs or QDs in an amorphous dielectric matrix connected by a defect tunnel junction to a thin-film Si cell.

To boost the efficiencies of silicon-based cells in a tandem and retain the other advantages of third-generation approaches, wider bandgaps in Si-based materials can be realised using quantum confinement in nanostructures. Such engineered bandgap material fabricated in a cell can be used as an element on top of a thin film bulk Si cell, as shown in Figure 9.5 [Cho et al., 2004; Conibeer et al., 2008]. For an AM1.5 solar spectrum the optimal bandgap of the top cell required to maximise conversion efficiency is ~1.7 to 1.8 eV for a 2-cell tandem with a Si bottom cell [Meillaud et al., 2006].

Thin-film techniques are used for nanostructure fabrication. These include sputtering and plasma-enhanced chemical vapour deposition (PECVD). The deposition is a variation of the multilayer alternating 'stoichiometric dielectric / Si-rich dielectric' process, shown in Figure 9.6, followed by an anneal during which Si nanocrystals precipitate, limited in size by the Si-rich layer thickness [Cho et al., 2004; Zacharias et al., 2002]. Nanocrystal or quantum dot size is therefore significantly more uniform. The matrix remains amorphous, thus avoiding some of the problems of lattice mismatch. The most successful and hence most commonly used technique is sputtering, because of its large amount of control over deposition material, deposition rate and abruptness of layers. For sufficiently close spacing

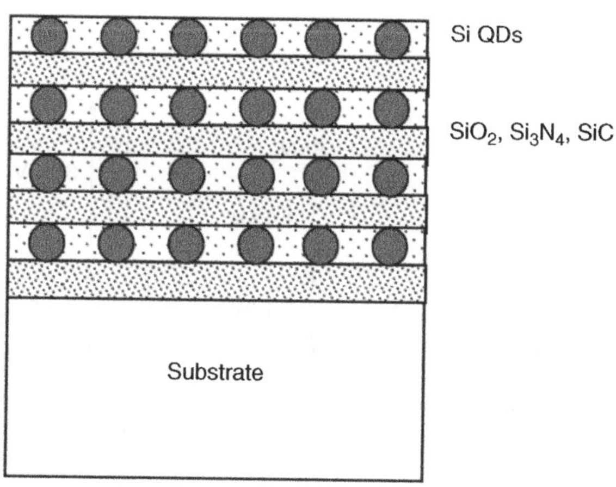

Figure 9.6 Multilayer deposition of alternating Si-rich dielectric and stoichiometric dielectric in layers of a few nm. On annealing, the Si precipitates out to form nanocrystals of a size determined by the layer thickness.

of QWs or QDs, a true miniband is formed creating an effectively larger bandgap as shown in Figure 9.5. For QDs of 2 nm (QWs of 1 nm), an effective bandgap of 1.7 eV results, ideal for a tandem cell element on top of Si.

Heterojunction PV devices consisting of multilayers of n-type Si QDs were fabricated on p-type c-Si wafers [Park et al., 2009; Song et al., 2007]. The resultant heterojunctions were used to evaluate the suitability of phosphorus (P)-doped Si QDs for photovoltaic applications. The n-type Si QD multilayer in a multilayer structure consisted of either 15 or 25 bilayers of SiO_2 and P-doped Si QDs with a SiO_2 thickness of either 1 or 2 nm. The nominal diameters of the Si QDs were 3, 4, 5 or 8 nm. The vast majority of the absorption is in the c-Si wafer, and the positive PV power recorded is as a result of this, but the devices do serve to test the applicability of P-doped Si QDs in forming a p-n heterojunction. The best device in this respect was the one with 3 nm QDs, with a very reasonable efficiency of 10.6% (V_{oc}= 556 mV; J_{sc}= 29.8 mA/cm^2, FF= 63.8%). The fact that this is somewhat comparable to a conventional p-n junction crystalline-silicon solar cell with a nontextured surface indicates that a reasonable quality junction has been formed [Park et al., 2009].

The first working PV device based on an all-Si QDs homojunction was reported in 2008, as would be used in the upper cells in all-silicon tandem solar cells [Green et al., 2008]. Since then progress has been made in increasing the open-circuit voltage (V_{oc}), the most critical parameter for these devices to indicate the high-efficiency output from higher bandgap in the structure [Perez-Wurfl et al., 2009, Perez-Wurfl et al. 2012; Green et al., 2009]. The highest voltage achieved is 492 mV, measured by illuminated I–V. The devices consist of p-n or p-i-n diodes of sputtered alternating layers of SiO_2 and SRO onto quartz substrates with in-situ Boron and Phosphorus doping. The top B-doped bilayers were selectively etched to create isolated p-type mesas and to access the buried P-doped bilayers [Cho et al., 2008; Hao et al., 2009].

Carrier-transport properties in such nanostructures are expected to depend on the matrix in which the silicon quantum dots are embedded. As shown in Figure 9.7 different matrices produce different transport barriers between the Si dot and the matrix, with tunnelling probability heavily dependent on the height of this barrier. Si_3N_4 and SiC give lower barriers

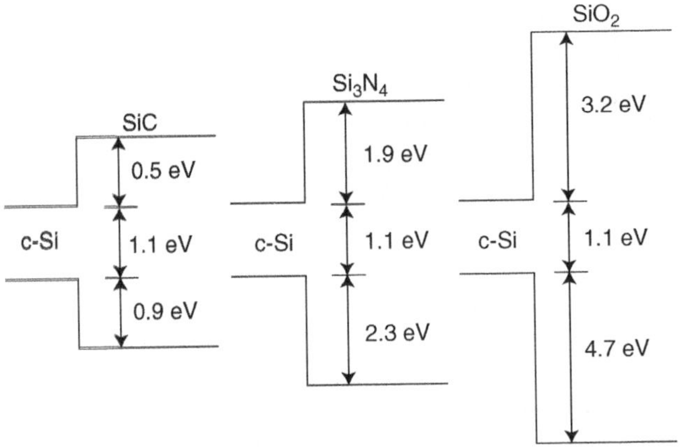

Figure 9.7 Bulk band alignments between silicon and its carbide, nitride and oxide.

than SiO_2 allowing larger dot spacing for a given tunnelling current. From the approximate relation of the tunnelling probability T_e through a square potential well:

$$T_e \approx 16 \exp\left\{-d\sqrt{\frac{8m^*}{h^2}\Delta E}\right\}$$

where ΔE is the energy difference between the CB edge of the matrix and the confined energy level of the quantum dots, d is the spacing between quantum dots and m^* is the effective mass of electrons in the matrix, the important parameter in determining the degree of interaction between quantum dots is $m^*\Delta Ed^2$. As barrier height decreases the barrier thickness for a given probability increases, thus requiring a lower dot density for a given conductivity or higher conductivity for a given dot density. As the dot size decreases ΔE also decreases, thus increasing T_e and enhancing the effect further for smaller quantum dots [Boer et al., 1990]. This suggests that QDs in a SiO_2 matrix would have to be separated by no more than 1–2 nm of matrix, while they could be separated by more than 4 nm of SiC. Fluctuations in spacing and size of the dots can be investigated using similar calculations. It is also found that the calculated Bloch mobilities do not depend strongly on variations in the dot spacing but do depend strongly on dot size within the QD material [Jiang et al., 2006, Jiang and Green, 2006]. Hence, transport between QDs can be significantly increased by using alternative matrices with a lower barrier height, ΔE. The spacing of dots would have to be closest in the oxide, nitride and carbide, in that order. Also alternative QD materials such as germanium and tin offer the possibility of lower temperature precipitation of quantum dots and the potential for bandgaps lower than that of silicon, which could be required for tandem cell elements under a silicon cell [Gao et al., 2008; Huang et al., 2009; Song et al., 2008]. Ge nanocrystal materials in SiO_2 have been fabricated with p-type characteristics [Zhang et al. 2011].

Heterodevices combining various combinations of these materials are also possible. One approach to this has been the use of Si nanocrystals embedded in SiO_2 matrix to give good quantum confinement but with SiN_x interlayers to give good transport between layers and to limit interdifusion between layers. These materials have been shown to have lower resistivities whilst maintaining quantum-confined energy levels [Di et al. 2011]. Several other heterojunction structures are possible combining two of these different nanostructured material types. Work on these is progressing and is likely to be the route to progress in this area.

9.2.2 Multiple-Exciton Generation (MEG)

Carriers generated from high-energy photons (at least twice the bandgap energy) absorbed in a semiconductor can undergo impact ionisation events resulting in two or more carriers close to the bandgap energy. This approach is an implementation of approach (b) to produce more carriers per incident photon and thus boost current, as is discussed in Section 2.4.1.1. In order to achieve this, the rates of photogenerated carrier separation, transport, and interfacial transfer across the contacts to the semiconductor must all be fast compared to the rate of carrier cooling [Nozik et al., 1980]. But impact ionisation has a vanishingly small probability in bulk materials because of restrictions imposed by energy and momentum conservation

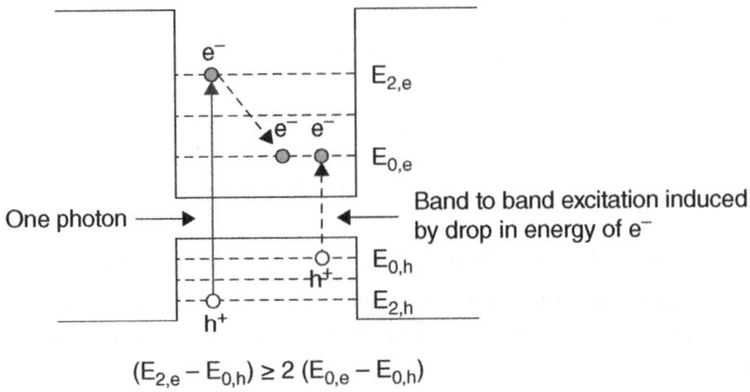

Figure 9.8 Multiple exciton generation in QDs: a high-energy photon is absorbed at a high energy level in the QD, which then decays into two or more electron–hole pairs at the first confined energy level.

and the extremely fast carrier cooling rate. However, it has been discovered that this process can be much more efficient in QDs [Schaller et al., 2004; Hanna and Nozik, 2006; Klimov, 2006]. Because of the localised nature of the electrons and holes in these confined systems it is bound excitons that are generated rather than free carriers. Hence, the phenomenon is known as multiple exciton generation (MEG). The exact mechanisms involved are not yet entirely clear, but they are likely to be related to the reduced requirement for conservation of crystal momentum in the small spatial volume of a QD. The mechanism is shown schematically in Figure 9.8. There is now a large amount of experimental evidence showing production of up to seven electron–hole pairs for the absorption of one high-energy photon, i.e. a quantum efficiency (QE) of seven, with high luminescence efficiencies. Until recently MEG had only been observed with absorption spectroscopy measurements and there is quite some controversy not only as to how it occurs but also whether it is repeatable [Pijpers et al., 2008]. The most likely explanation seems to be that the mechanism is very sensitive to the exact surface states on the QDs [Nozik, 2008A].

The ratio of the photon energy to the bandgap energy ($E_{h\nu}/E_g$) must be $\geq$ QE. MEG was first seen in PbSe QDs but has now been seen in quite a wide range of QDs fabricated from II-VIs and other materials, including Si, which is significant for possible future large-scale implementation. The formulation for the limiting efficiency of such a device is given elsewhere [Hanna and Nozik, 2006]. The efficiency is calculated in a similar way to a single-bandgap cell but with the current enhanced by an energy-dependent quantum yield term, QY(E). The value of this depends on the particular model used to simulate the data. It is always equal to one, up to a threshold energy above the effective bandgap energy, E_0. QY then increases to two, indicating the production of two electron–hole pairs above this energy. This threshold energy varies with material and is usually equal to $\sim 3E_0$, but is sometimes as low as $2E_0$, yielding efficiency limits of 36% and 42%, respectively, at one sun [Klimov, 2006].

The incorporation of MEG QDs directly in a solar cell should boost the current in the device through direct carrier multiplication. A Schottky device consisting of an all-inorganic metal/nanoparticle/metal sandwich solar cell was reported by Nozik and coworkers, which produces an exceptionally large short-circuit photocurrent (>21 mA cm^{-2}) and 2.1% conversion efficiency [Luther et al., 2008]. The functional PbSe nanoparticle film, deposited via layer-by-layer dip coating yields a high EQE of 55–65% in the visible and up to 25% in the infrared region of the solar spectrum, although not yet above 100% as observed in some of II-VI QDs solutions and thin films. This device produces one of the largest short-circuit currents of any nanostructured solar cell. However, this result has been augmented by results showing QE of 106% at photon energies of 3.5 eV for a PbSe QD solar cell [Semonin et al. 2011]. This very significant result demonstrates the real feasibility of this approach for boosting current in a device from high-energy photons and further significant improvement is likely to follow.

9.2.3 Intermediate-Band Solar Cells (IBSC)

The IBSC has an additional energy level (the intermediate band) within the bandgap of a single-junction cell such that this level absorbs photons below the bandgap energy in parallel with the normal absorption of above bandgap photons in the cell. Photons with less energy than the primary bandgap can be absorbed by transitions from the VB to the IB and from the IB to the CB. This is shown schematically in Figure 9.9 and is also an implementation of strategy (b), absorption of below-bandgap photons to produce carriers at the bandgap energy. The semiparallel operation of the direct band-to-band and transition via the IB processes, offers the potential to be much less spectrally sensitive than a tandem cell, but still has the potential to give high efficiencies. Such a device has the same limiting efficiency as a three-level tandem – 63% under maximum concentration, 48% under one sun, because it has the same number of energy thresholds. These calculations are expanded in Section 2.4.4.2. However, this calculation does not take into account spectral sensitivity and assumes ideal properties such as ideal photon selection, a potential problem with intermediate-level devices. Strictly to ensure complete photon selectivity, it is necessary to modify the absorption and emission ranges of the device such that the photon energy ranges do not overlap. This implies limited-width conduction and valence bands that would be very difficult to arrange in practice. Hence, without this rather artificial modification, there is a problem with photon selection, although the device nonetheless collects photons that would otherwise not be absorbed. However, note that the current now only has to be equal across the two lower energy levels, while the main current across the bandgap is independent. This reduces the spectral sensitivity and compensates to some extent for the reduced photon selectivity.

The IBSC has the advantage that the delocalisation of carriers in its continuous band means that these photons do not necessarily have to be absorbed by the same electron. This gives a

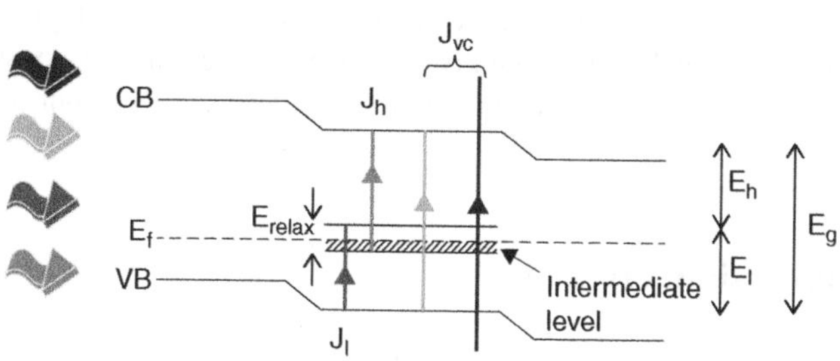

Figure 9.9 The IB solar cell: below-bandgap photons are absorbed by the two transitions to and from the intermediate level contributing to photocurrent, in parallel to normal operation of the cell.

much longer lifetime to the intermediate level, allowing much more time for absorption of the second photon. To maximise this advantage, the intermediate band should be half-filled with electrons, i.e. it should have a Fermi level at half the band energy, as illustrated in Figure 9.9, such that absorption of an electron from the valence band or emission of an electron to the conduction band are equally likely [Luque and Marti, 1997; Luque et al., 2006]. Formation of an intermediate band for an IB solar cell has been suggested in QD materials [Švrcek et al., 2004]. Three-dimensional confined states of the electrons in the potential wells of QDs may introduce isolated IB similar to the formation of minibands in QD arrays. Luque and coworkers have reported IB devices using the confined energy levels of an InAs QD/GaAs superlattice sandwiched by p and n GaAs emitters and grown by molecular beam epitaxy in the Stranski–Krastinov growth mode. In this structure the intermediate band is introduced by InAs QDs in GaAs [Marti et al., 2006]. These devices have demonstrated several of the indicators of true IB solar cell operation, including the existence of a split between the quasi-Fermi levels that corresponds to the conduction and intermediate bands, although an efficiency advantage has not yet been achieved [Luque et al., 2007]. Nonetheless, this seems likely in the near future, particularly if they are operated under concentration.

9.3 MODIFICATION OF THE SOLAR SPECTRUM

As an alternative to modifying the structure of a solar cell, another approach is to modify the spectrum incident on the cell to narrow its bandwidth and make it closer to optimal for a single-bandgap cell. This involves a limiting quantum efficiency (QE) of conversion of photons to electron–hole pairs in the cell not limited to unity. (The physics of such nonunity quantum efficiency processes is discussed further in Section 2.4.3.) For short-wavelength photons – at least twice the energy of the cell bandgap – this means a QE > 1. This 'downconversion' (DC) or 'quantum cutting' (QC) approach means that more electron–hole pairs are generated than the number of blue photons in the spectrum. It has been shown by detailed balance calculations that DC can boost the efficiency of a solar cell from 31% for a single junction to 36.75% under 1 sun [Trupke et al., 2002A]. For long-wavelength photons – with energy less than the cell bandgap – this would yield a QE < 1. Such 'upconversion' (UC) results in the combination of two or more below bandgap photons to give one electron–hole pair above the bandgap energy. Again by detailed balance a potential efficiency enhancement to 47.6% has been calculated [Trupke et al., 2002B].

These detailed balance efficiencies assume that the UC and DC devices are perfectly radiatively efficient and assume perfect photon absorption, materials and systems behaviour. This is valid in calculating the limiting efficiencies of solar cells employing such devices, but for real devices only much lower radiative efficiencies are currently possible. This can be critical in deciding which materials are suitable for UCs or DCs, but also in deciding between an UC or a DC approach.

9.3.1 Downconversion, QE > 1

A downconverting device (DC) must be placed in front of a standard cell and can boost current by converting a UV photon to more than one photon just above the bandgap of the solar cell – thus boosting the current. However, the DC does require that more lower-energy photons are emitted than high-energy photons absorbed, i.e. its quantum efficiency (QE) must be greater than 100%. Hence, there must be at least as many photons emitted at the lower energy as are absorbed at the higher, or else the DC layer will *decrease* the number of photons absorbed by the cell. (In fact, due to the partial transmission of photons from a low refractive index in air to a higher one in the DC layer, the QE must be greater than 100% in order not to be parasitic – a QE of about 110% for a refractive index similar to that for a Si solar cell at 3.6, [Würfel, 2005]. DC devices have been attempted experimentally but so far an overall QE greater than 100% has not been achieved, resulting in a downshifting of energies in, for example, Si nanocrystals, [González-Díaz et al., 2006], or porous silicon, [Švrček et al., 2004], which nonetheless can give a useful enhancement in spectral response at short wavelengths for some materials, due to better absorption of the down shifted photons. This effective narrowing of the bandwidth of the absorbed spectrum allows the solar cell to be designed more optimally with respect to absorption coefficient as a function of wavelength, junction depth and surface recombination, [Strumpel et al., 2007]. Similarly, luminescent downshifting layers based on luminescent dyes have been demonstrated to boost the short wavelength response of CdTe cells in which short wavelengths are usually attenuated in the CdS window layer, [Klampaftis et al., 2009]. True downconversion or QC as first suggested by [Dexter, 1957], requires absorption of a short-wavelength photon and re-emission of at least two photons of about twice the wavelength. In turn this requires an appropriately located intermediate energy half-way between the excited state and the ground state. Most work has focused on lanthanide materials because of their varied and discrete energy levels.

9.3.1.1 Lanthanide-Based DC

Praseodymium (Pr^{3+}) is a good choice because of its widely dispersed energy levels well matched for photon cutting, [Dieke, 1968], see Figure 9.10. The $^3P_2$, $^1I_0$ and P_0 levels at between 440 and 490 nm can absorb blue photons that can then radiatively recombine via the $^1G_4$ level at 1010 nm – at just greater than twice this wavelength – thus emitting two photons at just above the silicon bandgap, although nonradiative recombination via the other levels at longer wavelengths than $^1G_4$ is also likely. Experiments indicating such photon cutting have been carried out on Pr^{3+} embedded in various phosphors, [Meijerink et al., 2006], and also for other lanthanide-doped materials, [Wegh et al., 2002, Michels et al., 2002]. Transition metals with their partially screened 'd' shell electrons also have partially discrete levels and work on photon cutting in transition-metal-doped materials has also been carried out, [Ilmas, 1970, Berkowitz and Olsen, 1991].

9.3.1.2 Reverse Auger Generation

An alternative approach is to absorb a short-wavelength photon high up in the conduction band of various semiconductors. There is then the possibility of an impact ionisation event (i.e. reverse Auger recombination) in which the high-energy electron excites an additional electron to the conduction band, thus creating two or more electron–hole pairs at the bandgap energy, [Berkowitz and Olsen, 1991]. Luminescent recombination of these electron–hole pairs is then usually enhanced by choice of an appropriate doping level within the bandgap and an increased number of photons emitted. The quantum efficiency (QE) of such impact ionisation depends on the energy of the initial photon and is reduced by nonradiative thermalisation and recombination. QEs of greater than 200% have been achieved in some materials, usually

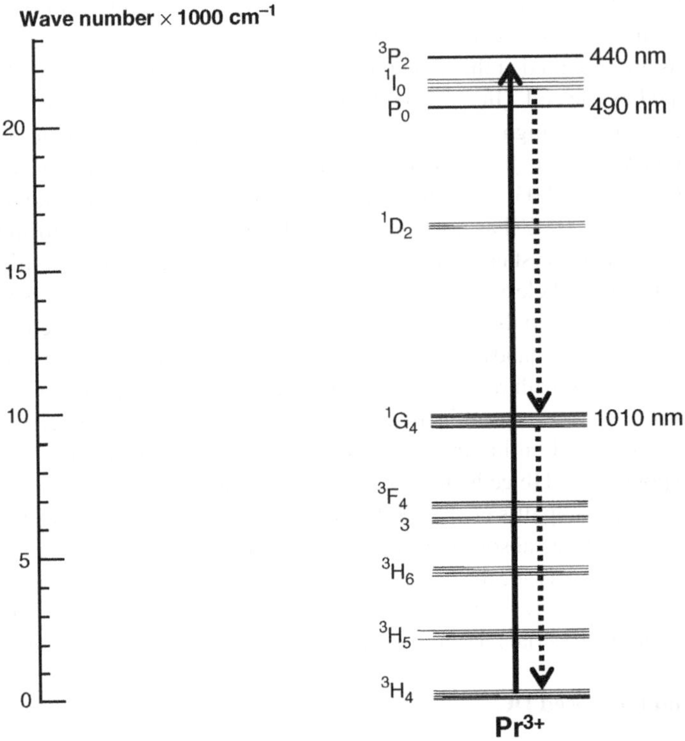

Figure 9.10 Energy levels for praseodymium (Pr) based on data from [Dieke, 1968]. Also shown is the possible photon cutting mechanism for a photon absorbed around 450 nm and emitted as two photons via the $^1G_4$ level. (Data based on Dieke, 1968. Copyright © John Wiley & Sons, Ltd, 1968.)

based on wide-bandgap oxides (at least 5 eV) doped with lanthanide or transition-metal atoms, but with incident photon energies of at least twice or three times the bandgap, [Michels et al., 2002, Ilmas and Savikhina, 1970]. Although QEs up to about 120–140% have also been achieved with the relatively narrow gap (3.4 eV) ZnS doped with Zn, Cu or Ag, but only with photons in excess of 20 eV, [Berkowitz and Olsen, 1991]. But at the incident photon energies required for this impact ionisation there are almost no photons in the solar spectrum. Also, the efficiency of the impact ionisation mechanism is very low. Hence, these are not good materials for downconversion for solar cells.

However, as discussed in Section 9.2.2, MEG in QDs has a dramatically higher efficiency for impact ionisation such that one or several bound electron–hole pairs can be generated from one incident high energy photon. For such a MEG material to be used directly as a DC with a single-bandgap solar cell, a few conditions are required. The MEG QDs would need to have an appropriate bandgap to illuminate a solar cell – for a Si solar cell Si QDs would be appropriate, and MEG has been shown in well-passivated Si QDs [Nozik, 2008]. But much more challenging would be the need for a high luminescent efficiency of the multiple excitons such that there would be a net QE greater than 100%. However, it is not yet clear whether such a luminescent efficiency from these materials is feasible. Certainly, the currently observable rate of MEG and subsequent Auger decay processes back to a single exciton at about 200 ps, [Beard et al., 2010; Nozik, 2008], and the long radiative lifetimes of these materials, at about 10 ns, would make efficient multiple photon emission unlikely.

9.3.2 Upconversion of Below-Bandgap Photons

An UC device is designed to absorb subbandgap photons in an UC layer behind a bifacial solar cell. This layer radiatively absorbs two or more long-wavelength photons and emits a photon of higher energy above the bandgap of the bifacial solar cell. Thus, the current in the device is boosted by photons that would not normally be absorbed. As the UC does not interrupt the incidence of photons on the front surface, even a very low efficiency of UC gives a small current boost and hence an efficiency increase.

For application to photovoltaics there are two broad possibilities. Either the current in a silicon solar cell can be boosted through the application of a simply applied thick-film upconverting layer on the back surface of a bifacial silicon cell. This approach would be reasonably simple to apply to cells in a production line and hence offers a near-term boost to efficiency. Although it would boost the current of the device the energy levels of around 0.8 eV for the intermediate level and >1.1 eV for the emission level (i.e. above the bandgap of silicon) are not optimum for the three-transition upconversion effect. Alternatively, a device in which both the intermediate and bandgap energies are optimised to about 1 and 2 eV, respectively, would give a higher limiting efficiency of just over 40%, for a symmetric device, [Ekins-Daukes and Schmidt, 2008]. The materials for this would both need to be tailor-made so this is not a near-term device, but does offer higher efficiencies.

The key to an efficient upconversion mechanism is for the transitional state – resulting from the absorption of one subbandgap photon – to be long-lived such that it lasts long enough to absorb a second low-energy photon and allow time for the reaction, boosting the energy of the state to a higher level to occur. If this second higher-energy state is above the bandgap, radiative emission from the state will boost the flux of above bandgap photons incident on the cell.

9.3.2.1 Singlet–Triplet Coupling in Organic Molecules

One such long-lived energy state can be achieved in the transfer of electrons from 'allowed' excited singlet states to 'forbidden' excited triplet states in some organic molecules (S–T transitions).[2] Exploitation of this for upconversion has been demonstrated by [Baluschev et al., 2007]. The spin-orbit coupling that can occur due to the heavy-metal atoms in some complexed molecules, such as porphyrins, can allow a mixing of singlet (S) and triplet (T) states, such as to transfer the excited electron in the first excited singlet state (S_1) to the excited triplet state (T_1) where it can have an extremely long lifetime (up to 100 μs). This is illustrated in Figure 9.11.

These transitions are strictly 'allowed' or 'forbidden' only for pure electric dipole transitions, but in real molecular systems these selection rules determine whether a transition will be fast or slow, respectively. Hence, if excited singlet and triplet states, as shown in

[2] These selection rules arise from spin conservation on absorption of a photon; the excited singlet state containing two electrons of antiparallel spin and hence able to absorb a spin 1 photon, whereas in the excited triplet state the two electrons are of parallel spins and hence not strictly able to absorb or emit a photon.

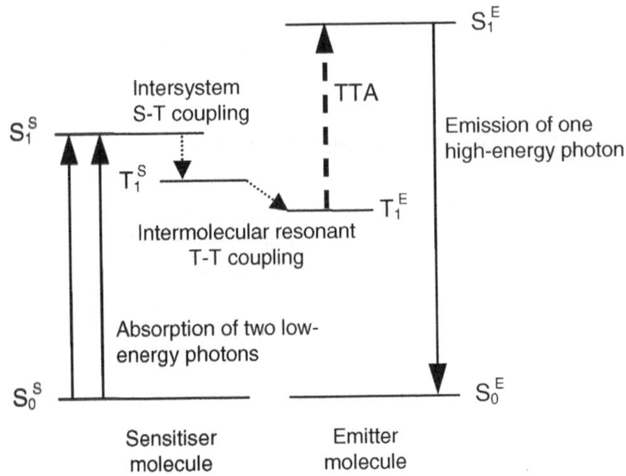

Figure 9.11 A simplified schematic illustration of the triplet-triplet annihilation (TTA) upconversion mechanism, [Baluschev et al., 2007; Ekins-Daukes and Schmidt, 2008; Cheng et al., 2010]. S and T are singlet and triplet states; subscripts 0 and 1 denote ground and excited states; superscripts S and E denote sensitiser and emitter states. Solid arrows are photon absorptions/emissions dotted arrows are interstate transitions and the dashed arrow is the TTA interaction. (Reproduced with permission from Conibeer, 2010. Copyright © Nova Science Publishers, 2010.)

Figure 9.11, are close in energy (or resonant) the transition is slow but reasonably likely. But similarly, the decay of triplet states is strictly forbidden and hence for a real molecular system they will be very slow making these states very long lived. From the $T_1{}^S$ state of the 'sensitiser' molecule very slow radiative emission can occur, giving rise to phosphorescence. However, if the molecule is placed close to another 'emitter' molecule that has a $T_1{}^E$ state close to resonance with $T_1{}^S$, very efficient population of $T_1{}^E$ by Dexter resonant energy transfer can occur. An important property of the emitter molecule is that, unlike the sensitiser, it does not have $S_i{}^E$ states close in energy to the $T_1{}^E$ state – molecules such as the anthracenes have this property. This very long-lived $T_1{}^E$ state can then absorb a second electron pumped by another $S_1{}^S$–$T_1{}^S$–$T_1{}^E$ transition. This then leads to an eventual triplet–triplet annihilation reaction (TTA) that produces an electron at the higher singlet state of the emitter ($S_1{}^E$) and a subsequent efficient radiative emission of a high-energy photon. This approach appears very promising for porphyrin–metal complexes that exhibit the required resonance in excited singlet–triplet energies, with high radiative quantum efficiencies recorded. The highest experimental efficiency so far is 16% for TTA UC [Cheng et al., 2010] importantly this being above the previously held maximum limit of 11%, due to the incorrect assumption that coupling to higher-order states other than triplets would reduce the overall efficiency. The longevity of the triplet state makes this possible because it makes the double absorption of below-bandgap photons only very weakly dependent on the illumination intensity as there is plenty of time to absorb the second photon. The S-T/TTA approach has also been suggested as a means to pump the excited singlet state of a molecular intermediate band solar cell, [Ekins-Daukes and Schmidt, 2008]. Here, the absorber material of a solar cell (perhaps a dye-sensitised cell) would replace the emitter molecule such that carriers are injected directly into the cell from low-energy photons as well as from those above the HOMO–LUMO level or bandgap of the absorber material.

The application of this approach to the boosting of a Si solar cell efficiency does not as yet look promising, because so far the TTA mechanism has only been demonstrated for absorption (and emission) of relatively high energy photons, all of which are above 1.1 eV and so would not boost the current in a silicon solar cell, [Ekins-Daukes and Schmidt, 2008]. It also seems probable that the lower intermediate and upper energy levels required for a Si cell (0.8 eV and >1.1 eV, respectively) are unlikely to be stable in organic materials. This is because the relatively localised electron–hole pairs resulting from photon absorption and the consequent large exciton binding energy, coupled with the weak bonds associated with these lower energy transitions, can easily lead to photodegradation.[3]

However, for the other PV UC application of a device with intermediate and upper energy levels tailored to give optimum efficiency, the high quantum efficiency of the S-T/TTA transition approach does offer great promise for a high-efficiency device. For instance, the upper energy of the rubrene molecule (an anthracene derivative) – in which TTA has been demonstrated – at 2.3 eV is close to the optimum for this type of UC device at 2.0 eV. It is therefore highly applicable for all of amorphous silicon (a-Si), organic solar cells (OPV) and dye-sensitised cells (DSSC). It has been demonstrated to boost the current in amorphous silicon cells [Cheng et al., 2012]. However, at present these require the TTA materials to be in solution, with clear practical problems. Attempts so far to fabricate solid-state devices have all resulted in very severe quenching of the effect. Significant improvement of the exact arrangement of sensitiser and emitter molecules is required in order to make such solid-state devices exhibit similar performance to solution.

9.3.2.2 Lanthanides for UC Devices

An alternative group of materials that give long-lived intermediate states are those based on the lanthanide (or rare-earth) group of elements, because of their narrow and suitably spaced energy transitions. The lanthanides have a valence shell of 4f electrons that are shielded by full outer 6s and 5p shells. Hence, their electronic transitions tend to remain similar to those of an isolated atom and fairly discrete. However, in a host material (or phosphor) the absorption and emission levels are broadened somewhat by the combination with the allowed phonon energies of the phosphor.

Several lanthanides have demonstrated upconversion, including Pr and europium (Eu), but this has usually been for laser applications, [Xie, 1993]. But erbium (Er) is the only lanthanide with energy levels at the values of interest for appropriate upconversion for a Si solar cell (i.e. around 1500 nm (~0.8 eV) for absorption and a little above 1100 nm (~1.1 eV) for emission), and that are also widely enough spaced that they would not be completely dominated by nonradiative recombination via the phonons in any host phosphor, [Dieke, 1968]. Hence, Er is the element that has been most widely investigated in this respect. This has included incorporation of Er into a Si matrix to engineer an impurity photovoltaic cell (IPV), [Keevers et al., 1997]. IPV is similar in concept to upconversion except that the upconverting species is embedded in the depletion region of the cell. It thus has the advantage that radiative emission from the upper level is not required as electron–hole pairs are injected directly into the cell above the bandgap, but has the severe disadvantage that not

[3] The partially delocalised nature of the organic species (e.g. benzene ring, porphyrins or fullerene molecules) used for this and most other organic electronic applications reduces this effect somewhat but does not give the same degree of delocalisation, and hence low exciton binding energy, as the completely dissociated electrons and holes in an inorganic semiconductor.

only does the device have to be very optically efficient at absorption of both the 1500 nm and the above-bandgap radiation but also very good electrically. It is very difficult to engineer both properties in a device, hence the lack of significant success to date. IPV is also closely related to Intermediate Band solar cells, as discussed in Section 9.2.2, except that the IB has a continuous density of states that allows for much longer intermediate lifetimes.

For investigation of true upconversion (i.e. in which the UC device is optically coupled but electrically isolated from the PV cell), Er has also been the main focus, [Shalav, 2006; Shalav et al., 2007]. The broadening of absorption/emission due to phonon energies in the phosphor, discussed above, is an important consideration in choosing the host phosphor. It would be desirable to have a wide absorption band but a narrow emission band, unfortunately the two are directly linked and a compromise is required (at least with the presently proposed mechanisms). The broadening of the otherwise discrete Er energy levels is determined by the phonon energies available in the host phosphor material. Large optical phonon energies are allowed in ionically bonded salts with very stiff lattice force constants. $NaYF_4$ has an average phonon energy of 45 meV (or 360 cm^{-1}) that is sufficient to give a broadening of 1480–1580 nm for Er absorption/emission. A very high upconversion QE of ~17% has been demonstrated with an Er^{3+}-doped $CaYF_4$ crystal, although only at laser intensities equivalent to several thousand suns, [Ivanova et al., 2009].

Upconversion can proceed by a few different mechanisms in Er atoms doped within a host. These were first described for lanthanides in general by [Auzel, 2004]. Figure 9.12 shows some of the energy levels for Er. With reference to Figure 9.12, and considering for the moment just the Er levels, all of these mechanisms start with GSA (ground-state absorption) in which a direct excitation of a single Er atom from $I_{15/2}$ (ground state) to $I_{13/2}$ (at 1523 nm) occurs. The difference between mechanisms comes in absorption of a second 1523-nm photon. This can either take place in the same Er atom (excited-state absorption, ESA), or by a second GSA in a nearby Er atom followed by a transfer of its energy to the first Er atom up to $I_{9/2}$ (at 810 nm). This energy transfer occurs for levels resonant in energy

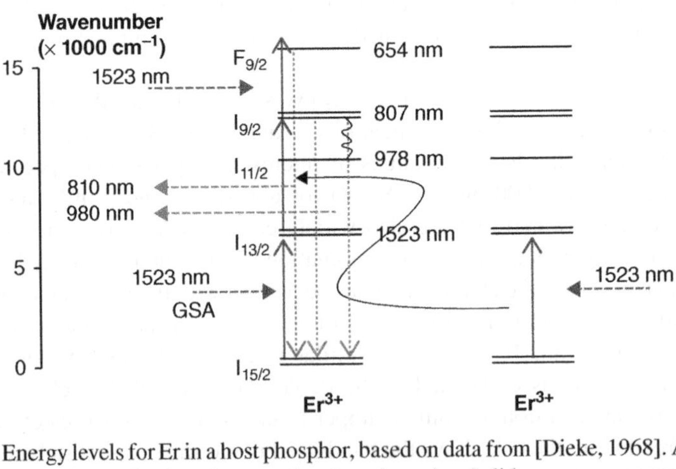

Figure 9.12 Energy levels for Er in a host phosphor, based on data from [Dieke, 1968]. Also shown are possible energy-transfer mechanisms between levels and species. Solid up arrows represent GSA; wavy lines rapid nonradiative relaxation; dashed arrows ETU; dotted down arrows the resulting upconverted luminescence; dashed horizontal arrows excitations; and solid nonvertical arrows resonant levels. (Data based on Dieke, 1968. Copyright © John Wiley & Sons, Ltd, 1968.)

via a dipole–dipole interaction mechanism or, i.e. by a delocalisation of the wavefunction. Actually, in order to match the energies between two Er atoms exactly for any of these three mechanisms, the levels need to be slightly broadened by emission or absorption of a phonon in the host material. This transfer of energy from one GSA Er atom to another is termed either energy transfer up-conversion (ETU) or cooperative state absorption (CSA), (with only a very subtle difference between the two that will not be discussed here). There can then be a direct radiative recombination from $I_{9/2}$ directly to the $I_{15/2}$ ground state by emission of an 810-nm photon or rather more probably a rapid nonradiative recombination down to the $I_{11/2}$ (980 nm) level by emission of about four phonons, followed by a radiative emission of a 980-nm photon down to the ground state. Either of these emitted photons are above the bandgap of Si and hence able to be absorbed by the solar cell and contribute to current generation. As ETU depends on excitation of nearby Er atoms, it is both Er concentration and 1500 nm light intensity dependent and it becomes the dominant mechanism above certain threshold values of both.

The response of an Er UC device, such as described above, has been measured by [Shalav, 2006], see Figure 9.13. It was found that the approximate optimum concentration for ETU is 20% Er replacing Y in $NaY_{0.8}Er_{0.2}F_4$. With the threshold intensity for the dominance of ETU occurring at a 1523-nm laser illumination power of about 1 mW, with the quantum efficiency increasing linearly up to 4% at about 5 mW laser power, beyond which point the Er excitation becomes saturated and there is no further improvement. However, this quite reasonable QE of 4% is only over the narrow absorption bandwidth of 1480–1580 nm,

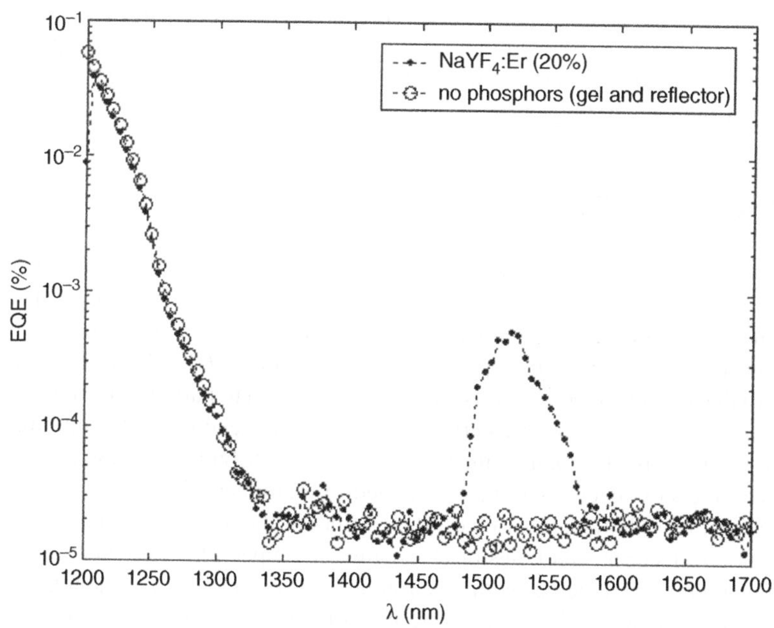

Figure 9.13 External quantum efficiency (EQE) of a bifacial silicon solar cell with and without $NaYF_4$:20% Er^{3+} phosphors on the back surface – [Shalav, 2006; Shalav et al., 2007] – showing the enhancement in current over a narrow range of incident wavelengths. Note that the log scale exaggerates this effect. (Reproduced with permission from Shalav, 2007. Copyright © 2007, Elsevier.)

hence the boost to the current for a solar cell although measurable, only amounts to a very small increase in the absolute solar cell efficiency – about 10^{-5}%. This is demonstrated in Figure 9.13 that shows the quantum efficiency of a bifacial silicon solar cell both with and without an upconverting Er-doped phosphor layer behind the cell. The clear but very small increase in spectral response over a narrow wavelength range – note the log scale for EQE – demonstrates this boost in current from upconversion of 1500-nm photons. Even if the QE was the maximum possible of 50% for this wavelength range, the absolute efficiency increase would only be 2.4%, [Shalav, 2006].

Due to the small number of photons in the solar spectrum over the absorption range of Er, the laser power quoted above for the maximum ETU QE translates to about 10 000 suns, although this is difficult to estimate because of the difficulty in measuring the illuminated area. This is a very high solar concentration ratio. Hence, for this reason and that of the narrow absorption bandwidth it is desirable to absorb a much larger proportion of the solar spectrum between 1100 and 1500 nm and hence boost the number of photons available for upconversion and hence the current in the solar cell. In addition, the inherent nonlinear aspect of upconversion means that efficiency will also be higher under concentration. There is a greater chance of the second photon absorption occurring if there are more photons incident on the cell. Hence, most work at present is either on sensitising upconversion in Er to a wider range of wavelength between 1100 and 1500 nm. This can be achieved by downshifting in a luminescent species such as narrow-bandgap nanocrystals that have their emission wavelength tuned to 1500 nm by their size. Or by increasing the local concentration of photons incident on the Er upconverter, which can be achieved through coupling of plasmonic modes in small metal particles with their plasmon resonance tuned to the 1480–1580 nm window of Er absorption. In addition, light trapping at these long wavelengths is entirely feasible and has been demonstrated to give a boost to UC, either in Rugate-type structures [Goldschmidt et al., 2010] or in the slowed light modes of distributed Bragg reflectors, DBRs [Johnson and Conibeer, 2012].

9.4 THERMAL APPROACHES

An alternative to specifically engineering multiple energy thresholds in a device or devices is to allow the photons to generate a thermal population of some sort in an absorber. The photon spectrum incident on the cell is essentially a thermal one, generated by the thermal emission from the surface of the sun at an approximate temperature of 6000 K, with an emissivity very close to that of a black body. If this energy is transferred to particles in the absorber this thermal distribution can also be transferred, with the 'excess energy' of the multiple energy levels of the incident photons maintained in the thermal distribution of these particles.

This thermal excess energy can be maintained in a number of different 'particle populations': the incident photons themselves; the carriers in the absorber (either with separate temperatures for electrons and holes or at a common temperature with electron-hole scattering); optical phonons together with electron–hole pairs; or acoustic phonons with fully thermalised electron–hole pairs; or acoustic phonons only with no energised electrons or holes. (The penultimate example is that for a normal solar cell, and the last example that for a fully thermalised material, probably with a zero bandgap, i.e. a metal.) The degree of irreversibility, and hence the entropy production, increases progressively through these examples. This is because for a thermal population described by a single Fermi temperature, it is not possible to collect the energy of each photon at its optimum chemical potential. And as the number of particles between which the energy is shared increases (an increase in the accessible microstates) the entropy and irreversibility increase.

9.4.1 Thermophotovoltaics (TPV)

A thermophotovoltaic (TPV) system consists of a narrow-bandgap photovoltaic cell (about 0.7 eV) that is illuminated by black- or greybody radiation from a hot source but at a lower temperature than the sun [Couts, 2001]. In order to give an advantage, thermal emission incident on the cell must be filtered by a selective emitter that only passes light just above the cell bandgap with photons at other energies above or below the bandgap, reflected back to the emitter. Hence, each photon's energy or excess energy respectively, is still utilised in reheating or maintaining the temperature of the emitter. This approach would normally use waste heat from an industrial process or similar and hence not be a solar cell, but it can be coupled to an emitter heated by solar thermal energy. In practice, the large number of different elements with their multiplying inefficiencies, and the need to thermally insulate some elements from others, makes the approach very difficult to optimise.

9.4.2 Thermophotonics

Thermophotonics is a variation of TPV in which the thermal source heats a luminescent diode rather than a broadband or metallic absorber as in TPV. This diode then illuminates a solar cell with a spectrum strongly peaked just above their common bandgap [Catchpole et al., 2003]. The advantage over TPV is that no additional selective emitter is required as the luminescent diode fulfills this role, (although its bandwidth is wider than an ideal selective emitter). It does, however, require a diode that has a very high luminescent radiative efficiency. Materials for such a device must be able to cool radiatively, i.e. such that they emit as much or more energy than they absorb as heat. Devices with almost the required radiative efficiency have been fabricated from III-V double heterojunctions [Catchpole et al., 2003]. These only have a temperature increase of mK when illuminated in a photoluminescent mode, with a measured external radiative efficiency of 96%. This is very close to the 98% required for cooling. However, for the diode structure required for thermophotonics the electroluminescent threshold of about 65% external efficiency is the more relevant. This more modest target is perhaps more attainable but does require the very pure properties of the undoped heterojunction device to be incorporated in a doped p-n junction.

9.4.3 Hot-Carrier Cells

The final approach for increasing efficiencies, strategy (c) is to allow absorption of a wide range of photon energies but then to collect the photogenerated carriers before they have a chance to thermalise. A hot-carrier solar cell is just such a device that offers the possibility of very high efficiencies (the limiting efficiency is 65% for unconcentrated illumination) but with a structure that could be conceptually simple compared with other very high efficiency PV devices, such as multijunction tandem cells. (Again the physics and limiting efficiency

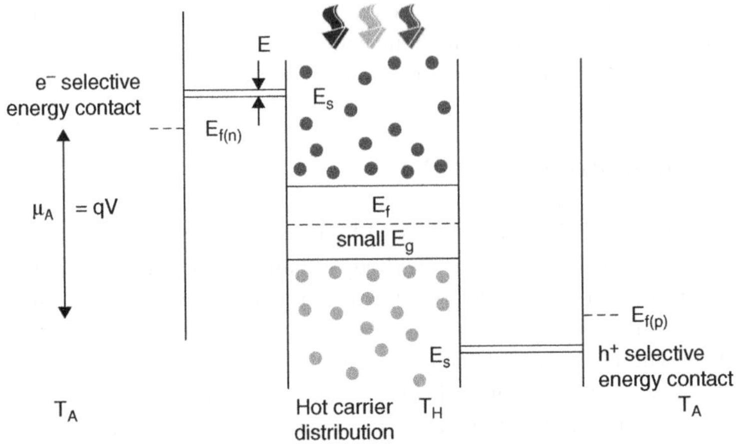

Figure 9.14 Band diagram of the hot-carrier cell that has two stringent requirements: firstly to slow the rate of photogenerated carrier cooling in the absorber and secondly to extract these 'hot carriers' over a narrow range of energies through the energy-selective contacts, such that excess carrier energy is not lost to the 'cold' contacts.

calculations for hot-carrier cells are outlined in Section 2.4.2.) For this reason, the approach potentially lends itself to thin-film deposition techniques with their attendant low material and energy usage costs and the ability to use abundant, nontoxic elements.

The concept underlying hot-carrier solar cells is to slow the rate of photoexcited carrier cooling, which is caused by phonon interaction in the lattice, to allow time for carriers to be collected while they are still at elevated energies ('hot'). This allows higher voltages to be achieved by the cell [Würfel, 1997; Ross and Nozik, 1982]. It thus tackles the major PV loss mechanism of thermalisation of carriers (1 in Figure 9.1). In addition to an absorber material that slows the rate of carrier relaxation, a hot-carrier cell must allow extraction of carriers from the device through contacts that accept only a very narrow range of energies (selective-energy contacts), as shown in Figure 9.14.

Slowed carrier cooling has been observed at very high illumination intensities via a phonon-bottleneck effect in which carrier energy decay mechanisms are restricted. Compounds with large mass difference between their anions and cations have a gap in their allowed phonon modes that can slow down these decay mechanisms and enhance the bottleneck effect [Conibeer and Green, 2004]. Examples are GaN and InN, with some experimental evidence for slowed cooling in the latter [Chen and Cartwright, 2003].

Whilst the III-nitrides look very attractive because of their large phonon bandgaps, the inclusion of the rare element Indium in a final technology is problematic. Analogues of InN with abundant elements, but also with wide phonon bandgap and narrow E_g include II-IV-VI compounds such as $ZnSnN_2$, IIIA nitrides such as LaN and YN, which should have large phonon gaps, and IVA nitrides such as ZrN and HfN that have measured large phonon bandgaps and are readily available. Bi and Sb compounds should also have large phonon gaps but are not abundant. Group IV compounds have large calculated gaps and small E_gs, as well as several other advantages [Conibeer et al., 2011]. Figure 9.15 shows calculations of phonon dispersions for group IV compounds that have large phonon bandgaps, sufficient to block Klemens' decay.

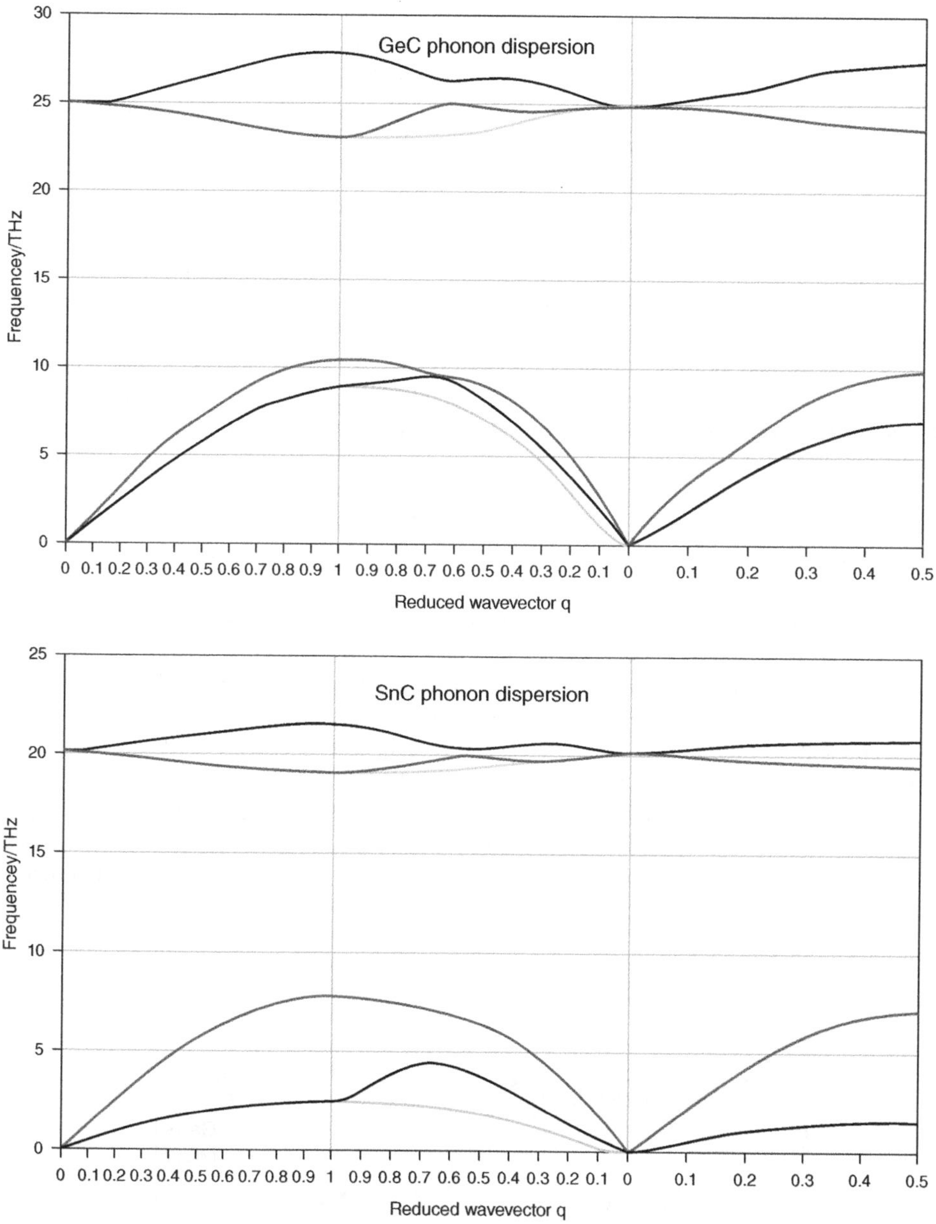

Figure 9.15 Adiabatic bond-charge calculations of phonon dispersions for group IV compounds. Phonon gaps increase as the mass ratio increases with those for GeC and SnC more than twice the acoustic phonon energy, large enough to block the primary phonon decay mechanism, Klemens' decay.

Theoretical work on replicating this effect by modifying the phononic band structures of QD nanostructure superlattices has shown that phonon dispersions of simple cubic (SC) SLs of core–shell QDs are expected to exhibit large phonon bandgaps for light-shell thicknesses on the order of a monolayer [Patterson et al., 2009]. Nanoporous materials with gas filled pores also show potential as candidate hot-carrier absorber materials. Fabrication of such highly ordered arrays from uniformly sized core–shell nanoparticles has been attempted experimentally using the Langmuir–Blodgett (LB) technique [Treiber et al., 2008]. This technique allows transfer of highly ordered monolayers onto a wide range of solid substrates such as glass or Si wafers [Huang et al., 2004]. By controlling the interspacing between adjacent particles, i.e. the shell thickness, optimised nanostructures for phonon bandgap materials may be achieved.

Low-dimensional multiple quantum well (MQW) systems have also been shown to have lower carrier cooling rates. Comparison of bulk and MQW materials has shown significantly slower carrier cooling in the latter. Figure 9.16 shows data for bulk GaAs as compared to MQW GaAs/AlGaAs materials as measured using time-resolved transient absorption by [Rosenwaks et al., 1993], recalculated to show the effective carrier temperature as a function of carrier lifetime by [Guillemoles et al., 2006]. It clearly shows that the carriers stay hotter for significantly longer times in the MQW samples, particularly at the higher injection levels by $1^1/_2$ orders of magnitude. This is due to an enhanced 'phonon bottleneck' in the MQWs allowing the threshold intensity at which a certain ratio of LO phonon reabsorption to emission is reached that allows maintenance of a hot-carrier population, to be reached at a much lower illumination level. More recent work on strain-balanced InGaAs/GaAsP MQWs by [Hirst et al., 2011; Hirst et al., 2012] has also shown carrier temperatures significantly above ambient, as measured by PL. Increase in In content to make the wells deeper and to reduce the degree of confinement is seen to increase the effective carrier temperatures.

The mechanisms for the reduced carrier cooling rate in these MQW systems are not yet clear. However, there are three effects that are likely to contribute. The first is that in bulk material photogenerated hot carriers are free to diffuse deeper into the material and hence

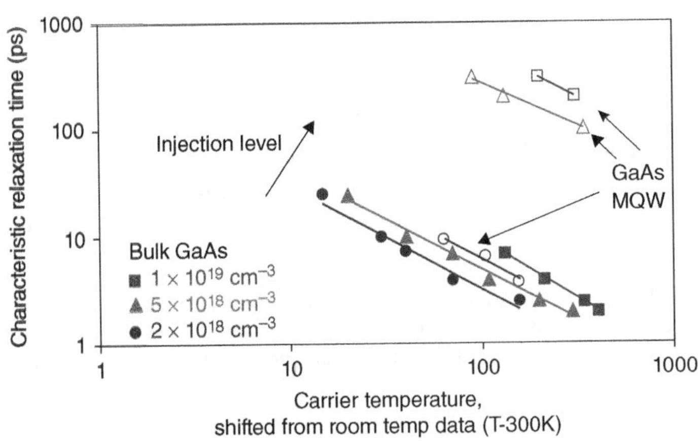

Figure 9.16 Effective carrier temperature as a function of carrier lifetime for bulk GaAs as compared to GaAs/AlGaAs MQWs: time-resolved transient absorption data for different injection levels, from [Rosenwaks et al., 1993], recalculated by Guillemoles et al. [Guillemoles et al., 2006].

to reduce the hot-carrier concentration at a given depth. This will also decrease the density of LO phonons emitted by hot carriers as they cool and make a phonon bottleneck more difficult to achieve at a given illumination intensity. Whereas in a MQW there are physical barriers to the diffusion of hot carriers generated in a well and hence a much greater local concentration of carriers and therefore also of emitted optical phonons. Thus, the phonon bottleneck condition is achieved at lower intensity.

The second effect is that for the materials systems that show this slowed cooling, there is very little or no overlap between the optical phonon energies of the well and barrier materials. For instance, the optical phonon energy ranges for the GaAs wells and AlGaAs barriers used in [Rosenwaks et al., 1993] at 210–285 meV and 280–350 meV, respectively, exhibit very little overlap in energy, with zero overlap for the zone-centre LO phonon energies of 285 and 350 meV [Colvard et al., 1985]. Consequently, the predominantly zone centre LO phonons emitted by carriers cooling in the wells will be reflected from the interfaces and will remain confined in the wells, thus enhancing the phonon bottleneck at a given illumination intensity.

Thirdly, if there is a coherent spacing between the nanowells (as there is for these MQW or superlattice systems) a coherent Bragg reflection of phonon modes can be established that blocks certain phonon energies perpendicular to the wells, opening up one-dimensional phononic bandgaps (analogous to photonic bandgaps in modulated refractive index structures [Conibeer and Green, 2004]. For specific ranges of nanowell and barrier thickness these forbidden energies can be at just those energies required for phonon decay. This coherent Bragg reflection should have an even stronger effect than the incoherent scattering of the second mechanism above at preventing emission of phonons and phonon decay in the direction perpendicular to the nanowells.

It is likely that all three of these effects will reduce carrier cooling rates. None depend on electronic quantum confinement and hence should be exhibited in wells that are not thin enough to be quantised but are still quite thin (perhaps termed 'nanowells'). In fact, it may well be that the effects are enhanced in such nanowells as compared to full QWs due to the former's greater density of states and in particular their greater ratio of density of electronic to phonon states that will enhance the phonon bottleneck for emitted phonons. The fact that the deeper and hence less-confined wells in [Hirst et al., 2012] show higher carrier temperatures is tentative evidence to support the hypothesis that nanowells without quantum confinement are all that are required. Whilst several other effects might well be present in these MQW systems, further work on variation of nanowell and barrier width and comparison between material systems, will distinguish which of these reduced carrier diffusion, phonon confinement or phonon folding mechanisms might be dominant.

Nanostructures such as QW and QD materials are also of importance in energy-selective contacts. Initial experimental progress has been made using double-barrier resonant tunnelling structures, with a single layer of Si QDs providing the resonant level [Jiang and Green, 2006; Conibeer et al., 2008]. A significant proof of concept has been achieved with negative differential resistance (NDR) observed at room temperature for a double-barrier quantum dot structure. The NDR resonance does not appear to be strong but nonetheless such a result at room temperature is very encouraging as evidence for 1D energy selection. Work now is focused on improving the quality factor for such NDR, using other barrier QD combinations and using QWs that, whilst not quite as effective in theory, are much easier to fabricate in high quality in practice.

Complete hot-carrier devices for proof of concept are likely in the near future. These are likely initially to combine double-barrier QW ESCs and MQW absorber in the same lattice-matched III-V system, most likely in the InGaN/InN system. A transition to other bulk and nanostructured thin-film-type systems of higher material abundance is anticipated subsequently.

9.5 OTHER APPROACHES

A few other schemes have been suggested to boost photovoltaic efficiencies that do not fit into the categorisations in the preceding sections. Their physics is not yet entirely proven but they may offer intriguing possibilities for much higher efficiencies, even if only in theory at present.

9.5.1 Nonreciprocal Devices

In Figure 9.1 one of the loss mechanisms is from radiative recombination (loss 4). In most devices this is assumed to be a minimum loss that cannot be reduced – for a cell at the radiative limit, i.e. no nonradiative recombination. This is necessary as a reciprocal device that can absorb solar wavelengths must also be able to emit those same wavelengths [Kirchoff, 1860]. However, as mentioned in Section 2.2.2 in this volume, it is possible that a nonreciprocal device could reuse some of this emitted radiation and boost efficiencies beyond the radiative limit. This is possible in theory because Kirchoff's law applies only to time-symmetric processes. Time symmetry can be violated in a detailed process if the deeper CPT symmetry is observed. Thus, either charge or parity must also be asymmetric for a process to be time asymmetric [Green, 2003; Weinberg, 2013]. A device that can be nonreciprocal in this time asymmetric way is known as a multiport circulator; with incoming light incident on port 1 emitted at port 2 but incident light on port 2 emitted at port 3, [Ries, 1983]. In principle, such a circulator can be used to redirect light emitted from a solar cell onto a second cell, light from this cell can then go through a second circulator to a third cell, etc. It has been shown that such an approach can boost efficiencies to 93%, the Landsberg limit, for an infinite number of circulators illuminating tandem cells with an infinite number of bandgaps [Brown and Green, 2002; Green, 2003]. The obvious practical difficulties of this are offset to some extent by the fact that most of the efficiency gain is obtained with the first circulator and, as we have seen, a tandem cell has most of its increase in efficiency for the first few layers. The nonreciprocal rotation of the polarisation of light by a magnetic field can be and is used to fabricate such esoteric-sounding circulator devices, for microwave and laser optics [Fujii, 1991]. However, the complexity of the components and the very small efficiency gains make such an approach applied to photovoltaics only appropriate for theoretical consideration.

9.5.2 Quantum Antennae – Light as a Wave

The idea of a quantum antenna is to use the wave nature of light rather than its particle nature [Bailey, 1972]. Incoming light waves oscillate electrons in an antenna that has dimensions such that excited oscillations are resonant for a particular wavelength of light. Each of these oscillations is then rectified for each antennae to give a DC output. The voltage is determined by the built-in voltage of the diode (V_{bi} = 2/3 of the bandgap of the diode semiconductor, depending on the radiative efficiency of the material). The current is determined by the number of electrons in the oscillation that are above the energy barrier determined by the diode built-in voltage. This number is a fraction of the total number of photons absorbed by that antenna, $N_p = (\sin f).V_{bi}$ (where f is the frequency of the light absorbed).

The short wavelengths of light mean that the devices have to be of the order of a few hundred nm. (For an ideal structure this is half the wavelength, but in practice probably slightly less because of penetration of the wave outside the antennae waveguide.) The broadband polychromatic solar spectrum also requires a wide range of antenna sizes to match all the wavelengths and its incoherent nature the need to arrange these antennae in two orthogonal directions of polarisation. In addition to these practical problems it has been suggested, based on wave theory, that the approach can only achieve 48% even under ideal conditions [Corkish et al., 2003]. Nonetheless, progress has been made on fabricating some of the small diode elements needed for such a device [Grover et al., 2013], although a proof of principle has yet to be demonstrated.

9.6 CONCLUSIONS

Third-generation PV aims to provide high conversion efficiency of photon energy with low manufacturing cost. The combined methodology of using multiple energy thresholds and low-cost processes with abundant nontoxic materials offers significant leverage in achieving this goal. However, the actual efficiencies and ease of optimisation depend on the different physical approaches. Efficiency, spectral robustness, and cost/ease of manufacture are important for a robust technology that can supply very significant increases in PV implementation.

Advanced nanomaterials offer great potential to optimise absorption, carrier generation, and separation. Low-dimensional quantum confinement provides bandgap flexibility in QW and QD materials. All-Si QDs devices fabricated by thin-film deposition technique have experimentally proved the application as upper cells in the all-silicon tandem solar cells. Efficient MEG has been observed in some semiconductor QDs due to the modified relaxation dynamics of photoexcited excitons. Devices designed to extract such multiple excitons have now shown QEs greater than 1. Work on intermediate-band and hot-carrier devices has also demonstrated that QD materials promise increased efficiencies and greater spectral robustness. Although these devices are still at early stages, implementation of such techniques could dramatically decrease cost per Watt with spectral robustness as they are compatible with conceptually relatively simple thin-film devices.

ACKNOWLEDGEMENTS

The author would like to thank the Australian Research Council for the support of his Future Fellowship; the Australian Renewable Energy Agency for support of Si QD and hot-carrier cell work; the LIMA project under Euro FP7 for support of Si QD work and the help and support of the Third Generation Photovoltaics group at UNSW for their contributions to the Third Generation projects and concepts discussed here.

REFERENCES

Aberle, A.G., (2006) 'Progress with polycrystalline silicon thin-film solar cells on glass at UNSW' *J. Cryst. Growth* **287**, 386–390.

Auzel, F., (2004), "Upconversion and anti-Stokes processes with f and d ions in solids", *Chem. Rev.* **104**, 139–173.

Bailey, R.L., (1972) *J. Eng. Power*, **94**, 73.

Baluschev, S., Yakutkin, V., Wegner, G., Miteva, T., Nelles, G., Yasuda, A., Chernov, S., Aleshchenkov, S., Cheprakov, A., (2007) "Upconversion with ultrabroad excitation band: Simultaneous use of two sensitizers", *Appl. Phys. Lett.* **90**, 181103.

Basore, P.A. (2006) 'CSG-1: manufacturing a new polycrystalline silicon PV technology', *4th World Conference of PV Energy Conversion*, p. 2089–2093.

Beard, M.C., Midgett, A.G., Hanna, M.C., Luther, J.M., Hughes, B.K., Nozik, A.J., (2010) 'Comparing Multiple Exciton Generation in Quantum Dots To Impact Ionization in Bulk Semiconductors: Implications for Enhancement of Solar Energy Conversion', *Nano Lett.* **10**, 3019.

Berkowitz, J., Olsen, J., (1991) *J. Lumin.*, **50**, 111.

Boer K. (1990) *Survey of Semiconductor Physics*, van Nostrand Reinhold, New York.

Brown, A.S., and Green, M.A. (2002) 'Limiting efficiency for current-constrained two-terminal tandem cell stack', *Prog. Photovolt.* **10**, 299–307.

Catchpole, K.R., et al., (2003) *3rd World Conf. on PV Energy Conversion* (Osaka, 2003).

Chen, F., and Cartwright, A.N. (2003) 'Time-resolved spectroscopy of recombination and relaxation dynamics in InN', *Appl. Phys. Lett.* **83**, 4984–4986.

Cheng, Y.Y., Khoury, T., Clady, R. G. C. R., Tayebjee, M. J. Y., Ekins-Daukes, N.J., Crossley, M.J., Schmidt, T.W., (2010) *Phys. Chem. Chem. Phys.*, **12**, 66–71.

Cheng, Y.Y., Fuckel, B., MacQueen, R.W., Khoury, T., Clady, R.G.C.R., Schulze, T.F., Ekins-Daukes, N.J., Crossley, M.J., Stannowski, B., Lips, K., Schmidt, T.W., (2012) Improving the light-harvesting of amorphous silicon solar cells with photochemical upconversion, *Energy Environ. Sci.*, Accepted Feb 2012, DOI: 10.1039/c2ee21136j.

Cho E.C., Cho Y.H., Trupke T., Corkish R., Conibeer G. and Green M.A. (2004) 'Silicon nanostructures for all-silicon tandem solar cells', *Proc. 19th European photovoltaic Solar Energy Conference*, p. 235–241.

Cho E.C., Park S., Hao X., Song D., Conibeer G., Park S.C., Green M.A. (2008) 'Silicon quantum dot/crystalline silicon solar cells', *Nanotechnology*, **19**, 245201–245204.

Colvard, C., Gant, T.A., Klein, M.V., (1985) "Folded acoustic and quantized optic phonons in (GaAl)As superlattices," *Phys. Rev. B*, **31**, 2080–2091.

Conibeer, G., and Green, M.A. (2004) 'Phononic band gap engineering for hot carrier solar cell absorbers' *Proc. 19th European Photovoltaic Solar Energy Conference*, p. 270.

Conibeer, G, et al., (2006) 'Phononic engineering for hot carrier solar cells including interface modelling' *Proc. 21st European Photovoltaic Solar Energy Conference*, p. 90.

Conibeer G., Green M.A., Corkish R., Cho Y.-H., Cho E.-C., Jiang C.-W., Fangsuwannarak T., Pink E., Huang Y., Puzzer T., Trupke T., Richards B., Shalav A., Lin K.-L. (2006A) 'Silicon nanostructures for third generation photovoltaic solar cells', *Thin Solid Films*, **511/512**, 654–662.

Conibeer G., Green M.A., Cho E-C., Konig D., Cho Y-H., Fangsuwannarak T., Scardera G., Pink E., Huang Y., Puzzer T., Huang S., Song D., Flynn C., Park S., Hao, X., and Mansfield D. (2008) 'Silicon quantum dot nanostructures for tandem photovoltaic cells', *Thin Solid Films* **516**, 6748–6756.

Conibeer G., Jiang C.W., König D., Shrestha S., Walsh T., Green M.A. (2008) 'Selective energy contacts for hot carrier solar cells', *Thin Solid Films*, **516**, 6968–6975.

Conibeer, G., Patterson, R., Aliberti, P., Xia, H., Huang, S., König, K., Puthen-Veettil, B., Shrestha, S., Green, M.A., (2011) "Hot Carrier cell absorber materials," *Proc. 26th European Photovoltaic Solar Energy Conference*, Hamburg, (2011).

Corkish, R. et al., (2003) *3rd World Conf. on PV Energy Conversion*, Osaka, (2003).

Couts, T. in Archer, M.D. and Hill, R. (eds) *Clean Electricity from Photovoltaics*, Imperial College Press, London, 2001.

Dexter, D., (1957) *Phys. Rev.* **108** (3), 630.

Di, D., Xu, H., Perez-Wurfl, I., Green, M.A., Conibeer, G., (2011) Improved nanocrystal formation, quantum confinement and carrier transport properties of doped Si quantum dot superlattices for third generation photovoltaics, *Progress in Photovoltaics*-10-174.R2, Accepted Sept 2011.

Dieke, H.H., (1968) *Spectra and Energy Levels of Rare Earth Ions in Crystals*, John Wiley and Sons, New York, Maryland, 1968.

Domine, D., Buehlmann, P., Bailat, J., Billet, A., Feltrin, A., Ballif, C., (2008) *Phys. Status Solidi: Rapid Res. Lett.*, **163–165**, DOI 101002/pssr.200802118.

Ekins-Daukes, N, Schmidt, T, (2008) "A molecular approach to the intermediate band solar cell: The symmetric case", *Appl. Phys. Lett.*, **93**, 063507.

Fujii, Y., (1991) *J. Light Wave Technol.* **9**, 1238.

Gao F., Green M.A., Conibeer G., Cho E-C., Huang Y., Perez-Wurfl I., Flynn C. (2008) "Fabrication of multilayered Ge nanocrystals by magnetron sputtering and annealing", *Nanotechnology*, **19**, 455611–455618.

Goldscmidt, J.C., Fischer, S., Löper, P., Krämer, K.W., Biner2, D., Hermle, M., Glunz, S.W., 'Upconversion to enhance silicon solar cell efficiency.' *Proc. 25th European Photovoltaic Solar Energy Conference and Exhibition* (September 2010) Valencia, Spain.

González-Díaz, B., Guerrero-Lemus, R., Haro-González, P., Borchert, D., Hernández-Rodríguez, C., (2006) *Thin Solid Films* **511–512**, 473.

Green, M.A. (2003) *Third Generation Photovoltaics: Ultra-High Efficiency at Low Cost*, Springer-Verlag, Berlin.

Green M.A., Conibeer, G., Perez-Wurfl I., Huang S., Konig D., Song D., Gentle A., Hao, X.J., Park S.W., Gao F., So Y.H. and Huang Y., (2008) 'Progress with silicon-based tandem cells using group IV quantum dots in a dielectric matrix', *Proc. 23rd European Photovoltaic Solar Energy Conference*, Valencia, 2008.

Green M.A., Conibeer G., König D., Perez-Wurfl I., Gentle A., Huang S., Song D., Hao X.J., Gao F., Park S., Flynn C., So Y.H., Zhang B., Di D., Campbell P., Huang Y. and Puzzer T., (2009) 'Progress with All-Silicon Tandem Cells On Glass Using Silicon Quantum Dots in Silicon-Based Dielectric Matrices', *Proc. 24th European Photovoltaic Solar Energy Conference*, Hamburg, 2009.

Green, M.A., Emery, K, Hishikawa, Y., Warta, W., Dunlop, E.D., (2013) Solar cell efficiency tables (version 41), *Prog. Photovolt.: Res. Appl.*, **21**, 1–11. www.sj-solar.com.

Grover, S., Joshi, S., Moddel, G., (2013) "Quantum theory of operation for rectenna solar cells", *J. Phys. D: Appl. Phys.* **46**, 135106.

Guillemoles, J.-F., Conibeer, G., Green, M.A., (2006) "On achievable efficiencies of Hot Carrier Solar cell absorbers based on reported experimental thermalisation properties," *Proc. 21st European Photovoltaic Solar Energy Conference*, Dresden Germany, 234–237.

Hanna, M.C. and Nozik, A.J., (2006) 'Solar conversion efficiency of photovoltaic and photoelectrolysis cells with carrier multiplication absorbers', *J. Appl. Phys.*, **100**, 074510–074517.

Hao X.J., Cho E.-C., Flynn C., Shen Y.S., Park S.C., Conibeer G., Green M.A. (2009) 'Synthesis and characterization of boron-doped Si quantum dots for all-Si quantum dot tandem solar cells', *Solar Energy Mater. Solar Cells*, **93**, 273–279.

Hirst, L., Führer, M., LeBris, A., Guillemoles, J.-F., Tayebjee, M., Clady, R., Schmidt, T., Wang, Y., Sugiyama, M., Ekins-Daukes, N., (2011) *Proc. 37th IEEE PV Specialists Conference*, Seattle, talk 927.

Hirst, L., Fuhrer, M., Farrel, D.J., Le Bris, A., Guillemoles, J.-F., Tayebjee, M.J.Y., Clady, R., Schmidt, T.W., Sugiyama, M., Wang, Y., Fujii, H., Ekins-Daukes, N., (2012) InGaAs/GaAsP quantum wells for hot carrier solar cells, *Proc. SPIE Photonic West* (Jan 2012).

Huang S., Minami K., Sakaue H., Shingubara S., and Takahagi T. (2004) 'Effects of the surface pressure on the formation of Langmuir-Blodgett monolayer of nanoparticles', *Langmuir*, **20**, 2274.

Huang S., So Y.H., Conibeer G., Green M. (2009) 'In situ formation of tin nanocrystals in silicon nitride matrix', *J. Appl. Phys.* **105**, 124303.

Ilmas, E., Savikhina, T., (1970) *J. Lumin.*, **1,2** 702.

Ivanova, S., Andria, C., Pellé, F., Guillemoles, J.F., Conibeer, G., (2009) "New efficient upconverters for c-Si photovoltaics", *Proc. E-MRS Spring Meeting*, Strasbourg, June 2009.

Jackson, E.D. (1955) *Proc. Conference on the Use of Solar Energy*, p. 122, Tuscon, Arizona, Oct. 31 – Nov. 1, 122; also US Patent 2,949,498, Aug. 16, 1960.

Jani, O., et al., (2006) *4th World Conf. on PV Energy Conversion* Hawaii, 2006, 20.

Jiang, C.-W., Conibeer G., Green M.A. (2006), 'Silicon quantum dots: application for energy selective contacts to hot carrier solar cells' *Proc. 21st European Photovoltaic Solar Energy Conference*, p. 168.

Jiang, C.-W., and Green M.A. (2006) 'Silicon quantum dot superlattices: Modeling of energy bands, densities of states, and mobilities for silicon tandem solar cell applications', *J. Appl. Phys.*, **99**, 114903.

Johnson, C.M., Conibeer, G.J., (2012) Limiting efficiency of generalized realistic c-Si solar cells coupled to ideal up-converters, *J. Appl. Phys.* **112**, 103108 (2012), http:dx.doi.org/10.1063/1.4766386.

Keevers, M., Saris, F., Zhang, G., Zhao, J., Green, M., (2007) *26th IEEE PVSC* (California, 1997) 215.

Kirchoff, G., (1860) *Annal. Phys.* **19**, 275.

King, R.R., et al., (2006) *21st Euro PVSEC,* Dresden, 2006, 124.

King, R.R., et al., (2007) *Appl. Phys. Lett.*, **90**, 183516.

Klampaftis, E., Ross, D., McIntosh, K.R., Richards, B.S., (2009) *Solar Energy Mater. Solar Cells*, **93**, 1182.

Klimov, V. (2006) 'Detailed-balance power conversion limits of nanocrystal-quantum-dot solar cells in the presence of carrier multiplication', *Appl. Phys. Lett.* **89**, 123118–123120.

Luque, A., and Martí, A. (1997) 'Increasing the Efficiency of Ideal Solar Cells by Photon Induced Transitions at Intermediate Levels', *Phys. Rev. Lett.* **78**, 5014.

Luque, A., et al., (2006) 'Intermediate bands versus levels in non-radiative recombination', *Physica B*, **382**, 320–327.

Luque A., Martí A., and Nozik A.J. (2007) 'Solar cells based on quantum dots: Multiple exciton generation and intermediate bands', *MRS Bull.*, **32**, 236–241.

Luque A., and Marti A., (2010) 'The Intermediate Band Solar Cell: Progress Toward the Realization of an Attractive Concept', *Adv. Mater.*, **22**, 160–174.

Luther J.M., Law M., Beard M.C., Song Q., Reese M.O., Ellingson R. J, and Nozik A.J. (2008) 'Schottky Solar Cells Based on Colloidal Nanocrystal Films', *Nano Lett.* **8**, 3488–3492.

Martı A., Antolín, E. Stanley C.R., Farmer C.D., López N., Díaz P., Cánovas E., Linares P.G., and Luque, A. (2006) 'Production of Photocurrent due to Intermediate-to-Conduction-Band Transitions: A Demonstration of a Key Operating Principle of the Intermediate-Band Solar Cell', *Phys. Rev. Lett.* **97**, 247701.

Meier, J., Dubail, S., Shah, A., (2002) *Sol. Energy Mater. Sol. Cells*, **74**, 457.

Meier J., Spitznagel J., Kroll U., Bucher C., Faÿ S., Moriarty T. and Shah A. (2004) 'Potential of amorphous and microcrystalline silicon solar cells', *Thin Solid Films* **451**, 518–524.

Meijerink, A., Wegh, R., Vergeer, P., Vlugt, T., (2006) "Photon management with lanthanides", *Opt. Mater.*, **28**, 575–581.

Meillaud F., Shah A., Droz C., Vallat-Sauvain E., Miazza C. (2006) 'Efficiency limits for single-junction and tandem solar cells', *Sol. Energy Mater. Sol. Cells*, **90**, 2952–2959.

Michels, J., Huskens, J., Reinhoudt, D.N., (2002) *J. Am. Chem. Soc.*, **124**, 2056–2064.

Nelson, J. (2003) in *The Physics of Solar Cells*, Imperial College Press, London.

Nozik A.J., Boudreaux D.S., Chance R.R., Williams F. (1980) in *Advances in Chemistry*, M. Wrighton (ed.), **184**, p. 162.

Nozik, A.J., (2008) "Multiple exciton generation in semiconductor quantum dots", *Chem. Phys. Lett.*, **457**, 3–11.

Nozik A.J. (2002) 'Quantum dot solar cells', *Physica E*, **14**, 115–120.

Nozik, AJ (2008A), Multiple exciton generation in semiconductor quantum dots, *Chem. Phys. Lett.*, **457**, 3–11.

Park S., Cho E.-C., Song D.Y., Conibeer G. and Green M.A. (2009) 'N-type silicon quantum dots and p-type crystalline silicon heteroface solar cells', *Solar Energy Mater. Solar Cells*, **93**, 684–690.

Patterson R., Kirkengen M., Veettil B.P., König D., Green M.A., Conibeer G. (2009) 'Phonon Lifetimes in Model quantum dot superlattices', *24th European Photovoltaic Solar Energy Conference, Hamburg*.

Perez-Wurfl I., Hao X., Gentle A., Kim D., Conibeer G., Green M.A. (2009) 'Si nanocrystal p-i-n diodes fabricated on quartz substrates for third generation solar cell applications', *Appl. Phys. Lett.*, **95**, 153506.

Perez-Wurfl, I., Ma, L., Lin, D., Hao, X., Green, M.A., Conibeer, G., (2012) "Silicon nanocrystals in an oxide matrix for thin film solar cells with 492mV open circuit voltage", *Solar Energy Mater. Solar Cells*, **100**, 65–68. DOI: 10.1016/j.solmat.2011.02.029.

Pijpers, J, et al., (2008) *Nanoletters* **8**, 1207.

Ries, H., (1983) *Appl. Phys.*, **32**, 153.

Rosenwaks, Y., Hanna, M., Levi, D., Szmyd, D., Ahrenkiel, R., Nozik, A., (1993) "Hot-carrier cooling in GaAs: quantum wells versus bulk," *Phys. Rev. B*, **48**, 14675–14678.

Ross, R.T. and Nozik, A.J., (1982) 'Efficiency of Hot-Carrier Solar Energy Converters', *J. Appl. Phys.* **53**, 3318.

Schaller, R.D. and Klimov V.I. (2004) 'High Efficiency Carrier Multiplication in PbSe Nanocrystals: Implications for Solar Energy Conversion', *Phys. Rev. Lett.*, **92**, 186601.

Semonin, O.E., Luther, J.M., Choi, S., Chen, H-Y., Gao, J., Nozik, A.J., Beard, M.C., (2011) *Science*, **334**, 1530–1533. DOI: 10.1126/science.1209845.

Shalav, A., (2006), "Rare Earth doped up-converting phosphors for an enhanced Silicon solar cell response", PhD Thesis, UNSW (Feb 2006).

Shalav, A., Richards, B.S., Green, M.A., (2007) Luminescent layers for enhanced silicon solar cell performance: Up-conversion. *Solar Energy Mater. Solar Cells*, **91**(9), 829–842.

Shockley, W. and Queisser, H.J., (1961) Detailed balance limit of efficiency of p-n junction solar cells', *J. Appl. Phys.*, **32**, 510–519.

Song D., Cho E.C, Conibeer G., Huang Y. and Green M.A. (2007) 'Fabrication and characterisation of Si nanocrystal/c-Si heterojunctions', *Appl. Phys. Lett.* **91**, 123510.

Song D., Cho E.-C., Cho Y.-H., Conibeer G., Huang Y., Huang S., Green M.A. (2008) 'Evolution of Si (and SiC) nanocrystal precipitation in SiC matrix', *Thin Solid Films* **516**, 3824–3830.

Strumpel, C., McCann, M., Beaucarne, G., Arkhipov, V., Slaoui, A., Švrček, V., del Canizo, C., Tobias, I., (2007) "Modifying the solar spectrum to enhance silicon solar cell efficiency-An overview of available materials", *Solar Energy Mater. Solar Cells*, **91**, 238–249.

Švrcek, V., Slaoui A., Muller J.-C., (2004) 'Silicon nanocrystals as light converter for solar cells', *Thin Solid Films*, 451-452, 384–388.

Takagi H., Ogawa H., Yamazaki Y., Ishizaki A., Nakagiri T. (1990) 'Quantum size effects on photoluminescence in ultrafine Si particles', *Appl. Phys. Lett.* **56**, 2379–2380.

Takamoto, T., Takamoto T., Kaneiwa M., Imaizumi M., Yamaguchi M. (2005) 'InGaP/GaAs-based multijunction solar cells', *Prog. Photovolt.* **13**, 495–511.

Treiber, L., Bumby, C., Huang, S., Conibeer, G., (2008) *23rd European Photovoltaic Solar Energy Conference*, Valencia, Spain (2008).

Trupke, T. Green, M.A., Würfel, P., (2002A) *J. Appl. Phys.* **92**, 1668.

Trupke, T. Green, M.A., Würfel, P., (2002B) *J. Appl. Phys.* **92**, 4117.

Trybus, E., et al., (2006) *J. Cryst. Growth,* **288**, 218.

Wanlass, M., Ahrenkiel P., Albin D, Carapella J., Duda A., Emery K., Friedman D., Geisz J., Jones K., Kebbler A., Kiehl J., Kurtz S., McMahon W., Moriarty T., Olson J., Ptak A., Romero M., Ward S. (2006) 'Monolithic, ultra-thin GaInP/GaAs/GaInAs tandem solar cells' *4th World Conference, PV Energy Conversion*, p. 729–732.

Wegh, R., Donker, H., van Loef, E., Oskam, K., Meijerink, A., (2000) *J. Lumin.*, **87–89**, 1017.

Weinberg, S., *Lectures on Quantum Mechanics*, (Cambridge University Press, 2013) p. 142: CPT symmetry invariance; Luders, G., Kong. *Danske Vid. Selskab Mat.-Fys.Medd.*, **28**, 5 (1954): *Ann. Phys.*, **2**, 1 (1957); Pauli, W., *Nuovo Cimento*, **6**, 204 (1957).

Würfel, P., *The Physics of Solar Cells*, (2009, Wiley-VCH, Weinheim, Germany).

Xie, P. R. (1993), S. *Appl. Phys. Lett.* **63** (23), 3125.

Yang J., Banerjee A., Lord K. and Guha S., (1998) Correlation of component cells with high efficiency amorphous silicon alloy triple-junction solar cells and modules, *Proceedings, 2nd World Conference on Photovoltaic Solar Energy Conversion*, Vienna, **1998**, 387–390.

Zacharias M., Heitmann J., Scholz R., Kahler U., Schmidt M. (2002) 'Size-controlled highly luminescent silicon nanocrystals: A SiO/SiO_2 superlattice approach', *Appl. Phys. Lett.* **80**, 661–663.

Zhang, B., Yao, Y., Patterson, R., Shresth, A.S., Green, M.A., Conibeer, G., (2011) Electrical properties of conductive Ge nanocrystal thin films fabricated by low temperature, *Nanotechnology*, **22**, 125204.

Concluding Remarks

Gavin Conibeer[1] and Arthur Willoughby[2]
[1]*School of Photovoltaic and Renewable Energy Engineering, University of New South Wales, Australia*
[2]*Faculty of Engineering and the Environment, University of Southampton, UK*

This book aims to present the latest developments in high-efficiency photovoltaics, contributed by experts in the respective fields.

The physics of solar cells and of advanced concepts as presented by Jean-Francois Guillemoles, gives a very useful insight into the underlying mechanisms required for solar cells and their limiting efficiencies. The descriptions of multiple energy threshold devices and their potentials to increase efficiencies above the Schockley–Queisser limit are particularly useful. An understanding of the physical limits is essential to guiding progress in design and fabrication of devices. This is particularly true for the multiple energy level devices with their more complex physical principles.

The characterisation of solar cell materials presented by Daniel Bellet and Edith Bellet-Amalric, ideally describes the characterisation techniques necessary for development of high-efficiency devices. In particular, the techniques for determining important crystal properties with X-ray and Raman techniques are most useful. It is not common to have such a focus on characterisation in a volume such as this, but its importance in determining and allowing control of the physical parameters of materials and devices is essential to significant progress.

The present status of crystalline silicon cells presented by Martin Green outlines the dominant position of crystalline technologies in the market, and the fact that this will continue for at least another decade. The developments in mono-, multi- and pseudo-monocrystalline cells and modules are allowing an approach to the fundamental limits for silicon, whilst also driving down the cost of production significantly. In particular, developments in material quality, light trapping and carrier extraction are allowing continuing improvements in the efficiency of cast cell materials and this is set to continue as the competiveness in the market continues to increase.

Ruud Schropp continues the theme of silicon solar cells but with amorphous and microcrystalline silicon materials. The inherent low material usage of thin-film deposition and the potential for cheap substrates make these technologies very competitive for cheap solar cell delivery. Efficiencies are boosted in double- and triple-junction micromorph and similar approaches. The differing morphology and defect density of amorphous materials and their characterisation presents challenges for strong light absorption and good carrier collection, but several p-i-n junction approaches with innovative characterisation techniques are proving successful in getting round these problems.

Solar Cell Materials: Developing Technologies, First Edition. Edited by Gavin Conibeer and Arthur Willoughby.

In the chapter on III-V solar cells, Nicholas Ekins-Daukes describes the large range of materials available in this group and their application to highly efficient crystalline cells. The roles of crystal quality, defects, hetero- and homojunctions are investigated. The very important area of multijunction III-V cells is explored with comparison of the several techniques used to balance the issues related to lattice matching and those of required bandgaps, with an excellent description of the route towards the highest efficiency solar cells in multijunction III-V materials.

In the next chapter on chalcogenide solar cells, Miriam Paire, Sebastian Delbois, Julien Vidal, Nagar Naghavi and Jean-Francois Guillemoles investigate the several different analogue materials of II-VI compounds. In a logical progression from CdTe to $CuInGaSe_2$ (CIGS) to Cu_2ZnSnS_4 (kesterite) cells, the increasing number of elements allows greater flexibility in the control over bandgap, optical absorption, lattice spacing and material abundance. This group of materials also has produced the record thin-film cell at just over 20% for CIGS cells. Whilst the control of stoichiometry and morphology become more of a challenge with the increasing complexity, the abundance and toxicity issues of some of the elements are largely addressed by the recent move to the kesterite group of materials. Very promising absorption, carrier collection and efficiency properties are possible with the potential to move on to tandem and other multiple junctions to boost efficiencies further.

In the chapter on printed organic solar cells, Claudia Hoth, Andrea Seemann, Roland Steim, Tayebeh Amin, Hamed Azimi and Christoph Brabec change direction to look at organic materials and the many different aspects required for a full photovoltaic product. The potential for these materials to produce really cheap solar cells depends on implementation of good materials and devices in a full product, preferably as a printed product. The crucial importance of morphology and the ability to characterise this in the fabrication of bulk heterojunctions that have all of good absorption, intimate mixing and good transport to contacts is investigated and put in the context of currently available and soon to be realised materials. The role of multiple-junction devices in further boosting efficiencies is also discussed and a case made for development of full OPV products.

In the final chapter on third-generation photovoltaics Gavin Conibeer addresses the potential of advanced concepts to boost efficiencies through the use of multiple energy thresholds. Linking back to several of the physical concepts in Chapter 2, the concepts of using quantum confinement in nanostructures to engineer semiconductor bandgaps, up- or downconversion of incident photons, or the capture of excess carrier energy usually lost as heat are investigated. Materials and practical approaches for each of these are explored and conclusions drawn on their near- or long-term viability.

This volume overall gives what we believe to be a good overview of the current status of photovoltaic development. It is not completely comprehensive with some technologies such as thin-film crystalline silicon and dye-sensitised cells not discussed in any detail, but all other technologies and to some extent their interaction with each other are covered. Comparisons can be drawn between the properties of bulk crystalline and thin-film materials particularly in terms of the light absorption and material usage of these approaches. The increasing move towards thinner and thinner layers of material and hence the importance of maximising absorption and hence light trapping makes such comparisons highly relevant. Similarly, the 'engineering' of solar cell materials' properties either alloying in chalcopyrites to modify the bandgap or for lattice matching in III-Vs or through quantum confinement in nanostructures, are all examples of areas in which there is an overlap of expertise in the quite different materials groupings and in their preparation methods. These synergies and shared expertise can continue to give new insights and lead to greater progress on efficiency improvement and reduction of cost, both in financial and environmental terms, of new and improved photovoltaic technologies. We expect this volume to become dated as these technologies move forwards, but the underlying principles and the approaches towards exploiting these principles outlined in this volume will remain to be built on further.

Index

Solar Cell Materials: Developing Technologies, First Edition. Edited by Gavin Conibeer and Arthur Willoughby.

Printed and bound by CPI Group (UK) Ltd, Croydon, CR0 4YY

16/09/2026

14963389-0003